The Oceanic Crust, from Accretion to Mantle Recycling

Springer
London
Berlin
Heidelberg
New York
Barcelona
Hong Kong
Milan
Paris
Santa Clara
Singapore
Tokyo

Thierry Juteau and René Maury

The Oceanic Crust, from Accretion to Mantle Recycling

Springer

Published in association with
Praxis Publishing
Chichester, UK

PRAXIS

Thierry Juteau, Professor of Petrology, Université de Bretagne Occidentale, France, and
René Maury, Professor of Petrology, Université de Bretagne Occidentale, France

Translated by Thierry Juteau and René Maury, Université de Bretagne occidentale

Original French edition, *Géologie de la croûte océanique*
Published by Masson, Editeur, Paris, © 1997

This work has been published with the help of the French Ministère de la Culture – Centre national du livre

Figures designed and translated by Bernadette Coleno and Jean-Luc Travers

SPRINGER–PRAXIS SERIES IN GEOPHYSICS
SERIES EDITOR: Philippe Blondel, M.Sc., Ph.D., Senior Scientist, Southampton Oceanography Centre, Southampton, UK

ISBN 1-85233-116-X Springer-Verlag Berlin Heidelberg New York

British Library Cataloguing in Publication Data
Juteau, Thierry
The oceanic crust, from accretion to mantle recycling.
— (Springer-Praxis series in geophysics)
1. Oceanography
I. Title II. Maury, Rene 551.4′6

Library of Congress Cataloging-in-Publication Data
Juteau, Thierry
[Géologie de la croûte océanique. English]
The oceanic crust : from accretion to mantle recycling / Thierry Juteau and René Maury.
p. cm. — (Springer-Praxis series in geophysics)
"Published in association with Praxis Publishing, Chichester, UK."
Includes bibliographical references and index.
ISBN 1-85233-116-X (alk. paper)
1. Submarine geology. 2. Earth—Crust. I. Maury, R. C.
II. Title. III. Series. QE39.J8713 1999
551.46′08—dc21 98-50711 CIP

Printed by MPG Books Ltd, Bodmin, Cornwall, UK

Cover design: Jim Wilkie
Typesetting: Originator, Gt. Yarmouth, Norfolk, UK

Printed on acid-free paper supplied by Precision Publishing Papers Ltd, UK

Preface

This textbook was first written in French for graduate students, and published in June 1997 by Masson S.A. (Paris). Its purpose is to discuss the structure, composition, spatial and temporal evolution of the oceanic crust, from the stage of its birth at oceanic spreading ridges, to its migration away from their axis and possible thickening when passing over hot spots, and finally to its recycling within the mantle of subduction zones.

The last 30 years have been marked by a fantastic international effort for exploration of the deep sea floor, especially the worldwide mid-ocean ridge system, thanks to the continuous development of new technologies and tools: high-resolution multibeam sonar systems and side-scan sonars (mapping and acoustic imagery of the ocean floor), sophisticated drilling ships (international DSDP (Deep-Sea Drilling Project) and ODP (Ocean Drilling Project) programs), deep submersibles, high performance geophysical tools, etc. These efforts have resulted in a much better knowledge of the (prominent) immerged part of the Earth's crust.

Our goals are twofold: (1) to present up-to-date results on the young and to a lesser extent fossil (ophiolites) oceanic crust and to describe its temporal evolution, with the help of examples taken from recent scientific studies, and (2) to discuss the models and hypotheses proposed to explain the formation of the ocean crust and its evolution through hydrothermal evolution, ageing and subduction, with deliberate emphasis on magmatic processes.

The idea of writing this textbook came progressively to us after multiple discussions with our graduate and Ph.D. students. There are excellent textbooks dealing with igneous petrology or igneous geochemistry, or both, but we got the feeling that a textbook devoted to the description of the crust of the present-day oceans and to its evolution and fate was still lacking. The volume and dispersion of specialised scientific literature on these topics convinced us of the necessity to try to summarise this broad and active field of research, with the purpose of helping students to understand the complex combination of geological processes operating in the oceanic crust and the underlying mantle.

We assume that the reader of this book will have some familiarity with the basic topics of mineralogy, igneous petrology, trace and isotope geochemistry. After three introductory chapters, the book leads its reader from the axial domain of oceanic ridges where the oceanic crust is generated (accretion processes, Chapters 4 to 9) to the oceanic intraplate domain, where intra-oceanic islands and oceanic plateaus are eventually added to the crust (Chapter 10), and finally to the subduction zones where the oceanic crust is dragged down within the deep mantle (Chapter 11). The last Chapter (12) deals with the cycle of the ocean crust at a global scale.

We have deliberately focused on the oceanic crust of the present-day oceans (including their oceanic islands and associated subduction zones), by describing a number of well-studied oceanic areas. We refer only briefly to the fossil relicts of the oceans of the past (ophiolites), since excellent texbooks already exist on these topics. The case studies we have selected were preferentially taken from our own experience and that of some French colleagues from the Brest scientific community, who provided us many documents and figures. This is an opportunity for non-French students to get away from excessive 'classical' textbook examples, for instance the Hawaiian hot spot and to discover its French Polynesian equivalents ...

We would like to point out that this book is written by two petrologists, whose own fields of research are the oceanic accretion processes (T.J.), and the island arcs and oceanic intraplate islands (R.M.), respectively, which indicates the limits of our expertise for treating such a broad subject. We hope that our necessary incursions in the fields of geophysics, isotope geochemistry or structural geology will not upset the specialists too much, and we warmly thank those of them who agreed to read our manuscript and suggested useful corrections. In particular, we gratefully acknowledge the help of Martial Caroff, Christian Coulon, Jérôme Dyment, Jean Goslin, Gérard Guille, Christophe Hémond, Yves Lagabrielle, Henriette Lapierre, Marcia Maia and Gaëlle Prouteau in providing us documents and/or reading chapters of this book, as well as Bernadette Coleno and Jean-Luc Travers for the tremendous work on the figures, and Marie-Pierre Schutz for typing the tables.

We would like to take this opportunity to thank the Springer-Praxis Geophysics Series Editor, Dr Philippe Blondel of the Southampton Oceanography Centre, for his careful review of this English-language edition, as well as our publisher, Clive Horwood (Springer-Praxis) and Neil Shuttlewood (Originator) and his typesetting team for the high-quality job they have done so efficiently.

Thanks to our graduate and Ph.D. students of U.B.O. (Université de Bretagne Occidentale, Brest): without them, this book would never have been written.

Finally, we heartily thank our wives Alicia and Marie-France for their patience and support during the long process of writing and translation of this book.

THIERRY JUTEAU and RENÉ MAURY

Contents

1

Oceans, continents, and the fundamental cycle of the oceanic lithosphere

Seen from space, the Earth's surface shows continental blocks decorated with high mountain ranges, separated by huge oceanic spaces, themselves dotted with volcanic islands and continent fragments. This present-day face of the Earth, as we know it today, is nothing but a transient image: at the geological time scale, the aspect of the Earth's surface is changing, continents and oceans move and are continually modified. These immense oceanic spaces of the 'blue planet' are ephemeral domains, whose lifetime takes generally less than 200 million years (Myr).

1.1 THE PRESENT-DAY FACE OF THE WORLD'S OCEANS

1.1.1 Surface and distribution of the world ocean

With a mean depth of 3700 m, the world ocean covers 361 million km^2, and is subdivided into three major oceans, the greater part of which is located in the southern hemisphere:

—The Pacific Ocean, by far the largest, covers 180 million km^2; that is, exactly 50% of the world ocean. It is also the deepest, with troughs deeper than 11 km.
—The Atlantic Ocean covers 106 million km^2; that is, 29% of the world ocean.
—The Indian Ocean covers 75 million km^2; that is, 21% of the world ocean.

As shown on the projection of Figure 1.1, the three oceans merge widely at the level of the 'Roaring Forties', and constitute effectively a unique world ocean from which continents emerge like giant islands. The Austral (Antarctic) and Arctic Oceans, whose boundaries are poorly defined, are generally incorporated within the three major ones, as are epicontinental seas like the North Sea for instance, the intra-mountain seas like the Mediterranean Sea, or the marginal seas like the Japan Sea.

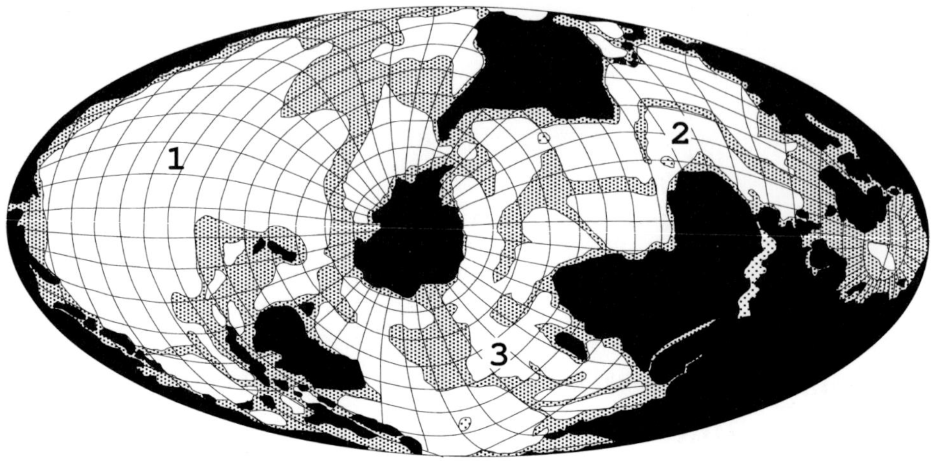

Fig. 1.1. The world ocean. This type of projection, centred on the Antarctic continent, allows visualization of the interconnection between the three major oceans at high latitudes in the Southern Hemisphere.

Figure 1.1 shows the unequal and grossly antipodal character of the distribution of land and sea: the northern hemisphere contains two-thirds of the emerged landmass, and the antipode of a continental point is most often marine (except for South America and Antarctica). In spite of this, the northern hemisphere has, however, more sea than land (53%). If the north pole were located in Britanny (western France), the contrast would be at a maximum between a terrestrial northern hemisphere and a southern marine hemisphere almost completely covered with water.

1.1.2 The fundamental duality of the Earth's crust

Figure 1.2 shows the frequency histogram and the hypsometric curve of altitudes and depths of the Earth's crust. The histogram shows a clearly bimodal distribution, with a maximum in the 0–1 km (altitude) range for continents, and a maximum in the 4–5 km (depth) for the oceans, with nearly three-quarters of the solid Earth's surface distributed between both maxima. This distribution reflects the duality of the Earth's crust, whose nature and thickness are fundamentally different for continents and oceans. The continental crust, with a thickness of 35 km on average, is mainly made of granitic and silica-rich metamorphic rocks (gneiss, micaschists, etc.). The oceanic crust, with a mean thickness of 7 km, has an overall basic composition: basaltic at shallow levels, gabbroic in depth. It results essentially from the emplacement and solidification of mantle-derived basic magmas in the axial zone of the oceanic ridges.

The oceanic crust (mean density 3) is denser than the continental granitic crust (mean density 2.7). This density difference is enough to explain the 4.5 km difference between the average depth of the oceans (3700 m) and the mean altitude of the

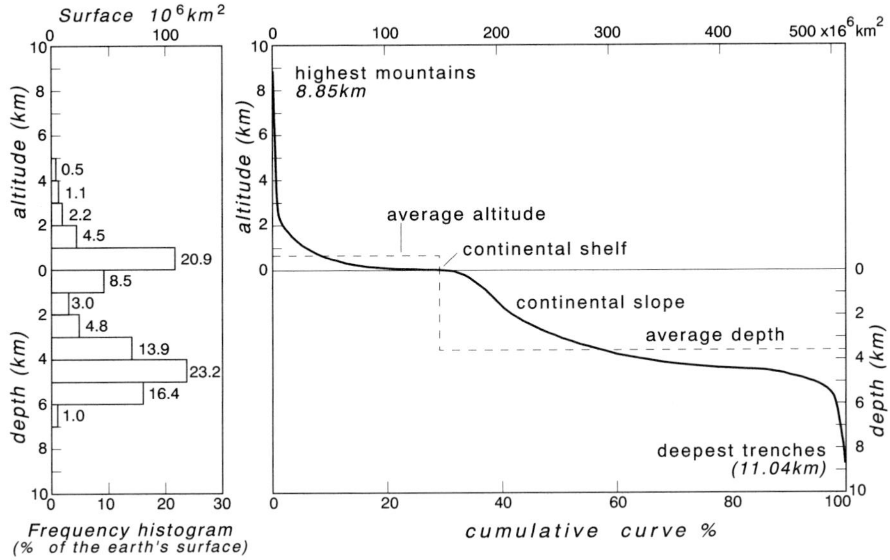

Fig. 1.2. Distribution of the Earth's surface levels. Left: frequency diagram. Right: cumulative curve of the frequencies, also called 'hypsometric curve'. This is NOT a profile of the Earth's surface. It is a curve which indicates the percentage of the Earth's surface located above any level. It can be used to calculate the percentage of the Earth's surface located below a given level, or comprised between two specific levels.

continents (800 m). This is a direct consequence of the isostasy principle: the external rocky envelope of the Earth tends toward the gravitational equilibrium by altitude adjustments between zones of different relative buoyancy.

The histogram and the hypsometric curve also show that 70 to 71% of the Earth's surface is located below the sea level. The maximum vertical offset of the terrestrial reliefs is 19.89 km; that is, 0.32% of the mean Earth's radius (6370 km).

1.1.3 The main physiographic features of oceanic basins

The map of the ocean floors (Plate 1 at the end of Chapter 11) has been known with reasonable precision for some 20 years. This map may puzzle the profane: the submarine reliefs are at least as complicated as the terrestrial reliefs! However, a more detailed examination identifies the important geological features.

The two most striking structures are the *worldwide oceanic-ridge system*, running across the three major oceans, and the *circum-oceanic system of deep trenches*, mainly circum-Pacific. Both types of seismically active structures reflect the mobility of the ocean floors. A new oceanic lithosphere is created continuously along the oceanic ridges by continuous seafloor spreading. The lithosphere thus created moves progressively away and symmetrically from the axial zone of the oceanic ridges, travels over hundreds to thousands of kilometers as a temporary ocean floor, and finally returns sooner or later to the deep mantle through the deep oceanic trenches marking the emplacement of the subduction zones.

Table 1.1. Surfaces covered by the main physiographic structures of the present-day oceans (after The Open University, 1989).

	Ocean			
	Pacific	Atlantic	Indian	World Ocean
Ocean surface (10^6 km^2)	180	106	75	361
Drained continental surface (10^6 km^2)	19	69	13	101
Average depth (m)	3940	3310	3840	3730
Continental shelf (%)	13.1	19.4	9.1	35.3
Continental slope (%)	2.7	8.5	5.7	5.3
Abyssal plains (%)	42.9	38.1	49.3	41.9
Intraplate volcanism (%)	2.5	2.1	5.4	3.1
Oceanic ridges (%)	35.9	31.2	30.2	32.7
Subduction trenches and island arcs (%)	2.9	0.7	0.3	1.7

Between oceanic ridges and subduction trenches are the immense sedimented abyssal plains, and also aseismic ridges, seamounts and submarine plateaus, some of them emerging to form intra-oceanic islands. Continental shelves, made of thick sediment accumulations, rim the continents (passive continental margins).

The surfaces covered by these various structures are summarized in Table 1.1.

1.2 SEAFLOOR SPREADING, FOUNDATION OF PLATE TECTONICS. OCEANIC CRUST, LITHOSPHERE AND ASTHENOSPHERE

Oceanic floors are mobile. According to plate-tectonics theory, the oceanic ridge represents an accretion centre, along which the seafloor-spreading process continuously occurs. In other terms, the ridge axis marks the trace of a deep fracture separating two divergent plates. A basaltic magma is continuously emplaced in the axial zone, whereas the adjacent plates separate.

The process responsible for magmatic production is the *adiabatic upwelling of the upper mantle* below the ridge: the upwelling is sufficiently fast (several centimeters per year) that the ascending mantle bodies, or 'mantle diapirs' (Nicolas, 1989, 1990), do not lose much of their heat and reach their melting point at about 60 km in depth (Figure 1.3). The magmatic rocks solidifying at the axis (basalts at shallow levels, gabbros in depth) form by cooling a rigid layer, the *oceanic crust*, about 7 km thick. The crust rests over a residual mantle which cools when moving away from the ridge axis, becomes rigid and thickens progressively at the bottom at the expense of the hot plastic mantle or *asthenosphere*, until it reaches its equilibrium thickness, between 70 and 100 km. Made of oceanic crust (upper part) and of upper residual mantle (lower part), this rigid layer constitutes by definition a *lithospheric plate* (Figure 1.3).

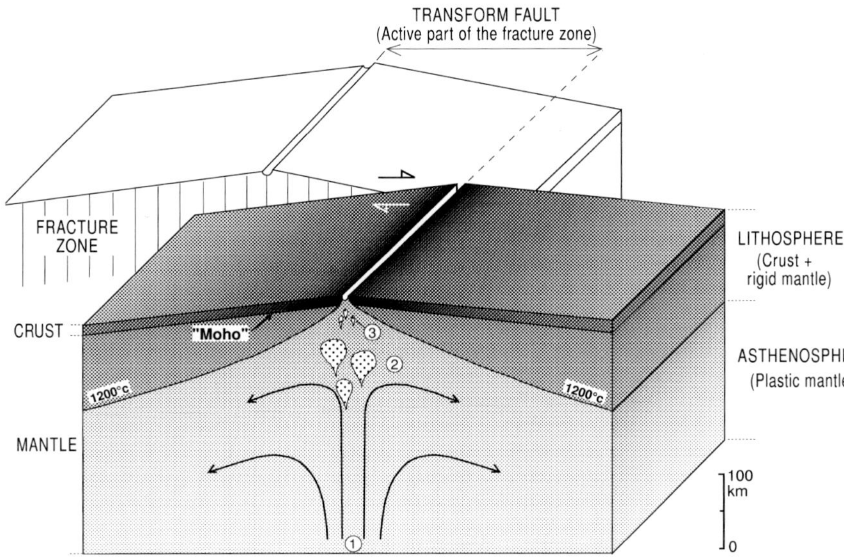

Fig. 1.3. Block-diagram showing the deep structure of the oceanic ridges (first plan). After Juteau (1993a). To the left are indicated the major lithological units, characterized by their mineralogical and geochemical compositions: these are the *crust* and the *mantle*, separated by the seismic Mohorovicic discontinuity, or 'Moho' in short. To the right are indicated the major physical units, characterized by their temperature and rheological (or mechanical) behaviour: these are the *lithosphere* (cold, rigid behaviour), and the *asthenosphere* (hot, plastic behaviour), separated approximately by the 1200°C isotherm. The lithosphere thickens progressively away from the oceanic ridge axis, at the bottom, at the expense of the asthenosphere. Below the ridge axis: (1) asthenospheric mantle upwelling; (2) partially molten mantle diapirs; (3) shallow accretion magma chambers. In the rear, the block-diagram shows an oceanic fracture zone and its tectonically active part, the *transform fault*, bounded by the crests of both ridge segments. Note the important vertical offsets on both sides of the fault, due to the thermal subsidence of the oceanic lithosphere with ageing.

The hypothesis of continuous seafloor spreading is today the foundation of plate tectonics. Thirty years of intense observation and modelling of the ocean floors have solidly confirmed it. Particularly, the hundreds of deep drillings done over the past 20 years in all the oceans of the world by the drilling ships *Glomar Challenger* (Deep Sea Drilling Project, or DSDP), and *Joides Resolution* (Ocean Drilling Programme, or ODP), have brilliantly confirmed the progressive thickening of the sediment layer and the more and more ancient age of the oceanic crust away from the ridge axis.

1.3 AGE OF THE OCEANIC CRUST AND KINEMATICS OF THE OCEANS

1.3.1 Magnetic anomalies, the growth records of the oceans

The map of oceanic magnetic anomalies constitutes a true image of the growth lines of the ocean basins (Plate 2 at the end of Chapter 11): it allows very precise reconstruction of the stages of growth of the oceans, and calculation of the oceanic opening rates and their eventual variations in the past.

During its cooling at the ridge axis, the basaltic magma acquires a thermoremanent magnetization when Ti-magnetite $(Fe, Ti)_3O_4$ passes below its Curie point (roughly 250° to 570°, according to titanium content, which lowers the Curie point temperature). The acquired magnetization is strong, of the order of several hundreds of nT (nano-Teslas), and parallel to the magnetic field of the epoch. We have known since the beginning of this century that the Earth's magnetic field has reversed many times during geological times, with a variable periodicity ranging from several tens of thousands of years to several million years. The first aeromagnetic surveys of the oceanic magnetic anomalies, realized to the South of Iceland or off the north-western American coast, revealed the now classical image of linear anomaly stripes, alternately positive and negative, parallel to the ridge axis and symmetrical with respect to it.

This remarkable pattern was then verified in all the oceans, and constituted a decisive argument in favour of the seafloor spreading hypothesis (Vine and Matthews, 1963). Admitting effectively that the basaltic layer is produced continuously at the oceanic ridge axis, it results that its magnetization must change in polarity at each inversion of the Earth's magnetic field. These magnetization contrasts of the basaltic layer are the source of the anomalies of the total magnetic field intensity recorded at the surface of the oceans (Figure 1.4).

As shown of Figure 1.4, each anomaly stripe created at the ridge axis is split in the middle by the following stripe, resulting in two half-stripes moving away symmetrically with respect to the ridge axis. The ages of the magnetic field reversals are well known today. The most important anomalies are numbered from 1 to 34 (starting from today) up to the 'magnetic rest' period of the Upper Cretaceous (no reversals for 35 million years), then M0 to M33 beyond this period, back to the Permian–Triassic period. The geomagnetic scale (Figure 1.5) and the map of the world ocean magnetic anomalies allows us to age the crust in any spot of the ocean with good precision.

It appears that the ocean floors are very young, since no oceanic crust older than 160 Myr was found in the present-day oceans. This surprising youth of the ocean floors contrasts with the age of the continents, where absolute ages as old as 3800 Myr were measured in various places.

1.3.2 Decrease in heat flow and seafloor altitude away from the ridge axis

The average temperature of the oceanic lithosphere is higher close to the ridge axis, where the heat flow is abnormally high, because of the presence of shallow magmatic reservoirs at more than 1000°C. Temperature and heat flow decrease with the age of the oceanic crust, hence with the distance to the ridge crest. Where the lithosphere is the hottest, it is also the least dense, and it 'floats' relatively high over the plastic asthenosphere. Calculations show that an oceanic lithosphere younger than 30 Myr floats over the asthenosphere, whereas an oceanic lithosphere older than 30 Myr is denser than the underlying asthenosphere and tends to sink spontaneously. When

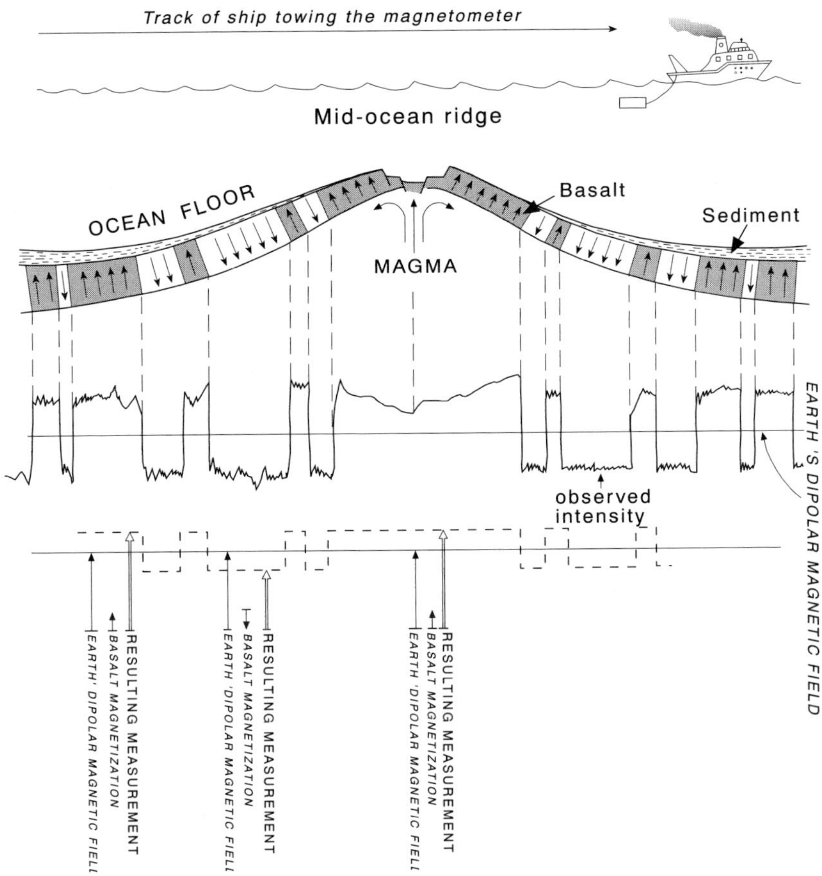

Fig. 1.4. Principle of creation and recording of oceanic magnetic anomalies (after Pomerol and Renard, 1995).

cooling, the lithosphere thickens at the bottom and the seafloor deepens (Figure 1.6), in order to maintain the isostatic equilibrium: this progressive subsidence is accompanied by burial beneath a sedimentary blanket whose thickness grows with age. The depth variation (Figure 1.6B) varies as the square root of the age until 70 Myr approximately. According to the Parsons and Sclater's formula (1977): $p = 2500 + 350t^{1/2}$, with depth p expressed in metres, and age t in million years. The simplest model to explain the variations observed is a model in which materials at high temperature are added at the margin of a plate cooling by simple conduction. The calculated contraction and the correlative density increase effectively explain the altitude decrease observed.

This simple and elegant relation between the age of the oceanic crust and its depth is precious, and allows estimation of the approximate age of any oceanic area by a simple glance at the bathymetric map.

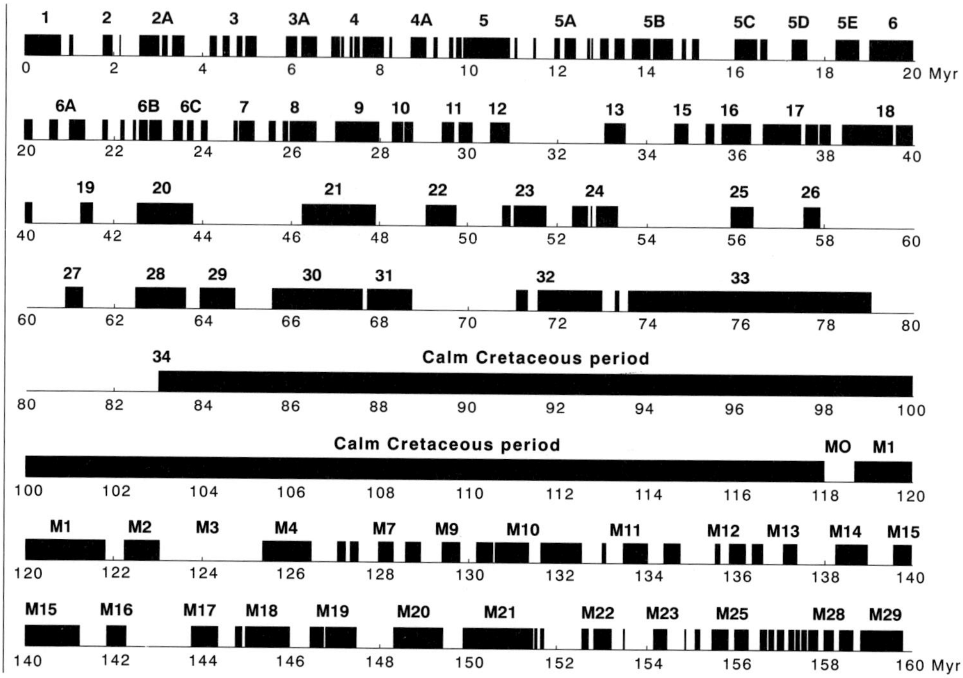

Fig. 1.5. Geomagnetic time scale, showing the reversals of the Earth's magnetic field during the last 160 Myr. In black: 'normal' magnetic periods. In white: 'reverse' magnetic periods. The ages are indicated in Myr, and the magnetic anomaly numbers increase backwards in time. They are labelled from 1 to 34 until the 'great Cretaceous quiet magnetic period', and then from M0 to M29 beyond this period.

The thickness e of the oceanic lithosphere increases with age according to a similar law: $e = 9.5t^{1/2}$ (Figure 1.6A).

1.4 THE AGE OF THE OCEANS AND THE EVOLUTION STAGES OF OCEANIC BASINS

1.4.1 Age of the crust and age of the oceanic basins

It is important to distingush between the *age of the oceanic crust* (cf. Plate 2 at the end of Chapter 11) and the *age of the oceanic basins.* The Pacific and Atlantic Oceans, for instance, exhibit identical crustal ages, particularly for the most ancient ages (around 160 Myr), but the Pacific Ocean is much older than the young Atlantic Ocean, born at the beginning of the Cretaceous during the break up of Pangea. The ancient history of the Pacific Ocean, traced back to at least 1 Gyr (10^3 Myr), has been simply erased by subduction of the pre-Jurassic oceanic lithosphere.

The examination of the map of the oceans reveals the present aspect of the various oceanic basins reached at different stages of their evolution. The complete life cycle

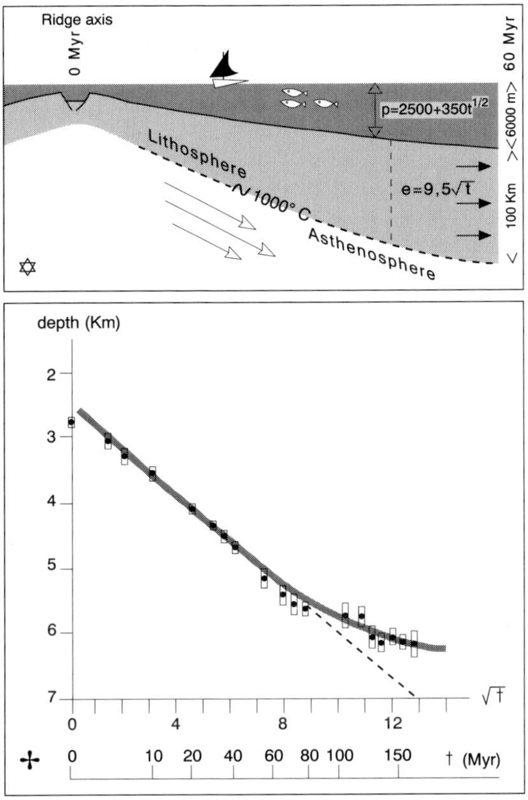

Fig. 1.6. (A) Depth variation of the ocean floor and thickening of the lithosphere with ageing of the crust. (B) Subsidence of the ocean floor with ageing of the crust (after Parsons and Sclater's data, 1977). Schemas inspired by Nicolas (1990, 1995). p: depth of the seafloor in km. e: thickness of the oceanic lithosphere in km. t: age of the crust in Myr (million years).

of an ocean should not be much longer than 200 Myr, since no oceanic crust older than 160 Myr is found in the present-day oceans, whatever their evolutionary stage. Now the oldest terrestrial rocks, found on the continents and dated at 3.8 Gyr, include both ancient marine sediments and relicts of very old oceanic crusts. We deduce from these facts that oceans were formed at the very beginning of the Earth's geological evolution, probably at about 4 Gyr. Supposing that the life cycle of a major ocean lasts about 200 Myr, it results that about 20 oceanic cycles may have occurred at the Earth's surface. This number is nothing but an order of magnitude, since the Earth's interior was much hotter in the remote past, and the recycling of the oceanic lithosphere was certainly a lot faster.

1.4.2 The Wilson cycle

Any oceanic basin begins by a stage of initial rifting, followed by a growth stage during which its surface increases and all its continental margins remain passive.

Later on, one or several of these margins begins to subduct and the oceanic basin shrinks, until the final continental collision which provokes its disappearance, leaving an *ophiolitic suture zone* in a continental domain (Figure 1.7).

Such a geological cycle, covering the opening and closing up of an oceanic domain with respect to its continental margins, is named the *Wilson cycle*, after the publication of the classical paper from Tuzo Wilson (1966), professor at Toronto University. In this paper, Wilson demonstrated not only the reality of the Pangea break up at the beginning of Mesozoic times, but also the existence of previous Pangeas that can only be explained by oceanic opening/closing cycles, implying important continental drift processes.

Each stage of this oceanic cycle is presently illustrated by one or several particular oceanic domains (Figure 1.7):

The *embryonic stage* (stages 1 and 2) can be observed in the great East African Rift, whose deep valley is developed over a major fracture in East Africa, and connects to the North with the world oceanic ridge system (Red Sea, Aden Gulf). The continental rupture is still incomplete, and the sea has yet to invade the rift, but one can already guess the contours of a future 'Somalian plate'.

The Red Sea of the Gulf of California provides the image of a *very young ocean in expansion* (stage 3), characterized by the active subsidence of its axial zone occupied by a young oceanic lithosphere, and by a thick terrigenous, evaporitic and pelagic sedimentary series, progressively building a shelf and continental slope along each margin.

The *mature stage* (stage 4) is illustrated by the present Atlantic Ocean, whose continental passive margins, still welded to the oceanic lithosphere, are now very far from the thermal influence of the mid-ocean ridge. This stage is characterized by a very important thermal contraction, with formation of immense abyssal plains adjacent to the ridge reliefs, and by the development of complete continental margins, with continental shelves, slopes and glacis. These margins remain parallel to the ridge track, which typically occupies the middle of the ocean. The lithosphere older than 110 Myr may begin to subduct locally (Antilla Trench, southern Sandwich Trench).

The *decline stage* (stage 5) involves the development of one or several subduction zones, island arcs and marginal basins. In spite of the presence of an active spreading ridge, the surface of the ocean shrinks inexorably. This is the stage of the present Pacific Ocean, whose very active ridge is apparently stimulated by the traction exerted by the numerous circum-Pacific subduction zones. The ridge is *excentred* and even collides with the American Continent at the 'Gulf of California'. The appearance of subduction zones in an ocean is probably the consequence of important geological events occurring in other parts of the globe (continental collisions, birth of new continental rifts, etc.).

The present Mediterranean Sea illustrates well the terminal stage of an ocean (stage 6). The immense Tethys Mesozoic Ocean began its closure phase at the time of the opening of the Atlantic, at about 160 Myr. Taken in a pincer movement between Africa–Arabia and India on one side, and the Eurasian continent on the other side, the Tethys Ocean closed progressively, subducting its lithosphere beneath Eurasia,

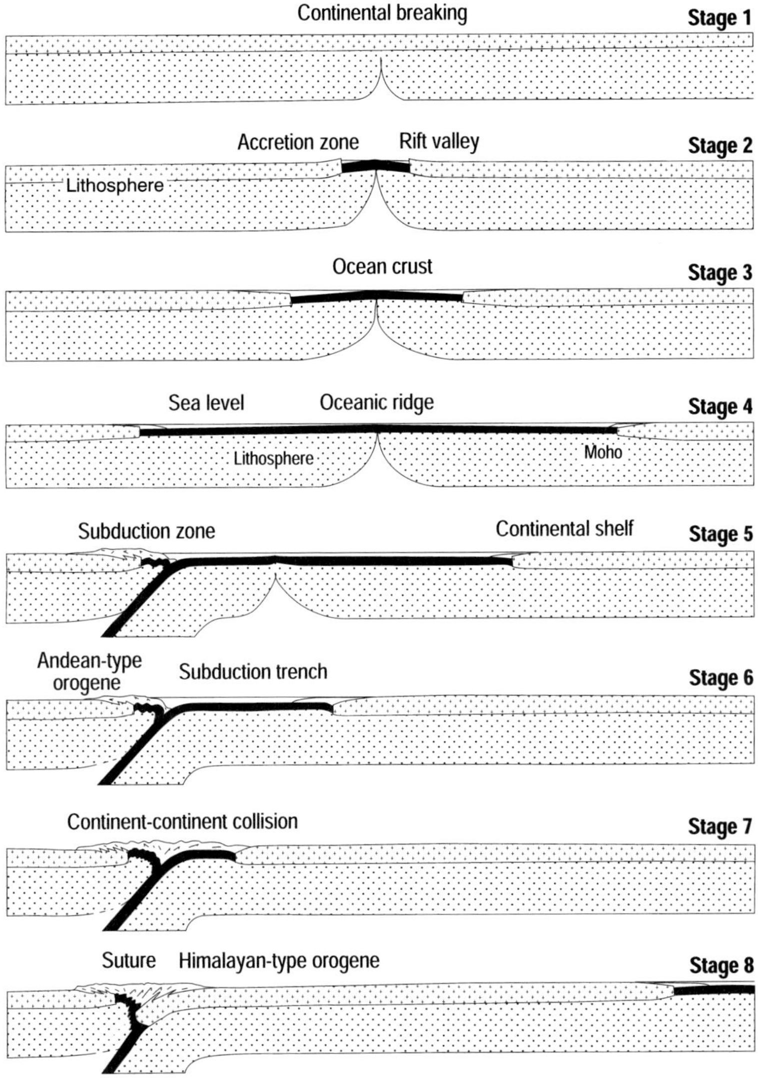

Fig. 1.7. Schematic illustration of a complete Wilson cycle, in the frame of lithospheric plate tectonics theory. This cycle, whose duration may be of the order of 200 Myr, comprises a stage of opening and *oceanic growth* (stages 1 to 4), with development of passive and symmetrical continental margins, and a stage of *oceanic closing* (stages 5 to 8), with the appearance of one or several subduction zones, and corresponding development of active, volcanized continental margins. The latest stage is a continental collision, resulting progressively in a Himalayan-type orogeny, in which an *ophiolitic suture zone* marks the emplacement of the disappeared oceanic domain. Comments in the text. Schema after Bird (1987).

including the ridge system which disappeared at the end of the Cretaceous. Some fragments of this lithosphere were tectonically obducted on the African, Arabic or Indian margins, forming numerous ophiolite nappes. The continental pincers are now in contact in several places, the bottom of the Mediterranean is uplifted and folded. Despite the fact that the ridge disappeared some time ago, small, ephemeral and transitory marginal or back-arc basins may have produced some recent lithosphere along limited accretion zones. The *collisional stage*, with residual *ophiolitic scar* (stages 7 and 8), may be observed in many places all along the Alpine-Himalayan mountain range (Indus suture zone for instance), where the continental collision provoked the upheaval of folded mountains at the emplacement of the ancient marine domain. This will be the destiny of the Mediterranean realm, if nothing comes and modifies the present movement of major plates.

This is how numerous oceanic basins developed and closed throughout the Earth's geological history, as attested by the ophiolitic sutures found in the major mountain range systems of all ages on the continents. Just as today, the oceans of a given epoch were probably competing, the youngest ones increasing their surfaces, the oldest ones shrinking, since it appears that the Earth's volume has not changed much during its evolution. In this case, the surfaces of the oceanic floor created at the ridge levels should be globally compensated by the destruction of identical surfaces in the subduction zones. No oceanic accretion without subduction, and vice versa!

1.5 THE FUNDAMENTAL CYCLE OF THE OCEANIC LITHOSPHERE

The oceanic crust constitutes the shallow part of the oceanic lithosphere. Its evolution in time and space, the subject of this book, occurs following a number of stages, constituting what can be called the fundamental cycle of the oceanic crust. These stages are as follows:

1.5.1 Extraction from the mantle: genesis of the oceanic crust

Created at the ridge axis by injection and solidification of magmas directly issued from the mantle, the new oceanic crust is made of a basaltic layer (about 2 km thick), laying over a gabbroic layer (about 4–5 km thick).

1.5.2 Basalt–seawater interaction

Once solidified, the oceanic crust is immediately percolated by seawater, with development of convective seawater cells in the crust resulting in the hydrothermal alteration and hydration of the oceanic crust. These mineralogical and chemical transformations are important: the prolonged seawater/basalt interaction results in recrystallizations with a range of metamorphic facies, ranging from zeolite to amphibolite facies. Polymetallic massive sulphides are precipitated by hot hydrothermal fluids on the seafloor, or in the crust at shallow levels, and the chemistry of seawater itself is modified.

1.5.3 Intraplate travel and meeting with hot spots

The oceanic crust thus mineralogically and chemically modified travels across the ocean floor over a period of tens of million years. During this travel, the crust will be charged with sediments, and submitted to various tectonic deformations and magmatic intrusions. Hot spots, in particular, will construct over the ocean floor huge submarine (seamounts) or aerial (volcanic islands) volcanoes. These overburdens result in the bending and sinking of the oceanic lithosphere and Moho, with metamorphic recrystallizations and remelting of the deep crust.

1.5.4 Return to the mantle

Sooner or later, the oceanic crust arrives at the level of a subduction trench, and will enter it, carrying the seamounts with it. During this progressive return to the mantle, the oceanic crust will be progressively dehydrated, with successive destabilization of the hydrated phases: clays and zeolites first, then chlorites, amphiboles, and finally micas. The oceanic crust recrystallizes first in the *amphibolite facies*, then in the *eclogite facies* (with phlogopite, and then anhydrous). At this stage, the oceanic crust has reached several kilometers in depth. It has been dismembered into blocks which deform plastically, forming a 'millefeuille' or a 'marble cake', contributing to the geochemical heterogeneity of the mantle.

During its descent, the crust may melt partially, forming *adakitic* magmas, or may provoke by its dehydration the melting of the mantle wedge of the overriding plate, with formation of tholeiitic to calc-alkaline magmas of the island-arcs or active continental margins. Secondary oceanic spreading zones may develop in back-arc basins.

1.5.5 Relict oceanic crust in a heterogeneous mantle

This heterogeneous mantle, several hundreds of million years later, may be eventually captured in an ascending mantle plume, rise up beneath an oceanic ridge and melt partially to create a new oceanic crust.

The tour is done. True foam of the mantle, the oceanic crust floats at the surface of the oceanic lithosphere until, by subduction and after several recrystallization episodes, it becomes finally denser than the surrounding mantle, and is incorporated by it as thin eclogitic layers, waiting for a new melting episode.

1.6 THE OPHIOLITIC EXCEPTION

In the vicinity of subduction zones, island-arcs or continental collisions, it may happen that fragments of oceanic lithosphere, instead of returning to the deep mantle, are thrust over continental margins or island-arcs, and incorporated in mountain-range systems. Slowly prepared by intense intra-oceanic slicings, the tectonic emplacement of these ophiolite nappes today allows detailed study *on land* of the structure and composition of fossil oceanic lithosphere fragments, precious witnesses of the oceans of the past.

These ophiolites often represent oceanic lithosphere fragments created not on the worldwide oceanic ridge system, but in marginal or back-arc basins near the continental or island-arc margins, along secondary and localized spreading zones. We shall describe the structure of ophiolites in Chapter 6.

2

Magma series from oceanic domains

Basalts are the dominant volcanic rocks in oceanic zones. They make up the upper part of the oceanic substratum, except in the zones where the upper mantle outcrops (Chapter 5), and they are only rarely associated with intermediate lavas resulting from their differentiation (ferrobasalts). In the emerged ridge sectors (Iceland), intermediate and evolved volcanic rocks are more common (icelandites, rhyolites), but basalts are still the predominant rock type, as well as in oceanic intraplate linear volcanic chains (Hawaii, Polynesia). The magmatic series from subduction zones are also generally derived from basaltic parent magmas, even if intermediate lavas (andesites) are sometimes more common in the surface than basalts. Magmas derived from sources other than the upper mantle are uncommon in oceanic areas (adakites from subduction zones). The knowledge of magmatic processes in the oceans is thus mostly based on the study of basalts, which also provide essential information about the compositions of their mantle sources.

2.1 TECTONIC SETTINGS OF OCEANIC BASALTS

The existence of links between the petrologic and geochemical characteristics of basalts and their tectonic settings has provided support to plate-tectonics theory. Initially based on major element chemistry, and especially on the notion of silica saturation (silica-saturated or -oversaturated tholeiites and silica-undersaturated alkali basalts (H. S. Yoder and C. E. Tilley, 1962), the distinctions between basaltic types have been refined with the help of incompatible trace elements and Sr, Nd, Pb isotopic ratios. In turn, numerous researchers tried, following J. A. Pearce and J. R. Cann (1973), to use the compositions of metavolcanic rocks to identify the tectonic setting of their emplacement, using 'discriminant' or 'geotectonic' diagrams. These diagrams are established from geochemical databases about young magmatic rocks from the ocean crust, island arcs and active continental margins, intraplate volcanoes, etc. Although this approach has initiated important progress in the knowledge of the geodynamic history of the Alpine and Himalayan chains, it also resulted in some confusing overlaps between the classifications of magmatic rocks derived from petrologic/geochemical features and those based on the usual tectonic setting of their emplacement.

There are five major types of modern oceanic basalts according to their various tectonic settings.

—The MORB (Mid-Ocean Ridge Basalts) correspond to the basalts emplaced along the axes of mid-oceanic ridges or sometimes from submarine eruptive centres located a few km away from these axes (off-axis seamounts). They make up most of the upper part of the 'normal' oceanic crust, as submarine flows and dykes, and represent the initial magmas from which oceanic gabbros crystallized.

—The BABB (Back-Arc Basin Basalts) are emplaced along the axes of the back-arc basin spreading ridges. They compose the upper crustal sequence of back-arc (or marginal) basins, the width of which is variable: 60 km (Okinawa Trough, behind the Ryukyu Arc), 450 km (northern part of the Lau Basin, behind the Tonga Arc), 700 km (North Fiji Basin), and up to 1000 km and more for marginal seas (e.g. the South China Sea).

—The OPB (Ocean Plateau Basalts) have been emplaced as thick piles of submarine lava flows in oceanic intraplate settings. The largest known examples are of Cretaceous age, e.g. the Nauru, Ontong Java and Manihiki plateaus in the Pacific, the Kerguelen plateau in the Indian Ocean, and the fragmented Caribbean plateau.

—The OIB (Ocean Island Basalts), also emplaced in oceanic intraplate settings, may constitue isolated seamounts or islands (St Helena, Réunion, the Kerguelen). More often, these islands and seamounts are grouped as linear chains (Hawaii-Emperor seamounts, Austral-Cook, Society) which in some cases straddle the ocean-continent boundaries (e.g. the oceanic islands of Annobon and Fernando Poo which define a single trend with Mt Cameroon volcanics). The OIB also occur in aseismic volcanic ridges like the Walvis Ridge in the South Atlantic or the Ninetyeast Ridge in the Indian Ocean.

—The IAB (Island Arc Basalts) are typical of island arcs and Andean-type active continental margins. Like the OIB, they display a rather wide range of petrologic and geochemical variations.

If we consider, in addition to typical oceanic domains, the intracontinental rifts, the above list has to be extended to the CTB (Continental Tholeiitic Basalts, also called trap basalts or flood basalts). These continental tholeiites show many similarities with MORB, but their emplacement through continental crust lead to chemical interactions with this crust. They are emplaced during the early stages of rift opening (Ethiopia, Kenya, South Kivu). They also occur in the continental Large Igneous Provinces (e.g. the traps of Deccan, India and Paraná, Brazil).

2.2 UPPER MANTLE MELTING

The high emplacement temperatures of basaltic magmas (ca. 1200°C) translate their deep origin (at least 60 km admitting an average temperature increase of 20°C/km in the crust and upper mantle), and their sources are thus located within the upper mantle. The Low Velocity Zone (LVZ), located at depths ranging from 70 to 120 km, and interpreted as a partially molten mantle zone, represents the most likely source of common basalts.

2.2.1 Composition of the upper mantle

Upper mantle rocks outcrop at the axes of slow-spreading oceanic ridges and within transform faults (Chapter 5), in Alpine-type chains (ophiolitic mantle: Chapter 6; high-temperature peridotite massifs: Chapter 11), and finally as very common peridotites and pyroxenites brought to the surface as xenoliths within alkali basalt magmas and kimberlites. The xenoliths from the latter, which contain diamonds stable at depths greater than 150 km (Chapter 11), represent the deepest Earth materials available for direct mineralogical and geochemical studies.

The two main kinds of mantle peridotites are the lherzolites and the harzburgites, respectively. The lherzolites are frequently found as xenoliths within basalts from continental rifts, oceanic and continental hot spots, and finally represent the dominant rock type in high-temperature 'orogenic' peridotitic massifs (which include the Lherz massif in the Pyrenean chain). They contain Mg-rich olivine and orthopyroxene (bronzite), calcic clinopyroxene of diopside type (>5 modal%), and finally an Al-rich mineral phase corresponding successively, under increasing pressures (Figure 2.1) to calcic plagioclase, Al–Cr spinel and Al–Mg garnet (pyrope).

Harzburgites make up most of the mantle sequence of ophiolites (with the exception of Alpine-type ophiolites; Chapter 6). They are also found at the axes of mid-ocean ridges and in transform faults. In these two latter kinds of occurrences, they have usually been converted to serpentines through hydrothermal alteration processes (serpentinites). When fresh, the harzburgites contain olivine, orthopyroxene slightly more calcic than that of lherzolites, low amounts of diopside (<5% and often <3% in volume), and, as the lherzolites, an Al-rich mineral (plagioclase, spinel or garnet depending on the pressure range).

There are several other petrographic types of mantle rocks. Dunites, usually associated with harzburgites, are poorer in pyroxenes (>90 modal% of olivine). Wehrlites are richer in clinopyroxene than the lherzolites with which they are often associated, together with pyroxenites (containing more than 40 modal% of pyroxenes). Peridotites containing undeformed Mg-rich amphibole (pargasite) and mica (phlogopite) occur as xenoliths within kimberlites and, exceptionally, into island arc calc-alkaline lavas (§11.3.3).

Clinopyroxene is the common peridotite mineral most able to host significant amounts of incompatible elements, i.e. elements which are not usually incorporated within crystal lattices and thus become concentrated within the liquid phase during petrogenetic processes involving solid–liquid equilibrium (e.g. partial melting or

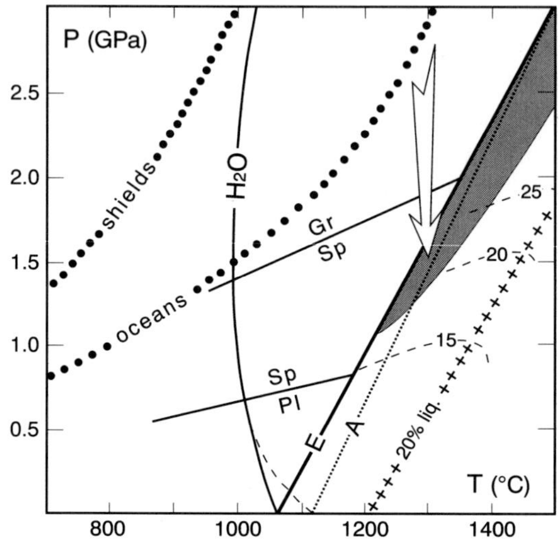

Fig. 2.1. Conditions of formation of basaltic magmas (after Jaques and Green, 1980, in Maury, 1993). Bold line: solidus of an enriched peridotite (lherzolite) in dry conditions (E) and in the presence of water (H_2O). Dotted line (A): dry solidus of a depleted peridotite. The line labelled '20% liq.' corresponds to a melt percentage of 20%; numbers 15, 20, 25 denote the percentages of normative olivine of the liquids. Below the solidus, the lines labelled Pl–Sp and Sp–Gr are the boundaries between the fields of plagioclase (Pl, stable at low pressures), spinel (Sp) and garnet (Gr) in peridotites. The curves labelled 'shields' and 'oceans' correspond to standard P–T paths in these areas.

fractional crystallization). Potassium is a typical incompatible element: despite its low concentrations in mantle rocks, it is more abundant in lherzolites than in harzburgites (Table 2.1).

Table 2.1 also allows comparison of the average composition of harzburgites (HZ) and lherzolites (LZ) with that of the theoretical primitive Earth's mantle (PM)

Table 2.1. Average compositions (oxide weight %) of the main components from the crust and upper mantle.

	PM	LZ	HZ	N-MORB	AN	OIB	CC
SiO_2	46.0	44.2	41.1	50.5	57.6	49.2	57.3
TiO_2	0.2	0.1	0.1	1.6	0.8	2.9	0.9
Al_2O_3	4.1	2.1	0.6	15.3	17.3	15.4	15.9
FeO*	7.5	8.3	10.6	10.4	7.1	6.8	9.1
MgO	37.8	42.2	46.3	7.6	3.6	9.2	5.3
CaO	3.2	1.9	0.2	11.3	7.2	9.6	7.4
Na_2O	0.3	0.3	0.3	2.7	3.2	3.4	3.1
K_2O	0.03	0.06	0.03	0.1	1.5	1.4	1.1

PM: primitive mantle; LZ: lherzolites; HZ: harzburgites; N-MORB: depleted ocean crust basalts; AN: arc andesites; OIB: intraplate island basalts from Central Pacific; CC: continental crust. Sources: PM, N-MORB, CC: Hofmann (1988); LZ: Maaloe and Aoki (1977); HZ: Harris *et al.*, (1967) AN: Gill (1981).

after segregation of the core but before the formation of the crust. MgO contents increase from PM to LZ and HZ. Magnesium is a compatible element which is preferentially incorporated in octahedral sites and is thus concentrated within early-crystallized iron–magnesium silicates (olivine, orthopyroxene, clinopyroxene) and consequently depleted in the liquids in equilibrium with them. On the other hand, Si, Al, Ca elements which tend to concentrate into early magmatic liquids, decrease from PM to LZ and HZ (Table 2.1). Thus, with respect to lherzolites, harzburgites represent incompatible element-depleted (and, in turn, compatible element-enriched) peridotites which can derive from the former through extraction of magma during one or several partial melting events. They are considered as residual peridotites, representative of the solid residues left in the mantle after partial melting followed by magma extraction. Lherzolites, which may represent the solid mantle prior to melting events, are considered as enriched (in incompatible elements, with respect to harzburgites) or less commonly fertile (i.e. able to produce basaltic magmas) peridotites. However, modern lherzolites are nevertheless depleted/residual with respect to the composition of the primitive mantle, because of the numerous melting events which affected the latter since the initial stages of planetary differentiation.

2.2.2 The genesis of basaltic magmas

The results of experimental studies of peridotite melting conditions are shown in Figure 2.1. In dry conditions (lack of free water and of water-bearing minerals) melting starts at 1100°C at low pressures, and 1200°C under 0.5 to 1.0 GPa depending on the depleted or enriched character of the mantle. Melting rates of ca. 20% require temperatures 200°C higher, unlikely to occur under normal Pressure-Temperature (P-T) regimes since the end of the Archean. Under relatively low pressures, basaltic magmas derived from relatively high partial melting degrees are silica-saturated or -oversaturated. For lower partial melting degrees and/or higher pressures, they become silica-undersaturated (shaded field in Figure 2.1). In presence of water (generally provided by the dehydration of amphibole and/or phlogopite from metasomatized peridotites from arc mantle), partial melting starts at significantly lower temperatures of ca. 1000°C, whatever the pressure in the 0.5–3 GPa range. Typical P-T paths (geotherms) of stable continental zones (shields) and oceans are shown in Figure 2.1. Whatever the pressure, the P-T path of shields never encompass the peridotite solidus, either in dry conditions or in the presence of water. Consequently, mantle melting does not occur in normal P-T conditions below stable continental areas.

On the other hand, the oceanic geothermal curve encompasses the solidus of mantle peridotites in the presence of H_2O. The suboceanic mantle will thus suffer partial melting at depth, provided it contains either free water or water-bearing minerals such as pargasite and/or phlogopite.

Figure 2.1 shows that, under normal P-T conditions, the geothermal curve of oceanic domains does not cross the dry peridotite solidus. This implies that, like the subcontinental mantle, the suboceanic mantle will only be affected by partial melting under specific conditions. These can include a temperature increase displacing towards the right the oceanic geotherm in Figure 2.1 (hot-spot conditions), or more frequently an ascent of deep and hot asthenospheric mantle as diapirs or plumes. If this ascent is rapid, it will occur without significant loss of heat (adiabatic ascent) as shown by the arrow in Figure 2.1. Such a situation allows the uprising mantle to cross the dry peridotite solidus, and to initiate partial melting. This process prevails in all tectonic settings allowing diapiric ascent of deep mantle, i.e. below oceanic and back-arc ridges and plume-related magmatic intraplate provinces (continental rifts, oceanic plateaus, hot spots).

2.2.3 Magma extraction and ascent

The conditions of magma extraction from the partial melting zone, usually assimilated to the LVZ, are still a matter of debate. The first critical point concerns the mode of extraction of the liquids, which can be described by two extreme models. In the batch melting (or equilibrium melting) model, the magmatic liquid is extracted from its source peridotites at the end of the melting process. Thus, its composition is homogeneous, and it averages the effects of the chemical variations that occurred during the successive steps of the melting process. Contrarily, the fractional melting model, involves successive extractions of the liquids during the melting process. This model implies liquid compositional changes: the early magmas are rich in incompatible elements, the later ones less so, the greater part of which have already been extracted from the source materials by the former liquids. These chemical variations are amplified by the fact that the source peridotite minerals do not behave similarly during the melting process. The clinopyroxenes, which melt easily, will contribute more than the other minerals to the composition of the liquids, especially during the early steps of the melting process. The olivines, which are rather refractory, will have an opposite behaviour. Thus the source minerals will not contribute to the melts in proportions consistent with their modal abundances (non-modal melting). More complex melting models have also been developed. The dynamic melting model, for instance, considers melting within an uprising diapir or plume, with increasing degrees of partial melting and progressive but incomplete extraction of the liquids during the ascent.

The uprise of basaltic magmas is controlled by their density (2.8–3.0 at 1200°C) which is lower than those of the host mantle peridotites (3.2–3.3). Consequently, the lighter magmatic liquids are pushed towards the surface by the effects of lithostatic pressure. The density contrast between the basaltic magmas and their host peridotites is considerably attenuated at depths of 200 km or more, which makes rather unlikely the extraction of magmas of deep origin, which are indeed rarely observed at the surface (kimberlites).

It seems intuitively difficult that melts resulting from very low degrees of partial melting could be easily extracted from the mantle, because partial melting starts at grain boundaries, and the corresponding early liquid patches are not connected to one another. There is thus a permeability limit below which magma extraction seems precluded or at least extremely difficult. This limit is usually taken as close to 1% of partial melting (Chapter 8). Fractal models allowing extraction of liquids for melting degrees as low as 0.1%, via a network of microchannels connecting triple-grain boundaries, have been developed, but are not yet supported by experiments or natural observations.

The range of effective melting degrees of mantle peridotites is not precisely known. Geochemical models lead to comparative estimations: for instance, alkali basalts from intraplate ocean islands derive from melting rates twice as small as the associated tholeiitic basalts (§10.2). Experimental studies suggest that common basaltic magmas derive from mantle peridotites through melting rates ranging from 3 to 15%. Much higher degrees (20–30%) would be involved only for unusual magmatic rocks (komatiites, § 10.3.4; boninites, § 11.4.2), and very low ones (1 to 3%) for the most alkali-rich mafic magmas (nephelinites, lamprophyres).

The most poorly known processes affecting basaltic magmas are those linked to their percolation within the lithospheric mantle and the mineralogical and chemical interactions occurring during their ascent. Some examples (Tubuai, §10.2.3) suggest that basaltic magmas derived from the melting of garnet peridotites can re-equilibrate with spinel peridotites when stored at their level in the lithospheric mantle. The 'subduction-related' trace element signature (negative Nb and Ta anomalies) of some post-collisional basalts could have been acquired during their ascent through subduction-modified upper mantle previously underlying a magmatic arc. Chromatographic models have been applied to the study of magma–mantle interactions, and it is widely admitted that both the subcontinental and the suboceanic lithospheric mantle can be extensively modified (and especially enriched in incompatible elements) through metasomatic reactions with percolating magmas of deep origin.

2.3 OCEANIC BASALTS AND MANTLE HETEROGENEITIES

The chemical and mineralogical diversity of basaltic magmas from oceanic domains has been known for a long time (Lacroix, 1927). It was first interpreted as traducing variable partial melting degrees of a theoretical homogeneous mantle source (e.g. the 'pyrolite' of D.H. Green and A.E. Ringwood, 1967), tholeiitic basalts deriving from partial melting degrees higher than those involved for alkali basalts. Radiogenic isotope (Sr, Nd, Pb) studies have indisputably demonstated that the main types of oceanic basalts derive from different sources. Indeed, radiogenic isotopes are not fractionated with respect to stable isotopes of the same element during partial melting (as well as during fractional crystallization). Thus, the isotopic ratios of basaltic magmas are similar to those of their source at the time of partial melting. Numerous studies have demonstrated that they are rather variable, suggesting that the mantle sources of common basaltic magmas are heterogeneous.

2.3.1 Isotopic compositions of oceanic basalts

The strontium and neodymium isotopic compositions of various types of young oceanic basalts is shown in Figure 2.2. ^{86}Sr and ^{144}Nd are stable isotopes, and ^{87}Sr

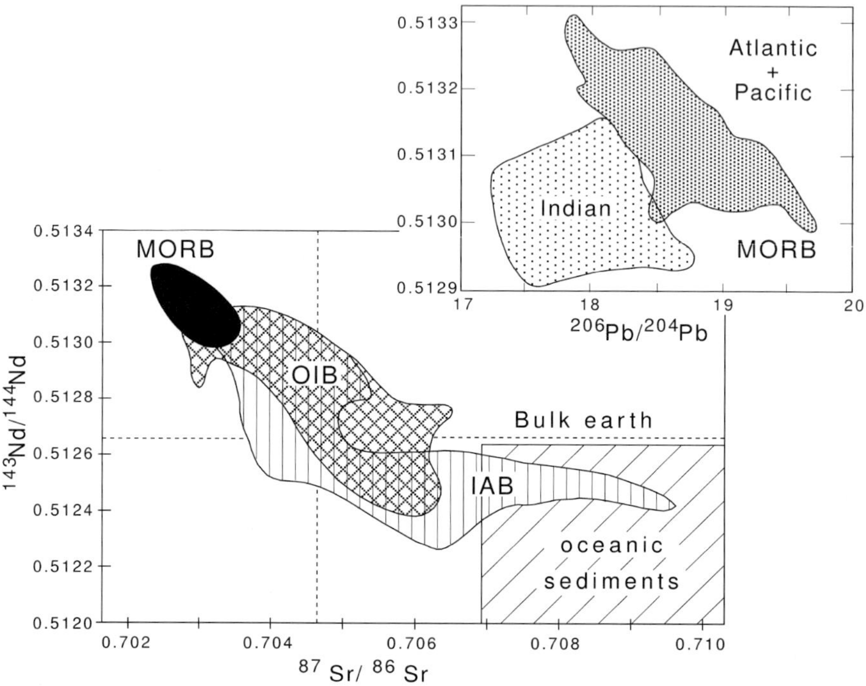

Fig. 2.2. Sr–Nd isotopic composition of oceanic magmas (simplified from Hawkesworth *et al.*, 1994). Black area: field of MORB; crossed hatches: field of OIB; vertical hatches: field of IAB; oblique hatches: field of oceanic sediments. The associated Nd–Pb diagram shows isotopic differences between MORB from the Indian, Atlantic and Pacific Oceans.

and ^{143}Nd radiogenic isotopes which derive from the radioactive decay of ^{87}Rb and ^{147}Sm, respectively. Rubidium is more incompatible than strontium and samarium slightly less incompatible than neodymium in mafic/ultramafic systems. A young magmatic rock derived from the partial melting of an ancient source, enriched in incompatible elements with respect to the bulk Earth, will thus be characterized by high ^{87}Sr/^{86}Sr ratios together with low ^{143}Nd/^{144}Nd ratios: it is the case for some OIB and IAB. The continental crust and the oceanic sediments display similar features with respect to the bulk Earth composition. On the contrary, the MORB are less radiogenic in Sr and more radiogenic in Nd than the bulk Earth, which implies that they derive from incompatible element-depleted mantle sources (see Hall, 1987 and Wilson, 1989 for further explanations).

The isotopic compositions of MORB are much less variable than those of other types of oceanic basalts, which suggest that they derive from a similar type of depleted mantle source. However, there are significant differences between the Indian Ocean MORB and the Atlantic and Pacific MORB (Figure 2.2). The former are characterized by ^{143}Nd/^{144}Nd and ^{206}Pb/^{204}Pb at a lower level than the others, which indicates that they derive from mantle sources with higher Nd/Sm and U/Pb ratios, i.e. enriched in the most incompatible elements with respect to that of the other MORB. This peculiarity is referred to as the DUPAL anomaly (Chapter 8).

2.3.2 Metasomatism or contamination?

The hypothesis according to which the isotopic composition of a basalt reflects that of its mantle source is only true if: (i) the analysed rock is fresh, i.e. devoid of significant secondary minerals formed during weathering, low-temperature interactions with seawater or hydrothermal alteration (Chapter 9), and (ii) if its parent magma has not been contaminated, i.e. modified through interactions with the lithospheric rocks during its ascent and occasional storage within magma reservoirs.

For instance, the identification of a 'sedimentary' geochemical imprint affecting an arc basalt, plotting close to the field of oceanic sediments (see the bottom right part of the Nd–Sr isotopic plot in Figure 2.2), may lead to very different interpretations. It may reflect the contamination of an IAB magma by sediments from the arc crust through which it was emplaced (crustal contamination). Alternatively, the mantle source of the basalt arc may be a depleted mantle similar to the source of MORB which has been modified by a chemical input from subducted sediments (mantle metasomatism). Finally, the arc magma may derive from the melting of the subducted ocean crust, involving its MORB and overlying sediments (melting MORB + sediments).

An elegant method proposed by James (1981) allows discrimination between these hypotheses by the coupled use of a radiogenic isotope ratio, e.g. strontium (or neodymium, or lead) together with another isotopic tracer, the stable oxygen isotope ^{18}O (Figure 2.3). The $^{18}O/^{16}O$ isotopic ratio is expressed as $\delta^{18}O$, which denotes the relative difference (as per mil) between the $^{18}O/^{16}O$ ratio of the studied sample and that of a standard, usually the Standard Mean Ocean Water (SMOW). The $\delta^{18}O$ ratio of sediments (+22) is much higher than that of the upper mantle and the magmas derived from its melting (+6). When oxygen is found in nearly equivalent concentrations in silicate rocks (where the cation charges are saturated by oxygen), the shape of the mixing curves describing the variations of $^{18}O/^{16}O$ and $^{87}Sr/^{86}Sr$

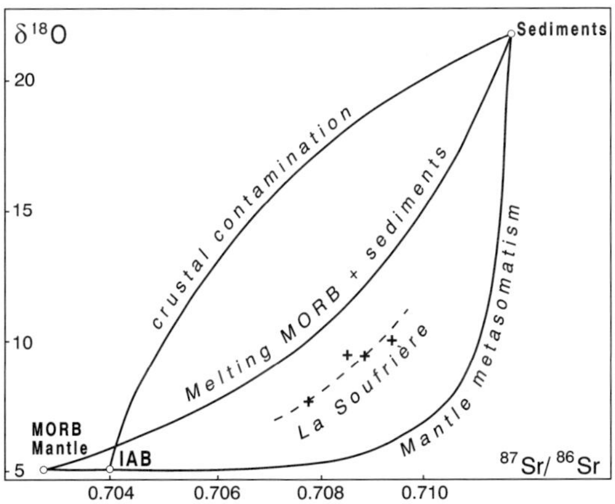

Fig. 2.3. Oxygen–strontium isotopic plot showing various mixing curves between mantle or basaltic magmas and sediments. (from Vidal *et al.*, 1991). See text for explanations.

isotopic ratios in the bulk mixes of IAB plus sediment, mantle plus sediment, or MORB plus sediment, will depend only on the concentrations of strontium in the mixing end-members. These are very variable. For instance, in the case of lavas from the St Lucia Soufrière, Lesser Antilles arc (Vidal *et al.*, 1991), Sr = 400 ppm in the sediments, 800 ppm in the IAB, 130 ppm in the MORB and 13 ppm in the postulated mantle source. Low amounts of Sr-rich sediment added to an Sr-poor mantle lead to a strong increase in the Sr isotopic ratio of the mix without changing its oxygen ratio significantly (concave-upwards 'mantle metasomatism' curve). The mix of MORB (Sr = 130 ppm) and sediments will be translated by a less-bent line ('melting MORB + sediments') because of the less-contrasted Sr contents of the two end-members. Finally, the contamination of an IAB very rich in Sr by less rich sediments will correspond to a concave-downwards mixing curve ('crustal contamination'). The position in this diagram of the analysed lavas allows the determination of the dominant petrogenetic mechanism, which in the case of St Lucia proves to be a combination of mantle metasomatism and melting processes.

2.4 THE DIVERSITY OF BASALTS FROM OCEANIC DOMAINS

The compositions of basalts from oceanic domains (including ocean ridges and basins, back-arc basins, oceanic plateaus and linear intraplate chains, and intra- and peri-oceanic volcanic arcs) integrate their whole geological history and that of their sources, namely: (i) their source features (depleted, enriched, metasomatized mantle rocks of variable mineralogy and age); (ii) the conditions of partial melting; (iii) the conditions of magma extraction from their source and its interaction with the upper mantle; (iv) their evolution during ascent through the oceanic or continental crust and their storage within magma reservoirs (fractional crystallization coupled or not with crustal contamination, magma mixing), and finally; (v) their chemical modifications during post-magmatic processes, e.g. low-temperature alteration by sea-water or interactions with hydrothermal fluids (see Chapter 9).

2.4.1 Classification

Basalts are mafic volcanic rocks rather poor in silica ($SiO_2 < 53\%$) and magnesium-rich ($MgO > 5\%$). Their average SiO_2 content is ca. 50% (Table 2.1), but some types of silica-undersaturated basalts contain only 40–45%. MgO contents of basalts are also very variable, and can reach up to 25%. Magnesium is a compatible element concentrated within olivine, which is the principal mineral of mantle peridotites, and crystallizes as early phenocrysts in most mafic magmas. A high MgO content in a basalt will indicate either that it represents a genuine product of mantle melting still in chemical equilibrium with peridotites (primitive basalt) or alternatively that it has mechanically concentrated olivine phenocrysts during its storage or emplacement at the surface. Quantitative models of partial melting of lherzolites show that 'reasonable' equilibrium melting degrees (less than 15%) produce liquids containing 12 to 8% MgO, which provides a criterion for the identification of primitive basalts.

Evolved basalts ($MgO = 8$ to 5%), also called low-Mg or Mg-poor basalts, have usually experienced fractional crystallization involving fractionation of olivine. Picritic basalts ($MgO > 12\%$) represent either magmas resulting from unusually high rates of partial melting, or alternatively derived from a source more Mg-rich than the lherzolites (e.g. harzburgites: Table 2.1), or finally highly porphyritic basalts which have accumulated olivine phenocrysts formed at the expense of other magmas ('oceanites' from Réunion and Polynesia). The classifications of basalts have for a long time been exclusively based on major element analyses, and more especially, following Yoder and Tilley (1962) on the concept of silica saturation. Most microlitic basalts contain interstitial glass of either rhyolitic, trachytic or phonolitic composition, which hides their silica-oversaturated or -undersaturated character from the petrographic analysis. Cross, Iddings, Pirsson and Washington (1903) proposed a method of recalculation (known as the CIPW norm) of the major element analyses of basalts to determine their silica-oversaturated (normative quartz plus hypersthene), -saturated (normative olivine plus hypersthene) or -undersaturated (normative olivine plus nepheline or nepheline + leucite) states. MORB, BABB, OPB and the vast majority of IAB are silica-saturated or -oversaturated. On the other hand, OIB include silica-undersaturated varieties (alkali basalts with less than 5% normative nepheline; basanites with more than 5%; nephelinites devoid of normative alkali feldspars) as well as silica-saturated or -oversaturated basalts (ocean island tholeiites). These two OIB types are often associated in a single volcano, and can present similar isotopic features. An efficient discrimination of various types of oceanic basalts is thus necessarily based on both major and trace elements, together with selected isotopic features.

Phenocrysts appear in basaltic magmas following a roughly constant order of crystallization at decreasing temperatures: (1) Cr–spinel (magnesiochromite) as tiny octahedra often found as inclusions within olivine phenocrysts, (2) Mg-rich olivine, (3) calcic plagioclase, (4) calcic clinopyroxene. In water-rich magmas (calc-alkaline and some alkaline basalts), clinopyroxene begins to crystallize before plagioclase, and titanomagnetite phenocrysts appear in fifth position.

Petrographic features of oceanic basalts provide some clues to their diversity. MORB are often subaphyric, IAB usually highly porphyritic (15–50 modal% of phenocrysts, plagioclases being the most abundant). IAB commonly contain orthopyroxene phenocrysts which are lacking in the other types of basalts. Plagioclase phenocrysts from alkali basalts (labradorite with An 50–70) are more sodic than those of tholeiites and calc-alkaline basalts (bytownite with An 70–90). Feldspathoids (nepheline, leucite and sodalite-type) cristallize as small interstitial patches in the groundmass of alkali basalts, and are often found as phenocrysts or microphenocrysts in the most silica-undersaturated types (nephelinites, leucitites). The alkali basalt groundmass commonly contains, in addition, interstitial alkali feldspars (soda sanidine, anorthoclase) which are missing from other basalts. Clinopyroxene compositions differ between tholeiitic and calc-alkaline basalts and alkali basalts. In the former, calcic clinopyroxenes are augites, accompanied in the groundmass by pigeonite (low-Ca clinopyroxene). In the alkali basalts, basanites and related rocks, clinopyroxenes are comparatively more calcic (diopsides).

2.4.2 Trace elements and diversity of basalts

Compatible and incompatible trace elements have contrasting behaviour during partial melting and fractional crystallization, and both types provide useful information on the petrogenetic processes affecting oceanic basalts. Compatible elements include Ni and Co (preferentially incorporated into olivine), Cr (into magnesiochromite and pyroxenes), Sc (into pyroxenes) and V (into titanomagnetite). These elements are present in fairly high concentrations in primitive basalts still in equilibrium with mantle lherzolites: 200–500 ppm Ni, 50–70 ppm Co, 500–800 ppm Cr. In evolved basalts, these contents are lower due to the fractionation of early phenocrysts (magnesiochromite, olivine, clinopyroxene).

Incompatible elements are characterized by bulk distribution coefficients (ratios between the concentration of a given element into the solid phase, i.e. the bulk phenocryst assemblage or the residual minerals, and that of the same element into the liquid in equilibrium with the solid mineral assemblage during partial melting or fractional crystallization) lower than unity and ranging usually from 0.01 to 0.3 or 0.4. The origins of incompatible behaviour (non-incorporation into the crystal lattice) are variable. Alkali and alkali-earth elements (Cs, Rb, Ba, Sr and the major element K) have ionic radii larger than 10^{-10} m, which prevent their admission into tetrahedral and octahedral sites of basaltic phenocrysts: they are called Large Ion Lithophile Elements (LILE), 'lithophile' referring to their high concentrations in the continental crust. A second group of incompatible elements is characterized by smaller ionic radii but high valencies (4 or 5: Nb, Ta, Zr, Hf and the major element Ti), and tend to attract oxygen anions in the silicate liquids. The anionic polyhedra surrounding these cations prevents them from entering tetrahedral and octahedral sites. They are called High Field Strength Elements (HFSE). Th and U are highly incompatible elements because their ionic radii are larger than 10^{-10} m and also their high valencies. Finally, the Rare Earth Element (REE) series, or lanthanids, contains 14 natural elements naturels, all trivalent (except Ce, sometimes tetravalent, and Eu, sometimes divalent). The REE chemical properties vary regularly from the Light REE (LREE: La, Ce, ...) to the Heavy REE (HREE: ... Yb, Lu), with a progressive decrease in ionic radii ($La = 1.14 \times 10^{-10}$ m; $Lu = 0.85 \times 10^{-10}$ m), so that LREE are more incompatible than HREE.

Incompatibility is a relative notion. Indeed, some elements are incompatible with respect to most of the minerals constituting the phenocrystic or residual assemblage, but can be selectively incorporated into one or a few of them. For instance, plagioclases have relatively high concentrations in strontium and europium, elements usually incompatible but which could thus have a compatible behaviour during melting or crystallization processes involving large amounts of plagioclases. Similarly, Ti and V, usually incompatible, are preferentially incorporated within titanomagnetite. In addition, equivalent degrees of incompatibility (i.e. similar bulk distribution coefficients) do not necessarily imply similar crystallochemical properties. Two crystallochemically equivalent elements like niobium and tantalum (Nb: valency 5+ and ionic radius 0.78×10^{-10} m; Ta: valency 5+ and ionic radius 0.79×10^{-10} m) have bulk distribution coefficients so similar that in most magmatic

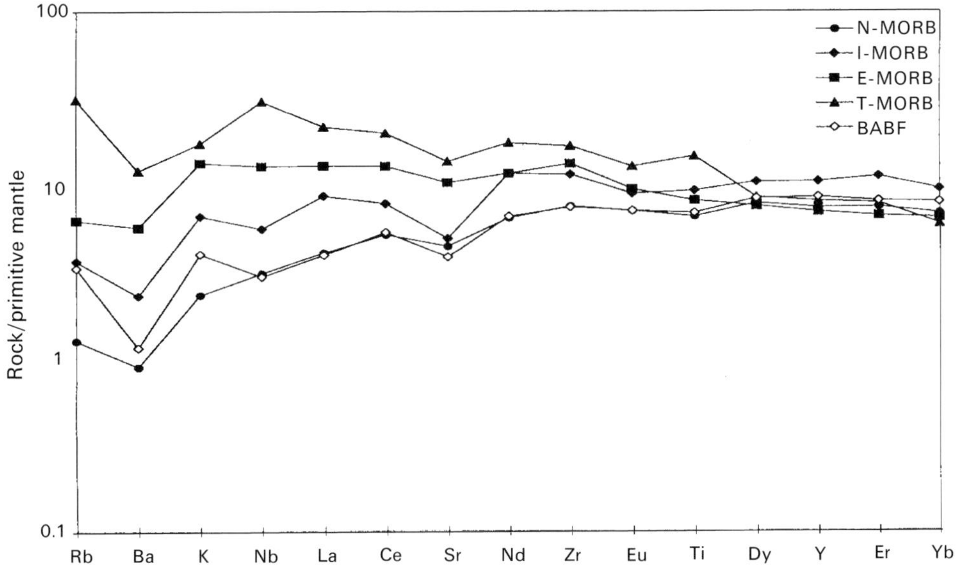

Fig. 2.4. Multi-element plots for oceanic crust basalts. Abbreviations as in Table 2.3; normalization to the primitive mantle (Table 2.2).

rocks their concentration ratios do not vary appreciably from the chondritic/primitive mantle ratio Nb/Ta = 17. On the other hand, other couples of elements with very different chemical properties may have similar bulk distribution coefficients (e.g. Ba and Th, or Ce and Pb).

The simplest graphic method representing the concentrations of a large number of incompatible elements is called the normalized multielement plot, or sometimes the spidergram (Figures 2.4 to 2.6). Elements are arranged along the x axis in order of decreasing incompatibility (or increasing bulk solid/liquid distribution coefficients) from left to right. According to the order of Table 2.2, these coefficients vary from ca. 0.01 (Cs, Rb) on the left part of the diagram to ca. 0.1 in its central part (Nd) and 0.3–0.4 on the right part (Yb, Lu). The variation in element concentrations as a function of their odd or even atomic number (Oddo–Harkins law) make their normalization to a reference geological material necessary. The composition of the primitive mantle (Table 2.2) is often used, but other reference materials can be more appropriate depending on the purpose of the plot (normalization to the average chondrite, average N-type MORB, or 'Normal' MORB, average continental crust, etc.). Whatever the choice, the parameters plotted along the y axis are the ratios (concentration of a given element in the studied sample)/(concentration of the same element in the reference material) using a logarithmic scale in order to accommodate very large variations. These normalized multi-element plots are used to compare the behaviours of a number of elements, and are widely used. However, they are usually completed by rectangular plots which enable precise study of inter-element correlations, essential for the approach of petrogenetic processes like partial melting, fractional crystallization and magma mixing.

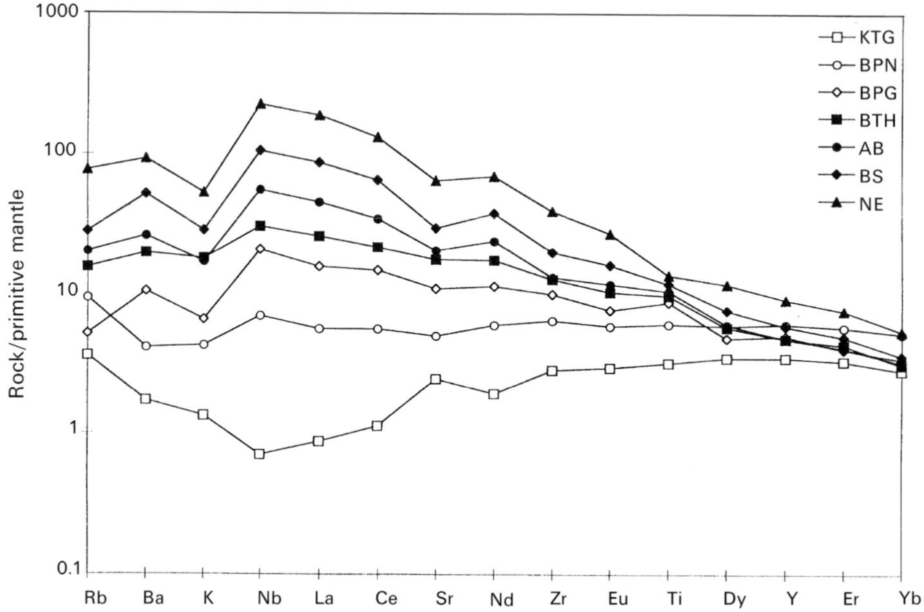

Fig. 2.5. Multi-element plots for basaltic lavas from intraplate ocanic settings (oceanic plateaus and linear chains. Abbreviations as in Table 2.4; normalization to the primitive mantle (Table 2.2).

2.4.3 MORB and BABB: basalts from oceanic accretion zones

Some representative analyses of these basalts are shown in Table 1.3. Their major-element compositions are remarkably constant: there are few significant differences between Recent MORB and Archean MORB, nor between Recent MORB and BABB. Within the MORB group, however, incompatible major elements K and Ti tend to increase from N-MORB to T-MORB.

These variations are consistent with those of incompatible trace elements (Figure 2.4). N-type MORB, which are the basalts most commonly found at ridge axes and in oceanic basins, are depleted in the most incompatible elements, and thus display a

Table 2.2. Incompatible element contents (as ppm) of the primitive mantle (after Sun and McDonough, 1989).

Cs	Rb	Ba	Th	U	Nb	Ta	K	La
0.0079	0.635	6.989	0.085	0.021	0.713	0.041	250	0.687
Ce	Pb	Sr	P	Nd	Sm	Zr	Hf	Eu
1.775	0.071	21.1	95	1.354	0.444	11.2	0.309	0.168
Ti	Gd	Tb	Dy	Y	Ho	Er	Yb	Lu
1300	0.596	0.108	0.737	4.55	0.164	0.480	0.493	0.074

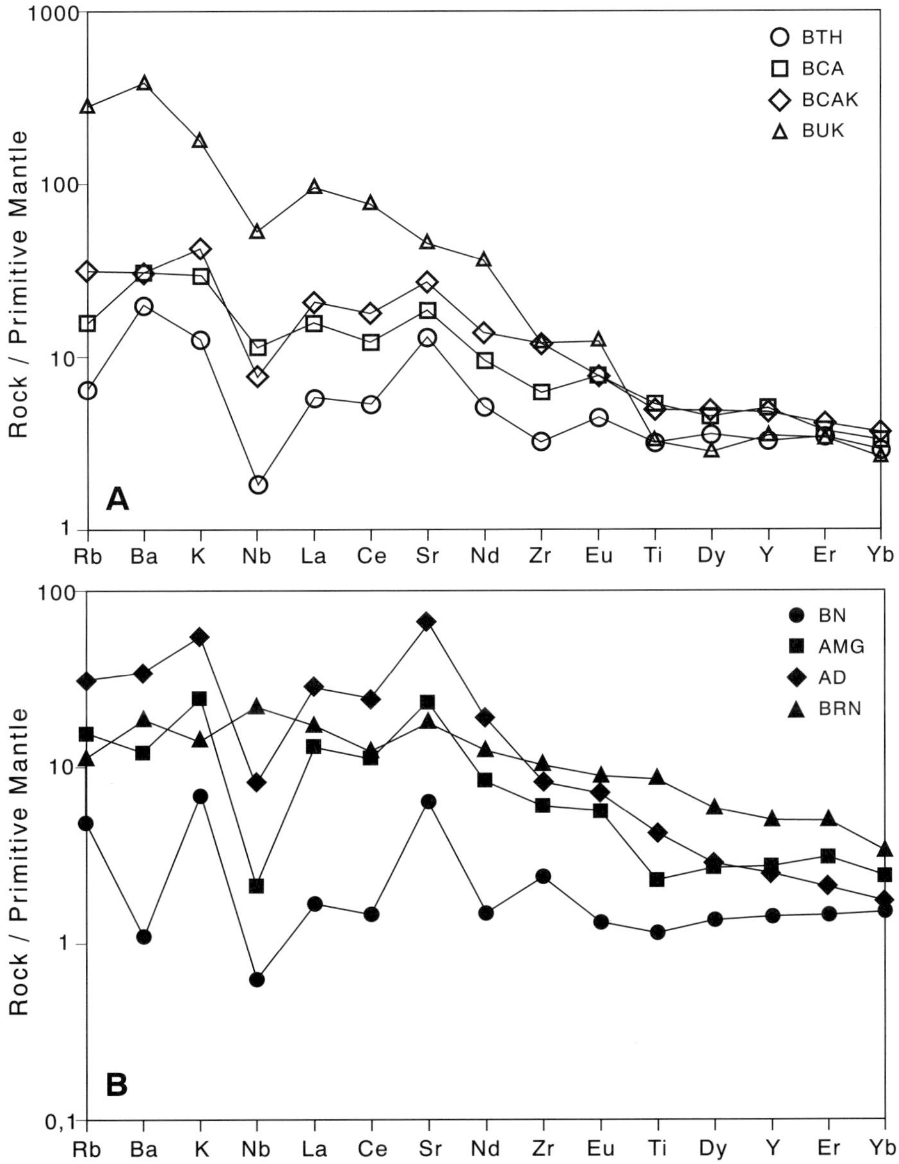

Fig. 2.6. Multi-element plots for arc lavas. Abbreviations as in Table 2.5; normalization to the primitive mantle (Table 2.2).

rather regular pattern with a positive slope in Figure 2.3. Corresponding concentrations vary from ca. 10 to 8 times the primitive mantle for HREE to nearly those of the primitive mantle for Rb and Ba. The weak negative Sr anomaly which characterizes most of the basalt patterns in Figure 2.4 traduces plagioclase fractionation effects, very commonly observed in MORB.

Table 2.3. Analyses of ocean crust basalts. Major elements as oxide weight %, trace elements in ppm.

Type	N-MORB	I-MORB	E-MORB	T-MORB	BABF	BABV	A-MORB
SiO_2	50.35	50.20	49.80	48.29	50.00	49.00	50.04
TiO_2	1.43	2.02	1.77	3.20	1.50	1.15	1.17
Al_2O_3	14.72	13.87	16.25	13.45	14.00	16.50	15.48
$Fe_2O_3{}^*$	11.30	13.67	9.78	14.15	13.20	10.14	14.10
MnO	0.18	0.23	0.17	0.20	0.21	0.14	0.23
MgO	8.65	6.91	7.55	6.59	7.10	8.55	7.78
CaO	11.20	11.70	11.00	10.76	11.57	12.40	9.37
Na_2O	2.45	2.46	3.08	2.75	2.36	2.30	1.54
K_2O	0.07	0.20	0.41	0.53	0.12	0.28	0.01
P_2O_5	0.13	0.21	0.25	0.34	0.15	0.05	0.10
L.O.I.	−0.83	0.00	−0.20	<0.01	−0.30	0.26	—
Total	99.65	100.80	99.87	100.25	99.91	100.77	99.82
Cr	365	188	250	—	124	274	—
Ni	152	63	93	47	62	115	118
Co	45	45	36	—	46	42	—
Sc	39	45	37	—	44	37	—
V	300	409	264	—	374	216	—
Rb	0.8	2.3	4	20	2.1	2.0	0.3
Ba	6.2	16	40	86	8	41	24
Sr	94	104	220	291	81	314	208
Nb	2.2	4	9.3	22	2.1	1.8	1.9
Zr	85	130	150	190	84	60	65
Y	34	48	32	37	39	20	21
La	2.8	6	9.0	15.1	2.7	3.8	3.5
Ce	9.2	14	23	35.8	9.5	8	9.3
Nd	8.8	16	16	23.9	9	7	7.9
Eu	1.2	1.5	1.6	2.16	1.2	0.85	0.82
Dy	5.8	7.7	5.6	—	6.2	3.0	3.8
Er	3.6	5.4	3.2	—	3.9	2.1	2.3
Yb	3.4	4.7	3.2	294	3.95	1.85	2.3

N-MORB: depleted ('normal') MORB from the Mid-Atlantic Ridge; I-MORB: 'intermediate' MORB with a flat REE pattern, Juan de Fuca Ridge; E-MORB: enriched MORB; East-Pacific Rise; T-MORB: transitional basalt (between MORB and OIB), Isabela Island, Galapagos; BAAF: back-arc basalt from the North Fiji basin; BAAV: back-arc basalt behind the Vanuatu arc; A-MORB: Archean MORB, for comparison. Analyses by ICP-AES, J. Cotten, Brest, except T-MORB (Geist *et al.*, 1995) and A-MORB (Ohta *et al.*, 1996).

There are just a few chemical differences between N-MORB and the North Fiji Basin back-arc basalt (BABB), the multi-element pattern of which is shown in Figure 2.4. The latter is slightly enriched in LILE such as Rb, Ba and K, and in addition shows a weak but significant negative anomaly in Nb with respect to neighbouring elements K and La. These features, which are still more apparent for the other example of BABB presented in Table 2.3 (Vanuatu), are typical of magmas formed above or within a subduction zone (§ 11.3).

The other types of Recent MORB differ from N-MORB, by far the most common, by their progressive enrichment in LREE (with respect to HREE), and more generally in the most incompatible elements. For instance, I-type MORB (for Intermediate) display nearly flat REE patterns and a slight depletion in the most incompatible elements. The E-type MORB (for Enriched) patterns are characterized

Table 2.4. Analyses of basaltic lavas from oceanic intraplate settings (oceanic plateaus and linear chains).

Type	KTG	BPN	BPG	BTH	AB	BS	NE
SiO_2	47.10	49.87	49.81	48.80	46.00	42.80	40.50
TiO_2	0.70	1.33	1.91	2.14	2.30	2.60	3.00
Al_2O_3	11.84	3.88	13.47	15.48	12.00	11.40	12.55
$Fe_2O_3{}^*$	12.71	13.39	11.31	10.92	12.60	14.10	15.90
MnO	0.18	0.20	0.18	0.15	0.18	0.21	0.29
MgO	18.04	6.80	8.80	6.79	8.90	10.55	8.02
CaO	9.67	10.52	10.59	12.45	13.80	12.30	9.80
Na_2O	0.84	3.51	3.55	2.25	2.30	3.40	6.52
K_2O	0.04	0.13	0.20	0.55	0.52	0.87	1.61
P_2O_5	0.05	0.11	0.18	0.31	0.34	0.50	0.94
L.O.I.	2.81	1.25	1.89	0.73	1.06	0.87	0.73
Total	101.18	100.99	99.99	100.60	100.00	99.60	99.86
Cr	827	119	413	315	370	560	166
Ni	684	79	167	110	136	232	128
Sc	24	49	41	31	38	30	16
Rb	2.3	6	3.3	10	13	18	49
Ba	12	29	74	140	185	365	650
Sr	52	107	233	378	440	635	1380
Nb	0.5	5	15	22	40	75	161
Zr	32	73	113	145	150	226	445
Y	16	28	23	22	22	27	42
La	0.6	3.9	10.9	18.0	31.5	60	129
Ce	2.0	10.1	26.6	39	62	117	234
Nd	2.6	8.2	15.6	24	33	52	95
Eu	0.5	1.0	1.3	1.75	2.0	2.75	4.65
Dy	2.6	4.4	3.6	4.3	4.5	5.7	8.8
Er	1.6	2.8	1.95	2.1	2.0	2.4	3.7
Yb	1.4	2.6	1.7	1.55	1.6	1.8	2.7

KTG; komatiite from Gorgona Island; BPN: basalt from Nauru plateau; BPG: basalt from Gorgona; BTH: tholeiitic basalt from Gambier, Polynesia; AB, BS and NE: alkali basalt, basanite and nephelinite from Tubuai, Polynesia. Analyses by ICP-AES, J. Cotten, Brest, except KTG, BPG (Kerr *et al.*, 1996) and BPN (Floyd, 1989).

by a slight negative slope traducing their enrichment in LREE, Nb and LILE. Finally, the T-type MORB (for Transitional) are intermediate between E-MORB and alkali basalts. Their composition plot near the boundary between the fields of subalkalic and alkali basalts defined by Miyashiro (1978). Like most of the latter, they show weak positive Nb anomalies with respect to K and La.

Despite their variable enrichment in incompatible elements, the MORB is the most homogeneous group of oceanic basalts. The dominant N-MORB compositions derive from the melting of upper mantle rocks depleted in incompatible elements, thus residual in character (harzburgitic mantle). The compositions of I-MORB, E-MORB and T-MORB reveal the increasing contribution to the sources of oceanic basalts, of one or several enriched components typical of oceanic intraplate magmatism (Chapter 10). This evolution also corresponds to increasingly enriched isotopic signatures.

2.4.4 OPB and OIB: oceanic intraplate basalts

Analyses of Table 2.4 and multi-element plots of Figure 2.5 show for this group a much greater variability than for MORB. Two main types of intraplate magmas can be distinguished: the OIB group, clearly enriched in highly incompatible elements, and the OPB group, either depleted or only slightly enriched in these elements.

The tholeiites from intra-oceanic islands show numerous similarities with the E-MORB, for major elements as well as for trace elements (Tables 2.3 and 2.4). The main differences between the two groups concern radiogenic isotope ratios, the sources of oceanic island tholeiites being as a whole more enriched than those of MORB (Figure 2.2 and Chapter 10). These lavas are also usually more porphyritic than the MORB. They often contain augite phenocrysts which are uncommon in MORB. Their groundmass contains Ca-poor clinopyroxene (pigeonite), but lacks any interstitial alkali feldspar or feldspathoid. They are either silica-oversaturated (quartz tholeiites) or silica-saturated (olivine tholeiites). In the large Hawaiian and Polynesian shield volcanoes, quartz tholeiite flows are usually overlain by olivine tholeiites which are in turn overlain by transitional basalts. These latter are characterized by a mixture of geochemical and mineralogical features of tholeiites and alkali basalts, respectively (e.g. silica-saturated character and occurrence of diopside phenocrysts). In alkali-silica diagrams, they plot very close to the boundaries between the fields of tholeiites and of alkali basalts (for instance, in Figure 2.9, the limit proposed by Miyashiro, 1978).

Many authors consider that the trace element compositions and Sr, Nd, Pb isotopic ratios of ocean island tholeiites and transitional basalts is relatively constant for a given shield volcano. However, the example of Nuku Hiva (§ 10.1 and 10.2) shows that important geochemical variations occur during the edification of such shields, especially during the final stages of their growth.

In ocean islands, the emplacement of tholeiites is followed by that of alkali basalts, which occurs usually at the end or after the shield-building stage (post-shield stage, according to the Hawaiian nomenclature). Detailed chronologic and geochemical studies, however, indicate that there is little or no time gap between the two episodes, and that lavas transitional between the two groups are often emplaced at this time.

Ocean-island/alkali magmas differ significantly from associated tholeiites. They are often highly porphyritic, and picritic lavas enriched by mechanical accumulation of 50 or even 60 modal% phenocrysts are common (olivine-rich 'oceanites', clinopyroxene-rich 'ankaramites'). Basanites usually contain easily recognizable feldspathoids, which occur only in small interstitial patches in the groundmass of alkali basalts. Most nephelinites are feldspar-free.

Incompatible trace-element concentrations increase from alkali basalts to basanites and nephelinites (Figure 2.5), usually proportionally to their degree of incompatibility. This feature has often been interpretated as traducing decreasing partial melting rates (§ 10.2.3). Positive niobium anomalies are also typical.

Figure 2.5 and Table 2.4 show that the major and trace element compositions of OPB are quite different from those of OIB. Although emplaced within an intraplate setting, they are much closer to MORB than to OIB. They are typical tholeiites, often subaphyric, with rare phenocrysts of olivine, calcic plagioclase and clinopyroxene, the groundmass of which shows textural features (subdoleritic, microlitic or quenched) similar to those of MORB (§ 7.5). However, picritic varieties rich in olivine and to a lesser extent calcic clinopyroxene are much more common in OPB. The famous Gorgona Island (§ 10.3) occurrence of phanerozoic komatiites (analysis KTG, Table 2.4) corresponds to OPB-type picritic-lava flows with very large skeletal olivines ('spinifex' texture). Trace-element variations in OPB are also nearly identical to those found among the MORB: multielement patterns range from depleted to flat and slightly enriched like the E-MORB (Figure 2.5). The Sr, Nd isotopic compositions of some depleted OPB plot within the MORB range, whereas those of enriched OPB are much closer to OIB signatures (Figure 10.21).

2.4.5 IAB, boninites, adakites: subduction zone magmas

The variability of arc magmas (Table 2.5 and Figure 2.6) is even greater than that of OIB. However, they display a number of common features. They are usually highly porphyritic (20 to 50 modal% phenocrysts), with in order of decreasing abundance calcic plagioclase (except in boninites), calcic clinopyroxene, olivine, titanomagnetite and orthopyroxene. These minerals, especially plagioclases, olivines and clinopyroxenes, often exhibit complex reaction/dissolution/crystallization textures ('cloudy' zones, glassy and fluid inclusions, complex zoning patterns) which suggest rather complicated crystallization histories.

From the point of view of trace elements, arc magmas are characterized by low HFSE contents with respect to LILE and LREE. Nb and Ta, and to a lesser extent Zr and Ti, display negative spikes in multi-element patterns, which are quite typical of IAB. Most of them also show selective enrichments in LILE, especially Rb, Ba and K, with respect to LREE. These features allow the IAB to be readily distinguished from MORB and OIB by their higher Ba/La, La/Nb and Th/Ta ratios.

Table 2.5. Analyses of arc lavas.

Type	BTH	BCA	BCAK	BUK	BN	AMG	AD	BRN
SiO_2	49.00	47.98	49.80	45.80	53.00	58.25	58.80	49.40
TiO_2	0.67	1.08	1.07	0.71	0.25	0.49	0.92	1.87
Al_2O_3	13.90	17.08	15.22	11.81	12.46	14.85	17.20	15.32
$Fe_2O_3{}^*$	9.79	11.66	8.55	9.22	7.54	7.62	5.81	11.18
MnO	0.17	0.19	0.14	0.16	0.14	0.13	0.07	0.13
MgO	11.62	7.45	9.60	9.98	13.08	5.64	2.61	7.04
CaO	13.10	11.48	8.55	11.87	6.70	8.96	7.00	9.07
Na_2O	1.73	2.48	3.09	2.63	2.76	3.19	4.48	2.86
K_2O	0.38	0.89	1.30	5.54	0.21	0.75	1.68	0.43
P_2O_5	0.13	0.21	0.30	1.07	0.04	0.10	0.44	0.29
L.O.I.	−0.03	−0.33	0.00	0.84	2.56	0.46	0.81	2.72
Total	100.46	100.27	98.95	99.63	98.74	100.03	99.82	100.31
Cr	620	102	426	390	786	185	13	220
Ni	130	63	236	99	296	38	16	158
Co	50	41	48	40	39	22	25	41
Sc	48	35	30	36	34	32	12	23
V	245	318	140	292	202	180	170	200
Rb	4.1	10	20	187	3.1	10	19.8	7.2
Ba	142	216	215	2790	7.6	85	243	132
Sr	280	395	578	980	136	505	1400	383
Nb	1.3	8	5.5	39	0.44	1.5	5.8	15.8
Zr	36	70	133	138	27	68	93	116
Y	15	23	22	16	6.5	12.4	11.4	23
La	4.0	10.8	14.6	68	1.15	9.1	19.9	12
Ce	9.5	22	32	141	2.6	20	44	22
Nd	7	13	19	50	2.0	11.2	26	17
Eu	0.75	1.3	1.3	2.1	0.22	0.95	1.2	1.5
Dy	2.6	3.3	3.6	2.1	1.0	2.0	2.1	4.3
Er	1.6	1.8	2.0	1.6	0.7	1.5	1.0	2.4
Yb	1.4	1.6	1.8	1.3	0.75	1.2	0.85	1.65

BTH: island arc tholeiitic basalt, Ambrym, Vanuatu; BCA: calc-alkaline basalt, Java; BCAK: potassic calc-alkaline basalt, Mindanao, Philippines; BUK: ultrapotassic basanite, Java; BN: boninite, Marianas; AMG: Mg-rich andesite, Vanuatu; AD: adakite, Mindanao; BRN: niobium-rich basalt, Mindanao. Analyses by ICP-AES, J. Cotten, Brest, except BN (Pearce *et al.*, 1992).

IAB also present distinct isotopic features. Most arc magmas (with the exception of some low-K basalts and the majority of adakites) are characterized by $^{87}Sr/^{86}Sr$ ratios higher than those of MORB together with lower $^{143}Nd/^{144}Nd$ ratios (Figure 2.2), and they are also more radiogenic in Pb. These peculiarities are consistent with a modification of their mantle sources through the input of a component of sedimentary origin (§ 11.3).

The main types of arc basalts can be identified on a K_2O/SiO_2 plot (Figure 2.11).

Potassium content increases from low-K calc-alkaline basalts (also called island-arc tholeiites) to medium-K calc-alkaline basalts (or 'normal' calc-alkaline basalts), then to K-rich calc-alkaline basalts and to absarokites (basalts from shoshonitic series). All these lavas are silica-saturated or -oversaturated. Higher K_2O content corresponds to silica-undersaturated compositions: K-rich island-arc alkali basalts and ultrapotassic basanites. These rocks contain leucite, often converted to analcite. They differ from OIB-type K-rich lavas by the negative Nb and Ta anomalies characteristic of their multi-element patterns (Figure 2.6a). The same diagram shows that highly incompatible trace elements behave like potassium: their concentrations increase from low-K series (with nearly flat or even slightly depleted patterns) to medium-K series (with moderately enriched patterns) up to strongly enriched ultrapotassic series.

The patterns shown in Figure 2.6a are representative of a vast number of IAB. Island arcs contain less common rocks which are nevertheless important for understanding the fate of subducted oceanic crust. Boninites and adakites represent the only non-basaltic primary magmas known to occur in truly oceanic settings. Both types are more silica-rich than basalts (SiO_2 = 53–60% for boninites, 56–70% for adakites) and have very low contents in HREE (one to five times the primitive mantle). On the other hand, boninites are Mg-rich (MgO = 8–20%) and depleted in all kinds of incompatible elements, whereas adakites are rich in Sr and other LILE (Figure 2.6b). The rather unusual petrogenesis of these arc magmas is discussed in § 11.3 for adakites and § 11.4 for boninites. Mg-rich andesites, still poorly known, display chemical characteristics intermediate between those of boninites and adakites (Figure 2.6b) and occur in the same tectonic settings (subduction of very young lithosphere or even of an active ridge). Niobium-rich basalts, the only arc lavas sometimes devoid of negative Nb and Ta anomalies (Figure 2.6b), are spatially, temporally and petrogenetically associated with adakites (§ 11.3).

2.5 MAGMA SERIES FROM OCEANIC DOMAINS

Although largely dominant, basalts and gabbros are far from being the only magmatic rocks occurring in oceanic domains. Although intermediate and evolved rocks are not commonly found at the axes of active ridges, as well as in oceanic plateaus and tholeiitic shields, they are widespread in island arcs (andesites) and within the alkalic series emplaced during the final stages of building of ocean islands (post-shield stages).

2.5.1 Processes of evolution

Fractional crystallization is the main process explaining the progressive transition from basaltic magmas to evolved rhyolitic, trachytic or phonolitic magmas. Its modes of occurrence within magma reservoirs underlying large volcanic edifices are schematically illustrated in Figure 2.7.

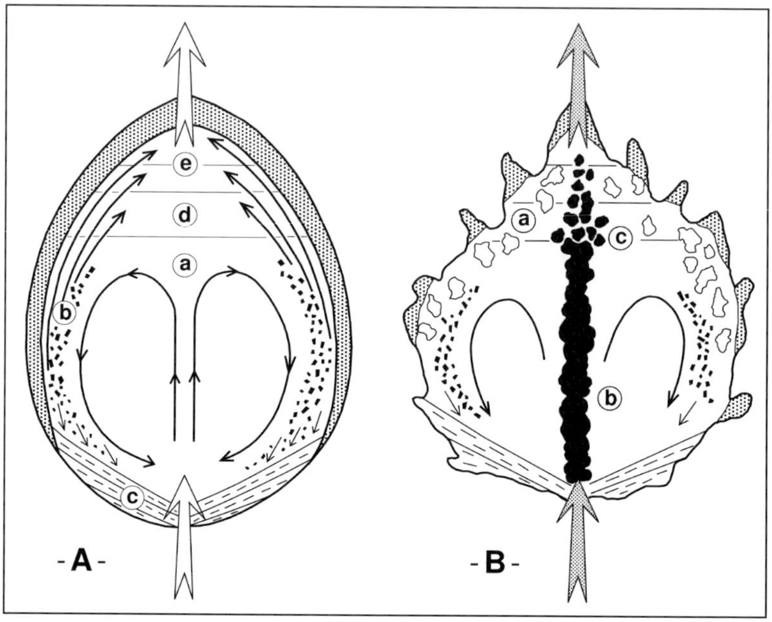

Fig. 2.7. Simplified models of magma reservoirs (Maury, 1993). A: reservoir evolving by closed-system fractional crystallization. B: reservoir evolving in open-system conditions. The shapes of the reservoirs are arbitrary. See text for explanations.

In a theoretical case of closed-system fractional crystallization (A), the magma reservoir is, at the initial stage, filled up by ascending primitive basaltic magma of supposed homogeneous composition. It is postulated that there will be no refilling of the reservoir or magma transfer to the surface (closed-system evolution). The cooling of the reservoir induces the progressive solidification of magma near the boundaries of the chamber (shaded pattern). The temperature gradient induces convective motions within the reservoir. The hot primitive magma from the bottom part of the reservoir uprises in its central part (a), then cools down during its descent near the walls (arrows). The phenocrysts formed at this stage (b) are dragged down by the convective motion, and accumulate near the bottom of the reservoir (c), forming layered cumulates. The residual liquids, initially in equilibrium with the phenocrysts at level (b), are less dense than the primitive magma. They will uprise near the walls of the reservoir (arrows), and form density-stratified liquid layers (d, e) in its upper part. The result is a two-level reservoir with a lower, hot, convective and basaltic part and an upper stratified part in which intermediate and evolved liquids are accumulated. With time the liquid volume decreases through solidification on contact with the walls and formation of cumulates. At the same time, the importance of the basaltic layer decreases because of the continuous formation of phenocrysts which are accumulated near the base of the reservoir and of intermediate and evolved liquids which concentrate near its top.

In a more realistic case of open-system fractional crystallization, shown in

Table 2.6. Analyses of lavas from a tholeiitic series. Example of the enriched series (with transitional features) of Alcedo volcano, Isabela, Galapagos (Geist *et al.*, 1995).

Type	MB	EB	FB	FA	IC	RD	RH
Phen.	PL + OL + CPX	PL + OL + CPX	PL + CPX + OL	PL + CPX + OL	PL + CPX + MT	PL + CPX + MT + AP	PL + CPX + MT + AP
SiO_2	48.61	48.66	49.21	51.70	53.78	64.05	70.92
TiO_2	2.60	3.25	3.72	3.11	2.76	0.97	0.42
Al_2O_3	14.18	13.70	13.40	13.39	12.90	13.67	13.04
$Fe_2O_3{}^*$	12.62	14.30	15.70	14.65	13.77	8.07	5.89
MnO	0.17	0.20	0.23	0.23	0.22	0.17	0.15
MgO	6.79	5.99	5.08	4.07	3.29	1.66	0.14
CaO	11.27	10.80	9.73	7.77	6.76	3.38	1.73
Na_2O	2.75	2.90	3.27	3.63	3.81	5.20	4.83
K_2O	0.41	0.45	0.59	0.90	1.11	2.20	2.69
P_2O_5	0.27	0.35	0.46	0.86	0.53	0.24	0.06
L.O.I.	<0.01	<0.01	0.35	0.14	<0.01	0.62	<0.01
Total	99.62	100.61	100.73	100.44	98.95	99.35	99.87
Rb	6		10	18	30	44	63
Ba	51	95	115	175	183	305	399
Sr	347	290	320	311	328	174	123
Nb	17	20	25	37	47	66	88
Zr	161	185	240	336	521	670	893
Y	30	28	36	63	70	85	102
La	12.7	14.9	17.9	29.1	33.4	48.5	60.6
Ce	29.4	38.2	47.5	68.9	78.3	109	132.7
Nd	19.5	23.6	30.3	44.4	48.2	59.1	68.5
Eu	1.83	2.20	2.57	3.65	3.80	3.92	3.56
Tb	0.85	1.12	1.35	1.80	1.87	2.19	2.55
Yb	2.39	2.98	3.57	4.88	5.62	7.20	8.93

MB: Mg-rich basalt; EB: evolved (Mg-poor) basalt; FB: ferrobasalt; FA: ferroandesite; IC: icelandite; RD: rhyodacite; RH: tholeiitic rhyolite. The equivalent nomenclature for plutonic rocks is: olivine-rich gabbro; olivine-poor ferrogabbro; ferrodiorite; monzonite; monzogranite; tholeiitic granite (plagiogranite). Abbreviations used for phenocrysts: AP: apatite; CPX: calcic clinopyroxene; MT: titanomagnetite; OL: olivine; OPX: orthopyroxene; PL: plagioclase.

Figure 2.7B, magma supplies to the surface can no longer be considered as negligible. Each event of magma outpouring is compensated by a refilling with hot basaltic magmas which are thus lighter than their previously resident equivalents. The reinjected primitive magmas will thus uprise in the axial part of the reservoir, forming a turbulent column (b) which, if it reaches the reservoir's summit, will be able to mix (c) with intermediate and evolved magmas. Assimilation of the reservoir wall rocks (a) will occur concomitantly with fractional crystallization and magma

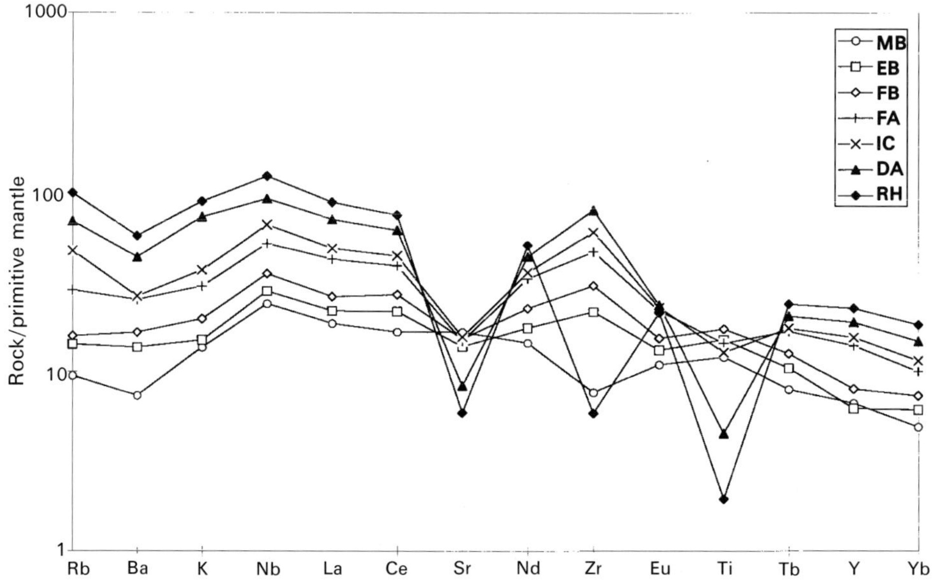

Fig. 2.8. Multi-element plots for Galapagos tholeiitic lavas. Abbreviations as in Table 2.6.

mixing. Assimilation and fractional crystallization (AFC) are often coupled processes, especially within reservoirs located in the arc crust (§ 11.2.1). The occurrence of crustal contamination can be demonstated using coupled isotopic ratios, e.g. strontium and oxygen (Figure 2.3).

Fractional crystallization coupled or not with assimilation can generate a number of variants, depending on:

—the size and the shape of the reservoir (§ 8.2a);
—the variable densities and viscosities of the residual liquids;
—the frequency of basaltic magma refilling events;
—and finally from the sequences of refilling/mixing/crystallization/eruption events in periodically refilled magma reservoirs.

In summary, each magma reservoir has its specific behaviour, and almost identical basaltic magmas stored within two (or more) neighbouring chambers can evolve in quite different ways.

2.5.2 Tholeiitic series

The compositions of successive liquids derived from the fractional crystallization of 'primitive' Mg-rich tholeiitic basalt are shown in Table 2.6 and Figure 2.8. Tholeiitic magmas, being water-poor, evolve under low oxygen fugacities. Consequently, iron-titanium oxides (titanomagnetite) phenocrysts crystallize at a late stage of evolution of these series (in icelandites, Table 2.6). Until their fractionation, iron and titanium content increases in successive liquids: ferrobasalts contain ca. 16% of total iron calculated as Fe_2O_3 and 3% TiO_2. The corresponding multielement patterns are

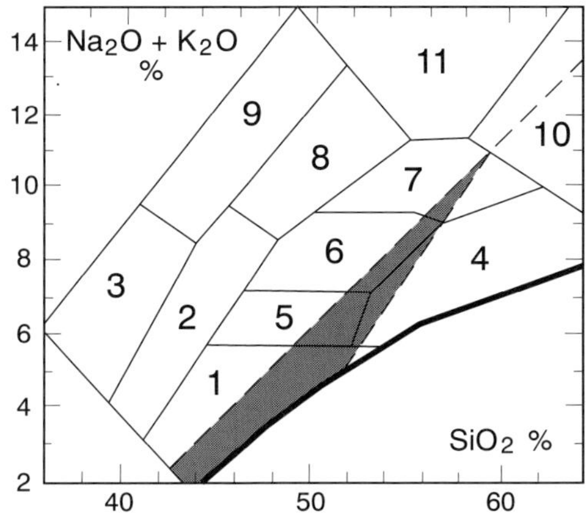

Fig. 2.9. Simplified nomenclature of volcanic rocks from alkalic series (Maury, 1993, modified from Cox *et al.*, 1979). Bold line: boundary between subalkalic and alkalic magma series (Miyashiro, 1978). The shaded field and the dashed lines separate the silica-undersaturated (left) and silica-oversaturated (right) domains. Mafic lavas: 1: alkali basalts; 2: basanites and tephrites; 3: nephelinites. Intermediate rocks: 4: trachyandesites; 5: hawaiites; 6: mugearites; 7: benmoreites; 8: phonolitic tephrites; 9: phonolitic nephelinites. Evolved rocks: 10: trachytes; 11: phonolites.

enriched (Figure 2.8), as the example shown is derived from T-type MORB, and the concentration of incompatible elements increases regularly from basalts to rhyolites. Exceptions to this enrichment pattern include strontium, fractionated by calcic plagioclases still the stage of evolved basalts, titanium, fractionated by titanomagnetite at the level of icelandites and finally zirconium, fractionated by zircon in the rhyolites.

The Galapagos tholeiitic series taken as an example evolves from basalts to rhyolites. Such a wide petrographic range is rarely observed at the level of oceanic ridges, where the most evolved volcanic rocks commonly dredged or drilled are ferrobasalts. Because of the iron increase during the early stages of the evolution of tholeiitic series, the densities of ferrobasaltic magmas are greater than those of primitive basalt magmas refilling the reservoirs underlying the ocean ridges (Chapter 8). Such primitive magmas will thus be able to uprise through these reservoirs without being affected by fractional crystallization and directly reach the ocean bottom. Ferrobasaltic magmas, which are mainly recovered from slow-spreading ridges and propagating rifts, would mostly reach the surface when primitive magma reinjections are not very frequent.

2.5.3 Alkaline series

A simplified classification of volcanic alkaline series, based on an alkali-silica plot, is shown in Figure 2.9. According to the level of silica-undersaturation of the primitive magmas (alkali basalts, basanites or nephelinites), fractional crystallization leads to trachytes or phonolites following paths approximately parallel to the dashed limits. Many ocean islands are characterized by a bimodal distribution of their major lava

Table 2.7. Analyses of lavas from an intraplate ocean island series. Example of Taiohae volcano, Nuku Hiva, Marquesas (Le Dez *et al.*, 1996).

Type	AB	HW	MG	BM	TR
Phen.	OL + CPX	CPX + PL + MT + AP	PL + CPX + AM + MT + MI + AP	PL + AM + CPX + MT + MI + FA + AP	FA + AM + BT + MT + AP
SiO_2	45.60	47.85	53.10	58.20	63.80
TiO_2	3.16	3.51	1.88	1.15	0.65
Al_2O_3	15.30	15.25	17.57	18.33	18.00
$Fe_2O_3{}^*$	12.20	12.22	8.12	5.42	3.08
MnO	0.17	0.16	0.20	0.16	0.06
MgO	7.13	5.64	2.77	1.26	0.47
CaO	9.03	8.08	5.20	2.95	1.53
Na_2O	2.83	3.24	5.39	5.18	6.15
K_2O	1.70	1.91	2.90	5.27	5.44
P_2O_5	0.56	0.61	0.86	0.34	0.20
L.O.I.	2.16	1.27	1.92	0.98	1.11
Total	99.84	99.74	99.91	99.24	100.50
Cr	190	165	4	16	3
Ni	89	123	2	7	7
Co	40	39	9	5	3
Sc	19	18	5	4	2
V	265	285	67	53	24
Rb	44	46	100	165	153
Ba	550	437	930	875	920
Sr	706	588	794	446	360
Nb	51	47	82	101	84
Zr	280	353	450	702	370
Y	31	64	40	35	28
La	46	60	79	97	79
Ce	96	108	161	162	137
Nd	47	73	73	63	48
Eu	2.70	4.60	3.70	2.35	1.05
Dy	6.2	11.1	7.8	6.1	4.6
Er	2.5	5.3	3.4	3.5	2.7
Yb	2.15	3.85	3.00	3.15	2.60
D.I.	34.06	39.69	63.33	76.58	87.59

AB: alkali basalt; HW: hawaiite; MG: mugearite; BM: benmoreite; TR: trachyte. Abbreviations used for phenocrysts: AM: amphibole; AP: apatite; CPX: calcic clinopyroxene; FA: alkali feldspar (sanidine-anorthoclase); MI: mica (phlogopite-biotite); MT: titanomagnetite; OL: olivine; OPX: orthopyroxene; PL: plagioclase; D.I.: Differentiation Index of Thornton and Tuttle (1960).

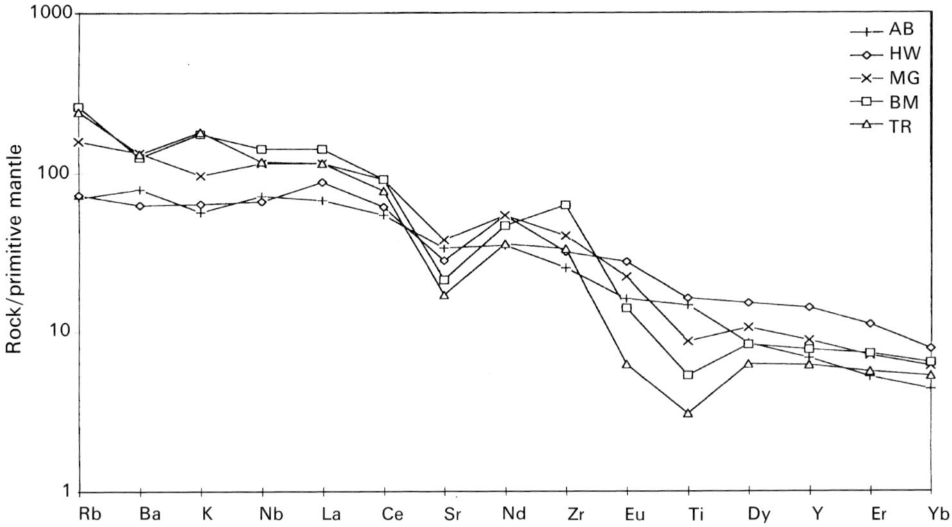

Fig. 2.10. Multi-element plots for lava from the alkali basalt series of Nuku Hiva, Marquesas. Abbreviations as in Table 2.7.

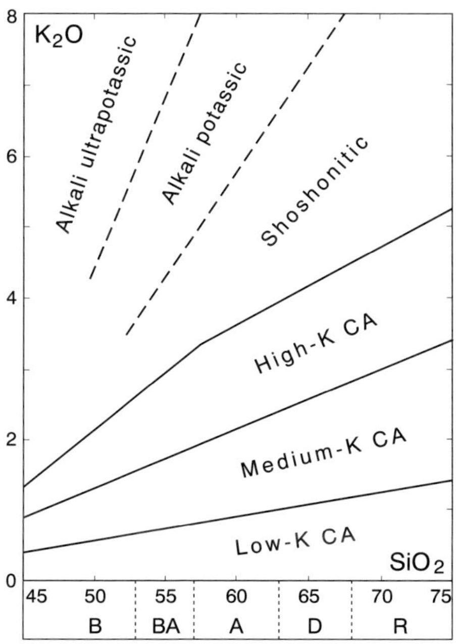

Fig. 2.11. Simplified nomenclature of volcanic arc lava (Maury, 1993). B: basalts; BA: basaltic andesites; A: andesites; D: dacites; R: rhyolites. This nomenclature is used for calc-alkaline series. The equivalents for shoshonitic series are: absarokites; shoshonites; banakites; latites; liparites. Figure 2.9. can be used for arc-type potassic and ultrapotassic alkali suites.

Table 2.8. Average analyses of lavas from the low-K (island arc tholeiitic) series of Martinique (Escalant, 1988).

Type	B	AB	A	D	R
Phen.	PL + OL + CPX	PL + CPX + OL	PL + CPX CPX + MTO	PL + MT PX + CXP	PL + MT OPX + CPX
SiO_2	50.90	54.59	59.68	66.00	71.11
TiO_2	0.83	0.89	0.88	0.62	0.38
Al_2O_3	18.68	17.55	16.10	14.70	14.33
Fe_2O_3*	9.91	9.34	8.23	6.33	2.47
MnO	0.19	0.20	0.19	0.13	0.05
MgO	4.86	3.4	2.54	1.0	0.34
CaO	10.34	8.46	6.10	3.44	2.59
Na_2O	2.62	3.33	4.05	4.91	5.26
K_2O	0.48	0.51	0.72	1.03	1.30
P_2O_5	0.12	0.15	0.17	0.18	0.09
Cr	50	17	10	8	5
Ni	19	9	6	7	7
Co	31	23	17	7	4
Sc	34	27	28	20	18
V	262	217	140	9	9
Rb	7	11	16	19	22
Ba	110	135	186	178	296
Sr	307	249	194	154	153
Nb	3.5	5	4	5.5	4.5
Zr	59	84	103	135	202
Y	21	27	32.5	41.5	34
La	5.1	8.1	7.9	10.3	10.8
Ce	12.3	18.8	19.8	28	25.6
Nd	11.6	15.9	16.0	—	9.3
Eu	0.9	1.1	1.2	1.6	1.6
Dy	3.0	–	5.8	—	4.9
Er	1.85	2.6	3.7	—	3.7
Yb	2.0	2.5	3.5	4.95	3.5

B: basalt; AB: basaltic andesite; A: andesite; D: dacite; R: rhyolite. Abbreviations as in Table 2.7.

types. Alkali basalts and trachytes, or basanites and phonolites, are predominant whereas intermediate compositions (mugearites, tephrites) are very uncommon or even not represented. This peculiarity is referred to as the Daly gap (§ 10.2).

The alkaline series shown as an example in Table 2.7 and Figure 2.10 is from the internal volcano (Taiohae) of Nuku Hiva, Marquesas Archipelago (§ 10.1). It is a mildly alkalic series, without noticeable Daly gap, ranging from alkali basalts to trachytes. The decrease of MgO is compensated by progressive increases of SiO_2, Na_2O and K_2O. In contrast with the tholeiitic Galapagos example, total iron and titanium decrease since the basaltic stage, due to titanomagnetite fractionation. Compatible elements Cr, Ni, Co, Sc concentrations also decrease rather abruptly in

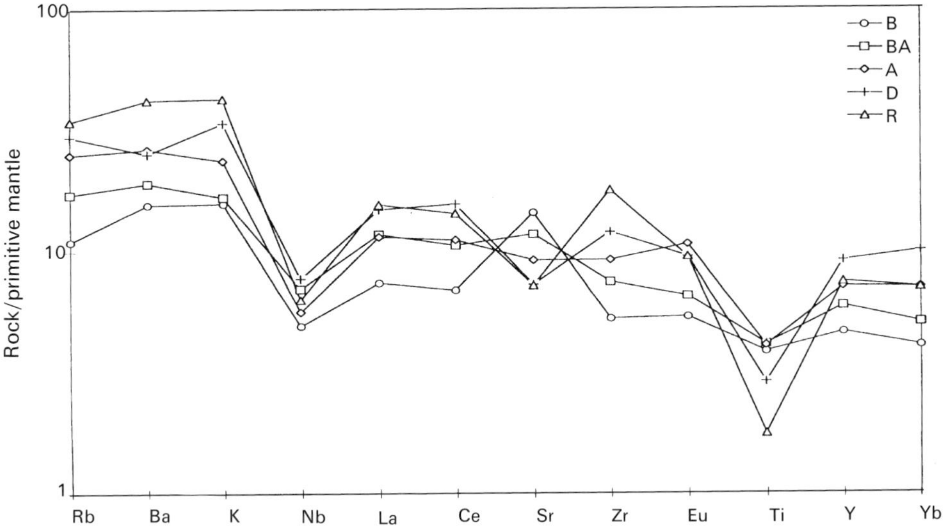

Fig. 2.12. Multi-element plots for the low-K arc suite of Martinique. Abbreviations as in Table 2.8.

the early steps, because of the importance of olivine and clinopyroxene fractionation. The following stages are dominated by plagioclase separation (negative anomalies in Sr and Eu), together with amphibole which fractionates some REE (Nd).

2.5.4 Magma series from subduction/collision zones

A simplified classification of these series, based on SiO_2 and K_2O content, is shown in Figure 2.11. The limits between the fields of the different series are nearly parallel to fractional crystallization paths involving the common phenocrysts of calc-alkaline magmas, i.e. calcic plagioclase, clinopyroxene, orthopyroxene, olivine and titano-magnetite.

The example of the low-K series (island-arc tholeiitic series) from Martinique, Lesser Antilles (Table 2.8 and Figure 2.12) shows that the variations of SiO_2, Na_2O and K_2O are rather similar to those observed in the Galapagos example. However, the initial basalts are more evolved, as evidenced by their low MgO, Cr, Ni, Co content. The decrease of total iron and titanium occurs at a late stage, at the level of dacites (negative Ti anomaly). This feature is typical of crystallization under low-oxygen fugacities. The important plagioclase fractionation which occurs after the early stages is marked by a clear decrease of Sr. The other LILE, i.e. Rb, Ba and K increase progressively, as well as the LREE. Although niobium concentrations are low with respect to its neighbours (negative Nb anomaly characteristic of arc magmas), this element has a typical incompatible behaviour during fractional crystallization.

The series taken as an example does not show peculiar features indicating the occurrence of magma-mixing or crustal-contamination processes. In many calc-alkaline volcanoes, these processes overprint, or even obliterate, the effects of fractional crystallization. For instance, the recent magmatologic evolution of Mt Pelée is dominated by magma-mixing events, which triggered the 1902 and 1929 eruptions. During these events, magma composition ranged from basalts to dacites, but most of the emitted volumes were intermediate between these two end-members (andesites containing numerous partly destabilized xenocrysts, originating either from the mafic or acid end-members).

3

Surveying and sampling the oceanic crust

The marine geologist working on a surface ship is separated from the ocean floor by a 'screen' of seawater several km thick. Peering through this screen is a permanent challenge to human ingeniosity: the progress of our knowledge about the oceanic crust has always been narrowly dependent on the evolution of oceanographic technology.

On land, detailed topographic maps, aerial photographs and satellite imagery today allow precise location on the Earth's surface to be plotted and permit the study of geological structures at all scales. The terrestrial geologist can take rock samples from judiciously chosen geological sections, and come back as often as needed to key outcrops.

Things are quite different at sea. The very first necessity is to have at hand an appropriate image of the ocean floor to be studied, which often means personally doing the bathymetric survey. The bathymetric information must be completed generally by geophysical surveys (seismic, magnetic, gravimetric, etc), which provide precious information about the subsurface structure, age, thermal state, etc. Finally, the marine geologist needs to collect rock samples, geologically reconstructed as precisely as possible.

The oceanic crust crops out directly on the ocean floor in the axial zone of the ridges, along a stripe several kilometres wide. Beyond this zone, it disappears progressively below the pelagic sediment blanket. The thickness of this sedimentary layer increases with distance to the ridge axis. The axial zone itself may be buried beneath mixed pelagic and terrigenous sediments, in the vicinity of continental margins (Gulf of California, Red Sea, etc.) Only major tectonic accidents allow the oceanic crust to appear away from the ridge axial zone (fracture zones and transform faults, intraplate deformations, etc.).

The study of the oceanic crust therefore comes up against very difficult technical problems.

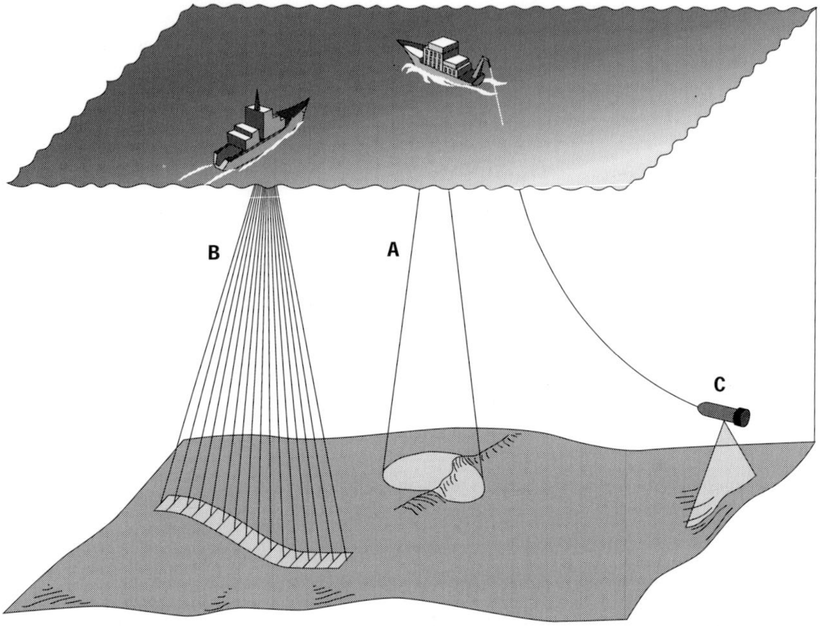

Fig. 3.1. The modern tools for acoustic imagery and mapping of the deep ocean floors. (A) *Echo sounder* with its divergent beam (apex angle of about 30°). (B) *Multi-beam echo-sounder*. A number of narrow acoustic beams are displayed side by side perpendicular to the ship's axis, sweeping a wide stripe of ocean floor. (C) *Side-scan sonar system*, towed close to the bottom behind the ship, in a torpedo-like casing. This device sends out fan-shaped signals that build up 'acoustic pictures' of the ocean floor (after a drawing from Averous, 1981).

3.1 MAPPING AND ACOUSTIC IMAGERY OF THE OCEAN FLOOR

The first, indispensable document for any study at sea is a detailed bathymetric map of the area to be studied. The marine geologist has today at his disposal a set of modern tools for cartography and acoustic imagery of the deep ocean floors (Figure 3.1). For mapping, *multibeam acoustic echo-sounders* are used. Since the First World War, *sonars* or single-beam echo sounders allowed general mapping of the ocean floor. Moreover, *side-scan sonars* also obtain acoustic images of the ocean floor.

3.1.1 Single-beam bathymetric sounders

The principle of sonar is simple, it is that of the echo. A high energy acoustic signal is emitted from the oceanographic ship toward the bottom, where it is reflected and comes back to the surface: this 'echo' is recorded, and the two-way travel time T of the sound is precisely measured. Knowing the velocity V of the sound in seawater, it is easy to calculate the depth P of the reflecting ocean floor (or of the reflecting object: a fish shoal, etc.): $P = VT/2$.

In an oceanographic ship equipped with a single-beam sounder, a sound vibration is generally emitted by a circular ceramic tile (or *transducer*) placed below the hull, which vibrates powered by an alternating electrical current. The acoustic frequencies used range generally between 5 and 30 kHz. When it receives the echo from the bottom, the ceramic tile vibrates again, creating by piezoelectric effect an electrical current, which is recorded. The tile is thus alternatively emitting and receiving ultrasounds. Of course an electronic system is necessary to manage the emission and reception times, to amplify the often very weak returned signal, etc. An important parameter is the angle of opening of the sound 'pencil', which depends on the ceramic tile diameter and on the emission frequency: the sharp slapping emitted by the transducer widens in a cone, all the more open with a small transducer and a deep sound. If the sound 'pencil' is narrow, it 'resonates' against a smaller area at the bottom, and the precision of the measurement is enhanced (Figure 3.2).

For graphical recording, each returned echo transmitted by the tile is transformed into a spark which burns a sheet of rolling paper, outputting a short black vertical dash, positioned higher or lower on the sheet according to the depth of the seafloor (Figure 3.3). As the sounds circulate in seawater at a velocity of the order of $1500\,\mathrm{m\,s^{-1}}$ (much faster than in the air: $331\,\mathrm{m\,s^{-1}}$), it is easy to convert the travel times into depths directly on the paper. Of course this is an average sound velocity, which may be modified by variation of seawater temperature or salinity, and also depends on the emission frequencies used. It may be necessary to make corrections: these are based on compilations of water sound-velocity profiles in various regions of the world's oceans (Matthew's Tables; Carter Tables). Inaccuracies in these tables may lead to appreciable errors in depth measurements: on a 3000-m-deep bottom, for instance, an error of 1% on the velocity signifies an error of 30 m on the measured depth.

Single-beam echo sounding reveals the mean-depth variations passing directly below the ship, and only beneath it. A number of details may be poorly expressed because of the width of the acoustic beam. When the opening angle of this is of the order of 30°, the surface resonated at 3000 m in depth is 1.5 km in diameter: bathymetric variations inside this surface will not be detected, and the recorded depth may be lower than the actual depth just below the ship (Figure 3.2). Only narrow-beam sonars, with an opening angle of only a few degrees, can solve these problems, and so are used for high-resolution surveys.

The tracks of the ships on the sea provide only linear, two-dimensional images of the bottom. These bathymetric profiles leave huge unknown zones in between the profiles, where important bathymetric features may therefore be overlooked. Since the 1920s, each cruise has added a bathymetric profile to the thousands of other profiles done, and patient compilation of these profiles has progressively revealed the world map of the ocean floors.

3.1.2 From single-beam sonars to multibeam-swath bathymetric systems: from the line to the boulevard ...

The dream of marine geophysicists, to evolve from line imagery to three-dimensional surface imagery, became reality at the end of the 1970s. The principle is simple: you

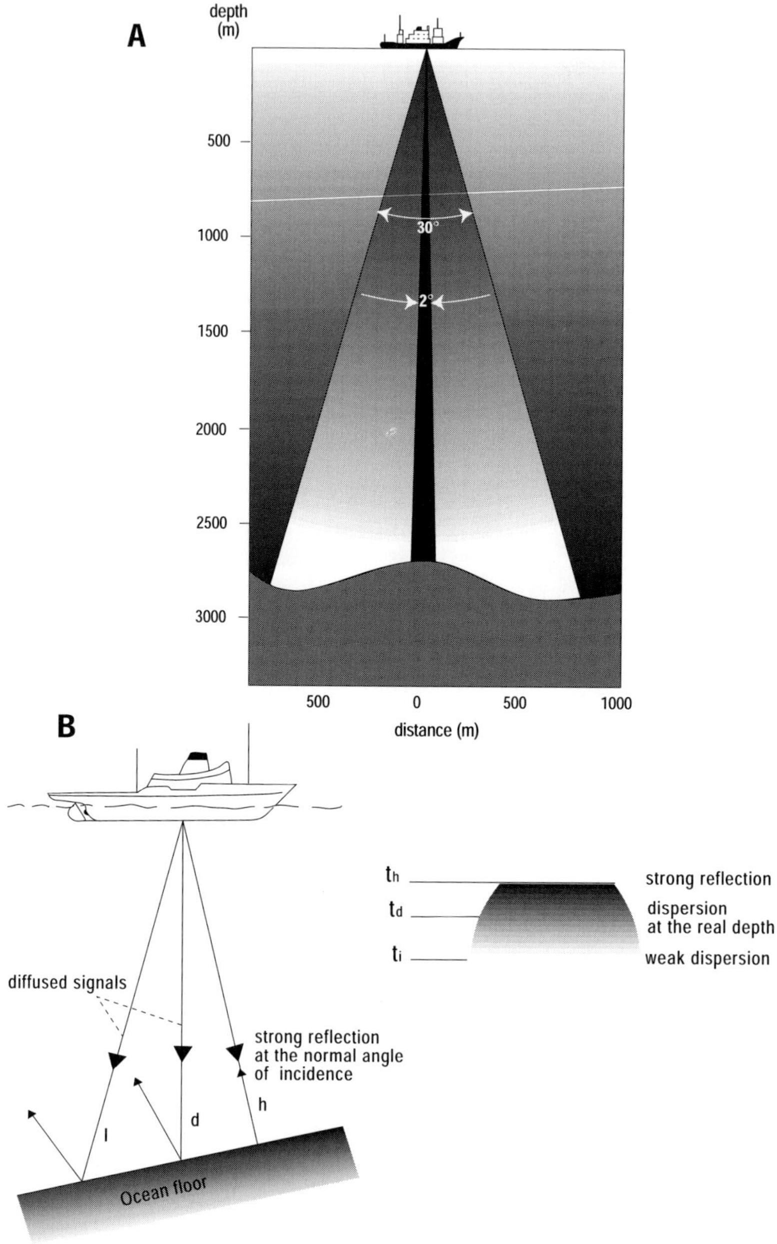

Fig. 3.2. (A) A classical echo-sounder has an apex angle of about 30°, resulting in a resolution of 1500 m at 3 km depth. High resolution, narrow beam (i.e. 2°) echo-sounders are necessary to map the submarine volcanoes of the oceanic ridge, which are usually less than 1 km wide. (B) Schema showing why errors in depth determinations are inevitable with poor-resolution conventional echo-sounders: in the case of a slope, the true depth *d*, directly below the ship, is masked by stronger reflexions at normal incidence, on pathway *h* (after The Open University, 1989).

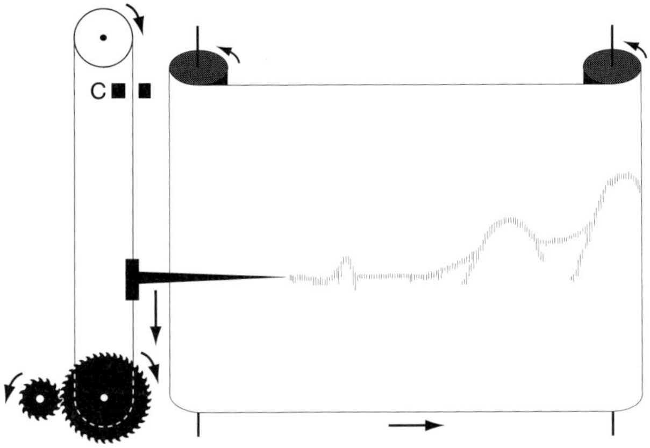

Fig. 3.3. *Principle of graphical recording of an echo-sounder*. The pen driven by a belt starts up the acoustic impulse toward the bottom every time it passes over contactor C. While the pen goes on, the acoustic signal is reflected on the bottom and bounces back to the receiver located in the ship's keel. At the moment of reception, the pen draws a small line on the rolling paper, at a position on the paper sheet corresponding to the length of answer time (i.e. the depth of the seafloor). Then the recording paper advances one step, ready to receive the next echo. The bathymetric profile is thus built point by point (after a drawing from Averous, 1981).

just have to place a number of sounding beams side by side arranged perpendicularly to the ship axis, in order to 'ensonify' a strip of the ocean floor on both sides of the ship (Figure 3.4). Carrying this out is technically very sophisticated, and demands powerful computer facilities aboard in order to manage and process the whole set of emission/reception signals.

The multibeam swath-bathymetric sounder was developed by the American Navy at the end of the 1970s, when advances in computer software allowed it. Commercialized afterwards by the American company GIC under the name of *Seabeam*, this system was installed on the French oceanographic vessel *Jean Charcot* as early as 1980. The French then provided the first precise survey of a segment of the East-Pacific Rise, and showed that the active volcanic zone was a narrow basaltic strip only 1–2 km wide. With the Seabeam system and its 16 juxtaposed beams, instead of a profile a strip is obtained whose width is approximately equal to two-thirds of the water depth. The data are processed on board by a powerful computer, and the submarine relief 'resonated' by the multibeam sounder is immediately converted to contour lines on a paper tape which unrolls following the ship's track. For a seafloor at 3000 m (following a ridge axis, for instance), the contour lines are drawn by the computer over a 2-km-wide area.

The document obtained aboard, however, is only a provisional document, which essentially gives the depths of the seafloor versus time. Long processing operations done on land will be necessary to draw the final bathymetric map. This is the 'replay' of the data, consisting in the conversion of time into distances, and in the relocation of the depth measurements on a spatial grid converted to international longitudes and latitudes. These operations include the resetting of the navigation, introducing in

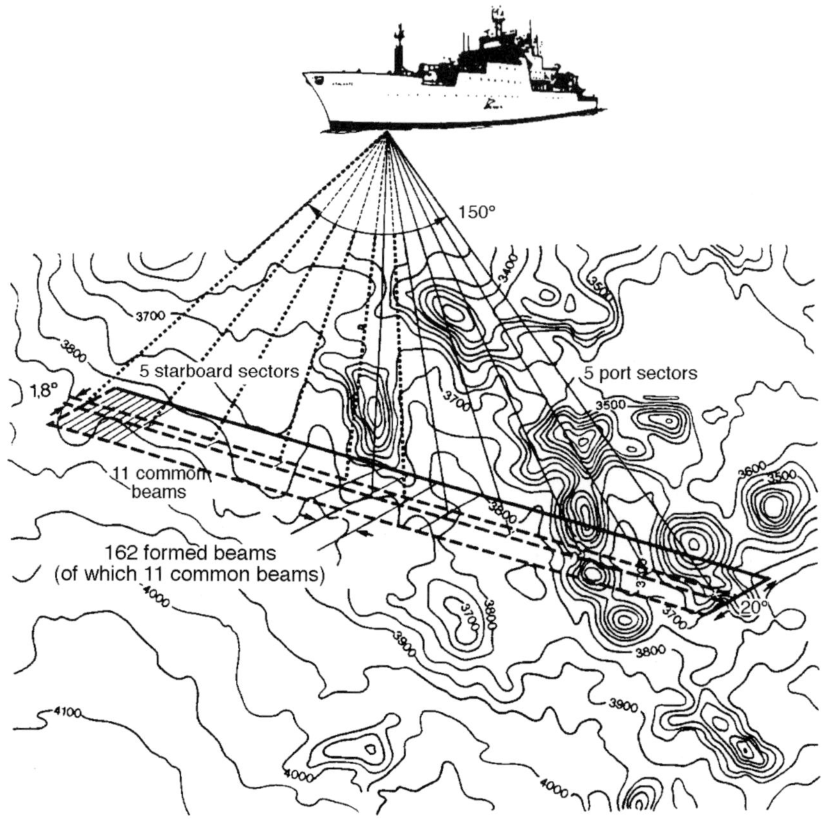

Fig. 3.4. Principle of the EM-12 dual multibeam echo-sounder, fitting out the French oceanographic ship *Atalante*. The bathymetric data are received in real time, then processed on board to obtain field numerical models (after a document from Ifremer).

the processing programme a number of satellite 'navigation points' (GPS, or Global Positioning System), and the drawing by computer of the contour lines, with the help of specialized software programs.

Today, a second generation of more improved multibeam swath-bathymetric systems has arrived on the market. The *Jean Charcot* and her Seabeam were replaced in 1990 by the *Atalante*, the *nec plus ultra* of the French oceanographic fleet, and her new EM-12 DUAL multibeam echo sounder, built by the Norwegian SIMRAD. Thanks to its 162 juxtaposed beams, the EM-12 system can carry out depth measurements between 100 and 10 000 m in a corridor with a width about seven times the water depth, with a maximum width of 20 km (beyond 20 km, the ray deviations of the sound waves due to salinity and temperature variations in the seawater column become too important). The *Atalante*'s powerful computers can restore in three dimensions, under the form of colored block-diagrams, any kind of relief mapped by its EM-12 sounder. To give an idea of the power of this system, suffice to say that it can map in 24 hours, to a depth of 5000 m, the surface area of two French départements.

3.1.3 The side-scan sonar, or the acoustic photography of the deep-ocean floors

Other techniques using ultrasounds have been developed. The side-scan sonar (Figure 3.5) consists in using the strength or *amplitude* of the reflected sound pulse, instead of measuring the travel time of the emitted sound. For deep seafloors, it is generally preferred to lower the transducer close to the bottom, in a torpedo-shaped vehicle towed behind the ship. The fan-shaped ultrasound pulse can then sweep the surface of the seafloor on both sides of the vehicle, and the recorded amplitudes of the signal reflected by the seafloor and all its reliefs draw a kind of 'acoustic photograph' of the bottom, with its 'acoustic shadows' outlining all the geological structures (sedimentary ridges, faults, rubbles and blocks, etc.), and sometimes man-made objects (ship, submarine or plane wrecks, etc.). The quality of the reflection is recorded in the intensity of the 'lightening' and the sharpness of the 'resonated' object: brilliant and neat images for anything sending back a clear and neat echo (steep hard rock cliff, huge rubble block, metallic wreck); grey and fuzzy images for anything absorbing a part of the signal and giving back a poor echo, for instance a heavily sedimented gentle slope.

Here also the surface of a seafloor strip is 'resonated' by the side-scan sonar, producing an 'acoustic picture', with its shadows reflecting faithfully the detailed nature of the bottom and its slope variations. By juxtaposing a number of strips side by side, a '*sonar mosaic-picture*' of a given area can be achieved, feeding back a lot of geological information on the bathymetric map (Figure 3.6).

Thanks to the development of new antenna design, both modes of using acoustic energy, sounding and scanning, can be combined in hybrid systems. The EM-12 system, for instance, comprises an optional 'side-scan imaging' function.

The side-scan sonars presently operated in various countries present notably different ranges:

The British GLORIA (Geological LOng Range Inclined Asdic) side-scan sonar (Figure 3.5B) scans the ocean floor along two strips of a 30 km width on each side of the ship. This device, towed at shallow depth (50 m), maps huge surfaces.

The American SeaMARC I and II and the English TOBI (Towed Ocean Bottom Instrument) have intermediate ranges, typically of the order of 5 to 10 km.

Finally the French SAR (Système Acoustique Remorqué, Figure 3.5A) is operated much closer to the bottom, for high-resolution surveys. With a swathe range of 750 m on each side with a 25 cm resolution, it is equipped with a 3.5 kHz sediment sounder (20° conical aperture, 80 m penetration).

Besides side-scan sonars, various kinds of deep-towed vehicles are used to take pictures of the ocean bottom and to measure various physical parameters (temperature, magnetic parameters, etc.).

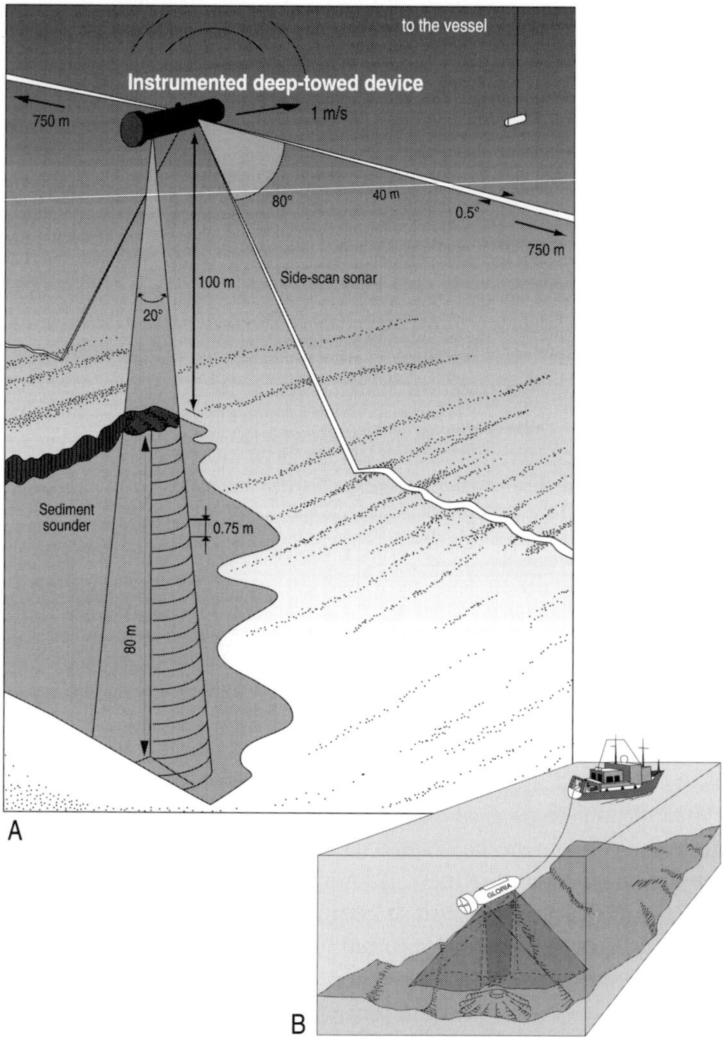

Fig. 3.5. *Principle of the side-scan sonar*. (A) The French SAR (for Systéme Acoustique Remarqué), towed close to the bottom, allows very high resolution surveys. Sweeping a stripe of 750 m on either side of the device with a resolution of 25 cm, it includes a sediment sounder (3.5 kHz, 20° conical opening, 80 m penetration) (after a document from Ifremer). (B) The British GLORIA side-scan sonar (for Geological Long Range Inclined Asdic) sweeps the ocean floor along two stripes, up to 30 km on either side of the ship, for depths of the order of 5000 m. Towed at 50 m below sea level, this tool is able to map huge surfaces (after a drawing from The Open University, 1989).

3.1.4 Seafloor mapping from space

For several years, altimetric satellites (GEOS-3, SEASAT, GEOSAT) have been measuring the altitude of the surface of the sea with respect to their orbits. These measurements show undulations on the seafloor surface that may reach amplitudes

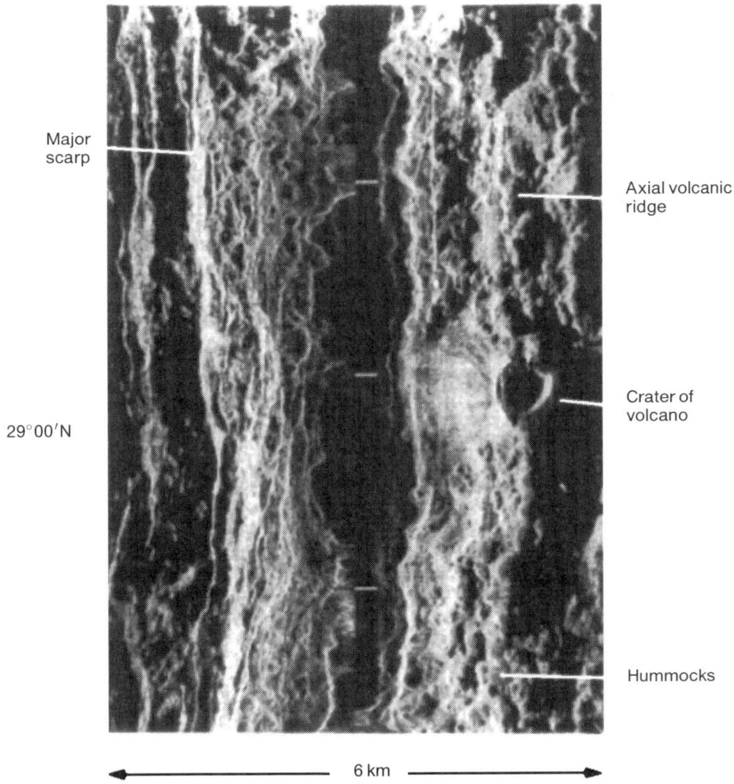

Fig. 3.6. Example of a mosaic assemblage of the Mid-Atlantic Ridge axial zone at 29°N, obtained by the British TOBI side-scan sonar, towed at about 500 m above the ocean floor. The pathway of the towed device is indicated at the centre of the image by a 'blind' stripe remaining black. The strip of ocean floor mapped by TOBI is 6 km wide, with a resolution of the order of 50 m. Light areas correspond to reflections, and dark areas to projected shadows. To the right, the axial volcanic ridge is clearly visible, bearing a large circular volcano 220 m high, with a caldera 600 m in diameter. The rest of the ridge is occupied by hummocky pillow-lava flows, affected in places by normal faults (upper right). To the left, major fault scarps bound the rift valley (after Smith and Cann, 1993).

of 200 m at the one 1000 km scale. These undulations reflect gravity anomalies due to heterogeneities in the distribution of mass in the earth. At short wavelengths (20 to 100 km), these anomalies are linked to the shallowest heterogeneities, that is, to the bottom topography.

The main submarine reliefs, although buried below several kilometres of water, thus have their 'signature' at the ocean's surface: one can 'see' and map them using altimetric measurements (Plates 3 and 4 at the end of Chapter 11). An oceanic ridge or a big submarine volcanic seamount, for instance, corresponds to a mass excess and is revealed by a several-metre-high positive undulation of the ocean surface; on the other hand, a deep trench, corresponding to a mass deficiency, is revealed by a negative undulation (a several-metre-low undulation), etc. The structural interpretation of the reliefs of the seafloor can thus be considerably improved (Figure 3.7).

The image of surface undulations of the world ocean, seen from space by altimetric radars, reproduces with incredible precision the whole oceanic ridge system, the subduction trenches, the submarine plateaus and seamounts presently known from classical oceanographic surveys (Plate 3 at the end of Chapter 11). The great advantage of altimetric satellites is in covering the oceans' surface *absolutely uniformly*, *whatever the latitude*. They allowed the discovery of numerous submarine reliefs at high latitudes (particularly in the Austral Ocean), inaccessible to oceanographic vessels. Two new extremely productive satellites, the first one European (ERS1), and the other one French–American (Topex-Poseidon) were recently put in orbit, and are presently refining the systematic survey of the worldwide ocean-floor

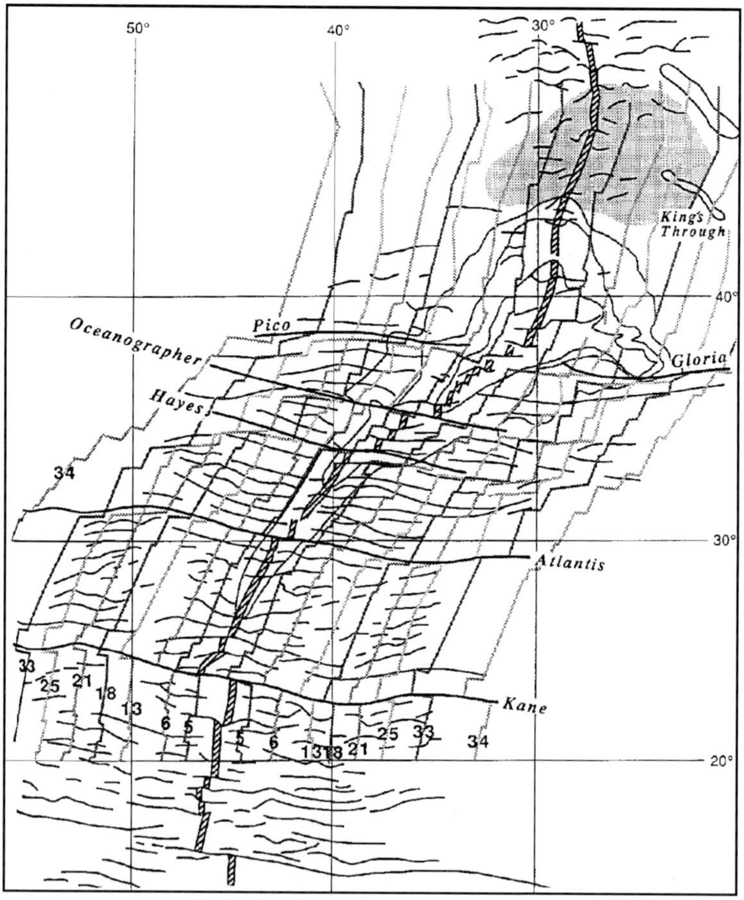

Fig. 3.7. Structural interpretation of Sandwell and Smith's (1994) gravimetric map by satellite altimetry for the Mid-Atlantic Ridge between 15°N and 50°N, and location of the main magnetic anomalies, indicated by their number (Klitgord and Schouten, 1986). The names refer to major fracture zones.

topography. Topex-Poseidon, for instance, has been measuring sea level since Summer 1992, with a precision greater than 5 cm. With one measurement station every 7 km, it covers the whole Earth's surface in 10 days.

The recent image given to scientists in October 1995 by David Sandwell, Walter Smith and collaborators combined the GEOSAT and ERS1 data (Plate 3). Both satellites together have covered the 350 million km^2 of the world ocean with a precision of 3 cm in geoid height, and an horizontal resolution of 10 to 15 km.

In summary, the last 80 years have been marked by a prodigious technical evolution for measuring water depth in the oceans, from long and static depth measurements below ships with a rope (before World War I), to the continuous, linear profiles (the *sonar* revolution), then to the cartographic strips, whose widths have evolved in 20 years from 2–4 km (*Seabeam* system) to 20 km (*EM-12 SIMRAD* system).

3.2 GEOPHYSICAL SURVEYS

Oceanographic ships are equipped to record a number of geophysical parameters. These geophysical surveys may use various methods. We mention here very briefly three main types of surveys, often routinely and simultaneously done aboard oceanographic ships.

3.2.1 Seismic surveys

Seismic surveys allow an image of the geological structures below the seafloor to be obtained. Acoustic energy sources are used for that purpose (air or water guns, boomers, sparklers, dynamite) emitting a signal in a range of frequencies between 10 to 200 Hz. The wave fronts, after travelling in seawater, penetrate the sedimentary layers or the hard rocks of the oceanic crust, then are reflected or refracted by geological discontinuities toward the surface, where they are recorded by one or several receivers (generally hydrophones, arranged along a seismic line trailing behind the ship (Figure 3.8). One may use either the *monochannel* (shallow penetration, good resolution), or the *multichannel* (deep penetration, variable resolution) seismic reflection method. The seismic refraction method, used to study deep structures, requires carrying out long seismic profiles, often done with two ships.

3.2.2 Magnetic surveys

Magnetic surveys allow magnetic anomalies of the ocean floor, to be mapped and identified in order to determine the age of the oceanic crust. The magnetic field is recorded by a magnetometer towed behind the ship.

3.2.3 Gravimetric surveys

Gravimetric surveys allow the distribution of mass deficiencies or excesses below the ocean floor to be detected. The oceanic gravimetric anomalies may also be mapped at higher scale by satellite altimetry (Plates 3 and 4, at the end of Chapter 11).

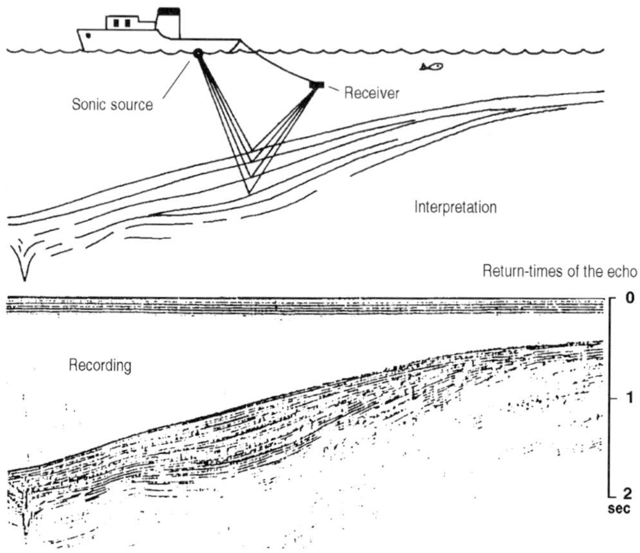

Fig. 3.8. Principle of a seismic profile recording. The structure of the oceanic substratum is studied along a profile by emitting acoustic impulses in seawater, which are reflected by the ocean floor and by some geological discontinuities of the substratum (boundaries between sedimentary layers, sediment/basalt contact, etc.). The record (below) shows the return time of the acoustic waves versus time, and may be interpreted, in a first approximation, as a geological section (top) (after a drawing from Van Andel, 1985).

3.3 SAMPLING METHODS

Once armed with the bathymetric map of the area he wants to study, and eventually acoustic images and bottom pictures, the marine geologist generally needs to obtain rock samples of his field of study. Here once again, several kilometres of seawater separate him from the ocean floor and from the rocky outcrops.

If the oceanic crust rocks are outcropping, they may be collected by dredging, by coring or directly using a submersible. If they lay beneath several hundreds of metres of sediment, the only way to sample them is by deep-ocean drilling.

3.3.1 Dredging

Historically, it was by using dredging techniques that the first hard-rock samples of the deep oceanic crust were collected. The first scientific dredgings were done probably around 1850, when American engineer Maury studied the reliefs of the Mid-Atlantic Ridge for the purpose of the first transatlantic telegraphic cables between Europe and the United States. But it was the famous English expedition of the corvette *HMS Challenger* (1872–1876) which systematically associated ocean-depth measurements with the sampling of the ocean floor by dredging. The rocks recovered were mainly manganese nodules, pelagic sediments, coral reef fragments and numerous volcanic rocks quite exclusively basaltic.

The principle of dredging is extremely simple: a strong welded steel frame, holding open a sturdy bag made of steel chain, is attached through hinged attachment arms to the heavy steel wire that lowers the dredge to the seafloor (Figure 3.9). The dredge is towed on the bottom, and the steel frame of this rudimentary device, armed with powerful teeth, pulls out rock fragments which accumulate in the metallic bag. The dredge must be sufficiently heavy so that its weight in water prevents it from bouncing off obstacles, and avoids getting stuck in the ground like a plough. The ship steams along at reduced velocity, towing the dredge with its whole mass, often over several kilometres. A tensiometer records the wire tension during the whole operation, in order to oversee the dredge work on the bottom.

It may happen that the dredge becomes stuck in the bottom, and that the wire reaches its rupture limit. In this case the ship comes back to a vertical position above the dredge and manoeuvers about, looking for the best angle to get free. If this fails to free the dredge, a clever mechanism is used to reach the rupture at the junction between the dredge and the wire, in order to safeguard the wire, which is much more difficult and expensive to replace than the dredge itself.

Dredging technology has not significantly improved since the 19th century, except on one point, which is the precise positioning of the dredge during its work on the bottom. This is done using an *acoustic pinger*, fastened to the dredge wire a few metres above the dredge itself. This battery-operated sound source gives the altitude of the dredge over the seafloor (using the echo of its sound impulse on the ground), and allows its x–y position to be calculated if two or three acoustic transponders have been launched close to the bottom in the area of dredging, by measuring the travel times of the sound impulses from the dredge to the transponder network. Setting up a transponder network, however, is time consuming and is not always done.

In summary, dredging is a simple and cheap method, which is still commonly used today. Its advantages are the volume of samples obtained (up to one ton per dredge), the possibility of obtaining huge samples (entire pillow lavas 50 cm in diameter for instance), its low cost and the ease of replacement of dredges lost on the seafloor (spare dredges are commonly carried onboard). Its disadvantages are evident: the samples arrive in total disorder on the deck! Moreover, the dredge track on the seafloor is often poorly positioned (when a transponder network is not set up). Another problem deals with *sampling bias*: it is clear that most samples pulled out from the bottom have been exposed on the seafloor to tectonic processes and to seawater action, often over long periods of time. Again, the large majority of dredged samples are taken from rubble talus accumulated at the feet of submarine cliffs. These samples are consequently abnormally altered or deformed with respect to 'normal' oceanic crust, reflecting their tectonic setting.

3.3.2 Coring

Coring methods are traditionally dedicated to sampling still-soft and water-saturated oceanic sediments. They consist in sending to the bottom a hollow metallic tube, sufficiently charged with cast-iron plates to be able to penetrate into the sediment

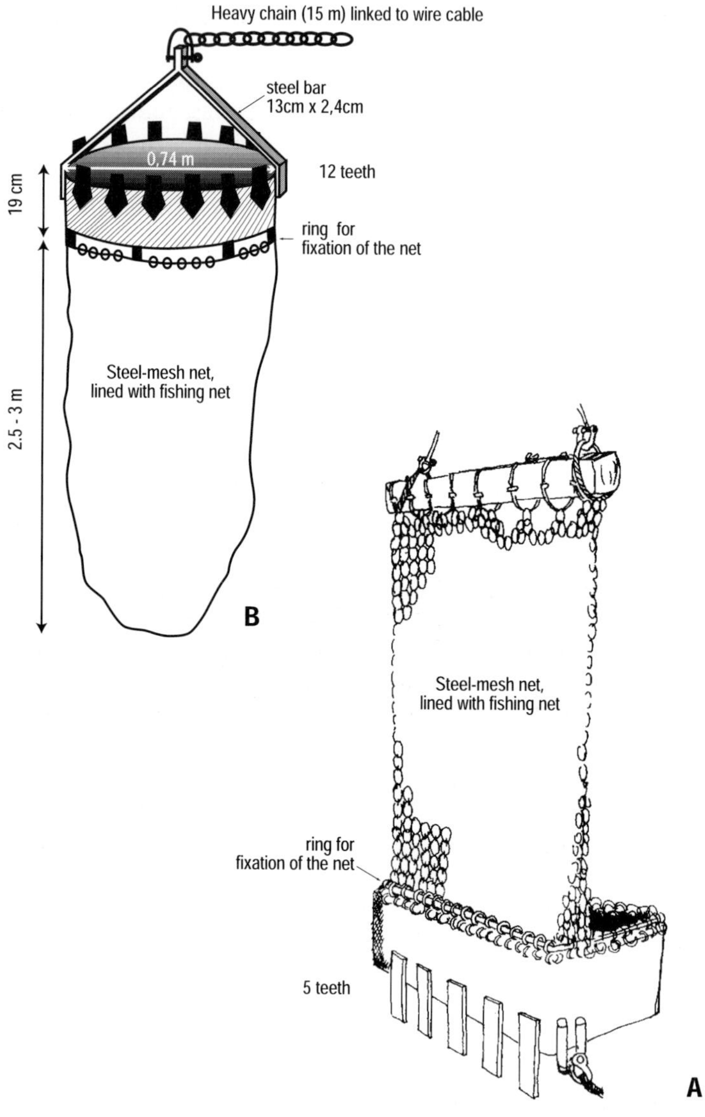

Fig. 3.9. Two kinds of dredges commonly used aboard French oceanographic ships. (A) Pautot dredging device (rectangular). (B) Lister dredging device (round). These dredges can take up to one ton of rock fragments from the ocean floor.

and sample a cylindric core 6–10 cm in diameter. The core barrels are commonly 5 to 29 m long, eventually 30 and up to 50 m in some cases. Of course, the hard rock of the oceanic crust cannot be sampled by this technique, because the core barrels bend or break when they hit them.

The core barrels, however, have often brought up fragments of fresh basaltic glass pulled out from the basaltic basement at the bottom of the sediment column. The American geochemist Charles Langmuir and his collaborators at the Lamont Doherty Geological Observatory have recently developed a method to sample basaltic glasses outcropping along the axial zone of the oceanic ridges by coring, smeared with a gluey resin. The weight of the core barrel splinters the pillow lava glassy crust, and numerous fragments of glass remain stuck to the resin at the core barrel tip. This is how Langmuir and collaborators have sampled fresh basaltic glasses all along some segments of the East-Pacific Rise to study their fine-scale geochemical variations.

3.3.3 Deep-oceanic drillings

In order to sample the oceanic crust buried below more than 50 m of sediment, away from the axial zones of the oceanic ridges, and also to core the hard oceanic crust over a several hundred metre area, it is necessary to operate at another scale, that of oceanic drillings, using the techniques developed by the petroleum industry. Only an oceanographic vessel specially built for this purpose can carry out deep oceanic drillings.

Drilling ship principle. A long drill string, made up of elementary 30-metre-long drill pipes which are screwed together, is lowered toward the bottom from a rigfloor located below the drilling tower (or derrick), in the middle of the ship (Figure 3.10). The drill pipe passes through a hole in the ship's hull (the 'moonpool'). The lower end of the drill string is equipped with a drill bit, designed to cut through the rocks encountered, and to allow a vertical cylindrical rock sample, the *core*, to pass through the inside of the bit into the hollow drill string. The bit drills by rotation of the whole drill string, rotated by a motor located at the top of the string, at the level of the drill tower. The drilling fluid is seawater, pumped from the ship down to the bit along the length of the drill string. Under pressure this water rises from the bottom of the hole between the string and the hole walls, carrying away the cuttings obstructing the bottom of the hole. Hollow metallic core barrels are lowered down by wireline through the drill pipe to bring to the surface the core samples, 6 cm in diameter and 10 m long.

Drilling and coring on land is not an easy task. To do it at sea from a floating platform is an extremely difficult undertaking, and requires overcoming an impressive number of technological challenges. The drill ship must be able to carry several kilometers of drill pipes in order to establish a hole entry on the ocean floor, to bring up regularly the core samples, to re-enter the drill hole after each change of bit, to compensate for vertical movements due to waves and heave, to remain at a position vertical to the hole without drifting by using a dynamic positioning system, etc.

The international programme. The cost of a scientific oceanic drilling programme is such that a single country alone cannot afford this burden. Launched in 1968 by a consortium of American oceanographic centers, the DSDP (*Deep Sea Drilling Project*) became international in 1974: the United Kingdom, France, Germany, then Japan and USSR entered one after another in the JOIDES (Joint Oceanographic Institutions for Deep Earth Sampling) Consortium between 1974 and 1976,

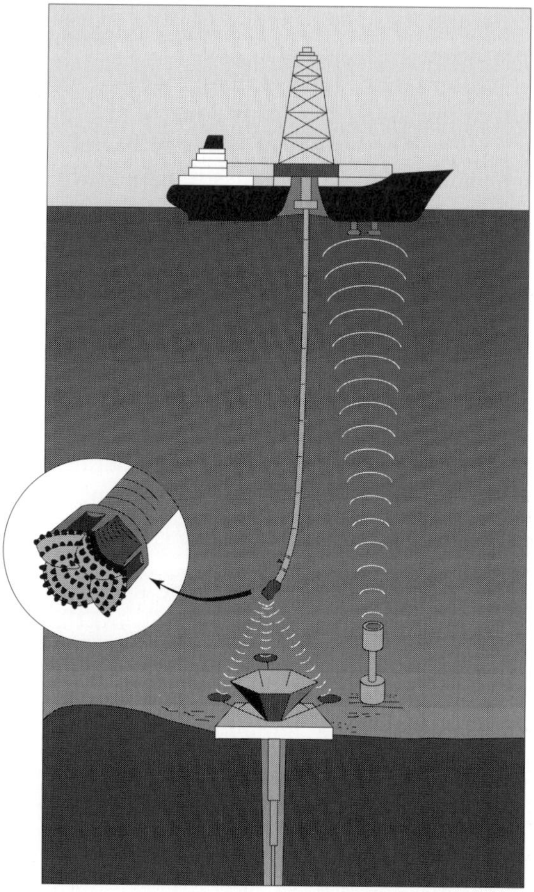

Fig. 3.10. Principle of the drilling ship. **(1) Dynamic positioning.** While the drilling ship is deepening a drill hole below several kilometres of seawater, it must remain exactly over the hole. In this situation, as illustrated here, its distance to an acoustic beacon dropped on the ocean floor just beside the re-entry cone is minimal. If the ship begins to drift, the distance to the beacon increases, and this information is recorded by the computer managing the dynamic positioning of the ship. When the ship reaches the drift tolerance, beyond which the drill string may be snapped, the computer triggers the lateral thrusters necessary to bring the ship vertically back over the drill hole. The figure shows an enlargement of the *drill bit* located at the drill string's tip: this is the tool which, by means of rotation and the weight of the drill string, pulverizes the rocks at the bottom of the hole. The drawing shows a classical four-conical headed bit, the cone surfaces being covered with nails made of extremely resistant tungsten steel. The rotation of the drill string, provided by a motor located at the top of the drilling tower, causes the rotation of the bit's four cones. These pulverize the rocks around a central cylinder 6 cm in diameter, which will become the drill core. The core is detached and raised up to the ship's deck every 10 m. **(2) Re-entering operations.** When the bit is worn out, it is necessary to raise the whole drill string (by unscrewing the 30 m-long pipes one by one), take off the used bit and replace it by a new one, then lower the string back and re-enter the drill hole. For this purpose, a sonar scanner mounted on the bit assembly emits sound signals that are echoed back from three reflectors spaced round the re-entry cone. The ship moves slowly over the cone, passing and passing again until the image of the three reflectors 'lit up' by the sonar form an equilateral triangle: the bit is then exactly over the hole, and the order is given to lower the drill string into the re-entry cone. This delicate operation may last several hours. Today, aboard the *Joides Resolution*, an ultra-sensitive and high resolution video camera moving along the drill string replaces the sonar. Direct vision of the bottom enables re-entry in a few minutes.

bringing their financial support. For France, the CNEXO (Centre National pour l'Exploitation des Océans) (IFREMER today) was the state establishment managing French participation. Each member country can develop its own DSDP structure, call for drilling proposals, put forward its delegates for the various scientific panels, and has the right to a quota of scientists and mission controllers aboard the drilling ship.

The first drill ship was the *Glomar Challenger*, well known in all the harbours of the world. The *Glomar* has steamed across all the oceans in all directions during her 96 cruises (called *Legs*) between 1968 and 1983. She covered 375 000 miles, drilled 1092 holes distributed in 624 sites, brought back 20 000 hard rock and sediment cores which, put together in line, would reach a length of more than 100 km . She drilled at all water depths, down to 7000 m in the Mariana Trough, and her penetration record is 1.7 km. The scientific cruise reports, the famous *DSDP Initial Reports* (familiarly called the 'blue books'), constitute a fabulous pluridisciplinary data bank, the world ocean archives.

Laid up in 1983, the *Glomar Challenger* was replaced by a new very capable American drilling ship, the *Joides Resolution*: 144 m long, 21 m wide, 17 000 tons displacement, equipped with a 61-m-high drill tower, able to drill down to 9000 m of water depth, The *Joides Resolution* is moreover a true floating research centre, with its 1100 m^2 of laboratories distributed over seven floors (Figure 3.11). The new ODP (*Ocean Drilling Program*) has succeeded the DSDP, with new partners (Canada, Australia, and a European consortium of the *European Science Foundation* representing 12 european countries). The *Joides Resolution* did her first Leg at the beginning of 1985 (ODP Leg 101), and the programme is still going on today, 13 years later.

Drilling the hard oceanic crust. From 1968 to 1974, the DSDP had no specific plan for the study of the oceanic crust, because the drillings were *single-bit holes*, pushed down until the bit wore out by abrasion. The operation, very simple on land, which consists in changing the bit, reoccupying the hole and continuing the drilling, was simply not possible at sea. All drillings were therefore ending rapidly in the first few metres of the hard and abrasive basaltic layer. The principle of the re-entry cone and its development at the beginning of the 1970s by the engineers of the Scripps Oceanographic Institution was a revolution for deep oceanic drilling operations: allowing re-occupation of a hole as many times as necessary, it was the start of a true 'oceanic lithosphere' programme. At the very first attempts, the re-entry system appeared to be operational: Leg DSDP 37 (1974), dedicated to drilling through the young oceanic crust of the Mid-Atlantic Ridge, close to the FAMOUS (French–American Mid-Oceanic Undersea Survey) area, was a brilliant success, with 583 m of penetration in the basaltic layer.

Starting from then and with the help of more advanced bits, several hundred-metre-deep holes were drilled and cored in Layer 2. Several of them penetrated more than 500 m into the Atlantic crust, but none of them reached 600 m. In the Pacific, penetration appeared to be much more difficult, and efforts were focused on a successful site, 200 km south of the Costa Rica rift; where Hole 504B reached the record penetration of 2111 m in seven DSDP/ODP legs, scattered from 1979 to 1993 (cf. Chapter 5).

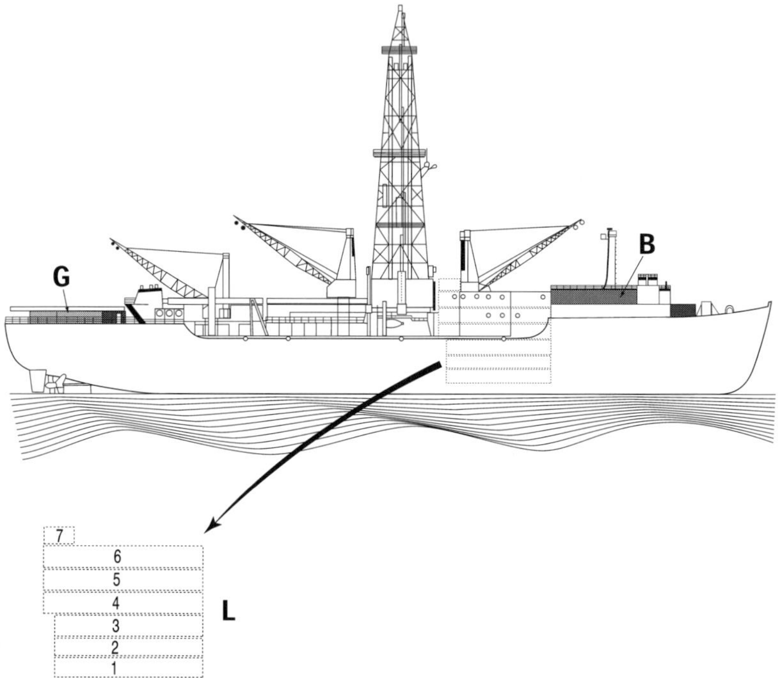

Fig. 3.11. Schema of the *Joides Resolution*, with her seven laboratory floors. G: Geophysical laboratory (reflection seismic and magnetic recordings during transits). B: Main deck, hosting the scientific library. L: laboratory block, distributed over seven levels: 1., storage of frozen cores and deep freeze; 2, storage of frozen cores, cold store; 3, photo lab, storage of electronic material; 4, computer rooms, offices of secretaries and chief scientists; 5, paleontology, electron microscopy, chemistry, petrology, thin sections, X diffractometry; 6, arrival of cores, study of physical properties of rock, palaeomagnetism, saw, description and sampling of cores, photographic bench for cores; 7, bottom-hole measurements (diagraphic recordings).

Bare rock drilling and access to the axial zone of the oceanic ridges. At the launch of the ODP programme, a new technological innovation, which allowed drilling directly into the bare crust, gained access to the axial zone of the oceanic ridges. In 1986 (ODP Leg 106), the first deployment of a *guide-base* (Figure 3.12), a device used to get rid of the 100 m sediment necessary to set up a classical drill hole, allowed drilling a hole directly at the very axis of the Mid-Atlantic Ridge, on the bare basalt of one of the axial volcanoes. This system was a disappointment when drilling 'zero age' basalts at the axis of the ridge, but gave remarkable results when drilling coarse-grained outcropping gabbros or peridotites. The axial zone of the ridges, formerly excluded, has been accessible to deep drillings for about 12 years.

Sampling characteristic. The rock samples obtained by drilling are cylindric cores, 6 cm in diameter (Figure 3.13). This core is broken during drilling into cylinder-shaped pieces, several cm to several decimeters long. Every 10 m, these cores are brought up aboard in a core barrel through the hollow drill string.

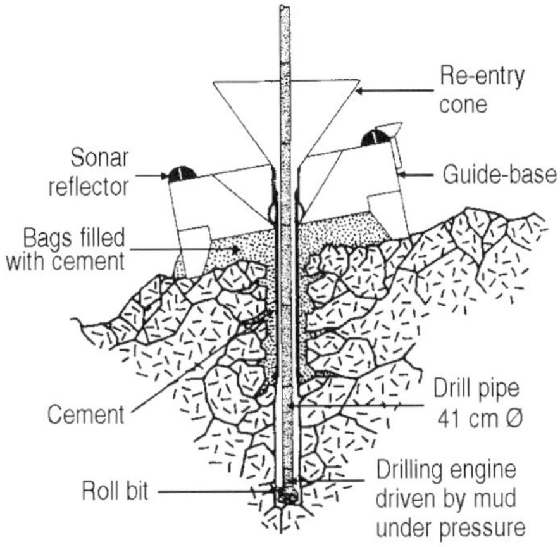

Fig. 3.12. Schema of a guide-base for bare-rock drilling. This device was conceived by the engineers of the Oceanic Drilling Program (ODP) to start a deep drilling on bare basaltic crust, devoid of a sediment blanket. It is made of a square-shaped metallic frame (6 m each side), resting on the ocean floor by four feet. This *guide-base* wears a re-entry cone and may accommodate a natural slope of 20°. Bags located below the guide-base are injected with dense, slow-setting cement. The bags inflate and take the exact shape of the terrain, providing a solid anchoring for the drill-hole casing head.

A very important parameter is the *recovery rate*, the ratio of the length of the recovered cores over the length drilled *in situ*. This ratio may range from 0 to 100% according to the difficulties encountered. Very high and close to 100% in the sediments, it drops frequently to 20–30% or less in the rocks of the hard oceanic basement. The recovery rate is higher in the aged crust: very poor (a few percent) in the very young crust of the ridge axial zone, it increases away from the axial zone, where the cracks and pores of the crust are filled with secondary minerals. For instance, the 110 Myr old crust of the Atlantic Ocean drilled to the south of the Bermuda rise during DSDP Legs 51 to 53 had excellent recovery rates (>80%). In gabbros and peridotites, the ratio may be either high (>80% in the gabbros of ODP Hole 735B), or low depending on the fracturing state and the hole stability.

With respect to other sampling methods, the rock samples provided by drilling present obvious advantages. First, they constitute an organized vertical succession, a geological section, more or less complete according to the recovery rate. This is the only sampling method able to provide rock samples not exposed on the seafloor, for instance sections of deep crust buried below hundreds of meters of oceanic sediment. The cores facilitate very detailed observation and analysis. A great number of fragile contacts (lavas/sediments, pillows/matrix, fault or shear zones, hydrothermal veinsystems, etc.) are preserved and can be studied in detail (Figure 3.13). The structural and lithological organization of the crust (magmatic contacts, magmatic, metamorphic and tectonic structures, crack and vein networks, etc.) can be analysed vertically over hundreds of meters.

Moreover, downhole measurements and more and more complete and sophisticated logging programmes are performed to complete the core studies, compensating for poorly or non-sampled intervals.

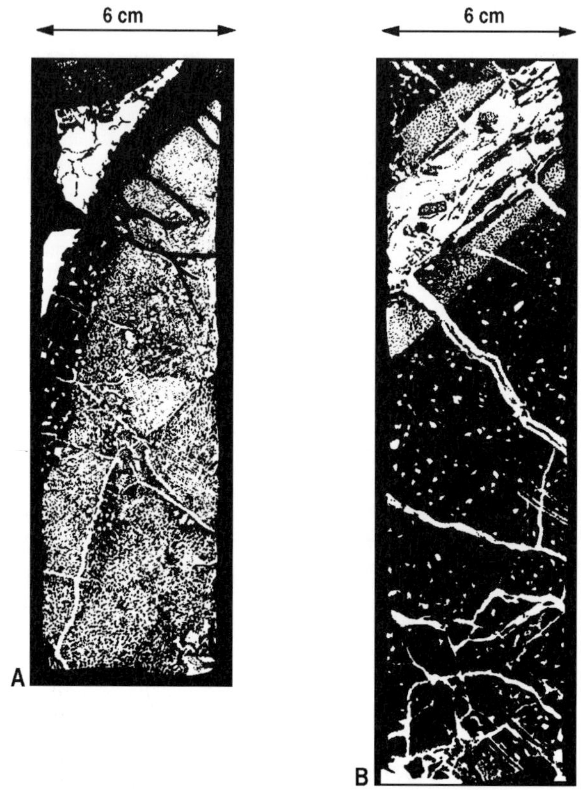

Fig. 3.13. Cores from DSDP Leg 51 (after pictures). Leg DSDP 51 drill holes, located to the north of Puerto Rico in the NW Atlantic, penetrated by more than 500 m the old Cretaceous crust of this ocean. They went through basaltic flows piled up over hundreds of meters, mostly made of pillow lava . The cores are 6 cm in diameter. They are sawn and split into two equal parts along their whole length, and nicely show here pillow-lava sections, with their typical radiating cracks and glassy margins. The drawing shows (A) *to the left*, fresh, non-altered basalts of Hole 417D: brilliant and black glassy margins, empty cracks, white carbonate material filling the spaces between pillows; (B) *to the right*, the altered basalts of Hole 417A: glassy margins transformed to clay minerals, cracks filled with secondary minerals. The delicate contact between the pillow-lava margins and the sediments or hydrothermal products cementing them are perfectly preserved. Any other sampling method would have destroyed this contact (after Juteau, 1993b).

The only major drawback of the drilled samples concerns the fact that their *in situ* orientation is lost, because the cores rotate during drilling. We know they are vertical, we know their top and bottom, and that is all. This may limit structural reconstruction. The orientation of the magnetization vector fossilized in the rocks may help to re-orient the core pieces.

All the cores taken since 1968 by DSDP and ODP are stocked in several lithotheques in the United States (Scripps Institution of Oceanography, Lamont-Doherty Geophysical Observatory, Texas A&M University), and can be observed *in situ*, or can provide core samples , for specialized studies.

Diagraphic measurements (*or loggings*) *and downhole measurements*. The sampling deficiencies, reflected by often poor recovery rates, can be compensated by

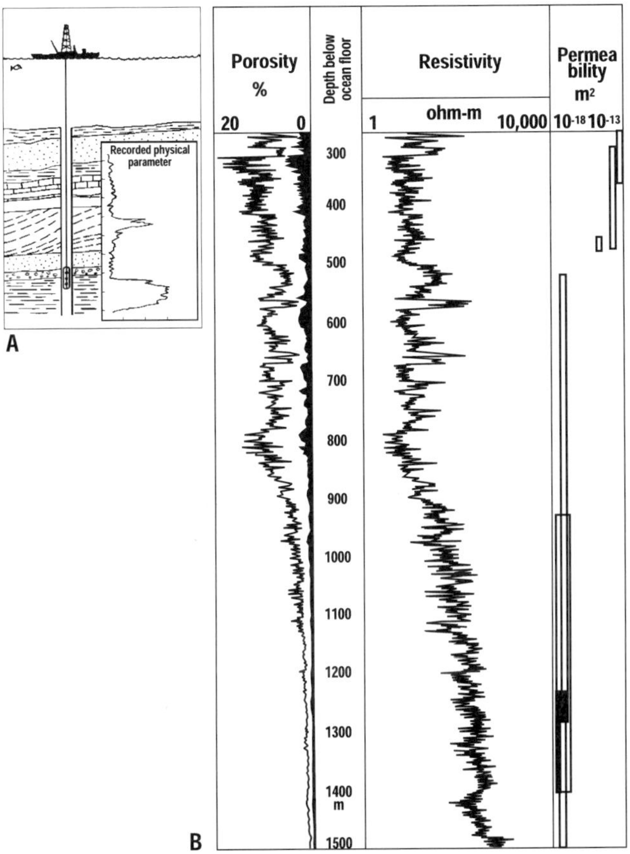

Fig. 3.14. Principle of diagraphic measurements or loggings. (A) At the end of drilling operations and before leaving the drill hole, an electronic wire cable wearing a series of logging probes, often emitting/receiving, is lowered into the hole. These tools measure a number of chemical and physical parameters versus depth in the crossed terrains. The recordings are done by slowly raising the cable from the bottom of the hole. Drawing after Biju-Duval (1994). (B) Example of a resistivity/porosity profile realized in DSDP/ODP Hole 504B, along with permeability measurements.

diagraphic measurements or loggings. These measurements consist in lowering probes (often emitting/receiving) attached to a wireline to record, all along the hole wall, physical or chemical parameters against depth (Figure 3.14A). One can thus record temperature or resistivity profiles (giving access to porosity, Figure 3.14B), V_p and V_s seismic velocities, or concentrations in various chemical elements in the wall rock, or else the 3D fracturing of the traversed terranes.

3.3.4 Direct sampling by manned submersibles

All the methods previously described involve taking samples from the seafloor to a surface ship. The marine geologist prefers generally to take his samples directly in the field, along a geological section, as on land. Manned deep submersibles offer this fascinating possibility (Figure 3.15). In France, deep sumersibles like *Cyana* (limited to 3000 m) or the *Nautile* (limited to 6000 m), thanks to their sophisticated mechanical arms, can take rock samples on the bottom and place them in a basket. Ten to twenty samples can thus be collected during a deep dive on the ocean floor, each of them being abundantly photographed *in situ* in their geological context before being removed. A diving cruise, when things are going well, will make on average 20 to 25 dives, and will haul on board 200 to 300 rock samples perfectly located on the bottom. The samples collected by submersible are most often picked

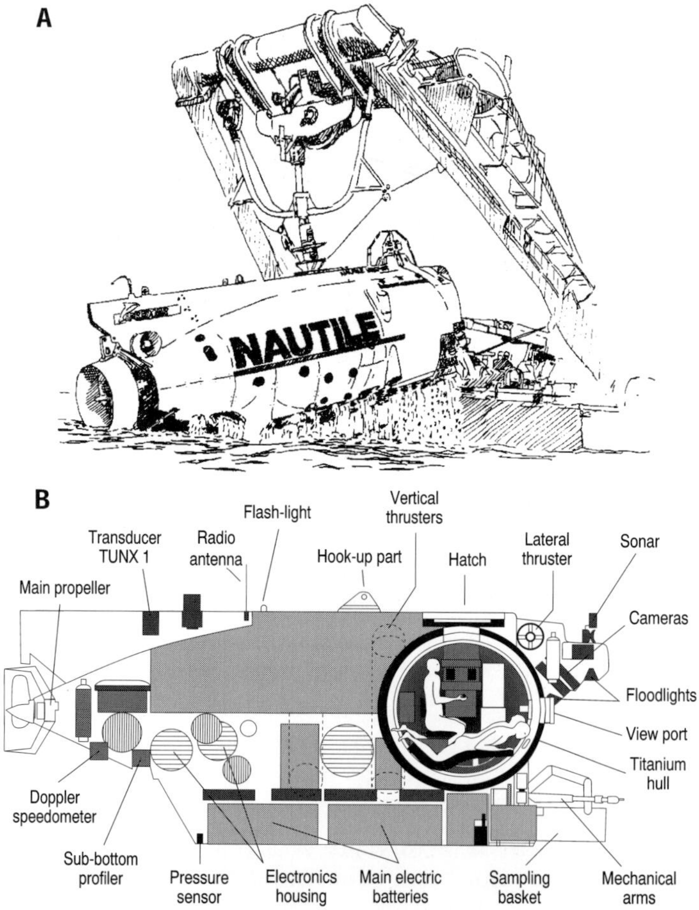

Fig. 3.15. Deep oceanographic submersibles. (A) The French submersible *Nautile* in action. (B) Schematic section of the *Nautile* submersible, built between 1979 and 1984 by CNEXO (Centre National pour l'Exploration des Océans, today Ifremer), and DCTN (Direction Technique de la Construction Navale), in the French National Navy dockyards in Toulon. This submersible can dive to 6000 m in depth (reaching 97% of the oceanic floor). Its weight is relatively low (18.5 tons) with respect to its diving capacity, thanks to a sphere made of a special titanium-rich alloy (2.10 m in diameter), housing the three passengers. The *Nautile* is easily operated from a relatively small-sized support-ship.

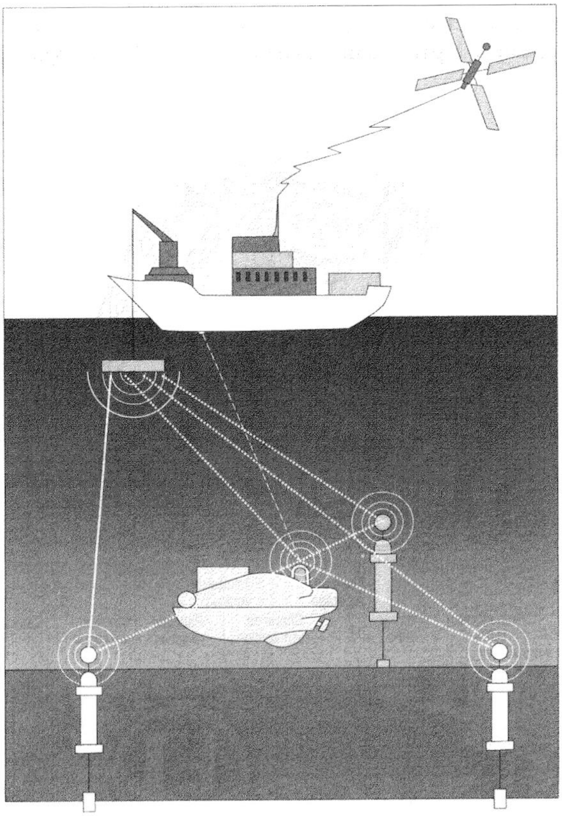

Fig. 3.16. Principle of the acoustic navigation of a deep submersible operating on the ocean floor. The support-ship of the submersible is positioned in the ocean using the civilian version of the American GPS system (Global Positioning System). Three acoustic transponders, each of them emitting on a different frequency, are dropped on the bottom before each dive. They are maintained at about 200 m above the bottom with the help of floating spheres, in order to avoid any interception of their signals by bottom reliefs. The mobile submersible, itself carrying an emitting/receiving transponder, interrogates the fixed transponders at regular time intervals. The answer times of the transponders, recorded by the submersible, are transmitted to the surface and processed by a computer on board the support ship. Moreover, an acoustic telephone is used to calculate the oblique distance of the submersible with respect to the ship. The computer calculates every 30 s the x, y and z (depth) coordinates of the submersible (after a document from Ifremer).

up on the ocean floor (rubble, etc.), since it is rarely possible to detach a rock fragment from the outcrop with the claw of the mechanical arm.

The crucial problem of precisely locating the submersible on the bottom is resolved to within 10 m, thanks to a navigation system using acoustic transponders on the seafloor (Figure 3.16).

Since the first dives in the axial zone of the Mid-Atlantic Ridge (the FAMOUS French–American programme, 1973–1974), deep-sea submersibles have played a major role in the exploration and detailed geological study of the oceanic ridges and associated fracture zones. Submersibles discovered massive hydrothermal sulphide deposits, active hydrothermal vents (the famous 'black smokers', cf. Chapter 9), outcrops of deep mantle rocks in the axial valley of the Mid-Atlantic Ridge, etc. They also enabled complete geological sections along the walls of the axial rift valley, or along the walls of the transform valleys in major fracture zones (cf. Chapter 5).

4

The worldwide oceanic ridge. Geographic extension, morphology and segmentation

The global oceanic ridge is the longest active volcanic range on Earth! It is also by far the principal source of volcanic and plutonic magmatic rocks at the Earth's surface, as shown in Table 4.1.

Table 4.1. Estimated production of the main magmatic province types (km^3/yr) (after Wilson, 1989).

Geodynamic context	Volcanic rocks (km^3/yr)	Plutonic rocks (km^3/yr)
Ocean ridge system	3.0	18.0
Subduction zones	0.4–0.6	2.5–8.0
Intraplate (continental)	0.03–0.1	0.1–1.5
Intraplate (oceanic)	0.3–0.4	1.5–2.0
Total production	3.7–4.1	22.1–29.5

This giant volcanic range is entirely submarine (except in rare places where the conjunction with a 'hot spot' (mantle plume) results in the emersion of the ridge, as in Iceland for instance).

4.1 THE RELIEFS OF THE OCEANIC-RIDGE SYSTEM. GEOGRAPHIC EXTENSION

The oceanic-ridge system extends over more than 60 000 km across the oceans of the globe and its reliefs, 1000 to 3000 km in width, occupy about one-third of the total surface of the oceanic floors. The crest of the ridge culminates at −2500 m in average,

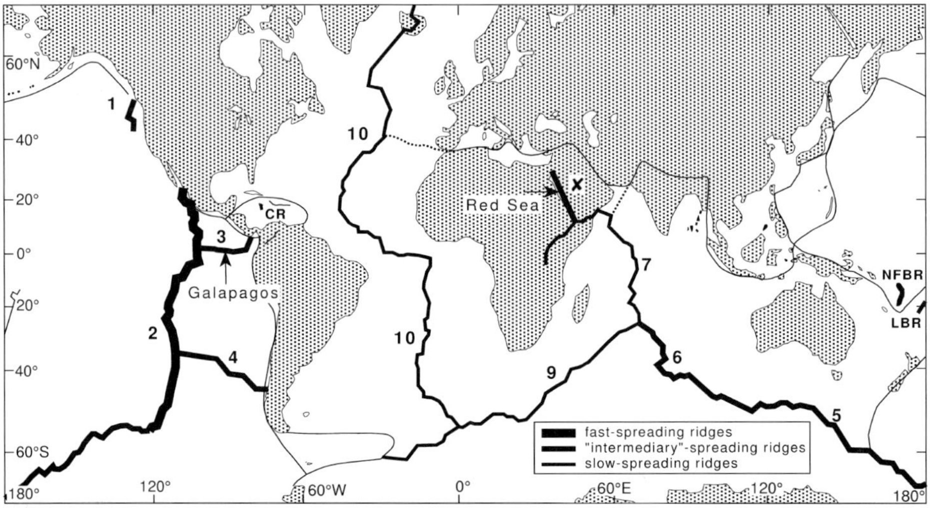

Fig. 4.1. Global tectonics of the Earth's surface (between 70°N and 70°S), showing the geographic extension of the world oceanic ridge system. The line thickness is proportional to the spreading rate (fast-, intermediate- and slow-spreading ridges): 1, Juan de Fuca Ridge; 2, East-Pacific Rise; 3, Galapagos Ridge; 4, Chile Rise; 5, Australia/Antarctica Ridge; 6, South-East Indian Ridge; 7, Carlsberg Ridge; 8, Red Sea Rift; 9, South-West Indian Ridge; 10. Mid-Atlantic Ridge; NFBR: North Fiji Basin Ridge; LBR: Lau Basin ridge; CR: Caiman Rift.

and stays therefore at 2.5 km above the large abyssal plains laying between the Ridge and the continental slopes.

The Ridge passes across the main oceans of the globe (Plate 1 at the end of Chapter 11 and Figure 4.1). In the young oceans, born in early Mesozoic times during the dislocation of the Pangea (Atlantic Ocean, Indian Ocean), the Ridge is equidistant from the continental blocks, whose 'passive' margins are still welded to the oceanic crust (Mid-Atlantic Ridge, East-Indian Ridge).

In the old Pacific Ocean submitted to active subduction along all its margins, the Ridge is offset eastward, and even contacts the north-American continent: penetrating into the California Gulf, it is relayed by the long San Andreas transform fault, which links the California Gulf to the northernmost segments of the East-Pacific Rise (Gorda rise, Juan de Fuca Ridge, Explorer ridge), located in the vicinity of the northwest-American coast. The opening of the Atlantic Ocean and the westward drift of the American plate resulted in the subduction of the eastern half of the Pacific Ocean, quite completely at the level of North America.

Moreover, the Ridge comprises several ramifications starting from localized triple junctions: Chile Ridge and Galapagos Ridge in the Pacific Ocean, Rodriguez Triple Junction in the Indian Ocean, Sandwich Islands Ridge in the southern Atlantic Ocean. In the Indian Ocean, the Carlsberg Ridge extends northward from the Rodriguez triple junction, turns westward to enter the Gulf of Aden (Sheba ridge), and enters the African continent by the Tadjura Gulf, to join the Afar Triple Junction, in conjunction with the East African Rift and the Red Sea. The latter, where oceanic accretion began some 4 Myr ago, represents the youngest oceanic domain in the world.

Finally, there are secondary accretion zones in many back-arc basins (North-Fiji Basin, Lau Basin, Woodlark Basin, Manus Basin in the south-west Pacific; Caiman Bift in the Caribbean Sea, etc.). In the south-west Pacific area, detailed mapping revealed the existence, in several of these basins, of oceanic ridge segments that may turn out to be more than 1000 km long (north Fiji basin for instance, Auzende *et al.*, 1988b, and Figure 4.1).

4.2 TRANSVERSAL MORPHOLOGY AND SPREADING RATE

In the axial zone, the morphology of the active zone presents important differences according to the *spreading rate*, the average velocity of opening calculated over several million years. This rate ranges from 1.5 cm/yr in the Red Sea, to about 18 cm/yr on the southernmost segments of the East-Pacific Rise (Figure 4.1). Macdonald (1982) distinguishes between three ridge types (Figure 4.2):

—The '*slow*'-*spreading ridges* (1–5 cm/yr), which show typically a deep axial valley or 'rift', about 30 km wide and 1 to 2 km deep, bounded by uplifted blocks (the 'rift mountains').

—The '*intermediate*'-*spreading ridges* (5–9 cm/yr), which show a slightly elevated axial plateau (example: the RITA (after the names of the RIvera and TAmayo fracture zones) area, East-Pacific Rise at 21°N).

—The '*fast*'-*spreading ridges* (9–18 cm/yr), which exhibit a dome-shaped axial structure (example: the East-Pacific Rise at 3°S).

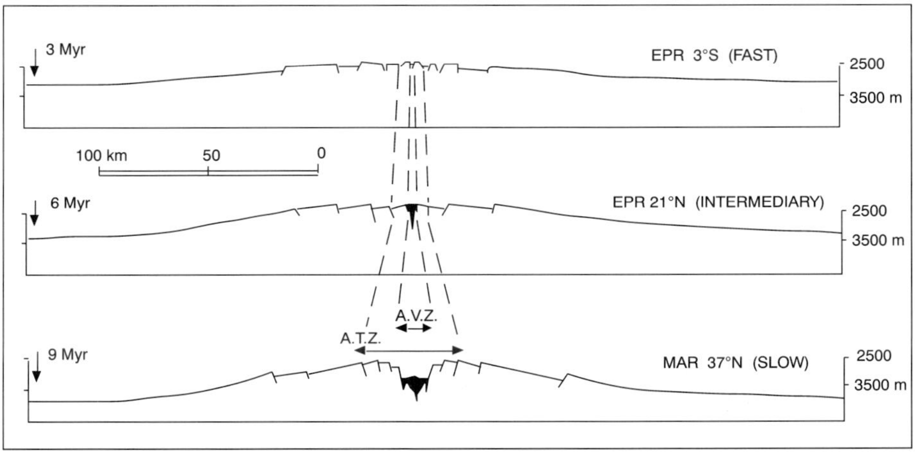

Fig. 4.2. Schematic transverse sections through the axial zone of three oceanic ridges with different spreading rates. This figure shows the influence of the spreading rate on the axial morphology, the width of the *active volcanic zone* (AVZ) and of the *active tectonic zone* (ATZ). EPR: East-Pacific Rise. MAR: Mid-Atlantic Ridge. Myr: million years (after Choukroune *et al.*, 1984). These general relationships have a number of exceptions (see text).

Recent studies have shown, however, that these general features must be tempered. For instance the approach of a major hot spot may modify considerably the morphological characteristics of the ridge (example: the Reykjanes Ridge, just south of the Iceland hot spot, presents the morphology of a fast-spreading ridge, without an axial valley, because of its hot thermal state). On the other hand, the axial morphology of a fast-spreading ridge may vary considerably along strike according to the volcano–tectonic cycle (see below).

Finally, one can distinguish, in summary, between two main oceanic-ridge types (the fast-spreading ones and the slow-spreading ones), since the intermediate- and fast-spreading ridges share many common features, whereas they both differ fundamentally from slow-spreading ridges in many aspects.

4.3 ALONG-STRIKE MORPHOLOGY: THE AXIAL SEGMENTATION OF THE RIDGE

4.3.1 First-order segmentation. Transform faults and fracture zones

The axis of the Ridge is not continuous, but is offset by numerous *transform faults*, generally (but not always) perpendicular to the ridge axis. The offset ranges from several tens to several hundreds of kilometres (Plate 1 at the end of Chapter 11). Transform faults are so called because they link the tips of two ridge segments, where the accretion movement is transformed to strike–slip movement. Seismically active, because the oceanic plates slide against each other along these faults, they follow a small-arc circle around the relative rotation pole between both plates (Figure 4.3).

Only the fault segment between the tips of the ridge segments is tectonically active (Figure 4.3; cf. also Figure 1.3). Beyond these tips, the extension of a transform fault is an inactive structure, called the *fracture zone*, extending through a single plate (and not at a plate boundary). Major fracture zones may develop over extremely long distances, up to 4000 to 5000 km in the Pacific Ocean. The strike changes observed along some fracture zones correspond to changes of direction in the relative movement of the plates.

An important consequence of this segmentation of the ridges is that the magma chambers feeding the axial volcanism of the ridges is necessarily interrupted at the level of the fracture zones.

4.3.2 Small-scale axial segmentation of the oceanic ridges, and notion of 'elementary accretion segment'

Until the 1980s, our knowledge of the Ridge was mainly based on transverse, widely spaced profiles, leading to a linear, bi-dimensional vision of the ridge structure. During the last 15 years, considerable effort has been made in studying the along-strike variations of the ridge morphology, and to obtain a true tri-dimensional image of the reliefs of the axial zone of the ridge. The systematic use of high-resolution

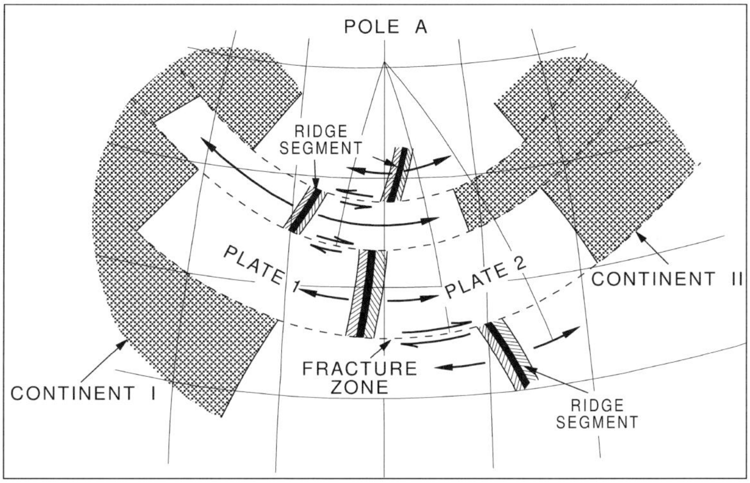

Fig. 4.3. The oceanic ridge is offset by numerous transverse fracture zones, the ridge crest being offset by several tens to several hundreds of kilometres. Each fracture zone comprises a tectonically active segment or *transform fault*, located between both ridge crests. The fracture zones are displayed following arcs of small circles centred on the rotation pole between two adjacent oceanic plates. The spreading rate is linked to the angular velocity and to the distance to the rotation pole: it increases progressively along the ridge, going away from the relative rotation pole. This latter is determined by tracing the large circles perpendicular to the fracture-zones: in the ideal case presented here, these large circles are all passing by pole A.

mapping devices (multibeam swath systems, side-scan sonars) has provided a new generation of bathymetric maps. With these high performance systems, our vision of the axial morphology of the ridges has become much more precise.

The style of the segmentation of fast- and slow-spreading ridges is notably different.

Fast-spreading ridges: example of the East-Pacific Rise

This 'young' oceanic ridge has been spreading for barely 11 Myr, following a ridge jump of several thousands of km, due to a general rearrangement of the Pacific spreading-ridge system. It presents an opening velocity increasing progressively southward, from 60 mm/yr at the entrance of the California Gulf, to 180 mm/yr, close to 22°S, in the vicinity of Easter Island. In this interval, it is offset by nine major fracture zones, cutting the ridge into 200 to 850-km-long segments (Figure 4.4).

Axial segmentation

The axial zone of the East-Pacific Rise, narrower and structurally simpler than the axial zone of the slow-speading ridges, has been mapped over long distances by single multibeam echo-sounding swaths, as early as the 1980s. The various segmentation types and axial morphologies were therefore described first on this ridge.

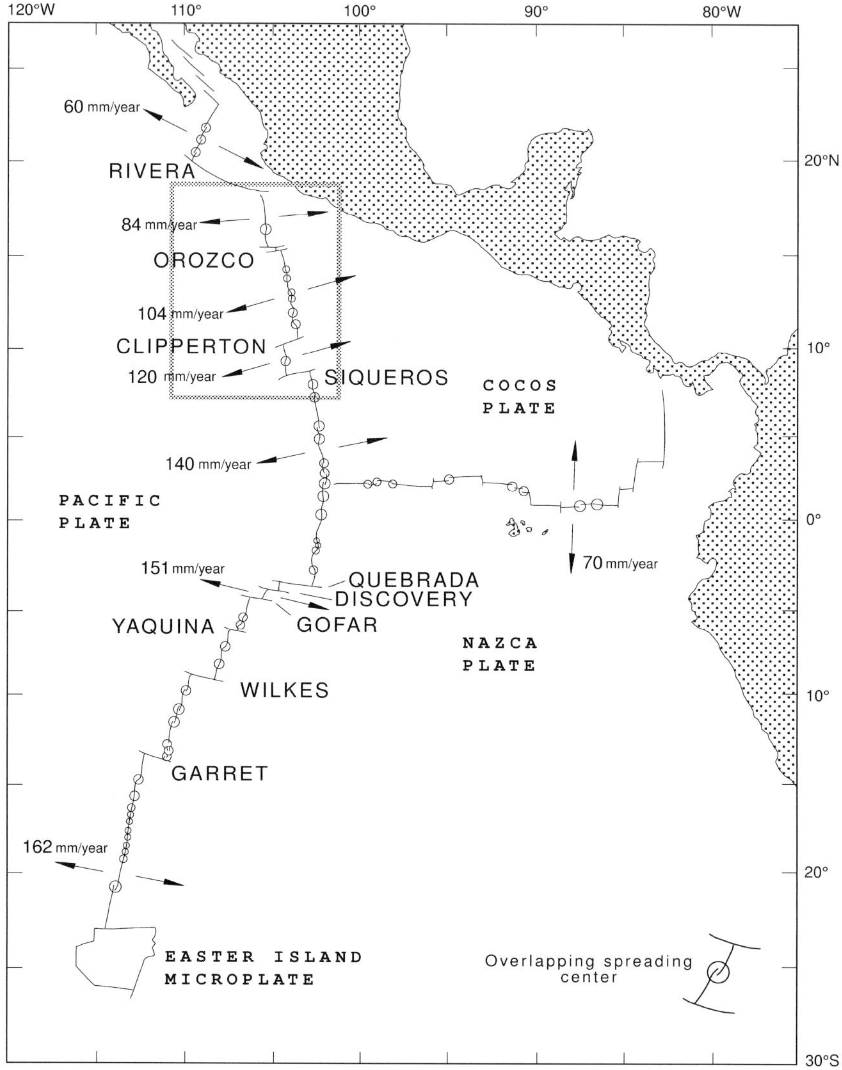

Fig. 4.4. General schema of the East-Pacific Rise segmentation between 30°S and 23°N, and of the Galapagos ridge, separating the Cocos and Nazca plates. The names of the main fracture-zones are indicated. The numbers refer to the present-day opening rate per year. The rectangle indicates Figure 4.5 domain. After Macdonald *et al.* (1986).

For example new detailed mapping surveys performed along the East-Pacific Rise between 8° and 16°N (Figure 4.5A) revealed the existence of an axial segmentation of the ridge, with a wavelength far inferior to that of the major transform faults. This segmentation is marked by progressive variations of the ridge axial depth, and by the existence of more or less pronounced axial discontinuities, which can be subdivided into four categories (Macdonald *et al.*, 1988; Figure 4.5B):

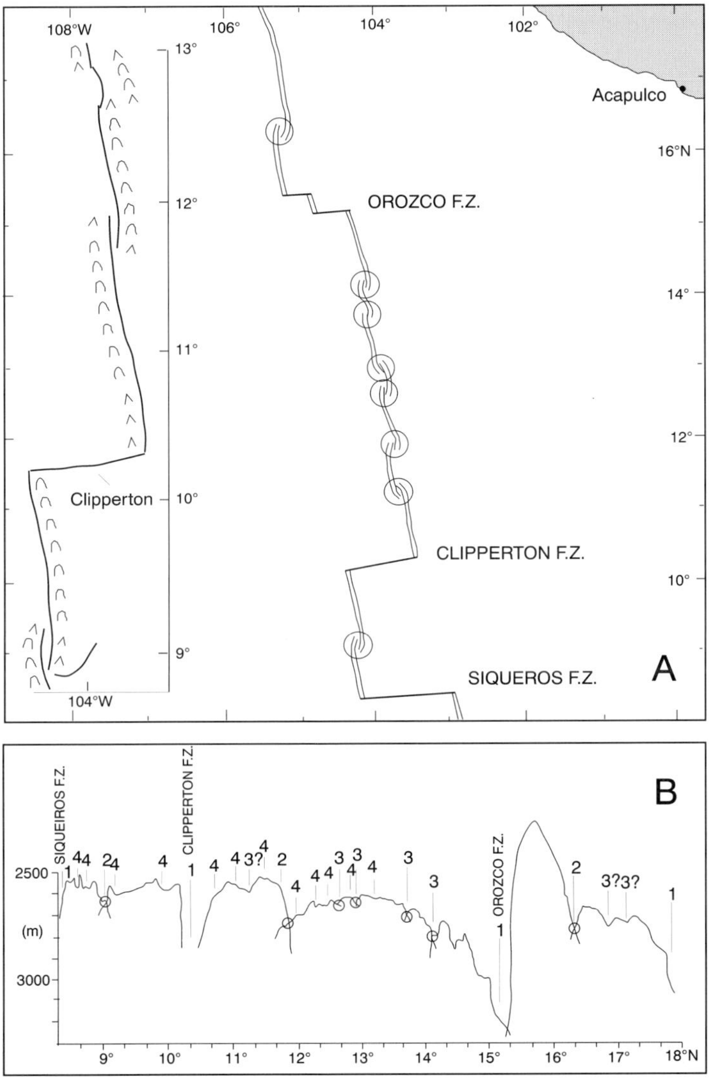

Fig. 4.5. (A) The axial segmentation of a fast-spreading ridge such as the East-Pacific Rise between 8° and 16°N is essentially determined by the main *fracture zones* (Orozco, Clipperton, Siqueiros, etc.), and by the *overlapping spreading centres* (circles). To the left, close-up of the area between 9° and 13°N, with indication of the axial morphology (wide dome or narrow triangle). (B) The axial bathymetric profile of this same ridge (with important vertical exaggeration) illustrates the various segmentation scales. The numbers refer to the four segmentation scale orders described in the text (after Macdonald *et al.*, 1988).

—The *first-order segmentation* is tectonically defined by the major transform faults. These are *rigid boundaries*, dividing the ridge into distinct and independant tectonic units, which persist for millions of years. Their typical spacing is 300 to 500 km. These structures are characterized by important depth anomalies (500 to 3000 m), and by frequent geochemical (FeTi-rich, very fractionated tholeiitic basalts) and mineralogical anomalies (highly phyric lavas).

—The *second-order segmentation*, with a 50–300 km typical spacing, is defined by shorter ridge offsets (3–5 km), characterized by a *non-rigid behaviour*: this is the case of the *overlapping spreading centres* (OSC) showing a large offset. These structures, discovered relatively recently thanks to the systematic mapping of the EPR axial zone, constitute non-rigid and temporary axial discontinuities. Marked by depth anomalies of several hundreds of metres, they evolve rapidly with time: one of the two branches wins rapidly over the other one and joins it, erasing the discontinuity, letting a scar remain in the 'structural grain' of the ocean floor. The ridge segments bounded by major OSCs show typically a domed along-strike profile, with a central 'topographic high' and depressed tips, with a vertical offset ranging from 100 to 400 m (Francheteau and Ballard, 1983).

—The *third-order segmentation* is defined by OSCs with a small offset (0.5 to 3 km), with depth anomalies of several tens of metres.

—Finally the *fourth-order discontinuities* include small axis jumps (<0.5 km), or small changes in the axis strike (1° to 5°). There is no bathymetric signature, but often a clear geochemical discontinuity.

Seismic structure of the East-Pacific Rise

Experiences of seismic tomography recently done at the East-Pacific Rise axis between 8° and 13°N, have shown the frequent existence of a strong reflector at about 1.5 km below the seafloor, interpreted as marking the top of an axial accretion magma chamber. Moreover, it appears that:

—Where the reflector disappears, the axial zone becomes narrower, and exhibits a triangle-shaped edifice, without axial graben. This occurs on relatively deep portions of the axis, often at the approach of a discontinuity (Figure 4.5A). In these zones, the magma chamber is probably less than 2 km wide, or does not exist any more.

—Where the reflector exists, the dome-shaped axial zone is wide and presents a well-expressed and extremely regular axial graben, along tens of km. This occurs on the topographic highs of the axial profile, away from discontinuities. In these zones, the axial magma chamber is inflated with magma and feeds an abundant volcanism. The resulting axial graben is due to a tectonic collapse similar to the circular calderas of aerial volcanoes.

This fundamentally segmented character of the East-Pacific Rise is thought to reflect the distribution of partial melting mantle sources (Figure 4.6A and B). These sources would be distributed in discrete places beneath the ridge axis, from which partial melting liquids would arise, feeding shallow crustal reservoirs. These magmas would inflate the crustal reservoirs, creating a local topographic high along the axial profile of the ridge. The local volcanic eruptions would be concomitant with along-

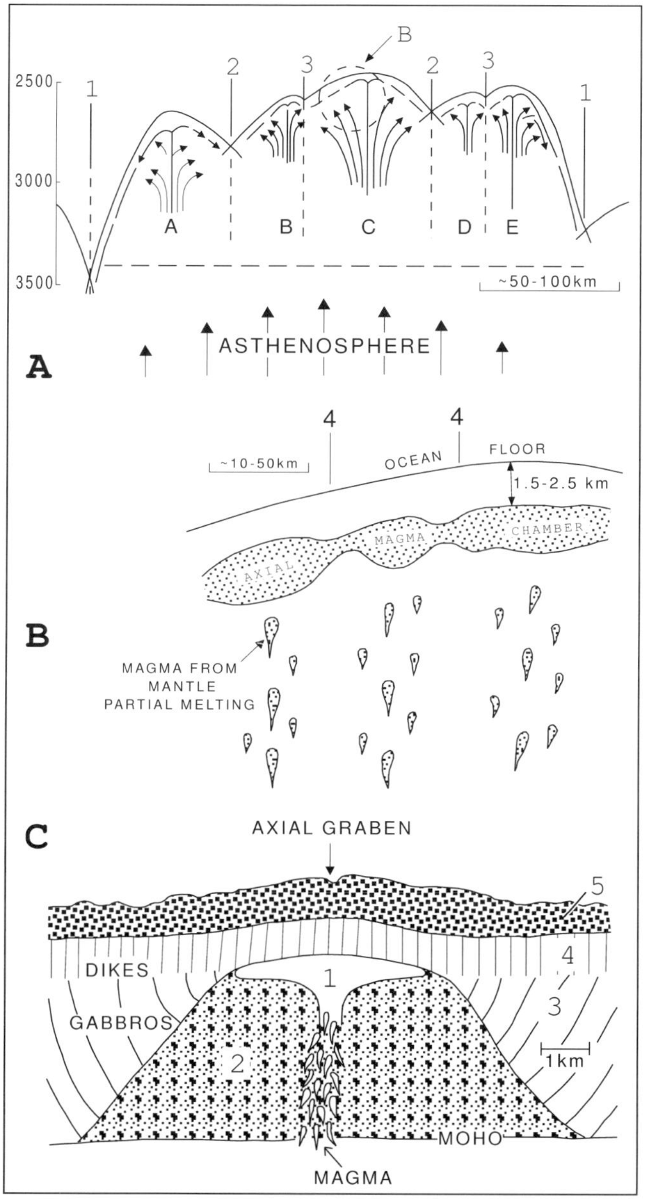

Fig. 4.6. (A) Scheme explaining how each elementary accretion segment may be fed, in its central part, by a discrete magmatic event (A, B, C, D or E). The magma would then migrate along-axis toward the segment tips. (B) Close up of area B, where geochemical discontinuities (segmentation of the fourth order) could be due to pinches of the magma chamber; (C) Transverse profile of the East-Pacific Rise at 13°N, based on acoustic tomography and gravimetric experiments: 1. Mushroom-shaped liquid pocket, 3–4 km wide, whose roof, located at 1.4 km below the seafloor, constitutes a vigorous acoustic reflector. 2. Hot crystal mush, about 6 km wide, containing several % of interstitial magma. 3. Gabbros, whose magmatic foliation steepens upsection. 4. Diabase sheeted dike complex. 5. Basaltic layer (sheet flows and pillow-lava flows (after Macdonald, 1989).

strike magma migrations, away from the source. In this magmatic model, the ridge axial discontinuities are located at the distal tips of the magma feeding zones, at the lowest points of the bathymetric profile, defining the tips of the ridge segments (notion of elementary accretion segments). The fact that the basalts dredged in these bathymetric lows are frequently evolved and fractionated strengthens this model.

Figure 4.6C presents a schematic transverse section of the axial zone in the middle of an accretion segment, based on seismic tomography and gravimetry experiments. The axial magma chamber is 3–4 km wide and hardly 1 km high. Its roof, located at 1.4 km below the seafloor, constitutes an excellent acoustic reflector. The magma pocket is 'perched' over a thick, very hot crystal-mush pile, widening out downward (6 km wide at the bottom). This crystal-mush pile, containing several percent interstitial magma, solidifies progressively into layered gabbros away from the axial zone.

Slow-spreading ridges: the example of the Mid-Atlantic Ridge

The axial zone of slow-spreading ridges, much broader than that of fast-spreading ridges, cannot be mapped during a single multibeam echo-sounding pass. Many more important mapping efforts are necessary to obtain a good image of the Mid-Atlantic Ridge segmentation over large distances.

Oceanic spreading in the north Atlantic between 10° and 60°N began at the end of the Liassic period, at about 180 Myr (Plate 2 at the end of Chapter 11). Its present-day opening rate ranges between 20 to 33 mm/yr all along the ridge.

Axial segmentation

A bathymetric section along the Mid-Atlantic Ridge axis between 10° and 60°N shows considerable variations in axial depth, obviously linked to the 'hot spot' distribution along the ridge (Figure 4.7). The shallower depths are located at the level of Iceland (emersion of the ridge) and the Azores (less than 1000 m at 38°N), and the maximal depths are observed between these major hot spots around 50°N, or else far away to the south of the Azores, between 16° and 23°N (3700 m to 3800 m). Besides these first-order undulations (with a wavelength of 700 to several thousands of km),

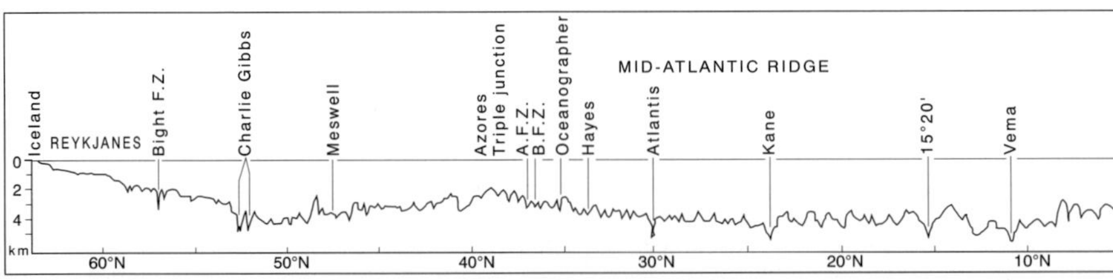

Fig. 4.7. Bathymetric profile of the Mid-Atlantic Ridge axis between 5° and 65°N. This profile shows *first-order undulations* at the 1000 km scale (Icelandic and Azores bulges), the bathymetric signature of the main fracture zones, and the small wavelength topographic variations (after Gente, 1987).

the axial topography presents depth variations of the order of 1000 to 2000 m, associated with important axial discontinuities.

As in the Pacific, the transform faults and associated fracture zones constitute first-order segmentation. They are associated with the most important horizontal offsets (>20–30 km) and age discontinuities (>2–3 Myr). They cut the Mid-Atlantic Ridge into segments of 200 to 1000 km long, which are offset along narrow transform fault zones.

Presently one of the best studied areas is the axial domain between 23°50′ and 30°40′N, including the Kane and Atlantis fracture zones in the north Atlantic. The mapping surveys done in 1988–89 over more than 800 km in this area allowed clear identification of 15 accretion segments between both fracture zones (*second-order segmentation*; Figure 4.8).

These accretion segments are 10 to 85 km long. They are offset by 'non-transform' axial discontinuities, whose morphology is quite different from that of transform faults. The horizontal offset at the level of these discontinuities ranges from 0 to 30 km. The axial bathymetric profile confirms the existence of a segmentation with an intermediate wavelength (10 to 100 km), with vertical offsets of 300 to 2200 m between the highest and the lowest points of the same segment (Figure 4.9). The axial discontinuities correspond to the lowest points of the profile, with the deepest ones corresponding to the transform offsets of the Kane and Atlantis fracture zones. In each segment, the inner floor of the rift presents a topographic high in the middle zone of the segment, and deepens progressively toward the tips. The segment centres are associated with the narrowest parts of the axial valley (3–15 km), which becomes progressively broader toward the tips (Figure 4.10). These morphological variations at a segment scale suggest along-strike variations in the magmatic and tectonic processes. As in the Pacific, the morphological analysis allowed identification of up to four segmentation orders.

This fundamentally segmented character of the Mid-Atlantic Ridge reflects probably, according to modern hypotheses, mantle upwellings focused below the central part of each accretion segment, at each segmentation scale. Figure 4.11 summarizes this 'fractal' vision of the mantle sources, with an identical scheme at the various observation scales: the '*hot-spot*' *scale* first (very long wavelength variations, 1000–5000 km), with voluminous deep mantle upwellings (Le Douaran and Francheteau, 1981); then the *fracture-zone scale* (long wavelength variations, 200–1000 km), with asthenospheric mantle upwellings between the fracture zones; and finally the *elementary accretion segment scale* (20–80 km), where focused asthenospheric mantle diapirs melt partially by decompression and produce magmatic events, feeding the central zone of the elementary accretion segments.

Off-axis evolution and 'structural grain' of the Mid-Atlantic Ridge

South of the Kane fracture zone, recent mapping surveys extended to the off-axis domain up to 10 Myr ago (Gente *et al.*, 1995; Durand *et al.*, 1995) demonstrated the presence of '*rhombic domains*' bounded by oblique linear structures forming topographic lows, indicating systematic variations in the length of the accretion

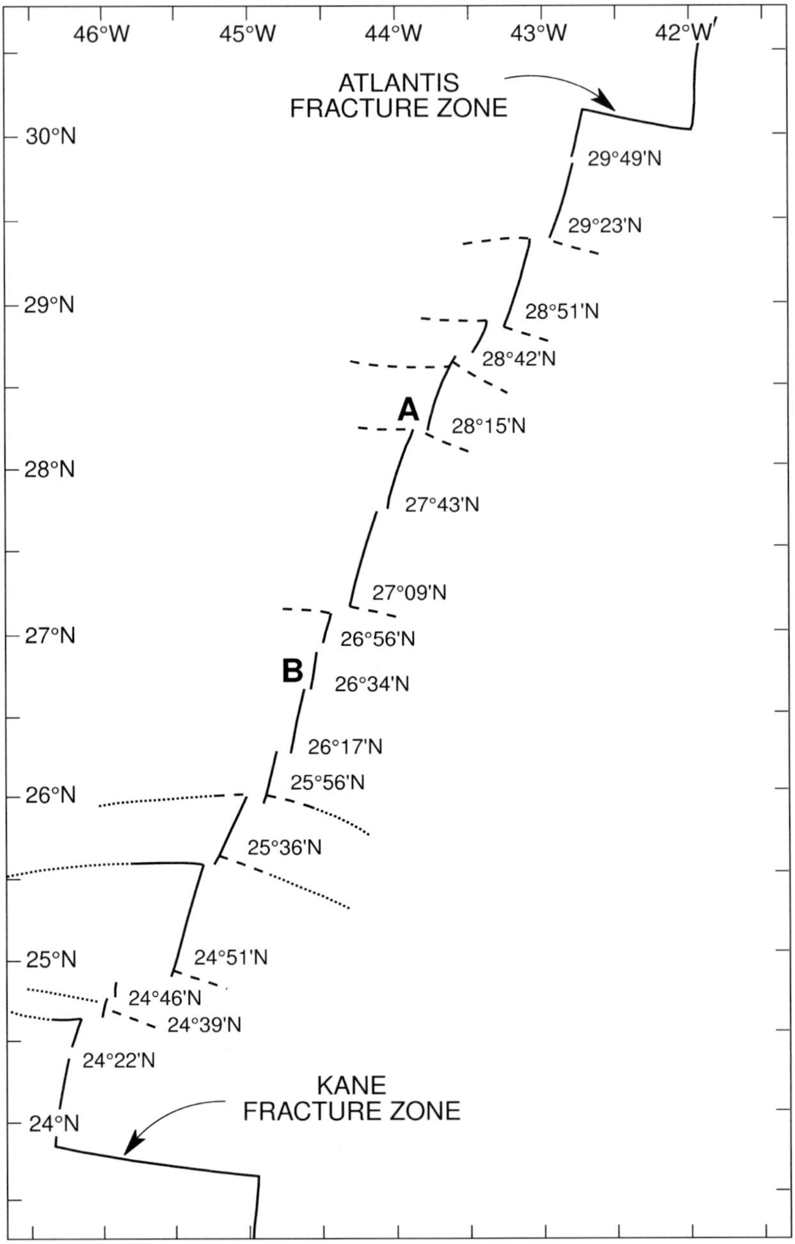

Fig. 4.8. Segmentation of the Mid-Atlantic Ridge between the Kane and Atlantis fracture zones, which correspond to the first-order segmentation scale. Detailed mapping by multibeam echo-sounder has revealed about 15 accretion segments, 20 to 120 km long, separated by 'non-transform' offsets of the axis (offsets <50 km, second-order segmentation scale), which are connected to off-axis lows (example A). The third-order segmentation corresponds to small offsets of the axis without off-axis extensions (example B), and the fourth-order corresponds to simple lows between the neovolcanic ridges, or between axial volcanoes (after Sempéré *et al.*, 1993).

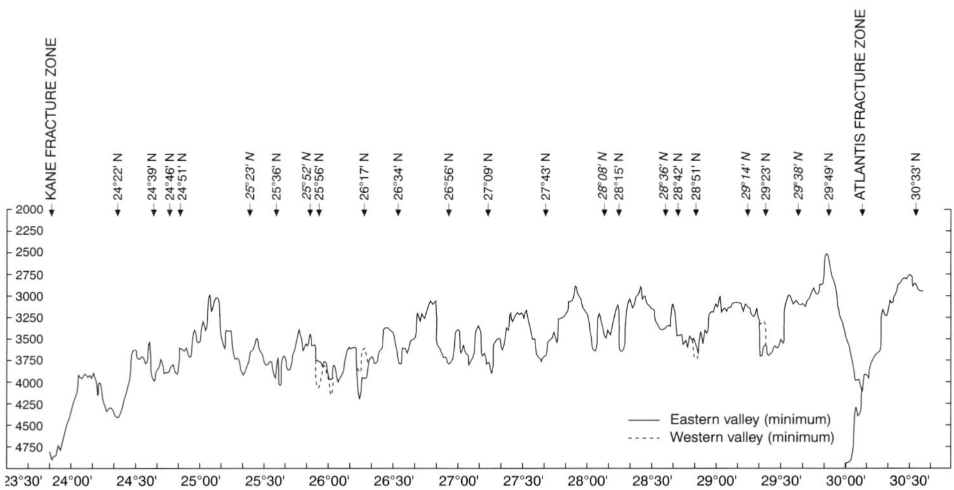

Fig. 4.9. Bathymetric profile along the Mid-Atlantic Ridge axis between 24° and 30°N, confirming the existence of an intermediate wavelength segmentation (20 to 100 km) (after Sempéré *et al.*, 1990).

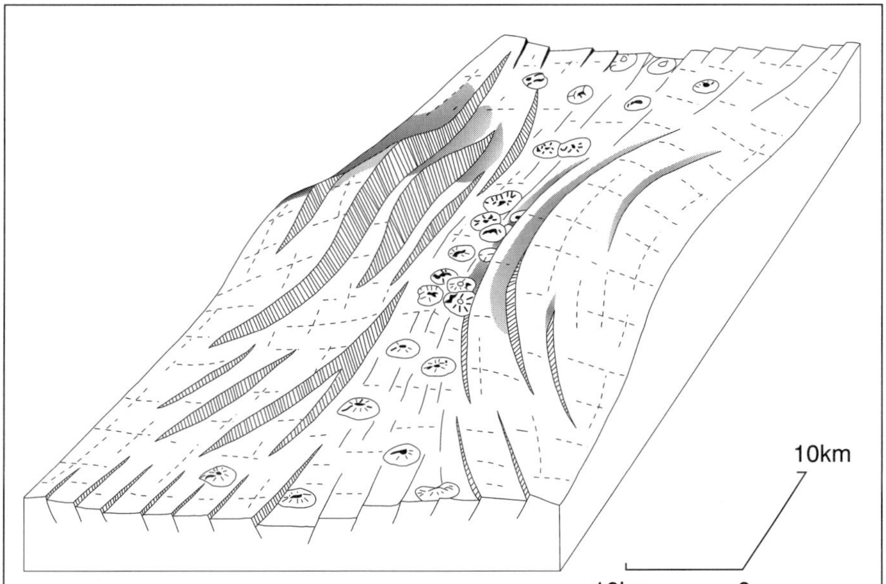

Fig. 4.10. Scheme of an accretion segment of the Mid-Atlantic Ridge, showing the variations in volcanic and tectonic activity from the centre to the tips of the segment. Both activities become looser and more diffuse at the segment tips (after Durand *et al.*, 1995).

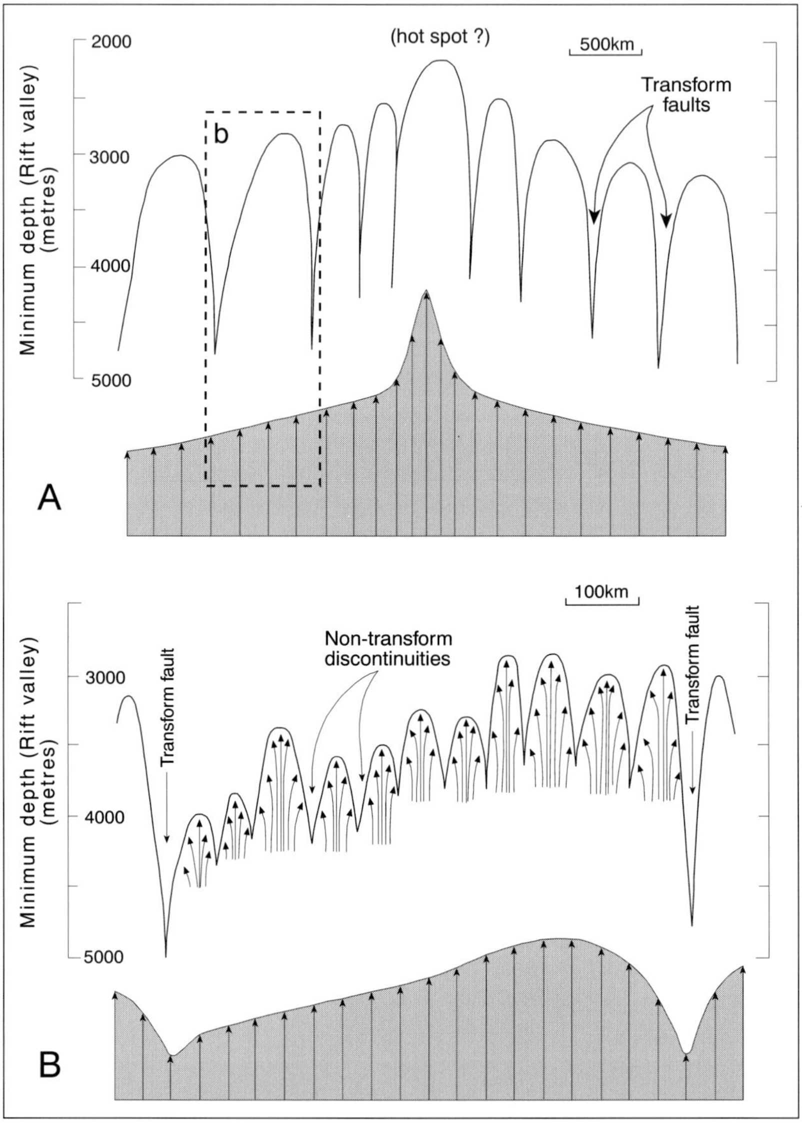

Fig. 4.11. 'Fractal' conception of the mantle upwelling distribution below the axial domain of the Mid-Atlantic Ridge, at the various segmentation scales. (A) Huge wavelength undulations dominated by deep mantle upwellings ('hot spots'). The rectangle corresponds to schema B domain. (B) Undulations corresponding to the wavelength of the fracture zones, with upwelling of asthenospheric mantle (after Sempéré *et al.*, 1993).

segments with time (Figure 4.12). Each segment may lengthen or shorten at the expense of its neighbours, with along-axis propagation velocities ranging between 0 and 25 mm/yr. The length of the same domain may vary between 20 and 120 km with time. Taking into account the average spreading rate and the transverse width of the

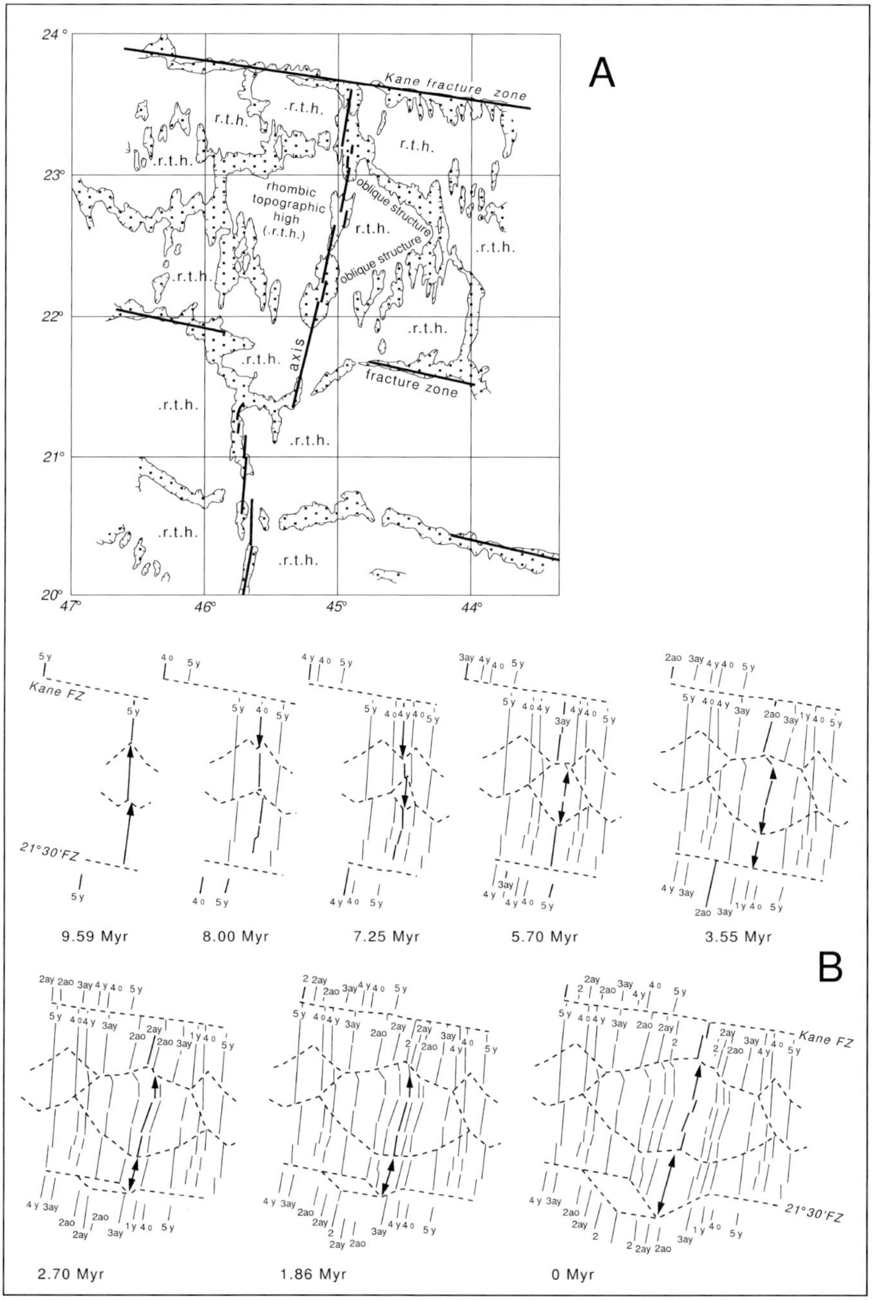

Fig. 4.12. (A) Schema showing 'rhombic domains' of variable lengths, bounded by linear, depressed structures oblique or perpendicular to the ridge axis (stippled). (B) Evolution of the Mid-Atlantic Ridge segmentation between 21°N and 24°N during the last 10 million years. Segments bounded by the oblique structures develop and then regress during a 'life-time' of about 3 to 9 Myr (after Gente, 1995).

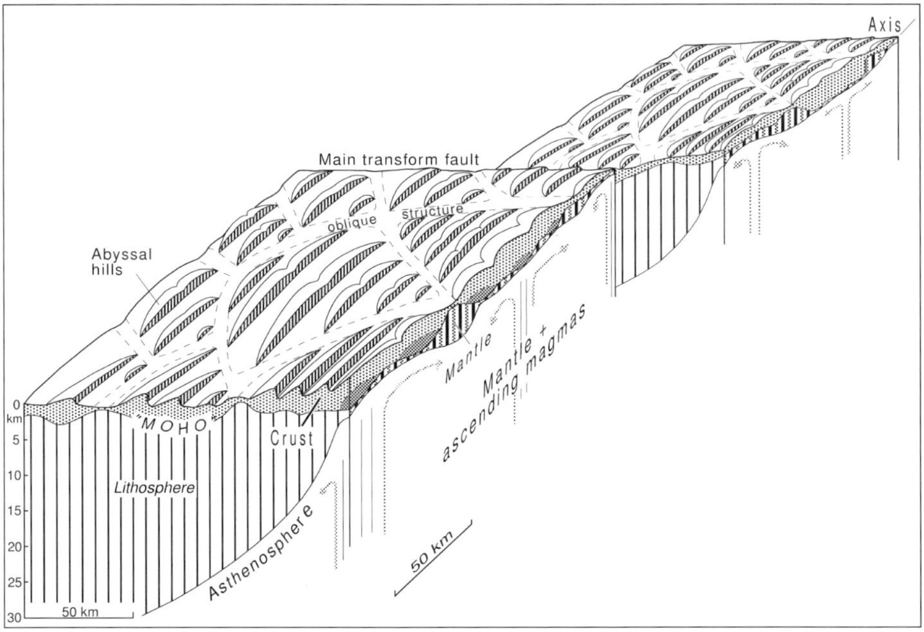

Fig. 4.13. Schematic block-diagram showing the axial segmentation of the Mid-Atlantic Ridge and the off-axis evolution of the various segments with time. Each 'rhombic domain' corresponds to a centre of asthenospheric mantle upwelling, which delivers its magmatic products during a period ranging between 3 and 9 Myr (after Gente *et al.*, 1995).

rhombic domains, one can calculate that the lifetime of a rhomboid domain ranges betwen 3 and 9 Myr.

The interpretation of these features is the following: any new asthenospheric mantle upwelling, when arriving below the lithosphere, creates a new segment (birth of a rhombic domain), whose initial dimension corresponds to the mantle diapir diameter (about 20 km minimum length of the segments) (Figure 4.13). This mantle flow spreads later on radially below the lithosphere, following preferentially two directions, respectively parallel and perpendicular to the ridge axis, up to a maximal distance of 120 km (maximal length and width of the rhombic domains). The competition between the various accretion segments suggests that these diapiric mantle upwellings are independant and 'desynchronized' along the axis.

This model implies that the lifetime of rhombic domains corresponds to the time necessary for a mantle diapir to deliver all its magmatic charge (3 to 9 Myr), which is in agreement with some theoretical models proposing that these upwellings occur as cells isolated in time and space.

The gravimetric map of the north Atlantic Ocean produced by Sandwell *et al.* (1995) (cf. Plate 6 at the end of Chapter 11) shows clearly that the 'structural grain' of the ridge, away from the axial zone, is composed of such juxtaposed rhomboid domains, and extends well beyond 10 Myr (Figure 4.14; cf. Figure 3.7).

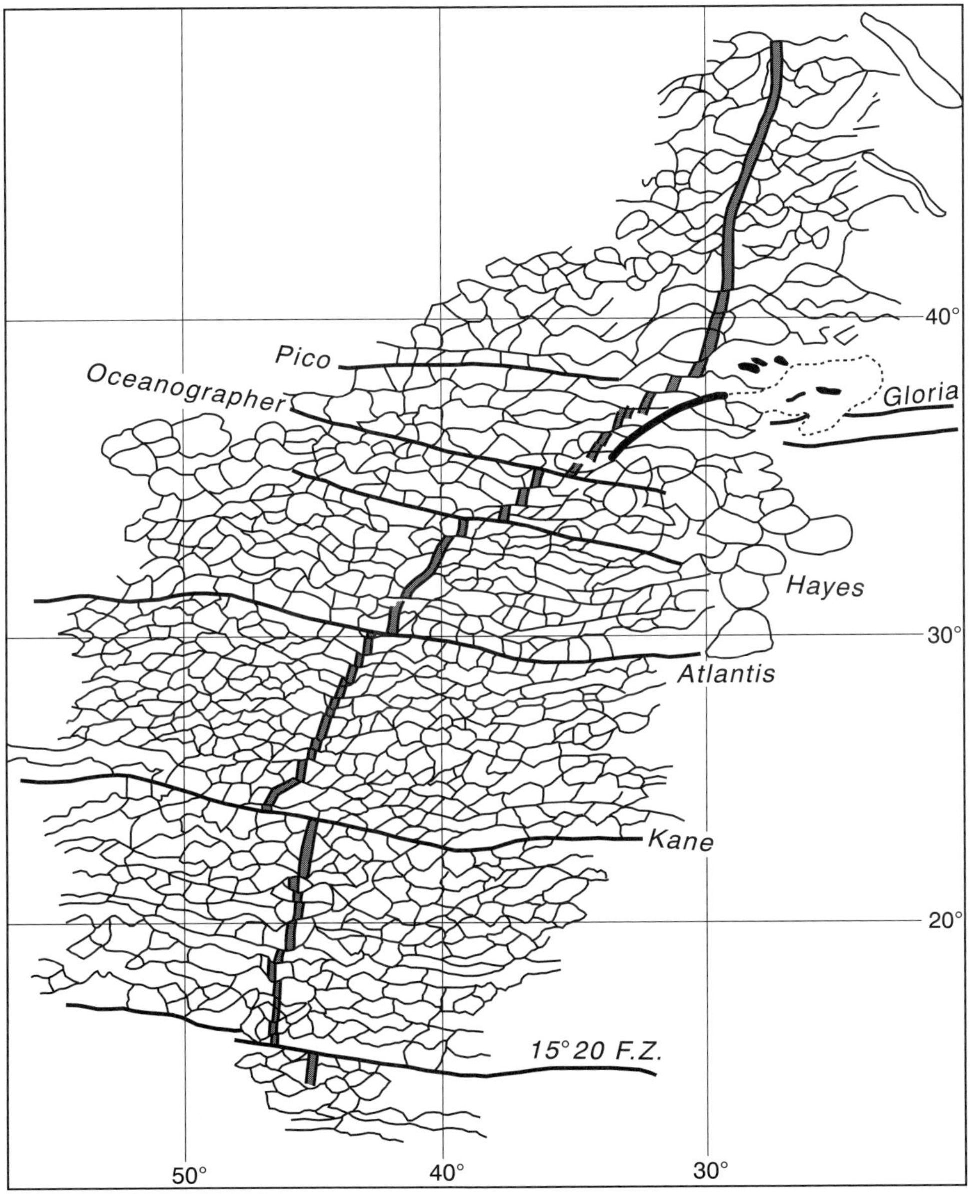

Fig. 4.14. Detailed interpretation of Sandwell *et al.*'s (1994) gravimetric map showing the persistence of the 'rhombic domains' on the flanks of the Mid-Atlantic Ridge, well beyond 10 Myr (after Gente, 1995).

5

Structure and lithology of the oceanic crust of the present-day oceans

Our knowledge of the lithological structure and composition of the oceanic crust is based on three main data sources: (1) the *geophysical data*, mainly the refraction and reflexion data; (2) the *geological sections and rock samples* taken directly from the ocean floors; (3) the *ophiolites*, fragments of fossil oceanic lithosphere from the oceans of the past, that can be studied on land.

We shall present the data in that order, beginning with presently known information about the environment of the present-day oceans (Chapter 5), and ending with ophiolitic data, for complementary information and comparison (Chapter 6).

The opposite approach, describing first a universal 'ophiolitic model' and then trying to fit the oceanic data to it, has often been used. This approach was certainly justified some 20 years ago, when the knowledge of the oceanic crust was still extremely fragmentary. Today, we know enough about the oceanic crust to propose lithological and structural models independently of ophiolitic models, even if numerous uncertainties remain.

5.1 SEISMIC STRUCTURE OF THE PRESENT-DAY OCEANIC CRUST

The seismic structure of the ocean floors has been studied for half a century. Contrarily to the marine geologist, the marine geophysicist makes the best of the existence of a seawater layer between his tools and the field of study. He can travel freely through the vast ocean without being hindered by the topographic accidents, or by political boundaries or private properties, unlike on land: the deep ocean belongs to everybody!

Large-scale surveys over vast surfaces are therefore a characteristic of geophysical studies at sea: bathymetric surveys by echo-sounding, mapping of the magnetic and gravimetric oceanic anomalies, seismic surveys, etc.

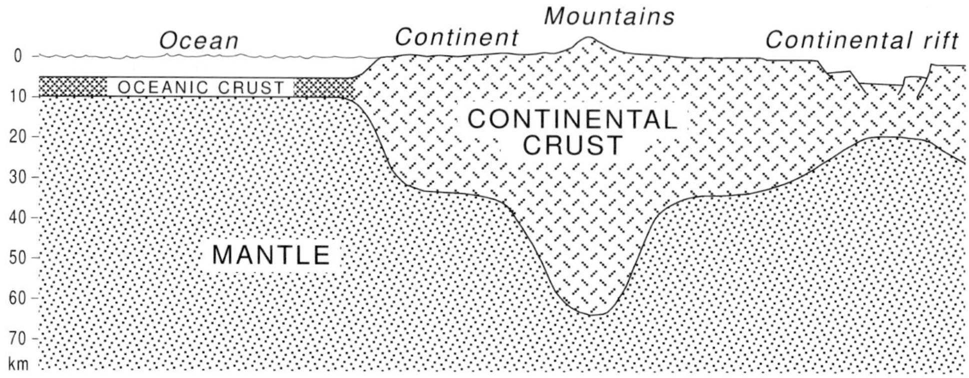

Fig. 5.1. Schematic section showing the contrast between the thin oceanic crust, constant in thickness (6 km), and the thick continental crust (35–40 km in average), whose thickness may vary drastically (from 60–70 km in the 'roots' of mountain ranges, to barely several km below continental rifts).

The thickness and general structure of the oceanic crust were first estimated using indirect seismological methods, revealing three major characteristics: the ocean crust is *thin*, its general structure is *layered*, and its thickness is *constant*.

5.1.1 General description

A thin oceanic crust

The first refraction seismic profiles through the oceans have shown the surprising thinness of the oceanic crust, compared with that of the continental crust. Whereas the latter is 35 to 40 km thick on average, with swells up to 70 km beneath some recent mountain ranges, the oceanic crust is hardly 6 to 7 km thick (Figure 5.1).

A layered oceanic crust

Despite local structural complexities, the oceanic crust presents a general downward increasing gradient of the seismic velocities.

Raitt (1963) was the first to propose a subdivision of the oceanic crust into three layers, each of them characterized by its own mean seismic velocity and its own

Table 5.1. Seismic structure of the oceanic crust, as proposed by Raitt (1963). Each seismic layer is supposed to have a uniform velocity.

Seismic layer	Velocity ($km\,s^{-1}$)	Thickness (km)
Layer 1 (sediments)	About 2	Variable
Layer 2 (basalts)	5.07 ± 0.63	1.71 ± 1.42
Layer 3 (gabbros)	6.69 ± 0.26	4.86 ± 1.42
Layer 4 (mantle)	8.13 ± 0.24	

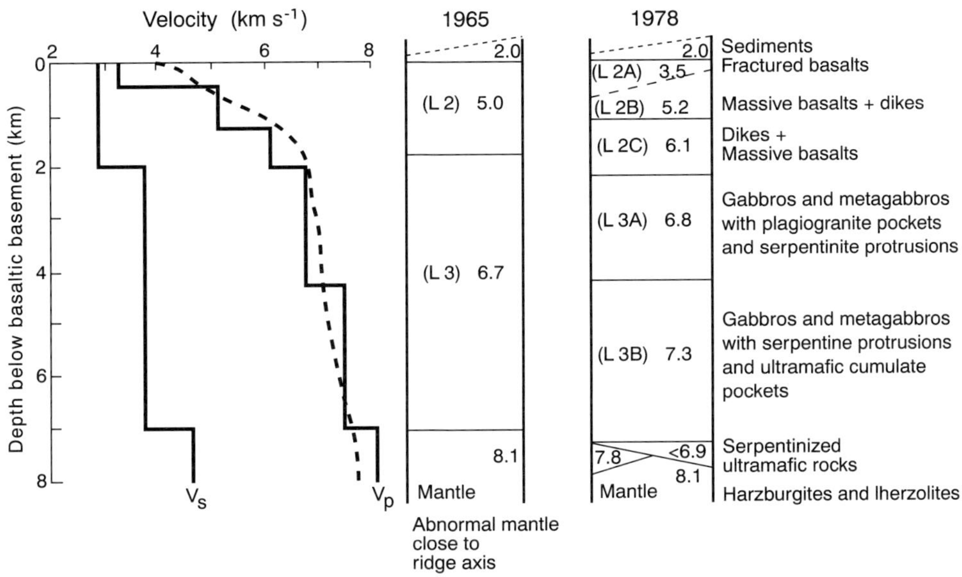

Fig. 5.2. Structure of the seismic velocities in the oceanic crust for P and S waves, and their interpretation in terms of layered crust models: comparison between Raitt's initial model (left column), and the more sophisticated model from the end of the 1970s (right column, with inferred lithology). The numbers indicate the V_p velocities in $km\,s^{-1}$. 'L' means 'layer'. The dashed curve indicates the progressive increase in velocity and the evolution of velocity gradients with depth, according to recent, more sophisticated inversion techniques (after Keary and Vine, 1990).

average thickness (Table 5.1). He noticed the remarkable fact that below layer 1 (made of a variable thickness of sediments), layers 2 and 3 present almost identical thicknesses and seismic velocity ranges in all oceans, whatever the spreading rate. By compiling the numerous data obtained by dredging and by drilling, and by comparison with the ophiolites, the petrographic nature of these two fundamental layers of the oceanic crust were assigned to *oceanic basalts* for layer 2, and to *gabbros* for layer 3.

This three-layer model was improved later (Figure 5.2): layer 2 was subdivided into three horizons (2A, 2B and 2C, Houtz and Ewing, 1976), and layer 3 into two horizons (3A and 3B, Maynard, 1970; Sutton *et al.*, 1971).

The main characteristics of layers 1, 2 and 3 and of their horizons may be summarized as follows:

Layer 1. This is the sedimentary layer resting over the basaltic basement. Its average thickness is 0.4 km. V_p velocities in water-saturated sediments (1.7 to $2.5\,km\,s^{-1}$) are barely superior to those of seawater ($1.5\,km\,s^{-1}$). Layer 1 thickens progressively away from the ridge, and thins until disappearance at the approach of the ridge axial zone. The general section of the sedimentary comumn of a half-oceanic basin, such as the north-eastern Atlantic basin for instance, typically shows

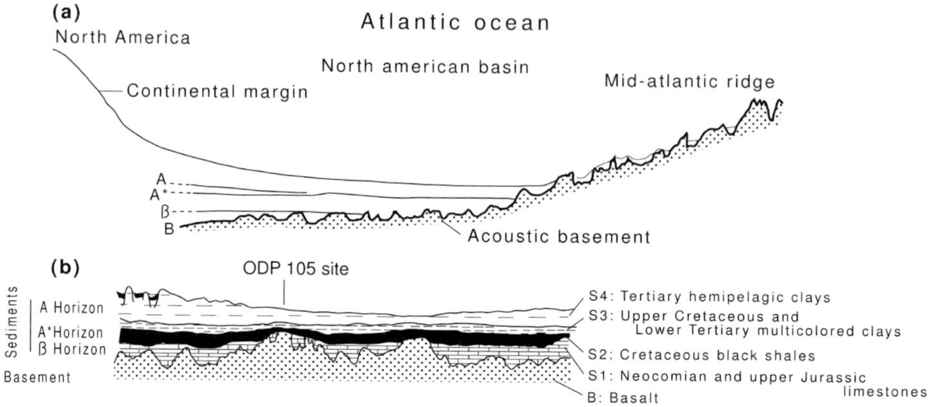

Fig. 5.3. (A) Schematic section of the north-west Atlantic basin, showing the attitude of the sedimentary layers, their thickness variation from west to east, and the main seismic reflectors of the sedimentary column. Note the roughness of the acoustic basement (layer 1/layer 2 contact). (B) Corresponding stratigraphic column, determined by deep drillings of the DSDP, and identification of the reflectors (after Edgar, 1974).

these variations in thickness and the 'transgressive' attitude of the beds over the basement of the ridge (Figure 5.3).

Layer 2. This layer, often called the 'oceanic basement' and representing the upper basaltic crust, is 2 km thick on average, and is characterized by a steep gradient of V_p velocities with depth, ranging from 2.7 to 6.3 km s^{-1}.

—The uppermost *sublayer 2A* presents V_p seismic velocities of the order of 3.6 km s^{-1}, decreasing eventually to 2.7 km s^{-1}, velocities far inferior to those predicted by laboratory measurements for basaltic samples. This layer is interpreted as a highly porous and abnormally low-density basaltic layer, due to the existence of numerous cavities and open fissures occupied by seawater, which 'slows down' the propagation of the seismic waves. This sublayer is well developed in the *young crust*, close to the axial domain of the ridge, where it presents the lowest velocities (2.7 to 3.7 km s^{-1}), and may reach at most 1 km in thickness. The velocities increase away from the axis, up to 4.5 km s^{-1}, and finally merge with sublayer 2B velocities. Sublayer 2A therefore becomes thiiner away from the ridge crest, and may even disappear when the crust is aged, between 20 and 60 Myr dependent on ocean (Houtz and Ewing, 1976). This evolution with ageing is interpreted as being due to the progressive filling of the pores, cracks and fissures of the upper crust by secondary (hydrothermal) minerals.

—*Sublayer 2B* presents higher velocities (4.8 to 5.5 km s^{-1}), in agreement with experimental data on basalts and metabasalts, and a thcikness of 0.5 to 1 km. It may represent either the deepest part of the young basaltic layer, or the whole aged basaltic crust, when sublayer 2A has transformed to 2B.

—*Sublayer 2C*, when detected, is about 1 km thick, with V_p velocities ranging from 5.8 to 6.2 km s^{-1}, indicating probably an important proportion of intrusive basic rocks (dikes and sills). Its thickness seems to be remarkably constant.

Layer 3. Often called the 'oceanic layer' and with an average thickness of 4–5 km, this layer represents the lower gabbroic crust, and is characterized by a weak vertical velocity gradient.

—*Sublayer 3A* remains very homogeneous in velocities (6.5 to 6.8 km s^{-1}) and in thickness (3 km) when going away from the ridge crest. This is certainly the most homogeneous and isotropic layer in the whole oceanic crust.
—*Sublayer 2B*, on the contrary (7.0 to 7.7 km s^{-1}, and 2 km thick), seems to thicken with ageing of the crust.

The crust–mantle transition. Several detailed studies of the crust–mantle transition have shown that the Mohorovicic seismic discontinuity may exhibit very variable characteristics, ranging from a sharp transition between crust and mantle velocities, to a true transition layer with inceasing gradient, or even a series of layers with alternating low and high velocities, over up to 2 km in thickness. Mapping the crust–mantle transition in ophiolites has revealed the same range of variations as the crsut–mantle transition zone on the scale of a single ophiolite massif (Reuber, 1988; Nicolas, 1989).

A constant thickness

Except in the vicinity of fracture zones or hot spots, the thin oceanic crust presents in all oceanic basins a remarkably constant seismic thickness, contrasting with the important thickness variations of the continental crust (Figure 5.1).

This constancy of the oceanic-crust thickness has been questioned. Several authors have presented arguments suggesting a thickening of the oceanic crust with age (Goslin *et al.*, 1972), explained either by off-axis plutonic intrusions (Christensen and Salisbury, 1975), or by a progressive serpentinization of the upper mantle with time (Lewis and Snydsman, 1979). Without excluding this latter process, which may explain some thickening of layer 3B in some cases, recent studies do not confirm the thickening of the ocean crust with age. Besides, a re-examination of the previous data does not favour the first interpretations.

The constancy of the seismic thickness of the 'normal' oceanic crust indicates that a fundamental process controls the generation of the igneous crust, whose thickness would therefore be independent of the spreading rates. Considering that the total thickness of the lithosphere is reduced to zero at the axis of an oceanic ridge, this means that all the magma necessary to produce the magmatic crust is extracted from the asthenospheric mantle, during its ascent and passive decompression. For example, if the spreading rate is multiplied by two, the asthenospheric volume ascending between the plates will be multiplied by two, producing double the amount of magma, which will fill a space twice as wide: the resulting solid crust will therefore retain the same thickness. On average, the oceanic crust would be the result of the partial melting of 10 to 15% of the asthenospheric source.

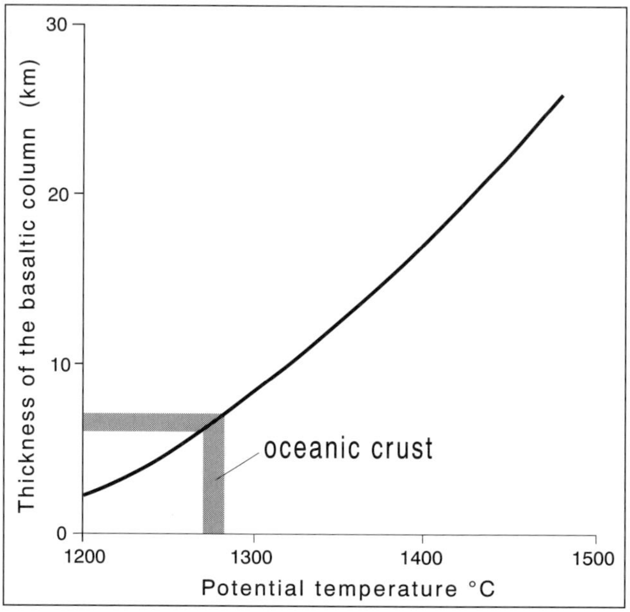

Fig. 5.4. Thickness of the magmatic column produced by adiabatic decompression of the asthenospheric mantle below an oceanic spreading centre, versus the mantle potential temperature (the temperature that the mantle would have if transported adiabatically to the surface, without melting). The 'normal' temperature of 1280°C results in a 'normal' crustal thickness of 6 to 7 km. Beneath a hot spot such as Iceland, the crustal thickness reaches 25–30 km, which indicates potential temperatures of more than 1500°C in the mantle; that is, about 250°C above the normal asthenospheric temperature (after White, 1991).

The volume and composition of the magma generated by decompression can be calculated using the equations developed by McKenzie and Bickle (1988), based on laboratory melting experiments. Extrapolating to terrestrial P-T (pressure-temperature) conditions, it is possible to model the 'incremental' partial melting of an ascending asthenospheric mantle. Almost all the magma produced flows rapidly toward the surface, leaving a negligible volume of primary magma trapped in the mantle mother-rock. The total volume of magma produced by decompression is extremely sensitive to the initial temperatire of the asthenospheric mantle. For instance a 100°C increase over the normal 1280°C potential temperature more than doubles the volume of produced magma (Figure 5.4).

A constant thickness of 6–7 km therefore confirms the global uniformity of the asthenospheric *potential temperature*, at $1280 + 20°C$ (the potential temperature being the temperature of the mantle if it was adiabatically raised up to the surface without melting). For White (1991), this confirms that the upper mantle convects vigorously and that plates move at relatively high velocities, preventing the maintenance of different temperatures over long periods below specific areas.

Figure 5.4 shows that in areas crossed by thermal plumes, such as Iceland, the volume of magma produced must be considerably higher. When approaching the hot spot, the temperature of the mantle increases, and consequently the thickness of the crust produced at the accretion centre axis increases also, reaching 25–30 km below Iceland, and indicating potential temperatures of more than 1500°C at the centre of the plume (that is, about 250°C above the normal asthenospheric temperature).

The limits of the model with homogeneous velocity layers

The model, with thick layers that are supposedly homogeneous (uniform velocities), gives a false impression of simplicity and of homogeneity of the crustal structure. From the time of the very first refraction data, it appeared that the oceanic crust was *locally* extremely variable in terms of crustal thickness and seismic velocity structure (at a scale of several tens of km). This heterogeneity was confirmed abundantly in many detailed geological and geophysical studies. The contradiction between the large and the small-scale interpretations is false, because:

- —Statistical compilations of hundreds of seismic profiles, on the scale of entire oceans, have smoothed out crustal heterogeneities, and created the illusion of large-scale homogeneous layers, introducing to the literature the concept of 'normal oceanic crust', with its layers whose thicknesses and velocities have been averaged and standardized.
- —The proposed thicknesses for the sublayers are close to the dimensions of the seismic waves used to explore the crustal structure. For the upper crust (<2 km), the wavelengths used are of the order of 500 m, and their resolution is hence limited to this dimension. For the lower crust (>2 km), the wavelengths reach 2 km, lowering still more the resolution of the crustal structure.

The velocity-gradient model

The model with homogeneous layers was replaced during the 1980s by a more sophisticated model with multiple thin layers with velocity gradients, based on new modelling methods (synthetic seismograms, inversion techniques, amplitude modelling, etc.). In this new concept (Figure 5.2, broken line; Figure 5.5), the oceanic crust consists in a discontinuous series of velocity gradients, the seismic velocities increasing with depth in each unit (Spudich and Orcutt, 1980; White, 1984, 1991). In the *upper crust* (layer 2) for instance, the seismic velocity increases from values as low as 3 km s^{-1} at the top to values close to 6 km s^{-1} at the bottom (half-way into the crust) that is, a steep gradient of 1–2 km s^{-1} per km in depth, mainly due to the *porosity decrease* with depth. In the *lower crust* (layer 3), the velocity gradient is much lower, of the order of 0.1 km s^{-1} per km in depth, reflecting a very homogeneous mineralogical composition, and an extremely reduced porosity, compared with that of the upper crust.

Absence of large-scale continuous reflectors

The oceanic crust does not show planar and continuous reflectors, mappable at regional scale, such as those observed in the sedimentary series. Reflectors have been

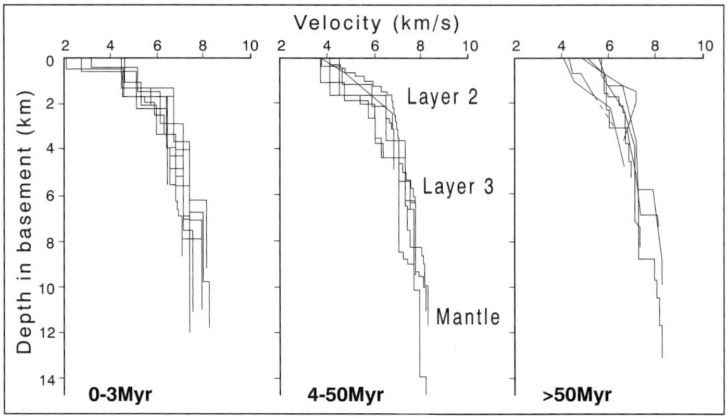

Fig. 5.5. V_p velocity profiles versus depth in the 'normal' Atlantic oceanic crust of various ages, far from fracture zones and hot spots (after White, 1984).

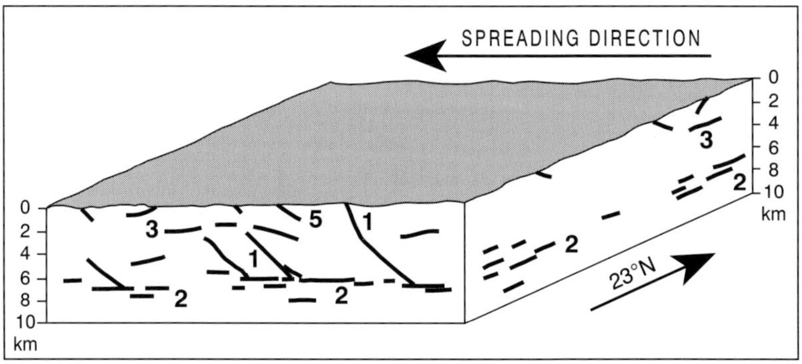

Fig. 5.6. Block-diagram showing various types of discontinuous reflectors observed on multichannel seismic profiles in the north Atlantic ocean, away from fracture-zones. 1. Oblique reflectors: traces of normal faults produced on the flanks of the accretion zone. 2. Flat reflectors in depth: Moho reflections, and planar intrusions in the lower crust (sills), or flat detachment faults, connected to shallow oblique normal faults. 3. Shallow flat reflectors (2-3 km in depth): lower limit of high-temperature hydrothermal penetration. After White (1991), simplified.

mapped only on short distances (several km in general). According to their attitude on the seismic profiles, they may be interpreted in various manners (tectonic features, magmatic contacts, etc., cf. Figure 5.6), but it is clear that there are no major and general discontinuities inside the oceanic crust, even between layer 2 and layer 3 for instance.

5.1.2 Lithological interpretation of the seismic profiles

The seismic velocities of the compression waves alone cannot be used to determine the nature and composition of the rocks, for the simple reason that extremely different rocks may present identical seismic velocities. Inversely, a similar material

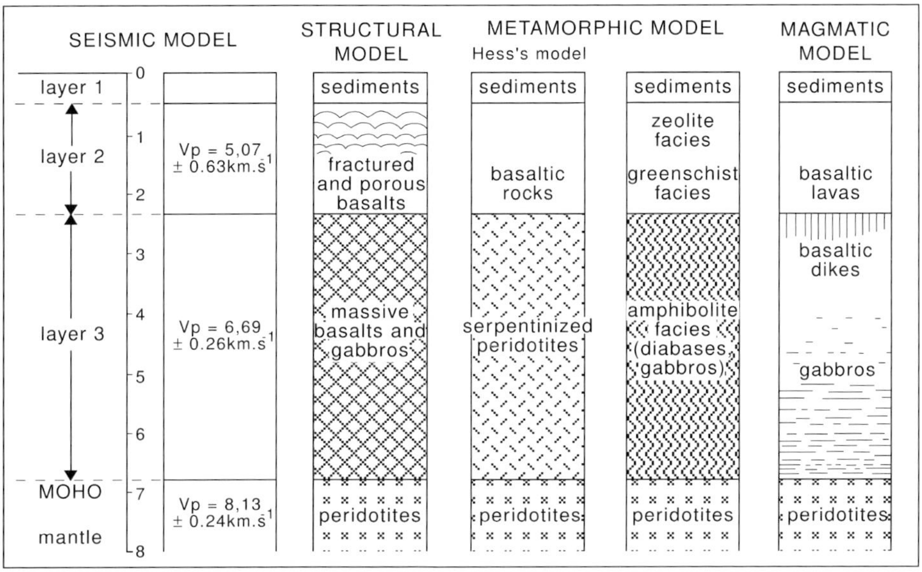

Fig. 5.7. Several possible models for the oceanic crust. Each of these models favours a single factor. By means of a judicious combination of these various models reality should be reached. Note that the serpentine model for layer 3 (Hess's initial model) should be discarded, because of experimental data on V_p and V_s (see text and Figure 5.9).

may present important velocity variations, according to its physical state (fracturing, porosity, thermal state, etc.).

The factors that may control variations in the crustal seismic velocities are: (i) physical factors, mainly porosity; (ii) the primary mineralogical composition of the rocks; (iii) the secondary mineralogical composition (low-temperature oceanic alterations, hydrothermal metamorphism). Figure 5.7 shows the various lithological models proposed for the oceanic crust, each of them emphasizing one of these factors:

The *structural model* considers that each velocity boundary corresponds to an important modification of the average porosity of the materials, ranging from non-compacted rocks, full of void spaces and open fissures in the shallow crust, to compressed, massive and poorly fissured rocks in depth. This model, based on *in situ* measurements done in DSDP (Deep-sea Drilling Project) and ODP (Ocean Drilling Program) drill holes, is presently the most popular.

The *magmatic model* supposes that the various seismic layers and sublayers of the oceanic crust correspond to primary (magmatic) lithological units such as: volcanic unit, sheeted dike complex, gabbroic sequence, etc. This somewhat naive and simplistic view is still presented in many textbooks, but is less and less credible.

The *metamorphic model* considers that each limit represents a change in the intensity of oceanic metamorphism, resulting in a change in mineral associations, ranging from the *zeolite facies* in shallow levels, to the *greenschist* and *amphibolite*

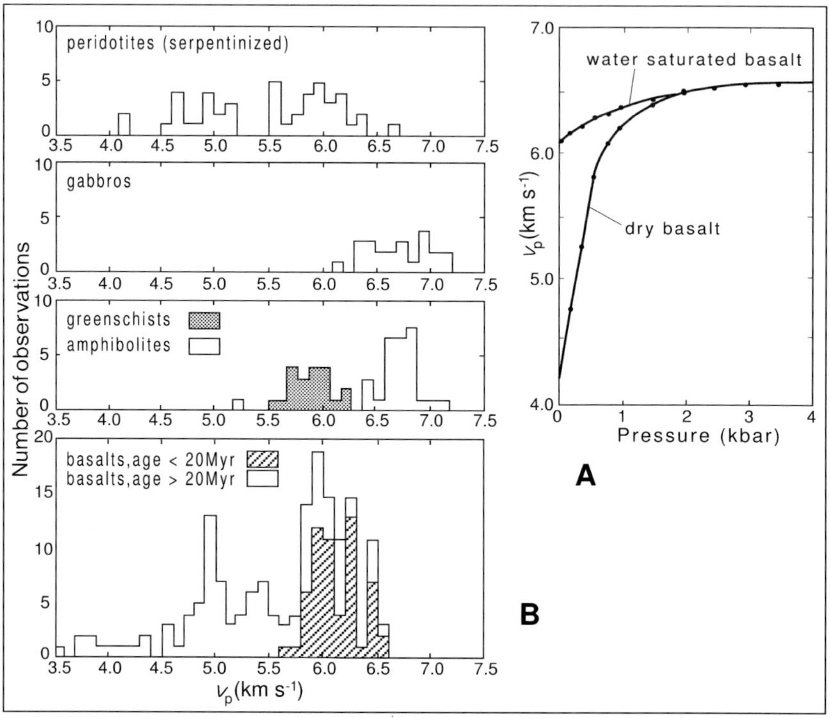

Fig. 5.8. (A) Compression-wave velocities (V_p), measured versus pressure in a basalt from the East-Pacific Rise, dried and saturated with water (after Christensen and Salisbury, 1975). (B) Compression-wave velocities measured in the laboratory at 1 kbar pressure, on various oceanic rocks saturated with water. The velocities in peridotites decrease with increasing serpentinization ratio (after Christensen and Salisbury, 1975).

facies in depth. Neglected for a long time, the 'metamorphic factor' is seriously taken into consideration after the results obtained in Hole DSDP/ODP 504B in the Pacific (see below).

An historically important variation of this model is the *serpentinite model* for layer 3, proposed by H. H. Hess as early as 1960, in which layer 3 is composed of deeply serpentinized mantle peridotites by seawater penetration and hydration. As we shall see, this proposition is considered again *pro parte* in recent models on layer 3 of the slow-spreading ridges (cf. Figure 6.18).

Of course, *only by carefully combining these various factors will the reality of the situation be reached.* In order to better evaluate their relative contributions, let us examine the results of laboratory experimental studies. Christensen and Salisbury (1975) were the first to measure systematically in the laboratory the velocities of compressional and shear waves in all kinds of oceanic materials. They took care to place these materials in their natural conditions; that is, under confining pressures of hundreds of atmospheres, and *saturated with water.* Figure 5.8A shows the V_p velocity variations with depth, obtained for a basalt of the East-Pacific Rise, on the

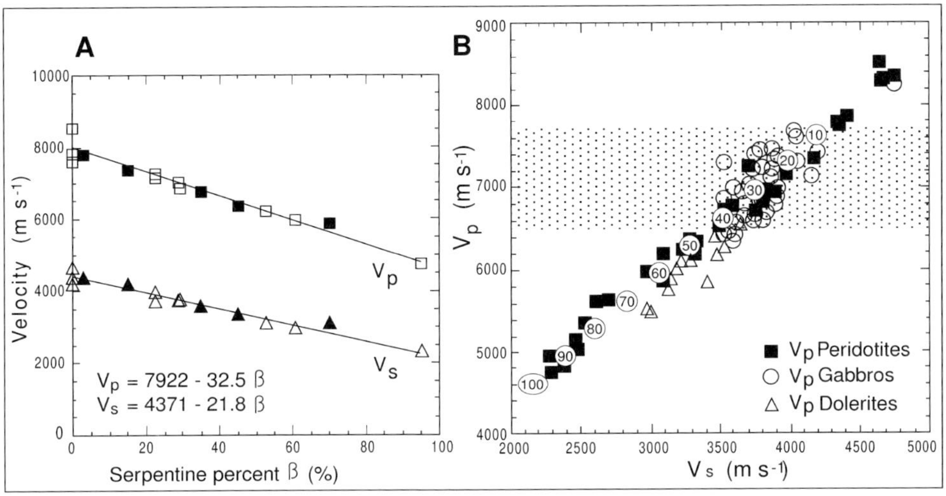

Fig. 5.9. (A) V_p (squares) and V_s (triangles) velocities measured in the laboratory on more or less serpentinized peridotites, plotted here versus the serpentine percentage β. Empty symbols correspond to Christensen's (1966) measurements, and black symbols to more recent measurements by Hören *et al.* (1996). Both series of measurements are perfectly complementary, and confirm the linear variation of V_p and V_s versus β (the linear equations obtained by linear regression are indicated). (B) Variations of V_p versus V_s, measured in the laboratory on oceanic peridotites, gabbros and diabases. Numbers in circles indicate the percentage of serpentine in peridotites. In the stippled area corresponding to oceanic layer 3, the V_p and V_s of gabbros cannot be distinguished from those of moderately serpentinized peridotites (10–40%) (after Hören *et al.*, 1996), slightly modified.

one hand in dry conditions, and on the other in the presence of water: itcan be seen that the velocities in a free atmosphere (dry conditions) are clearly lower than the velocities in the presence of water, and that the difference decreases with increasing pressure (hence with increasing depth); the two curves merge together at about 2 kbar, corresponding to 5–6 km in depth.

Figure 5.8B shows the result of these experiences. It appears that the V_p alone do not distinguish between the various models of Figure 5.7. If we consider V_p and V_s velocities simultaneously, two conclusions arise (Figure 5.9): (1) the model of a layer 3 formed exclusively of serpentinites, or of mantle peridotites more than 50% serpentinized, must be discarded, because the V_p and V_s of these materials are obviously too low; (2) on the other hand, the V_p and V_s of weakly serpentinized peridotites (10 to 40%) are exactly identical to those of gabbros (Horen *et al.*, 1996).

Therefore the V_p/V_s ratio allows improvement of the characterization of crustal materials, and rejection of some hypotheses. *In situ* measurements of V_s, however, are still scarce, because they are difficult to obtain in good conditions. Shear waves are not transmitted by liquids (hence by seawater), and it is therefore necessary to anchor hydrophones on the seafloor to record them. These *ocean bottom seismometers* (or OBS) are more and more used, and will certainly enables completion the catalogue of combined V_p–V_s records in the coming years.

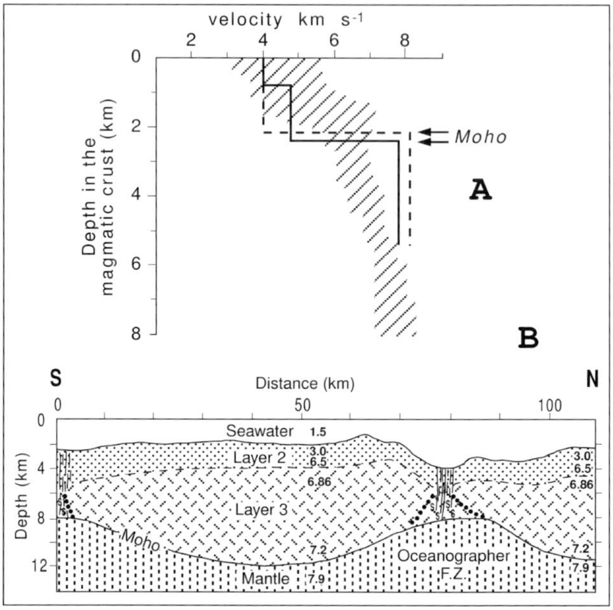

Fig. 5.10. (A) Crustal velocity profiles obtained along the Kane fracture zone (north Atlantic), by refraction seismic methods. The crust is hardly 2 km thick, and the crustal velocities are low up to the base of the crust, indicating a high porosity. The dashed zone corresponds to the envelope of velocities for the 'normal' north-Atlantic crust, away from fracture zones (after White, 1984). (B) Geological section parallel to an accretion segment, passing through two fracture zones. The schema shows the bathymetric variations of the ocean floor, and the thickness variations of the various layers. Fracture zones are characterized by a deepening of the bathymetry, a drastic reduction in crustal thckness, an almost complete disappearance of layer 3, an intense tectonic fracturing due to transform tectonic activity (vertical lines), and a serpentinization of the upper mantle (after White, 1984).

5.1.3 Seismic structure of the fracture zones

The study of fracture zones by wide-angle reflexion and refraction seismic methods, as well as by gravimetric methods, has shown strong anomalies in the crustal structure (Figure 5.10A). The crust is generally highly reduced in thickness, down to 1–2 km in some cases. This thinning is progressive at the approach of the reacture zone, over a distance of about 10 to 30 km, and occurs *at the expense of layer 3*, which often disappears completely below the fracture zone. Correlatively, as we have seen, the depth of the ocean floor increases (Figure 5.10B)

These data may be interpreted by admitting that the basaltic layer, tectonically intensely fractured and hydrothermalized, rests directly over an upper mantle strongly serpentinized *in situ* (Figure 5.10B). The thinness of the crust and the absence of layer 3 suggest an important deficit in the magmatic input, due to the distal position at segment tips, and to the 'cold-wall effect' resulting from the presence of an aged and cold lithosphere located in front of the accretion segments of the ridge, on bot sides of the fracture zone. Moreover, the low crustal velocities reflect an important porosity, explained by the accumulation of rubble in the deep, tectonically active transform valleys, and by the intense tectonic fracturing of the whole crust in the transform domain.

In such a context, seawater should be able to penetrate easily and massively down to the mantle, resulting in an important serpentinization of its upper part. The serpentinized zones tend to ascend, because they acquire a crustal density close to that of layer 3. However, gravimetric data show that this process remains limited, and do not confirm the idea of serpentinite diapirs ascending up to the seafloor in fracture zones, argued by Bonatti and Honnorez (1976; cf. Figure 5.20).

5.2 SAMPLING AND LITHOLOGICAL STRUCTURE OF THE PRESENT-DAY OCEANIC CRUST

Sampling of the oceanic crust, particularly its deep levels, is still very limited because of its giant dimensions (360 millions of km^2; that is, between 2.1 and 2.5×10^9 km^3).

Up to 1975, the only sources of information were dredged sample collections, with all the disadvantages linked to this type of sampling (approximative locations, samples in disorder, etc., cf. Chapter 3). These dredges were done essentially along the steep scarps of fracture zone valleys, which offer the only vertical offsets sufficiently important to expose the deepest parts of the crust and eventually the upper mantle.

Numerous igneous and metamorphic rocks were thus collected, including basalts, diabases, gabbros, serpentinites and more or less serpentinized ultramafic rocks, greenschists, amphibolites, mylonites, and all kinds of breccias, etc. It was of course very difficult to establish lithological sections from these dredgings. The only possibility was most often to estimate the relative abundances of the various rock types (Figure 5.19C–D). Moreover, the particular tectonic context of the fracture zones leave some doubts as to the representativity of these samplings concerning 'normal' oceanic crust.

From 1975, the advent of deep submersibles able to reach true geological sections along the ocean floor, and the possibility of re-entering the DSDP drill holes, gave new possibilities to study the deep crust (cf. Chapter 3).

The numerous data accumulated over more than 20 years have progressively shown considerable differences in crustal lithology and structure between fast- and slow-spreading ridges.

5.2.1 The continuous crustal tissue of fast-spreading ridges: the East-Pacific Rise and its ramifications

Hole 504B

The DSDP/ODP drillings in the basaltic crust of the Pacific have come across a lot of difficulties, and were, generally speaking, clearly less successful than in the Atlantic. It seems in particular that the abundance of fluid flows and fossil lava lakes (cf.

Chapter 7) in the basaltic column constitutes a redoubtable obstacle for drilling and coring.

An exceptional effort was, however, dedicated to Site 504, south of the Costa Rica rift (a branch of the East-Pacific Rise). Seven legs were performed between 1979 and 1993 (representing 14 months of drilling) to deepen Hole 504B down to a depth of 2111 m below the seafloor, with a record penetration of 1836.5 m in the ocean crust.

Description of Hole 504B. The site is located in the Panama basin, at 201 km from the Costa-Rica Rift axial zone, on the southern flank of the ridge, whose total spreading rate is 7.2 cm/yr. The age of the crust is 5.9 Myr, and the water depth is 3474 m (Figure 5.11).

The lithological section of Hole 504B is as follows (Figure 5.12):

(1) The *sedimentary cover* (274.5 m). The sediments are principally foraminifera and nannofossils bearing pelagic oozes, with some terrigenous clay beds (0 to 227 m), then cherts, silicified limestones and chalks of upper Pliocene age (227 to 274 m).
(2) The *pillow-lava zone* (571.5 m). It is made of alternating pillow-lava flows, hyaloclastites and flow breccias, and some massive flows. All lavas are relatively primitive tholeiitic baslts, containing 0 to 10% olivine and plagioclase phenocrysts.
(3) The *transition zone* (209 m). It includes pillow-lava flows and massive flows, cut by diabase dikes more and more abundant downward. The upper part of this zone is rich in brecciated pillows, and includes a mineralized zone, about 20 m in thickness, impregnated with hydrotermal sulphides ('*mineralized stockwork*').
(4) The *dike zone* (1056 m). It is entirely made of diabase dikes with steep margins, intersectin each other. The diabases are somewhat richer in Mg on average than the pillows, and are therefore issued from a somewhat more primitive magma than that of the extrusives.

Logging operatons and down-hole measurements done in Hole 504B allowed precise calculation of the physical characteristics of the crust, and to distinguish three zones:

—The first 100 m form a very *fractured unit*, showing low densities, thermal conductivities, electrical resistivities and seismic velocities (P and S waves), whereas porosity and permeability are correlatively high. This zone, whose cracks seem to be open and filled with seawater, corresponds well with the characteristics of seismic *sublayer 2A* (Figure 5.13), whose thickness here is greatly reduced, because of the age of the crust.

—From 100 to 650 m, porosity and permeability decrease, whereas the seismic velocities, electrical resistivity and density increase. The percentage of hydrated secondary minerals increases, correlatively to the porosity decrease. The seismic velocities correspond without ambiguity to *sublayer 2B*. Thi zone includes most of the pillow-lava section, and the upper part of the transition zone, with the stockwork zone. and density increase, as well as the Immediately below the stockwork, between 600 and 650 m, the physical properties change drastically. The electric resistivity, density and seismic

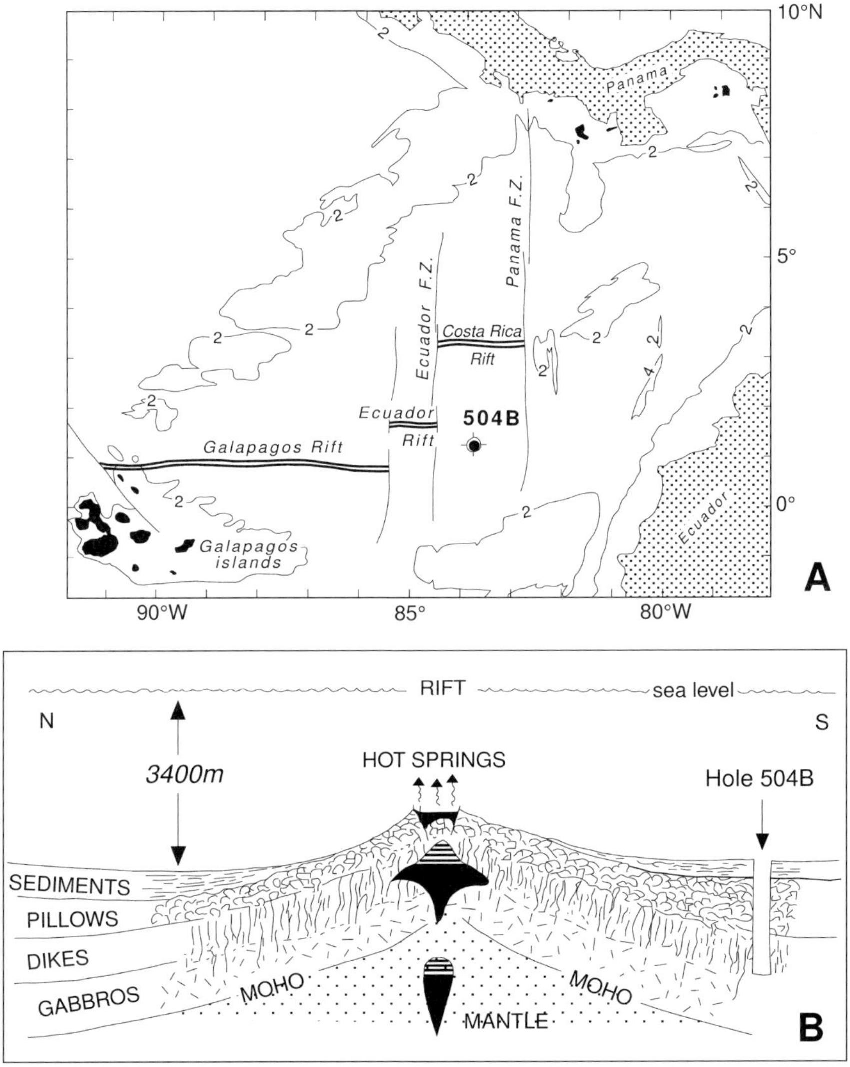

Fig. 5.11. (A) Location of Hole 504B on the southern flank of the Costa Rica rift. Bathymetry in thousands of meters. (B) Schematic section, transverse to the Costa Rica rift axis, showing the position of Hole 504B and the drilled lithological section. Inferred magma chambers are in black. Vertical scale not respected.

velocity gradients become steeper. The porosity, permeability, intensity of fracturing and percentage of secondary minerals decrease strongly. On the other hand, thermal conductivity increases. This 50 m zone is interpreted as the transition between sublayers 2B and 2C.

—From 650 to 1300 m: V_p velocities and resistivity increase regularly, whereas porosity decreases correlatively, and permeability drops by several orders of magnitude. This zone corresponds well with the characteristics of *sublayer 2C.*

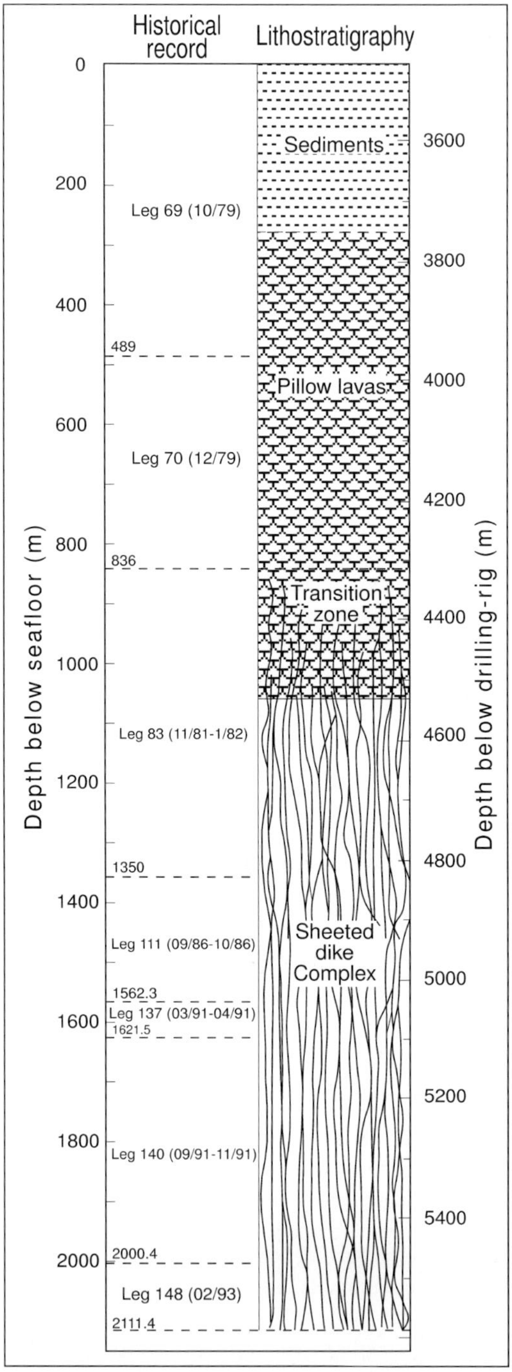

Fig. 5.12. General lithostratigraphic column of Hole 504B, with indication of the drilling historical record (after Ayadi *et al.*, 1996).

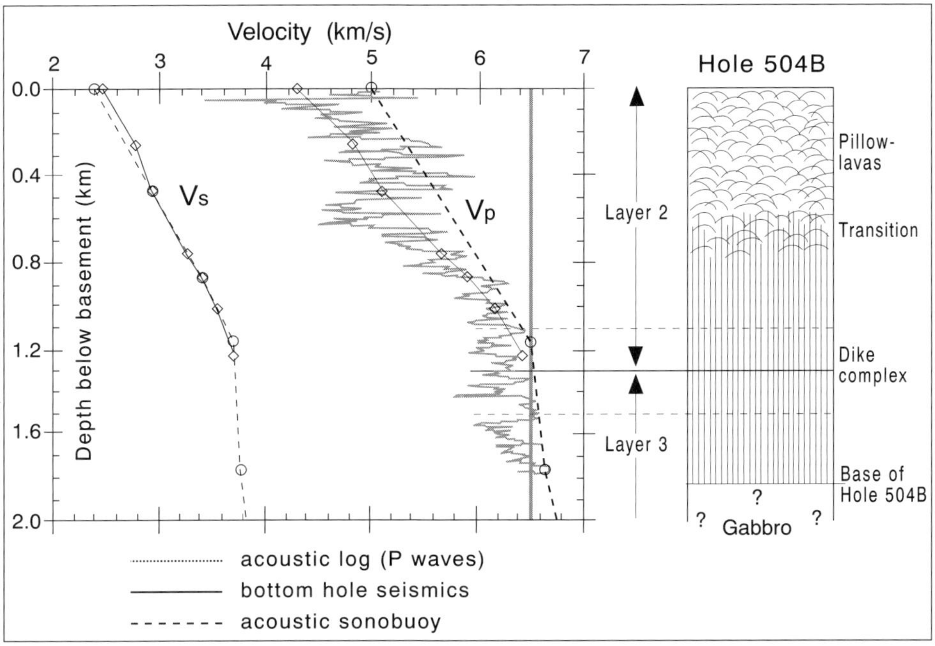

Fig. 5.13. Evolution of seismic velocities V_p and V_s with depth in Hole 504B, recorded by three different methods. Layer 2A is limited to the first 100 m. Layer 2B corresponds to the pillow lavas, whose initial porosity has been strongly reduced. Layer 2C corresponds to the upper part of the sheeted dike complex. The layer 2/layer 3 boundary is clearly located in the middle of the dike complex (after Detrick *et al.* , *1994*).

—From 1300 to 1856.5 m: the seismic velocity gradient shows a clear inflexion and becomes flat. The V_p velocities teach and pass 6.5 km/s. This is by definition the transition to seismic layer 3, right here in the middle of the dikes (Figure 5.13)! To the great surprise of a number of specialists, *the layer 2/layer 3 transition is not linked to a lithological or magmatic change* (*dikes/gabbros transition for instance*), *but to the variations of physical properties*, particularly of porosity (cf. Figure 3.14B).

Unfortunately, Hole 504B had to be abandoned before reaching the gabbros, which were not very far away. This section in the upper crust of the Pacific Ocean constitutes a reference point for intermediate- to fast-spreading ridges, but how can it be completed downward? Except along some fracture zones, places where the 'normal' deep oceanic crust of the Pacific is visible are very scarce. Outside of fracture zones, axial outcrops of diabases, gabbros and peridotites occur in three areas only. They correspond either to deep valleys (>500 m of water depth) linked to propagating rifts such as the Hess Deep, close to the Galapagos triple junction, or the Pito Deep, on the north-east side of the Easter microplate, or to abandoned rifts consecutive to a ridge jump, such as Mathematician ridge, an abandoned segment of the East-Pacific Rise, at 17–18°N in the Pacific.

The Hess Deep

The Hess Deep is located close to the Galapagos Triple Junction, at the junction between the East-Pacific Rise and the Cocos–Nazca Ridge. It is a deep topographic low (5300 m) located at the western tip of the Cocos–Nazca Ridge westernmost segment, barely 40 km from the East-Pacific Rise. This segment has been a westward propagating rift for about 6 Myr, as shown by the V-shaped rough crust formed by this ridge, whose topographic features are oriented east–west. It intrudes like a wedge into the smoother East-Pacific Rise crust, whose topographic features are oriented north–south (Figure 5.14).

The Hess Deep depression, bounded by east–west faults, cuts across the East-Pacific crust along important vertical offsets, reaching 3750 m in places. This 20-km-wide depression includes on its northern flank a 25-km-long and 8-km-wide intra-rift ridge, whose southern flank had provided dredged gabbros and serpentinized peridotites.

The Hess Deep is therefore a privileged area for observing the Pacific deep crsut below its basaltic carapace, in a non-transform context. Two recent diving programmes, one with the French *Nautile* submersible (Francheteau *et al.*, 1990, 1992), the other one with the American *Alvin* submersible (Karson *et al.*, 1992), measured precisely geological sections along the walls bounding the Hess Deep to the north and to the south, and on the flanks of the intra-rift ridge. Four lithological units were observed (Figure 5.15):

—An *upper volcanic unit*, relatively thin (100 to 200 m), formed mainly by pillow-lava basaltic flows.
—A *transition zone*, 50 to 500 m thick (200 m on average) consisting in pillow-lava flows, massive basaltic flows and sills, and basaltic dikes.
—A *sheeted dike complex*, at least 1200 m thick, oriented north–south parallel to the East-Pacific Rise.
—An *isotropic gabbro unit*, appearing at the base of the dike complex in places, whose thickness is unknown (it disappears beneath the rubble talus and sediments). Gabbronorites and some layered olivine gabbros were also observed in places.

Moreover, foliated serpentinized peridotites representing residual mantle (harzburgites, dunites) were observed and sampled at the bas of the southern flank of the intra-rift ridge, as awell as wehrlites and 'impreganted' plagioclase dunites representing probably the mantle-crust transition zone.

This intra-rift ridge has been drilled recently (Leg ODP 147), in two sites, the first one on outcropping gabbros (Site 894), the second on outcropping serpentinized peridotites (Site 895). The difficult drilling conditions on a bare crust did not allow deep penetration, but a 'multiple shallow-hole strategy' on both sites allowed recovery of an important collection of plutonic rocks. The gabbroic section recovered in Hole 894G includes olivine gabbros, gabbros and gabbronorites. The very hererogeneous textures and grain sizes, the increase in the degree of differentiation downward, and steep magmatic foliation, indicate that this section corresponds

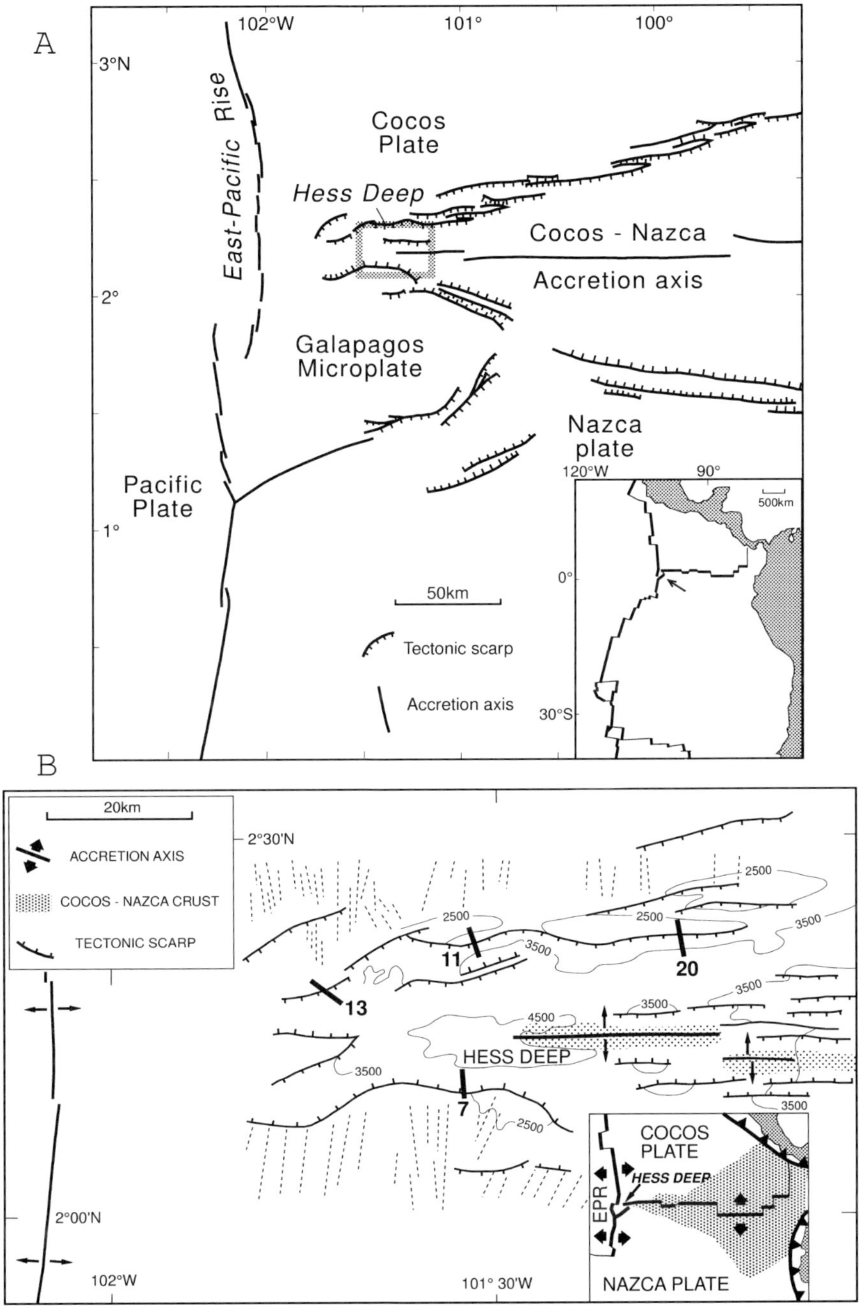

Fig. 5.14. (A) Structural map of the western part of the Cocos–Nazca ridge and the East-Pacific Rise, showing the location of the Hess Deep, in the region of the Galapagos triple junction. (B) Structural map of the Hess Deep and location of several *Nautile* dives along the northern and southern walls (after Karson *et al.*, 1992 and Francheteau *et al.*, 1992).

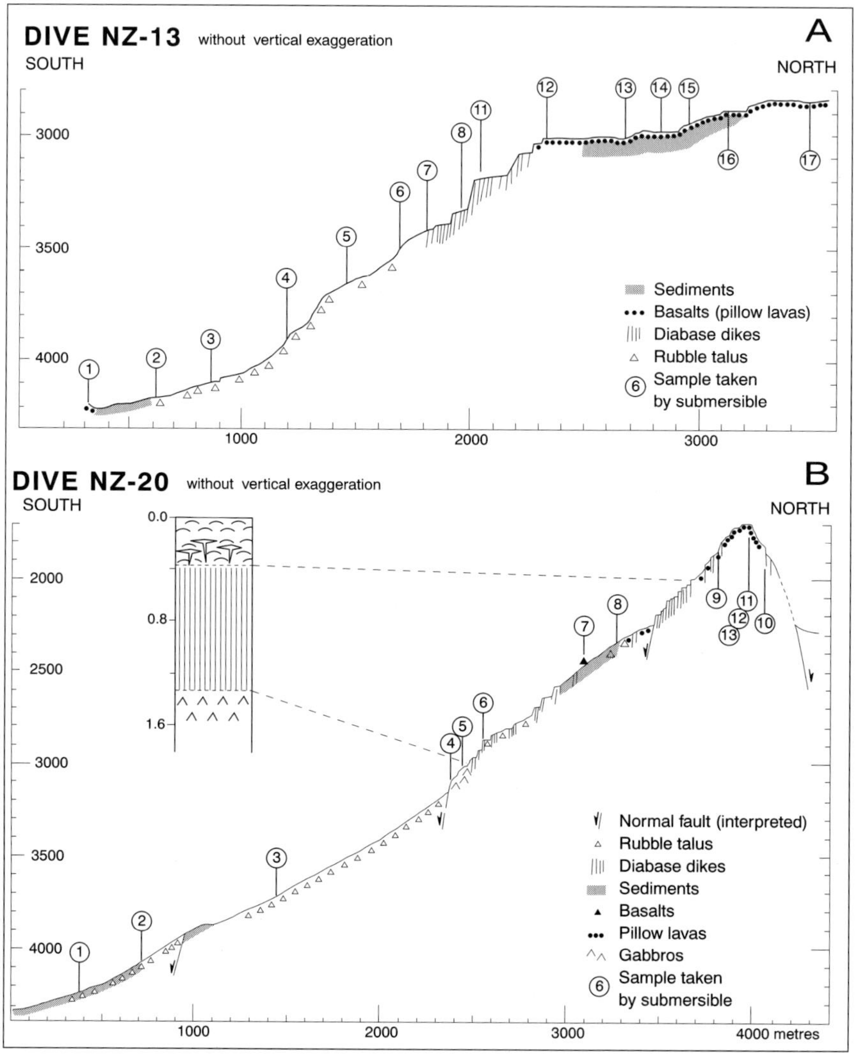

Fig. 5.15. Geological sections realized by the French submersible *Nautile* on the northern flank of the Hess Deep. See location of the dives on Figure 5.14 (after Francheteau *et al.*, 1992).

to the roof of a magma chamber. Holes 895A to F allowed recovery of well foliated spinel harzburgites, dunites impregnated with interstitial plagioclase and clinopyroxene, and some gabbros.

Though discontinuous and incomplete in the deepest parts (lower crust), the synthetic section of the flanks of the Hess Deep confirms and completes downward that of Hole 504B. The interpretative geological section through the Hess Deep (Figure 5.16) confirms and completes, presents a 4–5-km-thick oceanic crust, characterized by a thin extrusive layer, a robust sheeted dike complex (more than

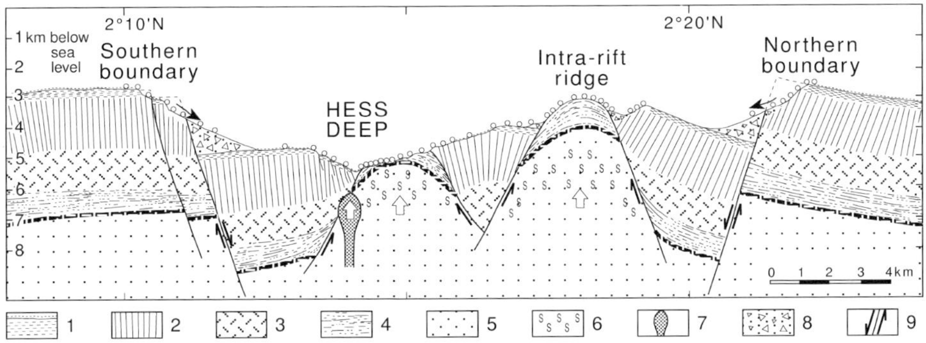

Fig. 5.16. Interpretative geological cross-section through the Hess Deep, drawn according to the bathymetric, geological and morpho-structural data. 1. Basaltic flows, eventually covered by a thin blanket of pelagic sediments. 2. Sheeted dike complex (100% diabases). 3. Isotropic gabbros. 4. Layered gabbroic cumulates. 5. Residual peridotites (upper mantle). 6. Serpentinized upper mantle. 7. Dike-like magma chamber, in line with the Cocos–Nazca ridge tip. 8. Rubble talus. 9. Normal fault (after Francheteau *et al.*, 1990).

1 km thick), and a 2–3-km-thick gabbroic layer, including isotropic gabbros at the top (where the dikes of the sheeted dike complex are rooted), then gabbronorites and layered gabbros at the bottom, grading downward to wehrlites and plagioclase dunites, and finally to the foliated harzburgites and dunites of the upper mantle. *The gabbroic layer remains mainly unknown*, as only its basal and summital parts were observed, exceeding several tens of metres in both cases.

According to Francheteau *et al.* (1992), the continuous section of upper crust observed in the Hess Deep confirms the data from Hole 504B and fits the model of upper crust deduced from seismic refraction experiments done in the axial zone of the East-Pacific Rise at 9°N (Figure 5.17; cf. also Fig. 7.20 and 21).

The only other places showing similar sections in the Pacific are: (1) the Pito Deep, a deep depression (5850 m) located at the tip of a propagating rift bounding the Easter microplate, near 23°29′S and close to the East-Pacific Rise (EPR); (2) Mathematician Ridge. Moreover, several sections observed in the transform valleys of the EPR or Galapagos rift fracture zones have shown everywhere the presence of a robust sheeted dike complex beneath a more or less thick extrusive carapace, resting itself over gabbros and mantle harzburgites. In the north-eastern Pacific, the Blanco fracture zone located between the Juan de Fuca and Gorda ridges has revealed recently a thick layer of diabases (650 m visible) beneath 800 m of basaltic extrusives (Juteau *et al.*, 1995).

Petrologically, the gabbros recovered in the south-eastern Pacific exhibit a wide range of compositions, from very Mg-rich troctolites to highly evolved ferrogabbros, passing by olivine gabbros and gabbronorites (Figure 5.18). These variations of composition can be explained by the crystal fractionation processes (Constantin, 1995). Figure 5.18 shows that the gabbros from fracture zones (Garrett, Terevaka) include ferrogabbros (sometimes extremely evolved) and no troctolites, whereas the gabbros from the propagating rifts (Hess Deep, Pito Deep) show more restricted

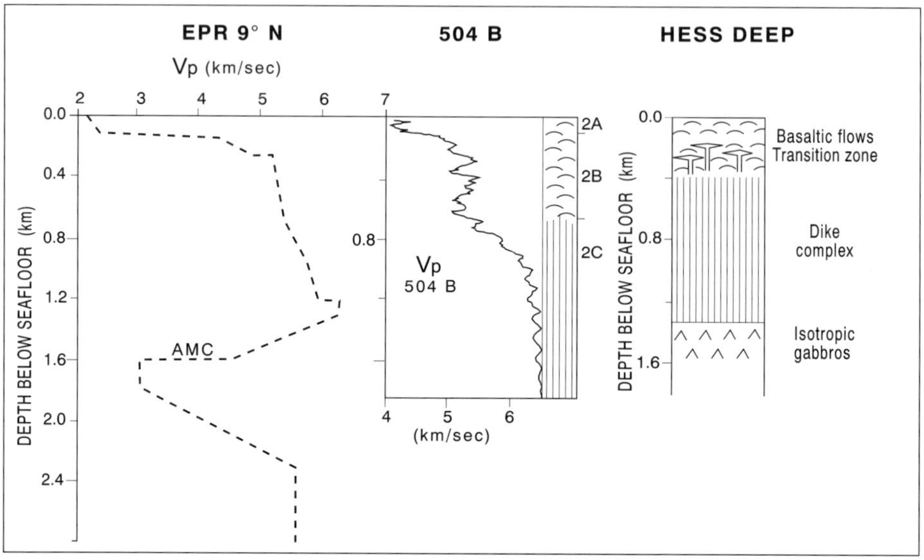

Fig. 5.17. Synthetic log of the Hess Deep upper crust. Comparison with the data from Hole 504B (lithology, V_p velocities and seismic layers), and with the model for the upper oceanic crust inferred from ESP seismic profiles in the axial zone of the EPR at 9°N (axial profile showing the roof of the magma chamber at 1.6 km below the seafloor (after Vera *et al.*, 1990). AMC: Axial Magma Chamber (drawing from Francheteau *et al.* (1992).

compositions ranging from troctolites to gabbronorites, sometimes to weakly evolved ferrogabbros. These differences reflect different crystallization conditions: 'colder' thermal regime, thinner crust and ephemeral small magma pockets evolving in closed systems in fracture zones; hotter thermal regime, thicker crust and periodically fed magma chambers in the propagating rifts.

These various sites have provided residual mantle peridotites whose compositions are in the range of the *most depleted harzburgites of the oceanic domain*, indicating a high melting rate (about 25%). In the mantle–crust transition zone, these peridotites are impregnated with magmatic dikes and veins (clinopyroxenites, gabbros, ferro-gabbros, etc.), and are often deeply modified by the basalt/harzburgite interaction.

In summary, the oceanic crust of the intermediate- to fast-spreading East-Pacific Rise, including the Galapagos ridge and the Juan de Fuca and Gorda ridges, is thick and continuous, including a continuous volcanic layer of variable thickness (100 to 800 m), overlying a well-developed and robust diabase sheeted dike complex (more than 1 km thick everywhere), and a gabbroic layer 3 at least 2–3 km thick, only the top (over 150 m) and the very bottom of which were observed. A transition zone rich in wehrlites and 'impregnated' plagioclase dunites marks the passage to the residual mantle, typically made of depleted harzburgites and dunites.

Emphasis must be given to the fact that *most of layer 3 in the Pacific Ocean was never observed or sampled, and remains presently unknown.*

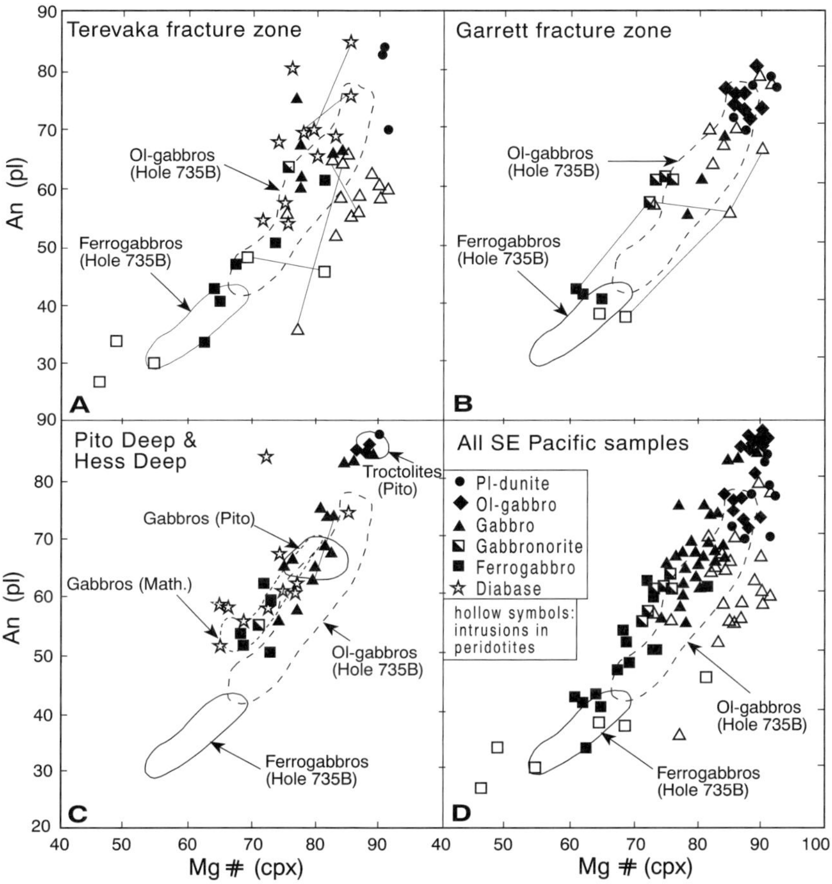

Fig. 5.18. Mg# (Mg/Mg + Fe) variations in the clinopyroxene versus An (anorthite %) variations in the plagioclase of oceanic gabbros from the south-east Pacific, and comparison with ODP Hole 735B gabbros (South-West Indian Ridge). Fracture zones (Garrett, Terevaka) exhibit highly evolved ferrogabbros, and no troctolites, whereas the Hess Deep and Pito Deep have given abundant troctolites and olivine gabbros, and very few ferrogabbros (after Constantin, 1995).

5.2.2 The discontinuous tissue of slow-spreading ridges: the Mid-Atlantic and South-West Indian Ridges

The Mid-Atlantic Ridge

A complete section of the oceanic crust along the southern wall of the Vema fracture zone

Located at about 10°45′N, the Vema fracture zone is one of the major faults of the equatorial Atlantic (Figure 5.19A). The left-lateral offset of the Mid-Atlantic Ridge along this fault is 320 km. The tectonically active part of this fracture zone consists of a deep transform valley oriented east–west, reaching 5100 m in depth, and 10 to 20 km in width. This valley is bounded by two fault scarps whose crests are between

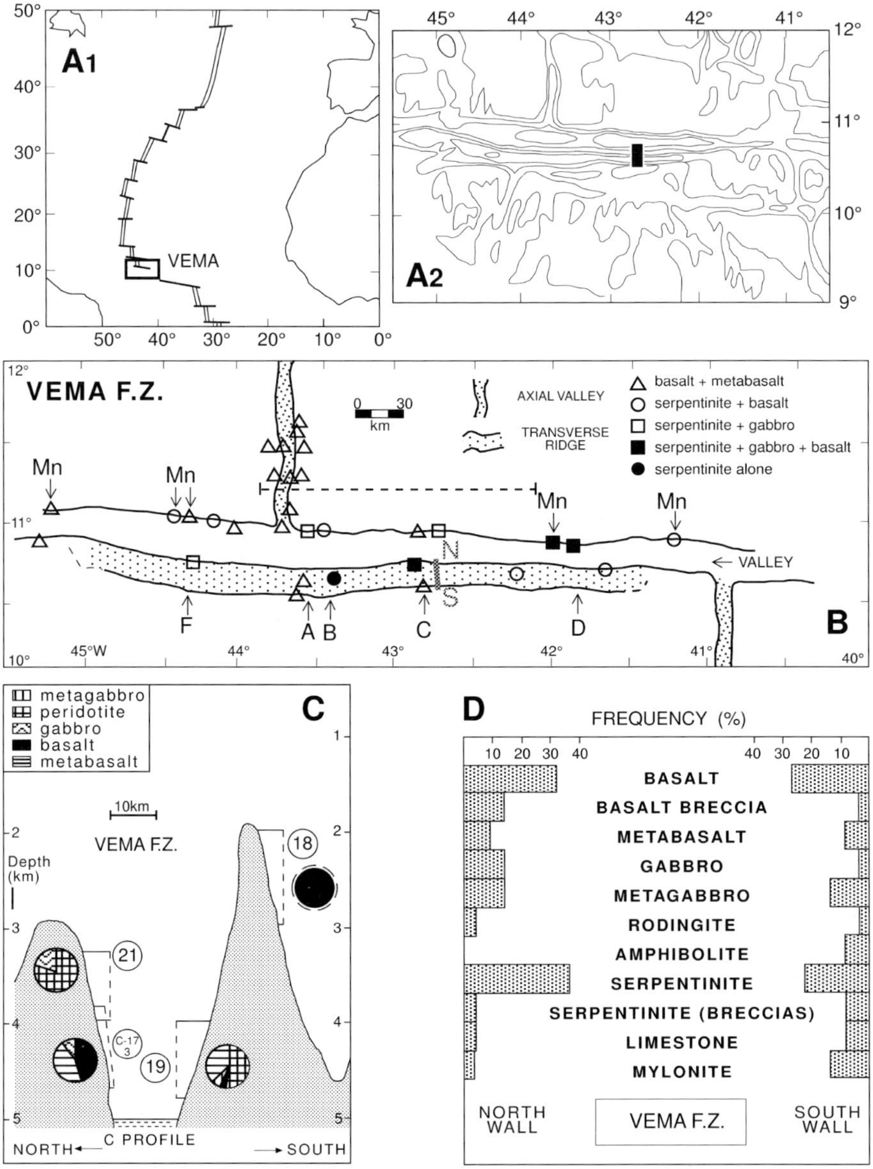

Fig. 5.19. Location of Vema fracture zone (A1), and simplified bathymetry of the transform valley and transverse ridge (A2). (B) Simplified morphology of the fracture zone at 11°N, and location of various dredge hauls (A, B, C, D, F, etc.), realized during the 1960s and the 1970s. The arrow pointing to 'C' indicates the emplacement of the section of Figure C. The sites marked with 'Mn' along the northern wall of the fracture zone exhibit basalts encrusted with a thick manganese crust (>1 cm). (C) North–south profile along section C. The vertical exaggeration is 20. The locations of four dredge hauls are indicated. The rock types collected in each dredge are indicated, with their relative percentages ('camembert' slices). The crustal structure is impossible to restore. (D) Frequency diagram of the various rock types met in the whole set of dredges along the northern and southern walls of the Vema fracture zone (after Bonatti and Honnorez, 1976).

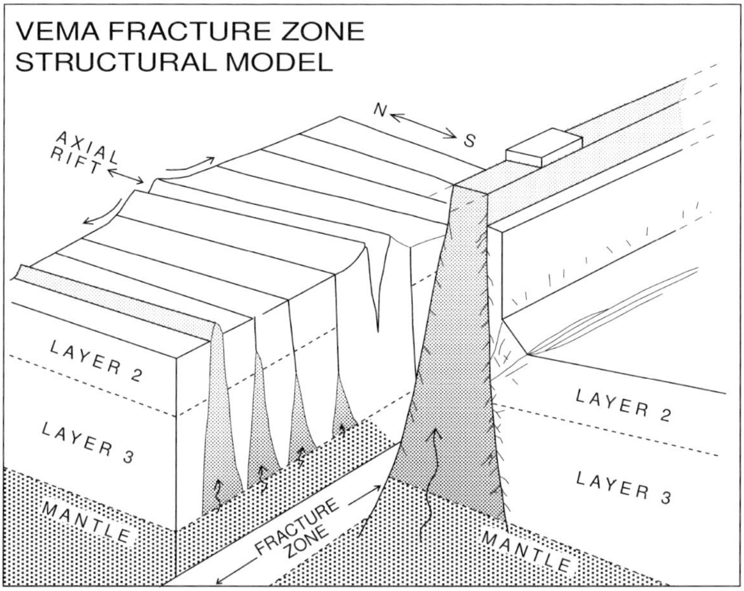

Fig. 5.20. Structural model of the Vema fracture zone, based on various collections of dredged samples proposed by Bonatti and Honnorez in 1976. The abnormally high transverse ridge located to the south of the transform valley is mainly made of mantle-derived serpentinite, interpreted as a vertical 'diapiric intrusion', raising blocks of crust. The margins of the intrusion are strongly tectonized. The northern wall of the fracture zone constitutes a 'window' in the normal oceanic crust. To explain the frequent occurrence of serpentinites at shallow levels in the crust, the authors propose the existence of vertical intrusions of serpentinite bodies, elongated parallel to the ridge axis.

2000 and 4500 m above the bottom of the transform valley. The southern wall corresponds to an abnormal topographic high called the transverse ridge, whose summit culminates at hardly 550 m below sea level (Figure 5.19B).

During the 1970s, systematic dredgings along the northern and southern walls of the transform valley provided in bulk a varied collection of magmatic and metamorphic rocks: basalts and metabasalts, gabbros and metagabbros, rodingites, amphibolites, serpentinized peridotites and serpentinites, mylonites, and varuous kinds of sedimentary rocks (Bonatti and Honnorez, 1976).

Despite good enough control over the dredging depths, any reconstruction of a lithological section was impossible. Only the proportions between the various kinds of rocks could be given, in the form of 'camembert slices', or of frequency diagrams (Figure 5.19C and D). In spite of these diffivculties, Bonatti and Honnorez (1976) proposed a model in which diapiric extrusions of serpentinites of mantle origin constitute the main part of the transverse ridges with their abnormally high topography, like the southern wall of the Vema fracture zone (Figure 5.20). The northern wall would be made of 'normal' crust injected with thin and vertical serpentinite intrusions, parallel to the ridge axis.

During the cruise to the Vema fracture zone in 1988, five dives of the *Nautile* submersible enabled completion of two geological sections of this southern wall

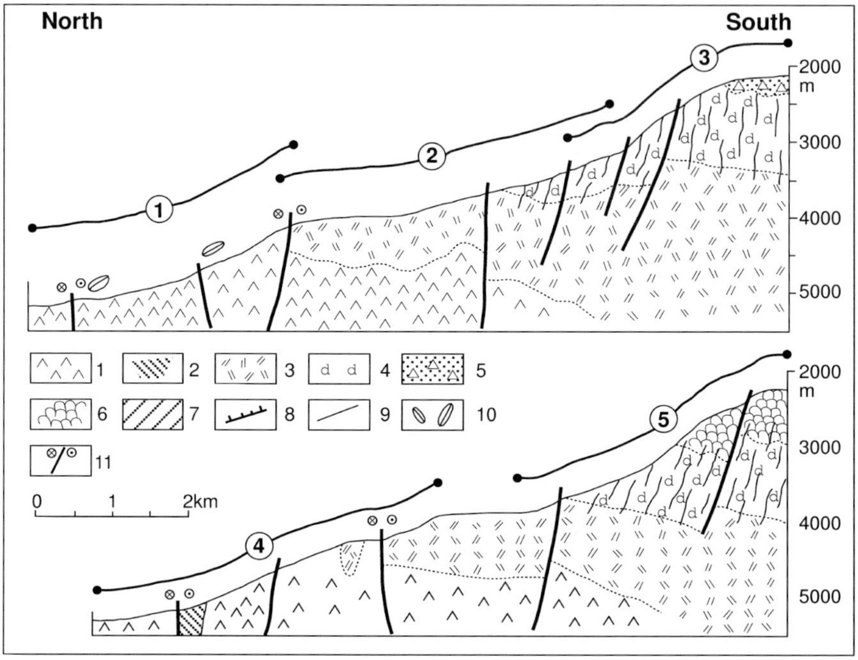

Fig. 5.21. Two synthetic geological cross-sections of the southern wall of the Vema fracture zone, with a spacing of 5 km, after the observations made with the *Nautile* submersible in 1988 (five dives, numbered 1 to 5). The location of these two dives is indicated by the bold line N–S in Figure 5.19B. 1. Serpentinites. 2. Amphibolites. 3. Gabbros. 4. Sheeted dike complex (100% diabases). 5. Basaltic breccias. 6. Pillow lava. 7. Consolidated sediments. 8. Major normal fault. 9. Minor fault. 10. Clam occurrences (giant shells). 11. Shear sense of transform faults (after Auzende *et al.*, 1988a).

between 5160 m and 2180 m that is, a vertical scarp of about 3000 m, showing complete exposure of the oceanic crust (Auzende *et al.*, 1988a) (Figure 5.21). Both parallel sections, separated by 5 km and quite identical, begin at about 5100 m on the bottom of the transform valley with discontinuous outcrops and breccias of mantle peridotites (serpentinized harzburgites). Then non-layered, massive gabbros appear from 4200 m to 3700 m, over about 500 m thick. These are evolved FeTi-rich ferrogabbros, at the top of which isolated diabase dikes appear, followed by a spectacular sheeted dike complex made of mutually intrusive vertical diabase dikes oriented north–south, suggesting that they were emplaced at the ridge axis. This unit is about 700 to 1000 m thick. The section ends upward with basaltic breccias for the western section (dive 03), and with pillow-lava flows for the eastern section (dive 05). These basalts constitute a volcanic layer covering the top of the transverse ridge.

This section is famous, because it is unique in the Atlantic. It is not fundamentally different from that of the Hess Deep, with the difference that *the crust exposed here is much thinner*, hardly 2.2 to 2.5 km in thickness. The main lithological units of the Pacific crust are there: a gabbro layer lies over the mantle peridotites, and are overlain by adiabase dike layer forming a true dike complex, supporting a volcanic layer mainly made of pillow-lava flows.

But is this section representative of the Atlantic crust? Today, it can be doubted.

During the very first 'crustal' drilling in the Atlantic (Leg DSDP 37, 1974), the results were not those expected. After drilling 50 m of basalts, the drill hole crossed 65.5 m of peridotite and gabbro breccias, resting over deeper basalts. The drill hole had obviously penetrated through a rubble talus made of blocks of deep rocks, indicating that these mantle and deep crustal rocks were outcropping along submarine cliffs in the vicinity.

During the following years, the dredgings, drillings and submersible studies done along the Mid-Atlantic Ridge confirmed this fundamental fact: that in many places of this ocean ridge, upper mantle peridotites, associated with deep crustal gabbros, are outcropping in the axial zone, or close enough to the axis to be covered by lava flows. DSDP/ODP drill holes have shown the frequency of this phenomenon for various ages of the crust (Figure 5.22).

It is clear today that a great majority of deep rocks from the crust and upper mantle collected in the oceanic environment come from *slow-spreading ocean basins* (Figure 5.23): Atlantic Ocean, Indian Ocean (SW Indian ridge), and also back-arc basins (Cayman rift in the Caribbean Sea for instance).

These observations were interpreted as, in a slow accretion context, long-duration 'amagmatic' stages of pure tectonic extension, alternating with relatively brief stages of magmatic production. During these amagmatic stages, the magmatic budget is low and the crust, particularly thin, is easily dismembered by extensional tectonics, facilitating access of mantle rocks and gabbros on the seafloor.

The gabbros of the MARK area

The Kane fracture zone and its eastern intersection with the Mid-Atlantic Ridge (zone MARK, for Mid-Atlantic Ridge Kane) have been intensively mapped and studied over the last 20 years. Between 1981 and 1992, four diving cruises with the submersibles *Alvin* and *Nautile* enabled acquisition of a number of geological sections in this area, and confirmed the existence of outcrops of deep rocks (gabbros and serpentinized peridotites). They are found in particular along the western wall of the MARK area axial valley, on the *intersection massif* (inner corner), and all along the southern wall of the Kane fracture zone (FZ) transform valley, over 80 km long. ODP drillings have taken core samples of the outcropping peridotites and gabbros of the MARK area, first along the western wall of the axial valley (Leg ODP 109, Hole 670A), then recently on the eastern flank of the intersection massif (Leg ODP 153, Sites 920 to 924).

The general setting is the following (Figure 5.24): the axial valley, oriented 20°N, is bounded by two asymmetrical walls, the western wall being more elevated than the eastern one. Its average depth is about 4000 m. The inner floor of the axial valley shows a neovolcanic axial ridge 500 m high and 3–4 km wide, (*Snake Pit Ridge*), at the top of which an important hydrothermal activity with 'black smokers' was discovered in 1985 (Leg ODP 106). At the intersection with the fracture zone, the axial zone deepens to form a *nodal basin*. This is a circular structure whose depth exceeds 6000 m, typical of the ridge/fracture zone intersections in the Atlantic.

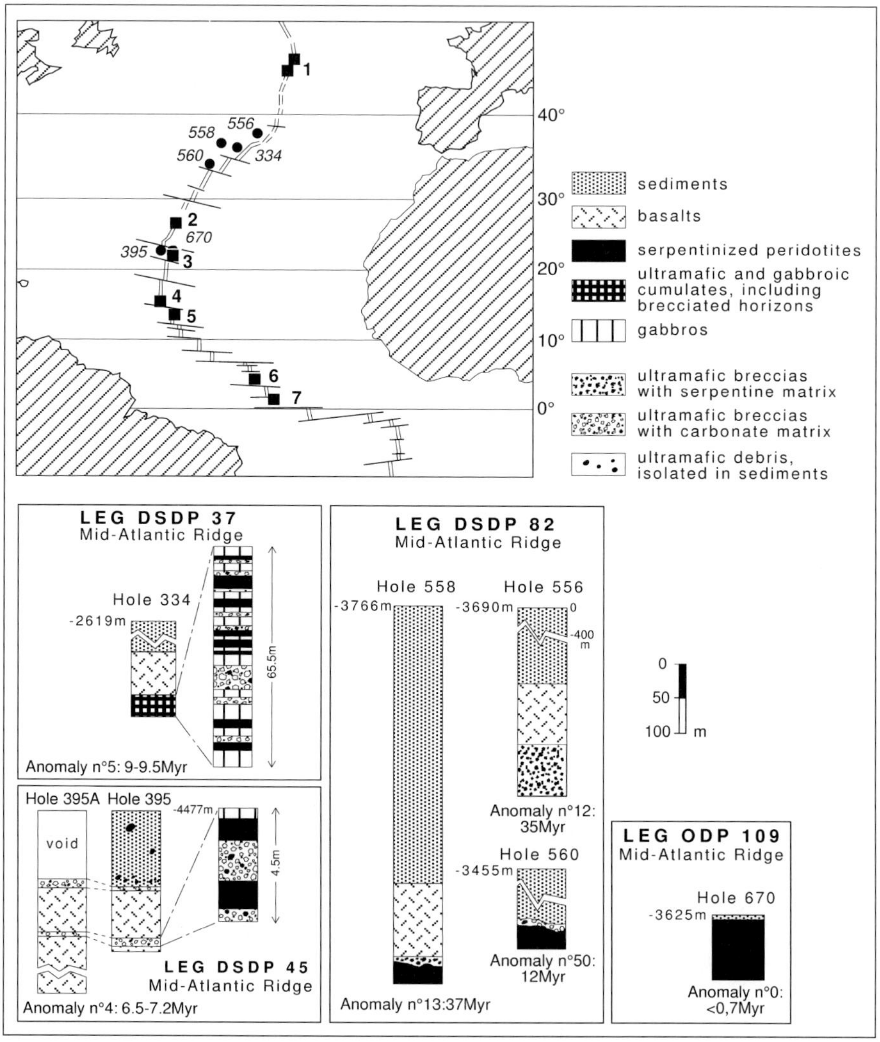

Fig. 5.22. Mid-Atlantic Ridge. Location of sites where variously serpentinized mantle peridotites, with or without associated gabbros, were dredged (sites 1 to 7 on the map) or drilled (DSDP/ODP hole numbers indicated on the map), and schematic illustration of the drilled columns (after Juteau *et al.*, 1990 and Lagabrielle and Cannat, 1990).

The transform valley, oriented 110°N, has an irregular geometry, with a succession of ridges and deep basins (4500 m in average). Its average width is 6–8 km. The northen wall of the fracture, made of aged crust, presents a topography of ridges and depressions oriented north–south. The southern wall, over a length of 80 km starting from the intersection zone (nodal basin), is made of a young crust formed at the MARK area axis, ansd consists of a succession of four massifs presenting a steep margin convex toward the transform valley, separated by deep

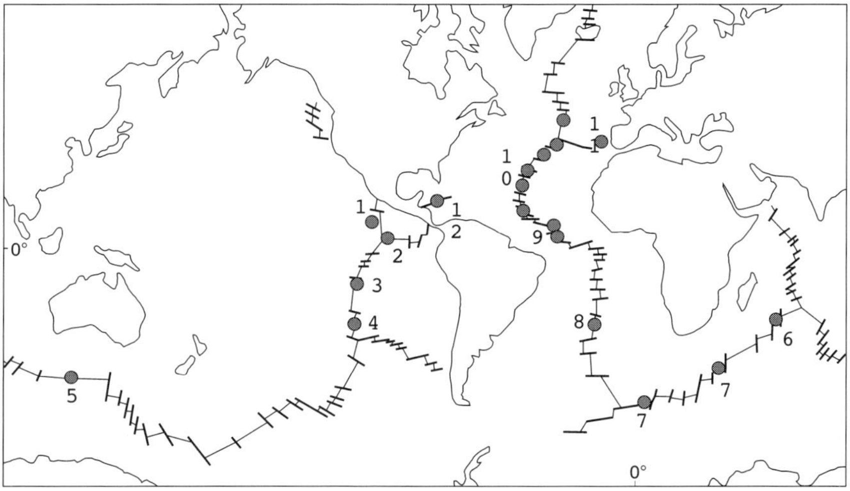

Fig. 5.23. Location map of areas where deep rocks (gabbros and mantle peridotites) were collected in the oceanic domain. **Pacific Ocean**: 1. Mathematician Ridge. 2. Hess Deep. 3. Garrett fracture zone. 4. Easter microplate (Pito Deep and Terevaka fracture zone). **Indian Ocean**: 5. Australia–Antarctica discordance zone. 6. Atlantis II fracture zone (ODP Hole 735B). 7. Other fracture zones. **Atlantic Ocean**: 8. Southern Atlantic. 9. Major fracture zones of the equatorial Atlantic. 10. Northern Atlantic, Mid-Atlantic Ridge. 11. Gorringe Bank. 12. Cayman Trough (after Mével and Cannat, 1991, Cannat, 1993 and Constantin, 1995).

and narrow depressions oriented north–south. The easternmost massif, or *intersection massif*, culminates at less than 1200 m, presenting therefore a vertical offsst of nearly 5000 m with the bottom of the adjacent nodal basin.

About 20 dives with both submersibles enabled compilation of a relatively complete geological map of the *intersection massif* (Figure 5.25A and B). On its eastern flank, massive gabbros, often intensely schistosed, associated in places with serpentinized peridotites, crop out from the bottom of the nodal basin to 2000 m. Normal faults oriented N–S to NW–SE and dipping E to NE affect the gabbros. On the northern flank, starting from the bottom of the transform valley, isome basaltic outcrops and rubbles appear first (pillow lavas and isolated diabase dikes), cut by right-lateral strike-slip faults oriented E–W. The whole slope above, up to 2500 m, is occupied by highly cataclastic metagabbros metamorphosed in the greenschist facies, associated here also with some serpentinites. These metagabbros are intersected by huge mylonitic zones developed along EW-trending fault planes, dipping 35 to 50° northward, toward the transform valley (Figure 5.25B). Above 2500 m the top of the massif consists of a basaltic blanket up to about 1400 m (pillow-lava flows, intersected by isolated basaltic dikes). The whole western flank of the intersection massif is occupied by a thick basaltic series, down to the transform valley bottom (pillow lavas). *No sheeted dike complex was observed: the lavas rest directly over the tectonized and metamorphosed gabbros.*

In summary, the *intersection massif* constitutes a tectonic window in which outcrop, over more than 3 km in height, the deep levels of the crust, associated

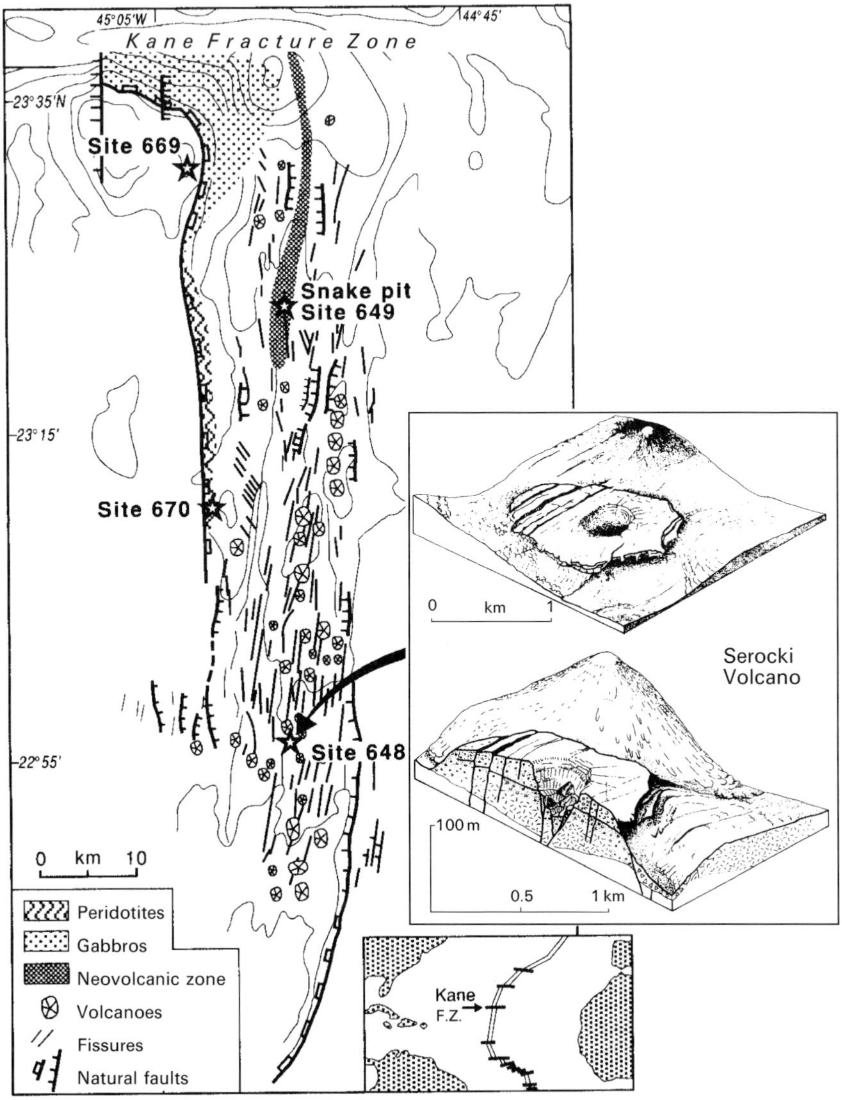

Fig. 5.24. General geological sketch of the MARK area (Mid-Atlantic Ridge Kane), and its intersection with the Kane fracture zone, with location of ODP drill sites.

with serpentinized upper-mantle rocks, strongly tectonized and metamorphosed. The three other massifs going westward, exhibit very similar sections, with alternating mantle peridotites (more or less serpentinized) and gabbros, with their basaltic blanket (Figure 5.25C). Serpentinites and metagabbros outcrop along huge ductile faults dipping 45° toward the transform valley. The mantle rocks are cut by numerous gabbro dikes, and everywhere a flat basaltic blanket covers the plutonic rocks, without showing any dike complex in between.

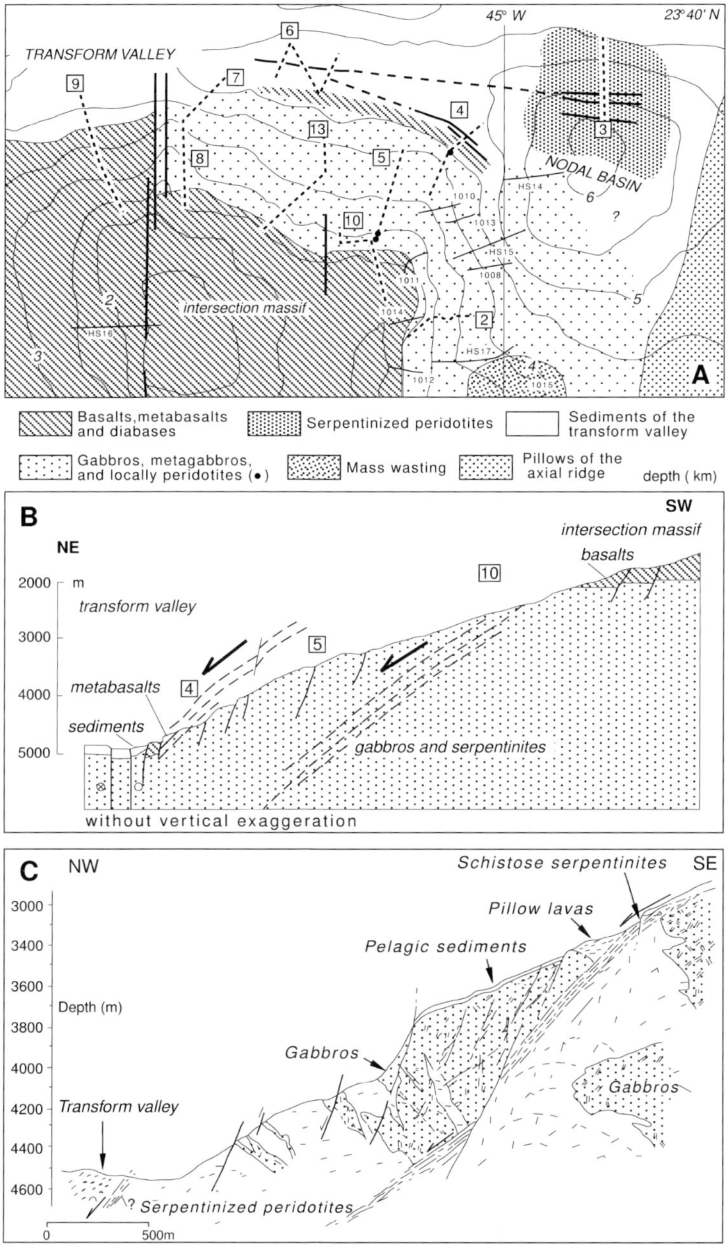

Fig. 5.25. (A) Geological map of the eastern intersection massif of the Kane fracture zone with the Mid-Atlantic Ridge, established after the data from four deep diving cruises. Dash lines: *Alvin* dives (1981, 1986). Thin continuous lines: *Nautile* dives (1989, 1992). The main observed faults are in bold lines. (B) Geological cross-section through the intersection massif, SW of the nodal basin. Tight, discontinuous dash lines indicate the main ductile faults (after Auzende *et al.*, 1994). (C) Geological cross-section of the southern wall of the Kane fracture-zone valley at 45°15′W. Tight, discontinuous dashed lines indicate the main ductile faults (after Auzende *et al.*, 1994).

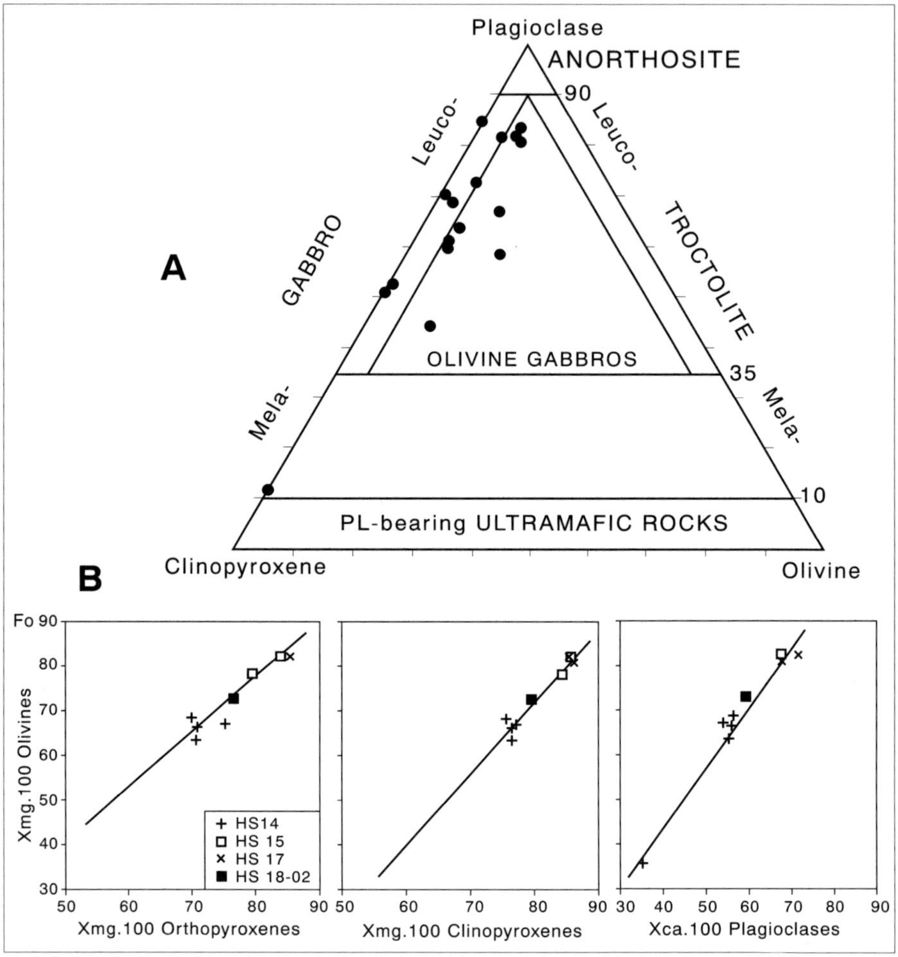

Fig. 5.26. (A) Modal composition of gabbros from the MARK area (*intersection massif*). (B) Correlations between the XMg = Mg/(Mg + Fe) ratios of pyroxenes, the anorthite % of plagioclases, and the forsterite contents of associated olivines, in gabbros of the MARK area (*intersection massif*) (after Marion, 1992).

Petrology of the gabbros. Petrographically, the gabbros of the massif intersection show a wide range of facies Marion, 1992): olivine gabbros, two pyroxene gabbros, very evolved ferrogabbros rich in Fe–Ti oxides, brown amphibole and apatite, lucogabbros and plagiogranites (rich in quartz, albite and zircon). A brown amphibole of magmatic origin is generally present in most of these facies (Figure 5.26A).

Olivine gabbros and clonopyroxene gabbros are dominant in the sampling. Two pyroxene gabbros and ferrogabbros are much less abundant. Plagiogranites appear as cm-thick veins in the gabbros, and are clearly late magmatic products.

The chemical composition of the primary minerals, represented for instance by the anorthite content of the plagioclase, and the XMg = Mg/(Mg + Fe) for the ferromagnesian phases, exhibit large variations in differentiation degree of the rocks, ranging from very Mg-rich olivine gabbros (Fo 80–85, An 75–80), to very evolved Fe-augite-bearing ferrogabbros (Fo 55–35, An $<$ 40), rich in ilmenite and apatite. From one sample to another, the chemical variations of olivine (Fo) are well correlated to the variations of the other primary phases (Figure 5.26), confirming the impression of a magmatic suite having evolved by crystal fractionation processes. Moreover, the most evolved facies were sampled very close to the Kane fracture zone. This evolution toward very differentiated end-members is characteristic of a magma chamber evolving in a closed fractionation system.

The study of secondary crystallizations in the gabbros of the intersection massif shows the effects of at least two successive episodes of transformations in the deep crust, linked to the circulation of hydrothermal fluids. A first episode occurred at high temperature, when the permeability of the gabbros was still very low. The penetration of seawater occurred along localized ductile shear zones, with seawater circulating along the grain boundaries. The parageneses associated with this first episode (diopside, plagioclase, brown to green amphiboles) suggest interaction temperatures compatible with the lower amphibolite facies (about 500–600°C), as well as a low water/rock ratio. The second episode is associated with an intense brittle fracturing, which considerably increased the permeability of the gabbroic mass, and allowed a masive introduction of seawater in the crust. The volume of the corresponding secondary hydrated phases is important. Their nature and composition suggest temperatures compatible with the transitional to greenschist facies (500 to 300°C), with a high water/rock ratio.

Figure 5.27 illustrates the present-day concepts concerning the crustal structure of a slow-spreading oceanic ridge, based mainly on the geological and geophysical studies performed in the MARK area. This along-strike section across two adjacent accretion segments shows that the segment tips are characterized by a thick lithosphere (10 km) and a thin, discontinuous crust, with a seismic Moho at 4 km, whereas the segment centres are characterized by a thinner lithosphere (2 km minimum), and a thicker, lithologically 'normal' magmatic crust (6 km). The magmatic crust is therefore supposed to be continuous, thick and probably 'layered' in the central zone of the segments, frequently fed with new magma; it becomes progressively thinner and discontinuous toward the segment tips, poorly fed with magma and dismembered by numerous strike-slip and normal oblique faults, favouring the appearance of deep-mantle rocks on the seafloor.

The South-West Indian Ridge

In the Indian Ocean, the South-West Indian Ridge (SWIR), oriented east–west, presents a very slow opening rate (1.6 to 2 cm/yr). Morphologically, it presents a deep axial valley and is offset by numerous north-south fracture zones, with a 50-km spacing on average. This ridge has provided a lot of deep rock samples through dredgings (gabbros, serpentinized mantle peridotites) representing close to 50% of the dredged samples.

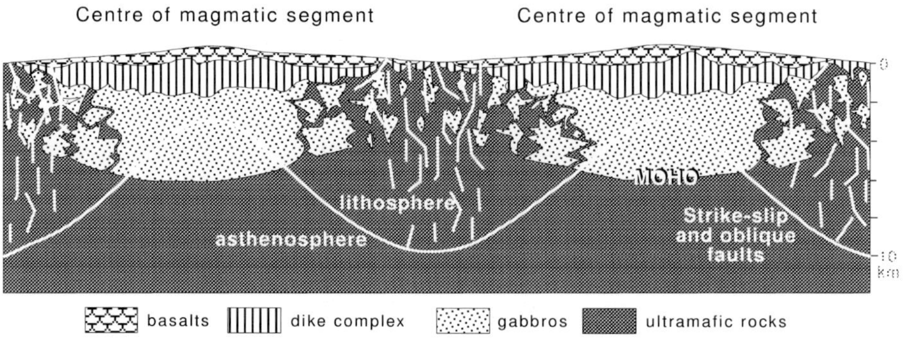

Fig. 5.27. Schema illustrating present-day concepts concerning the crustal structure of a slow-spreading ridge (Mid-Atlantic type, after Cannat *et al.*, 1995).

The gabbros of ODP Hole 735B

During Leg ODP 118 in 1989, a 500-m-deep section was drilled in the layer 3 gabbros of the Indian Ocean. 435 m of gabbros were cored and recovered on the eastern flank of one of these SWIR fracture zones (Hole 735B).

These gabbros crop out on the seafloor along the eastern wall of the Atlantic II fracture zone, at 90 km south of its intersection with the SWIR axis, with an estimated crustal age of 11 Myr (Figure 5.28). These gabbros are interpreted as having crystallized in a ridge/FZ intersection context, comparable with that of the gabbros of the massif intersection in the MARK area in the Atlantic.

The gabbros are petrographically very varied, covering a wide range of facies, from poorly evolved troctolitic and olivine gabbros, to evolved FeTi oxide-bearing ferrogabbros and gabbronorites, and finally to very evolved leucogabbros rich in quartz, zircon and apatite, and to trondhjemites. This variability is reflected in their chemical composition, which spreads widely following the tholeiitic trend in the MFA diagram for instance, suggesting a closed fractionation system of crystallization, in ephemeral and small magma pockets (Figure 5.29A).

Figure 5.30 gives the lithological log of Hole 735B, which comprises six major lithological units. Most of the section consists in a unique olivine gabbro body, intrusive at its top in a medium- to coarse-grained gabbronorite. At the bottom, the olivine gabbro body is cut itself by troctolites and troctolitic gabbros, probable precursors of a major troctolitic intrusion located at the base of Hole 735B (Dick *et al.*, 1991). The section is cut in various places by small microgabbro intrusive bodies, products of the crystallization of interstitial magmas which migrated through the olivine gabbro before its complete consolidation.

The whole section has suffered a '*syntectonic differentiation*', process during which the deformation and compaction of a rigid but still partially melted gabbro resulted in the extraction of a fraction of the interstitial liquid out of the olivine gabbro, and concentrated it in the ductile shear zones. The chemical exchanges, oxide precipitations and trapping of liquids during their migration at the end of the tectonic deformations have transformed the gabbros affected by shear zones to ferrogabbros.

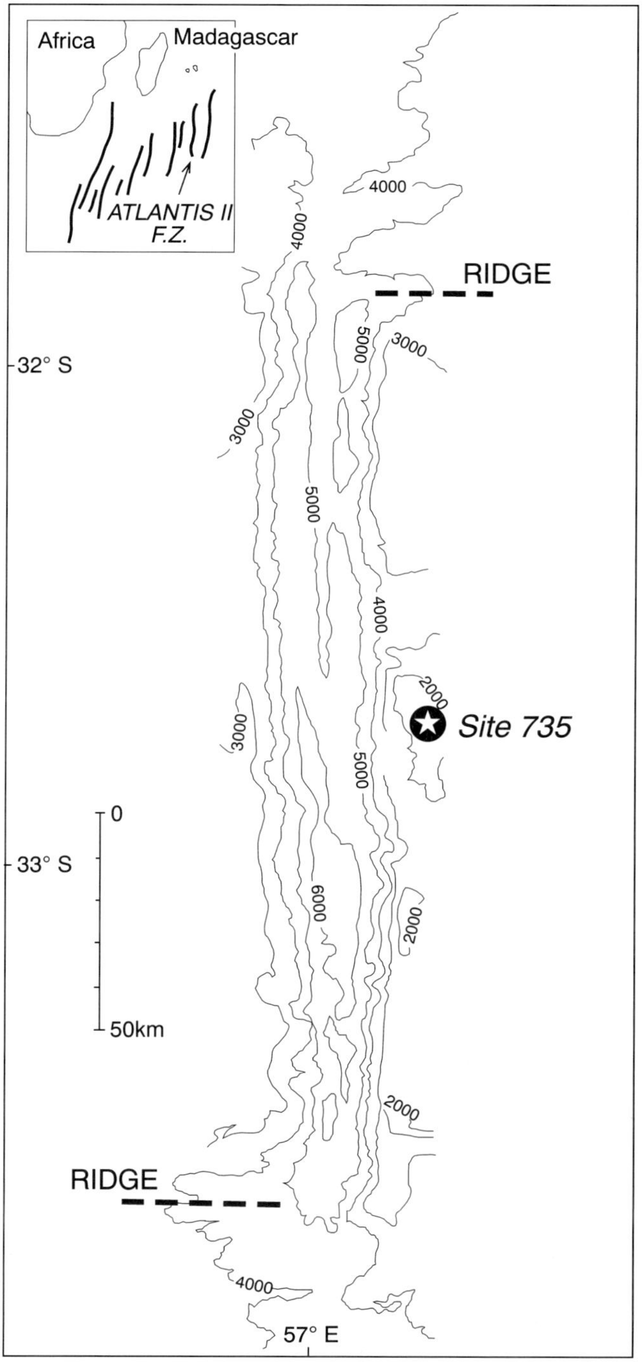

Fig. 5.28. Location of ODP Site 735, on the eastern flank of the Atlantis II fracture zone (south-western Indian ridge).

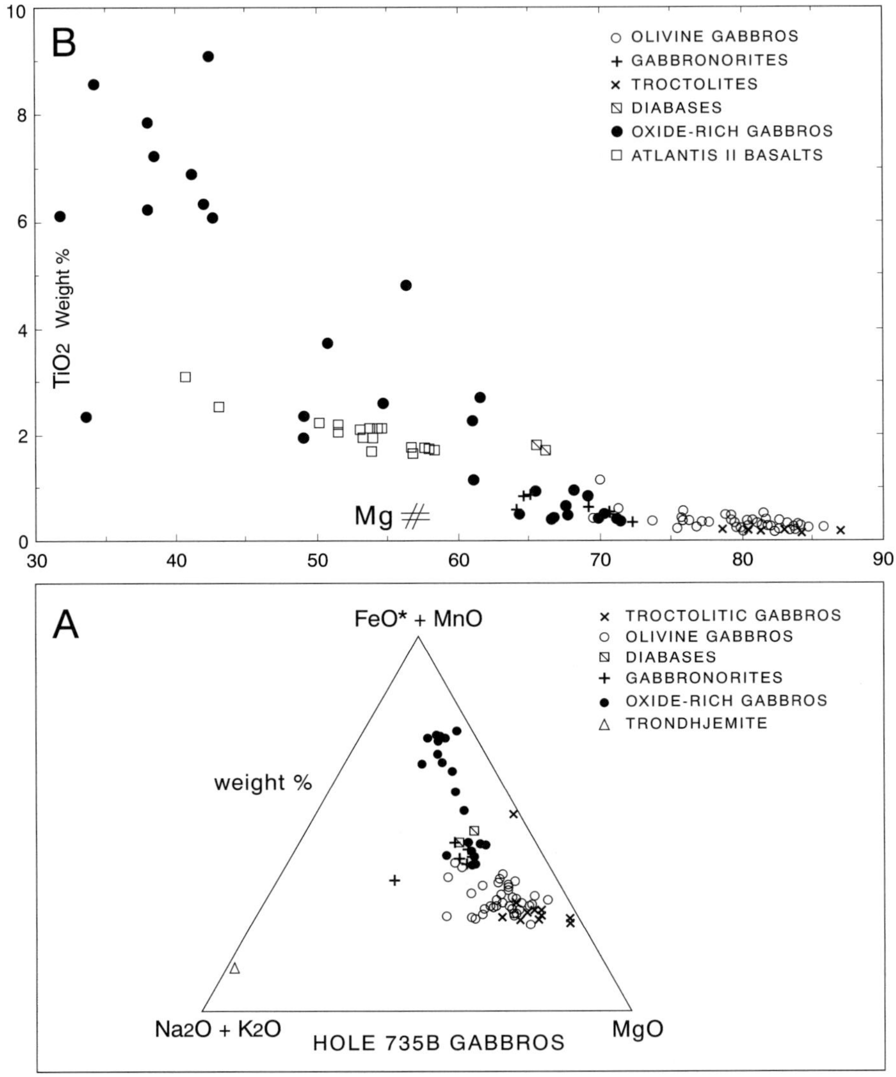

Fig. 5.29. Chemical variability of Hole 735B gabbros, and comparison with the basalts dredged in the Atlantis II fracture zone. (A) Ternary MFA diagram (all Fe as FeO). (B) Plot of TiO_2 content (in weight % of whole rocks), versus the Mg#, Mg = 100 Mg/(Mg + Fe (after Dick *et al.*, 1991) (simplified).

The last magmatic event consists in intrusive trondhjemite veins, formed by crystal fractionation from the interstitial liquids.

High-temperature metamorphism and deformation. 30% of the 435 m of cores were plastically deformed at high temperatures corresponding to the crystallization, in the foliation, of metamorphic assemblages of the *granulite*, then of the *amphibolite facies* (Figure 5.31). These high temperatures suggest that the gabbros were deformed in

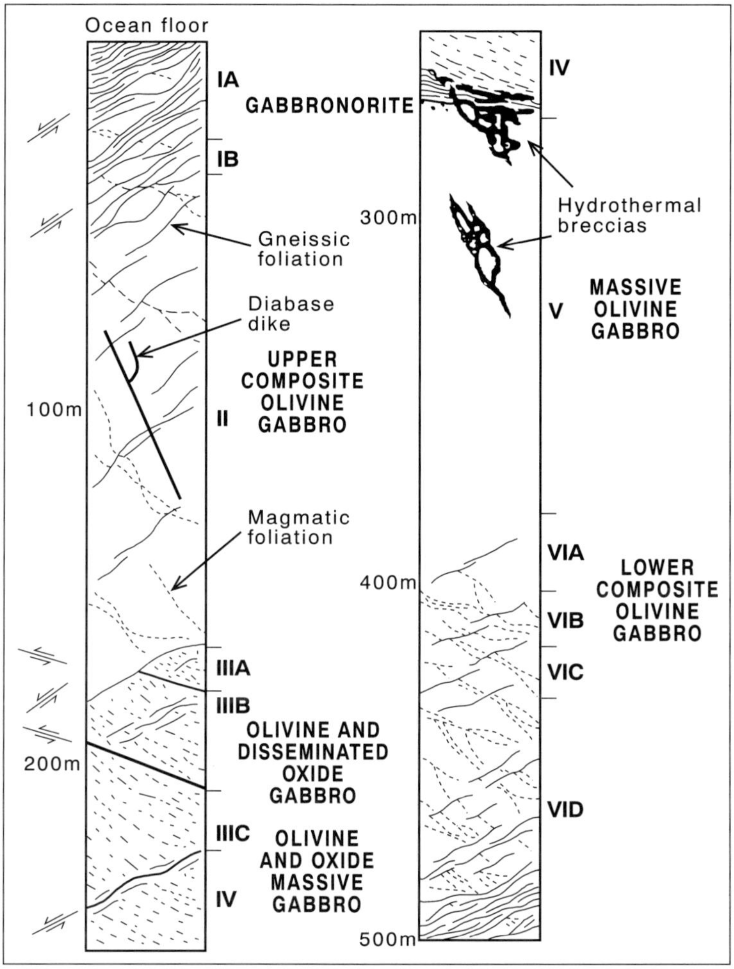

Fig. 5.30. Lithological log of Hole 735B gabbros (after Dick *et al.*, 1991).

the immediate vicinity of the ridge some 11 Myr ago. Some deformed gabbros are intersected by intrusive, non-deformed leucogabbros; confirming that the deformations were developed close to the magmatic pockets, during their differentiation. Moreover, both the non-altered, non-deformed gabbros and the metamorphosed, deformed gabbros exhibit identical and positive remanent magnetic inclinations, suggesting a magnetization during a same reverse magnetic episode. This is another indication of a short time interval between the initial crystallization of the gabbros in the axial zone of the SWIR, and their deformation.

The hydrothermal circulations which allowed the development of a secondary amphibole are linked to the formation of two types of structure susceptible of creating porosity: (a) the *ductile shear zones* at high temperatures; (b) a *fracture network*, developed at lower temperatures.

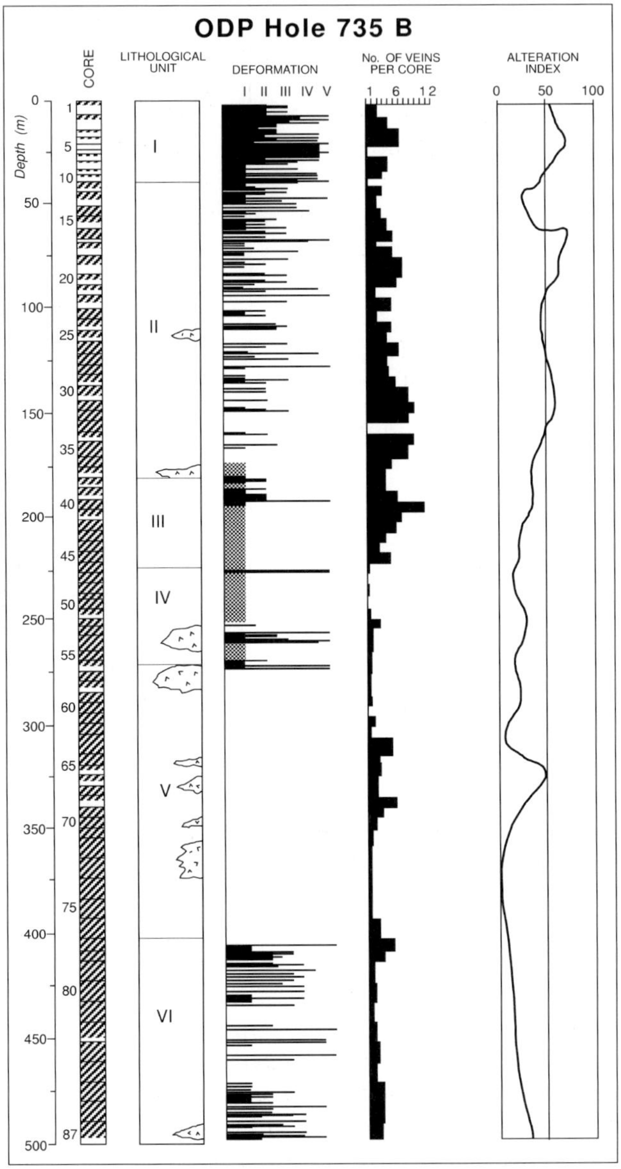

Fig. 5.31. Metamorphic and structural log of Hole 735B gabbros (after Mevel and Cannat, 1991). To the left, depths below the seafloor and core numbers. *Lithological units*: as in Figure 5.30. *Deformation*: the authors have defined five texture types (deformation increasing from I to V), ranging from non-foliated and weakly recrystallized rocks (I), to extremely deformed and foliated mylonites (V). *Number of veins per core*: hydrothermal veins were identified and counted on board. This parameter gives an idea of the intensity of hydrothermal circulation, which does not always coincide with the intensity of deformation. *Alteration index*: this index is estimated from the volumic content (measured by point counting under a microscope) of secondary amphibole replacing the primary clinopyroxene. It is high in the shear zones and in intervals rich in hydrothermal veins. There is no simple relationship between depth and intensity of metamorphism.

The earliest deformations show quite anhydrous high-temperature assemblages, where the primary magmatic minerals are granulated, forming plagioclase, clinopyroxene, orthopyroxene, Ti-hornblende and/or olivine neoblasts with triple junction, with identical compositions as the minerals of the non-deformed gabbros. At this stage, where equilibrium temperatures are estimated at 850–900°C (granulite facies), there are not yet any chemical exchanges with a fluid phase.

The following stage is marked by the penetration of external fluids, with development of syn-kinematic amphiboles in the shear zones. Fracturing at the ridge crest has penetarted below the brittle–ductile transition, allowing the penetartion of hydrothermal fluids in the hot rocks, and provoking hydration reactions at high temperature. This stage of plastic deformation and hydrated metamorphism resulted in the formation of foliated amphibolites at temperatures of the order of 700–600°C. The composition of the amphiboles ranges from pargasitic to edenitic hornblendes, then to actinolitic hornblende, according to the temperature, whereas the secondary plagioclase evolves toward andesine (Figure 5.32).

A fractured surface developed later on during the cooling of the crust, with penetartion of hydrothermal fluids and formation of vertical hornblende and sodic plagioclase veins at temperatures of 500–600°C (cf. Figure 9.10). The pyroxene and plagioclase around the veins have been replaced by green amphibole and sodic plagioclase respectively. Far from the veins, in the non-deformed gabbros, the static metamorphism is characterized by the formation of coronitic reactions around the primary mafic minerals, with development of Fe–Mg amphiboles, talc, phlogopite, pargasite and chlorite.

In summary, the gabbros of layer 3 in the Indian Ocean, in the Hole 735B section,do not present the image of a basic magma differentiating by gravity in a huge magma chamber, as the Skaergaarg intrusion in Greenland, for instance. The layered gabbros, so well developed in some ophiolites, are completely lacking. These gabbros are characterized, on the contrary, by successive multiple intrusions, forming small-sized intrusive bodies, evolving rapidly in closed crystal fractionation systems. The crystallization is accompanied by important deformations, reflecting the permanency of an active extensional tectonics during the whole consolidation period, then the cooling of layer 3 (Figure 5.33).

Finally, the episodic and irregular character of magmatic production at slow-spreading ridges results in the creation of a discontinuous and very irregular 'crustal tissue'. This is the main difference with the crust of the Pacific Ocean, where nothing similar -up to now- was observed: the high frequency of the volcanic eruptions and the sustained magmatic production have created there an homogeneous oceanic crust with a roughly constant thickness.

On the other hand, geophysical methods have never been successful, up to now, in identifying any seismic reflector at the Mid-Atlantic Ridge axis, that may be assimilated to a magma chamber roof. This constatation is in agreement with the conclusions of the specialists who valculate the thermal budget of the ocean ridges: below a certain accretio rate, the heat loss by conduction and hydrothermal convection of seawater in the crust is such that it results in the complete

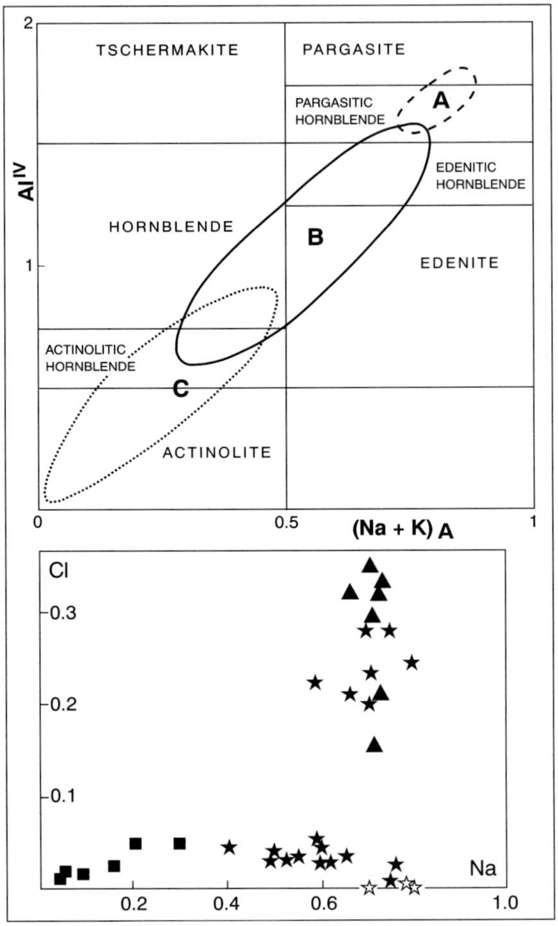

Fig. 5.32. Variations of secondary amphibole compositions in Hole 735B gabbros, plotted in the AlIV versus (Na + K) diagram. (A) Field of brown amphiboles associated with granulated pyroxenes in early shear zones. (B) Field of green amphiboles from shear zones and associated syn-kinematic veins. (C) Field of amphiboles from post-kinematic veins, or pseudomorphoses of clinopyroxene (static metamorphism) (after Mevel and Cannat, 1991).

consolidation of the magma chambers: these are therefore ephemeral in time, discontinuous in space, and probably are less than 2 km wide.

5.3 CONCLUSIONS ON THE STRUCTURE AND LITHOLOGY OF THE PRESENT-DAY OCEANIC CRUST

The data set out above may be summarized by the following points:

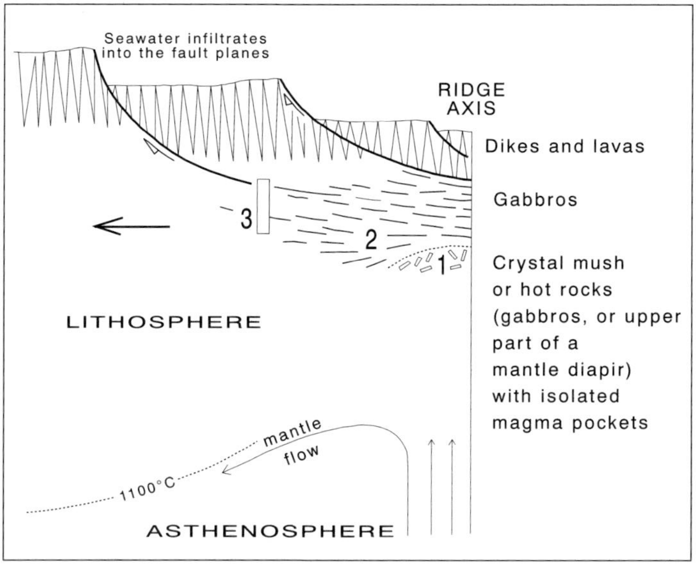

Fig. 5.33. Interpretative schema of the tectonic evolution of Hole 735B gabbros, in the axial zone of the South-West Indian Ridge. Phases 1, 2 and 3 are discussed in the text. The rectangle just to the right of number 3 represents the drilled section (500 m). The lithosphere/asthenosphere boundary corresponds to the 1100°C isotherm (basalt solidus) (after Cannat *et al.*, 1991).

5.3.1 Seismic structure

Seismically, the oceanic crust is thin (6–7 km), of constant thickness, and exhibits a layered structure, controlled essentially by the physical properties of the crust with depth (porosity, thermal state, etc.), much more than by the lithological composition.

The upper crust, essentially volcanic (basaltic flows and dikes), constitutes layer 2. This is subdivided into sublayer 2A (young, fractured and porous crust, with a high velocity gradient), Sublayer 2B (aged and hydrated oceanic crust, where most of the pores are filled), and sublayer 2C (deep upper crust, transition to layer 3).

The lower crust, mainly gabbroic, constitutes layer 3. This one includes sublayers 3A and 3B, whose velocity gradients are much lower (of the order of 0.1 km s^{-1}), reflecting a very low porosity.

There are no continuous planar reflectors at a regional scale in the oceanic crust.

The task of identifying the materials constituting the oceanic crust from geophysical data only is very difficult. The physical properties and thermal state of the crust, as well as its hydrothermal and metamorphic transformations, are parameters perhaps more important than the original magmatic structure and lithology. Taking into consideration the V_p/V_s ratios may certainly help to refine the lithological diagnostic, and to reject some possible hypotheses.

At the approach to the fracture zones, an important decrease in crustal thickness is observed, principally at the expense of layer 3, which may disappear completely.

The depth of the ocean floor increases correlatively. Layer 2, intensely fractured and hydrothermalized, rests directly over a strongly serpentinized upper mantle. The serpentinized zones have a tendency to ascend, but gravimetric data show that this process remains limited.

5.3.2 Lithological structure

The overall sampling of the oceanic crust still remains extremely fragmentary with respect to ist dimensions. The deep levels of the crust, in particular, are widely unknown. Complete sections of the crust remain exceptional.

These necessary reservations being made, two main types of crustal structure clearly appear:

(1) Fast- to intermediate-spreading oceanic ridges show a *continuous crustal tissue*, characterized by a thin volcanic layer (several hundreds of metres), a robust sheeted dike complex (>1 km), and a several-km-thick gabbroic layer, resting over a *harzburgitic* upper mantle. The seismic Moho, in this case, corresponds to the transition between gabbros and mantle peridotites (whose upper level is represented by impregnated dunites, cf. Hess Deep and Pito Deep). This crustal structure is explained by the vigorous and sustained magmatic activity along fast-speading ridges.

(2) Slow-spreading oceanic ridges (<6 cm/yr) show a *discontinuous crustal tissue*. An intermittent magmatic activity, alternating with an intense tectonic activity, often results in the uprise of the serpentinized upper mantle to the ocean floor, even away from fracture zones. A true, continuus gabbroic layer probably does not exist, and is rather represented by multiple, km-sized magmatic bodies intruding into the upper mantle. The sheeted dike complex, though observed in places (Vema FZ), is generally lacking. The residual upper mantle is *lherzolitic*, or quite lherzolitic. In this case, the significance of the seismic Moho becomes problematic: we discuss this point in the next chapter (cf. Figure 6.18).

Whatever the spreading rate, individual accretion segments of the Ridge exhibit a common along-strike structural organization. The crust is thicker in the central zone of the segment, and becomes progressively thinner toward the tips, whereas the ocean floor deepens.

6

Structure and lithology of fossil oceanic crust in ophiolites

Ophiolites are *fragments of fossil oceanic lithosphere*, which are found in most mountain ranges of the Earth, either in continents or in island arcs. Their ages are extremely variable, ranging from more than two billion years for the most ancient ones in the northern part of the Quebec province (Canada), to hardly 2–3 Myr for the youngest ophiolites, in the Taitao peninsula (southern Chile).

Collision ranges such as the Appalachians, the Urals or the huge alpine orogen *sensu lato* are particularly rich in ophiolite massifs, distributed along sinuous and narrow belts over thousands of km.

6.1 OPHIOLITES, MARKERS OF OCEAN CLOSURES

What is the geodynamic significance of these *ophiolite belts*?

As we have seen in Chapter 1, the normal fate of oceanic lithosphere is to return sooner or later to the deep mantle, through the subduction zones. Only in particular geotectonic circumstances can oceanic lithosphere fragments escape their destiny and be thrusted onto a continental or island-arc margin, being sometimes uplifted at several km in altitude in a mountain range.

These particular circumstances generally imply the progressive compression and closure of an ocean basin, followed by a stage of detachment and overthrusting of ocean basement slices (ophiolite nappes), either onto a passive continental margin, or in an active margin (Figure 6.1A). In this latter case, the overthrusting is provoked by the blocking of oceanic subduction, because of the arrival of a continental block in the subduction zone. If the overriding plate is itself oceanic, the result will be that a piece of oceanic lithosphere will be tectonically thrust over the continental margin of the subducted plate: this process is called *obduction* (Coleman, 1971). A variety of other cases are possible (Figure 6.1B), but the main fact is that the tectonic emplacement of ophiolites always indicates the disappearance of an oceanic domain

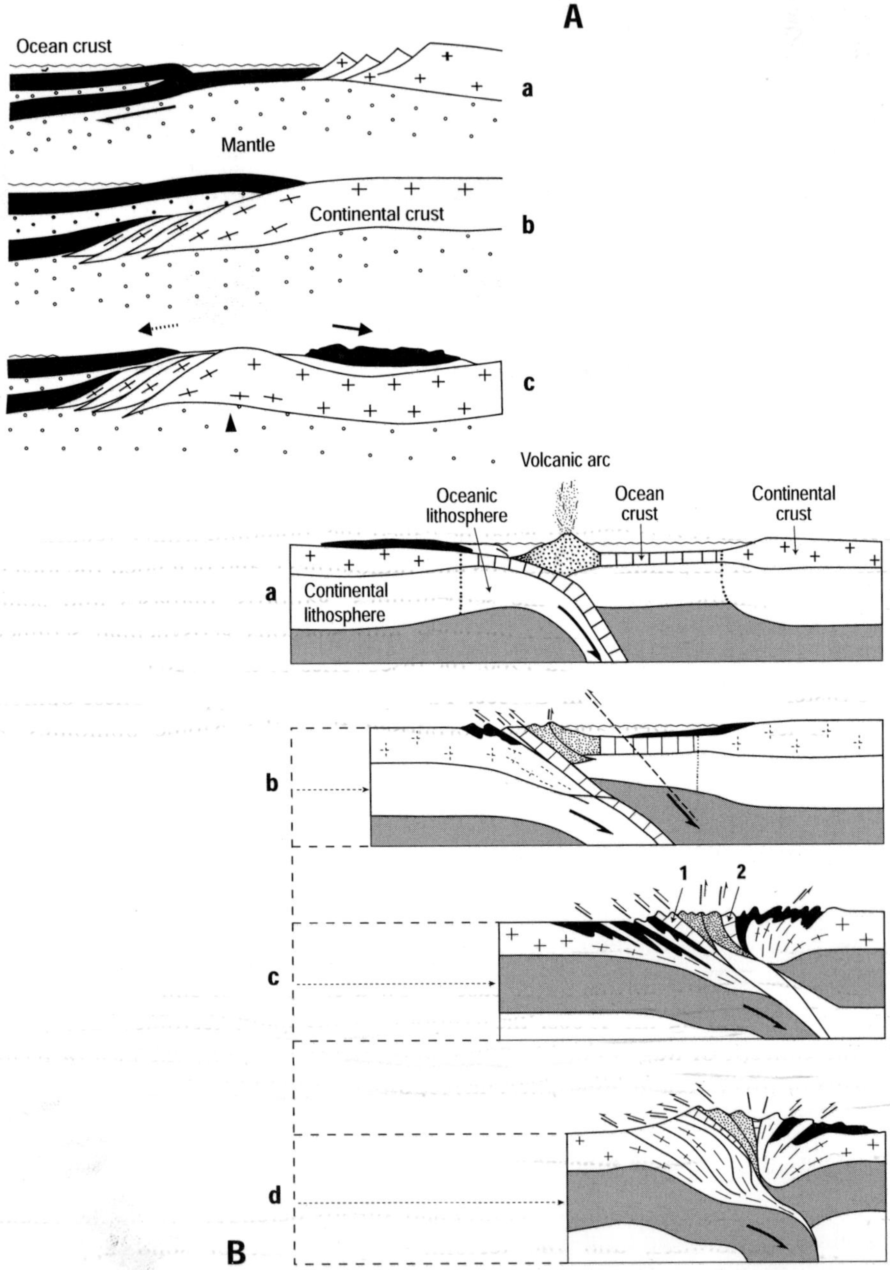

Fig. 6.1. (A) Obduction over a passive margin (Tethyan ophiolites): a, stage of intra-oceanic detachment and tectonic slicings; b, stage of obduction over the continental margin; c, gravity slidings after isostatic rebound (after Nicolas, 1989). (B) Obduction over an active continental margin, by blocking a subduction zone: a, an oceanic basin is closing up by subduction, ahead of an island arc; b, c and d, successive stages of the continental collision, resulting in the formation of two ophiolite belts, corresponding to the oceanic domain and to the back-arc basin respectively (1 and 2 on schema c) (after Mattauer, 1985).

in a context of plate convergence: *ophiolite belts are the markers of ocean closings, and indicate the emplacement of disappeared oceanic basins.*

6.2 HISTORICAL RECORD

6.2.1 From Steinmann's laccolites to the huge submarine flows

Ophiolites or 'snake stones' are well known from terrestrial geologists, since the term was apparently created by the French geologist Brongniart in 1813, to designate a serpentine matrix containing various minerals, often associated with volcanic rocks, gabbros and red cherts.

During the whole 19th century and the beginning of the 20th century, the term 'ophiolite' has designated an association of serpentinites, gabbros and spilites, with or without associated cherts or radiolarites, mainly described in the Ligurian Apennine (Italy) and in the French–Italian western Alps. These ophiolite outcrops were faulted, folded and metamorphosed, and not easy to interpret. In a famous paper, Steinmann (1927) defined what he called the 'ophiolite trinity' (consisting in an association of serpentinites, gabbros and radiolarites), and defended the idea of a cogenetic magmatic series for the serpentinites, gabbros, diabases and spilites, differentiated in huge 'laccolites', intruded into siliceous geosynclinal sediments.

Then came, between 1930 and 1960, the discoveries of the large ophiolite massifs of the eastern Mediterranean in Greece, Turkey, Syria and Cyprus. These ophiolites are much less tectonized and metamorphosed than the Alpine ophiolites, and allowed a more detailed study of the internal structures of the ophiolitic assemblage, particularly the spectacular magmatic layerings visible in the gabbro and ultramafic rocks. The model of *huge submarine flows* became very popular (at least in Europe: Routhier, 1946, Dubertret 1953, Brunn, 1960; Figure 6.2), and was adopted by the promoters of the geosynclinal theory (Aubouin, 1965).

At that time, the predominant ultramafic part of the assemblage was considered as an internal magmatic differentiation in an intrusive (Steinmann) or extrusive (Brunn) magma body; in this latter case below a carapace of chilled basaltic lavas (pillow lavas). During the 1960s, the irruption of the plate tectonics theory swept away the concept of huge ophiolite flows, and imposed rapidly the idea of *tectonic fragments of fossil oceanic lithosphere* incorporated in the mountain ranges.

6.2.2 Ophiolite = oceanic lithosphere

In a precursory paper, De Roever (1957) had already defended the mantle origin of 'alpinotype' peridotites, and the tectonic emplacement of solid upper-mantle fragments. At the end of the 1960s, several detailed studies on the ophiolites of Greece, Cyprus and Turkey concluded that tectonic emplacement of oceanic lithosphere fragments was caused by two radically different sequences: a lower *mantle sequence* made of upper-mantle peridotites, characterized by high-temperature plastic deformations (the *tectonites*), and an upper *crustal sequence*, made of

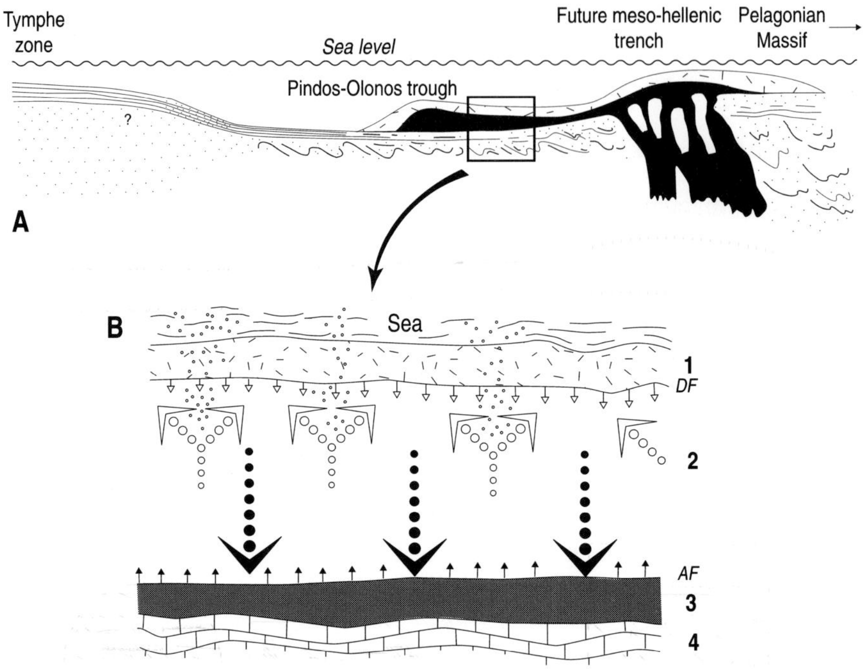

Fig. 6.2. The ophiolitic model of huge submarine outpourings. This model, very popular in Europe before the emergence of plate tectonics theory, postulated that ophiolites were the result of the emplacement of huge magmatic outpourings on the seafloor, delivered through major tears that develop on the flanks of 'geosyncline' basins. (A) Hypothetical section of the Pindos ophiolitic extrusion in Greece. (B) Solidification process of an ophiolitic extrusion. 1. Doleritic and microlitic crust (DF: descending front of solidification). 2. Magma body submitted to magmatic differentiation (soluble and volatile products are ascending, whereas cumulus crystals are accumulating on the floor). 3. Layered mafic and ultramafic cumulates (AF: ascending front of solidification). 4. Sedimentary substratum (after Brunn, 1960).

weakly deformed mafic magmatic rock, including a cumulative gabbroic series (the *cumulates*).

At the beginning of the 1970s, Coleman introduced the term *obduction* to describe specifically the tectonic thrusting of oceanic lithosphere over a continental margin, whereas Dewey and Bird (1970), in a series of excellent papers, explained the formation of mountain ranges and the existence of ophiolitic sutures within the new frame of plate tectonics.

6.2.3 The 'ophiolite manifesto' (1972)

Deep divergences still existed between European geologists, familiar for a long time with the notion of ophiolitic assemblage, and American geologists, who were still mapping separately the mafic/ultramafic coarse-grained assemblages and the submarine volcanic series (pillow lavas), even when they were closely associated in the field. A special Penrose Conference meeting dedicated to the definition of ophiolites

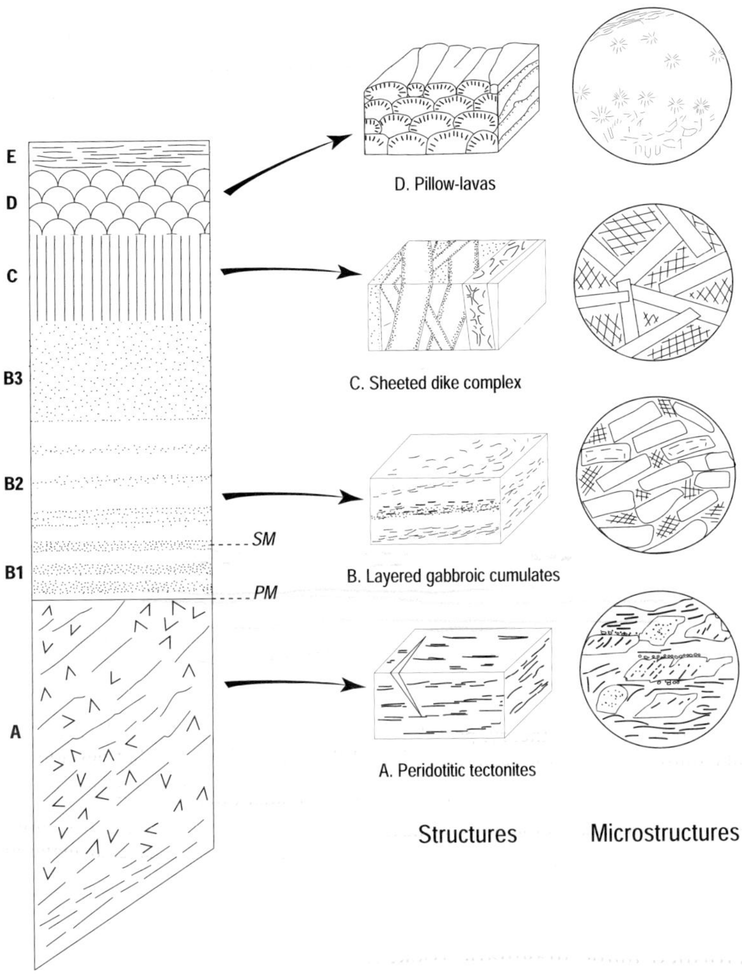

Fig. 6.3. The standard 'ophiolitic log', as described in the 'ophiolite manifesto' (Penrose Conference, 1972). A: Mantle residual peridotites (tectonites). B1: layered ultramafic cumulates. B2: layered gabbroic cumulates. B3: isotropic gabbros. C: sheeted dike complex (100% diabase dikes). D: basaltic flows, mainly pillow lava. E: pelagic sediments. PM: 'petrological Moho'. SM: 'seismic Moho' (after Caron *et al.*, 1989).

was held in 1972, at the end of which a common definition of the term ophiolite was proposed as follows (Anonymous, 1972):

> The term *ophiolite* is employed by the members of the present Conference to design a peculiar assemblage of mafic/ultramafic rocks. It should not be used on maps as a rock name or as a lithological unit. In a complete ophiolite assemblage, the type-facies are met in the following order, from bottom to top (Figure 6.3):
>
> —The *ultramafic complex*, consisting in harzburgites, lherzolites, dunites in variable proportions, generally with a tectonic fabric.

—The *gabbroic complex*, more often with cumulate textures, including commonly peridotite and pyroxenite cumulates, and generally less deformed than the ultramafic complex.
—The mafic *sheeted dike complex*.
—The basaltic *volcanic complex* generally made of pillow lavas.

Other facies commonly associated are:

—An overlying sedimentary series, including typically cherts, pelagic limestones and thin interbedded clays;
—Concentrations of podiform chromites, generally associated with dunites;
—Intrusive and extrusive Na-rich plagiogranites.

The authors of this definition added the important following remarks: (a) faulted contacts between mappable units are frequent, and complete sections may be missing; (b) an ophiolite assemblage may be incomplete, dismembered or metamorphosed, and in this case should be called 'incomplete, dismembered or metamorphosed ophiolite assemblage'; (c) although ophiolites are generally interpreted as fragments of oceanic lithosphere (oceanic crust + upper mantle), the use of the term ophiolite should be independent of its supposed origin.

This definition has survived rather well with time, and remains valuable today in its overall text, despite the fact that over the two following decades, the wide variety and diversity of origin of ophiolites has become conspicuous.

6.2.4 Ophiolites and their diversity of origin

Only one year after the 'ophiolite manifesto', the problem of the original geodynamic setting of ophiolites was raised by the Japanese geochemist Miyashiro (1973), who questioned the origin of the Troodos ophiolite in Cyprus, using very simple diagrams based on correlations between some major and trace elements (such as Si, Fe, Mg, Ti), and claimed that the Troodos lavas present all the characteristics of an arc volcanism, and have nothing to do with MORB-type basalts (mid-ocean ridge basalts). Other geochemists confirmed this diagnostic later on, and extended it to other ophiolites, interpreted as having formed in a 'supra-subduction' context (Pearce, 1980; Pearce *et al*., 1984).

We know today that a number of back-arc ocean basins are submitted to vigorous seafloor spreading processes (see Chapter 3), and comprise true oceanic ridges, mapped sometimes over thousands of km in length. We also know that marginal and inter-arc basins, and even fore-arc basins, may exhibit ocean-spreading centres, and therefore produce new oceanic lithosphere. The fate of these basins is to be tectonically welded, sooner or later, to the neighbouring continental margins. It is then reasonable to think that in the past a number of ophiolites obducted on the continental margins were issued from marginal or back-arc basins, and not from wide open oceans like the Atlantic or Pacific Oceans. When fossil magmatic arcs are thrust together with ophiolites having the same age over a continental margin, as in California for instance, the 'supra-subduction' origin of the ophiolites is quite clear.

6.3 CRUSTAL STRUCTURE OF OPHIOLITES: THE HOT AND LOT TYPES

There are about 150 ophiolite complexes of all ages presently known, among which about 40 have been mapped and studied with some detail (Nicolas, 1989). These ophiolites bear precious witness to the seafloor basement of the oceans of the past, and offer the possibility of studying in detail the structure and composition of the oceanic crust and associated upper oceanic mantle, along more or less continuous sections on the field.

The lithological structure of ophiolites, based on these forty reasonably known ophiolites, may be classified into two main types, identified by their *crustal structure* and by the *petrographic nature of their residual mantle peridotites* (Figure 6.4):

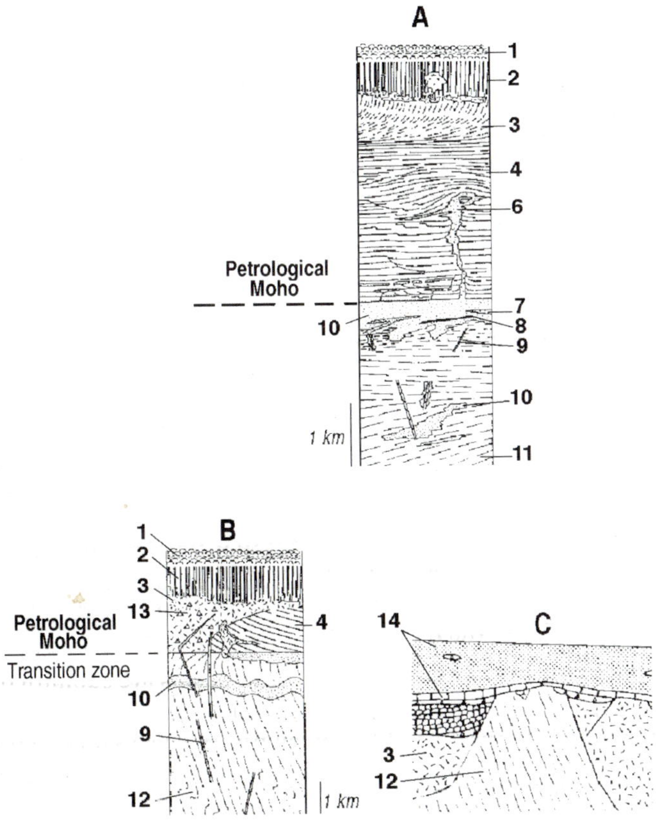

Fig. 6.4. HOT- and LOT-type ophiolites. (A) Schematic log of the Oman ophiolite (harzburgitic type, with thick and continuous crust). (B) Schematic log of the Trinity ophiolite (lherzolitic type, with thin and discontinuous crust). (C) Schematic structure of the Ligurian ophiolites (lherzolitic type, with very discontinuous crust). 1. Basaltic pillows. 2. Sheeted dike complex. 3. High-level gabbros (isotropic or foliated). 4. Layered gabbros. 5. Plagiogranites. 6. Wehrlitic intrusions. 7. Gabbroic sills. 8. Podiform chromites. 9. Gabbroic dikes. 10. Residual dunites. 11. Foliated harzburgites. 12. Foliated lherzolites (adapted from Boudier and Nicolas, 1985, Nicolas, 1989 and Lagabrielle, 1987).

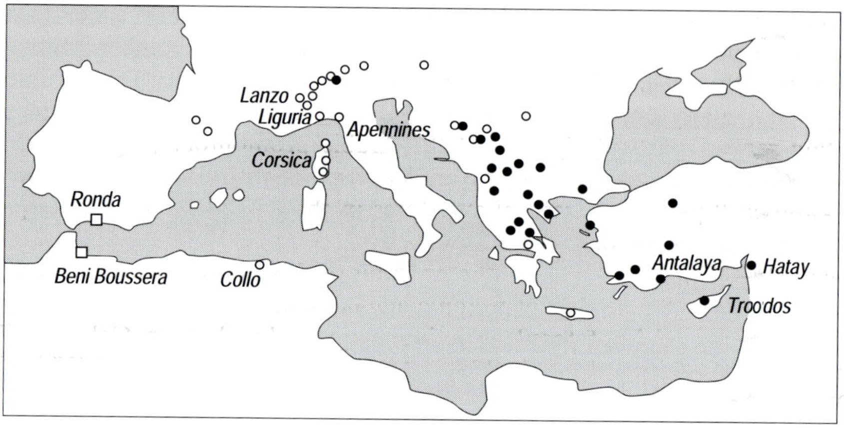

Fig. 6.5. Simplified map of the peri-Mediterranean domain, showing the distribution of the lherzolitic massifs (squares), LOT-type ophiolites (empty circles) and HOT-type ophiolites (black circles). This distribution suggests the existence of an eastern province, rich in harzburgitic ophiolites, and a western province, rich in lherzolitic ophiolites (after Nicolas and Jackson, 1972), slightly modified.

—The ophiolites having a thick and continuous crustal section, and a mainly harzburgitic mantle sequence: this is the harzburgitic type, or HOT (for Harzburgitic Ophiolite Type).
—The ophiolites having a reduced, eventually discontinuous crustal section, and a mainly lherzolitic mantle sequence: this is the lherzolitic type, or LOT (for Lherzolitic Ophiolite Type).

This fundamental distinction was proposed for the first time by Nicolas and Jackson (1972) for the peri-Mediterranean ophiolites, which can be subdivided into an eastern province with harzburgitic ophiolites, and a western province with lherzolitic ophiolites (Figure 6.5). The transition between both types occurs in the Dinarides (Albania, Serbia), where they form two parallel belts.

Actually there is a large variety of ophiolite types, intermediate between the end-member types, represented for instance by the Oman ophiolite (pure and typical harzburgitic type), and by the Ligurian ophiolites (pure and typical lherzolitic type). Both end-member types are described here, which will aid understanding the differences between the Pacific-type (fast-speading ridges) and the Atlantic-type (slow-spreading ridges) oceanic crust.

See Nicolas (1989) for more complete structural descriptions.

6.4 OPHIOLITES WITH THICK AND CONTINUOUS CRUSTAL SEQUENCE, AND HARZBURGITIC MANTLE: THE OMAN OPHIOLITE

The Oman ophiolite, or Semail nappe, is a fragment of the Cretaceous Tethyan oceanic lithosphere, obducted over the Arabic continental margin at the end of the

Upper Cretaceous. It forms a 500-km-long and 50–100-km-wide crescent (Figure 6.6B, 6C). Due to its geographic location on the eastern flank of the Arabic promontory, this giant slice escaped the Alpine continental collision, which has dismembered many other Tethyan ophiolites outcropping all along the Alpine mountain ranges, particularly around the Arabic promontory (Figure 6.6A). The Oman ophiolite presents not only a complete section of the crust, but also exhibits remarkably well-preserved internal structures in the crustal and mantle sequences, only weakly affected by obduction tectonics.

The reader will find excellent syntheses about the history of research in Oman, the geological context, the models of tectonic emplacement, internal structures of crust and mantle, or models of fossil accretion magma chambers in Coleman (1981), Ceuleneer (1986), Lippard *et al.* (1986), Nicolas (1989), Juteau and Whitechurch (1989).

Here only the lithological and petrological aspects of the Oman ophiolite are described, likely to aid understanding the lithological structure of the crust in present-day oceans. We begin with the description of the various lithological units of the Semail nappe, as they appear vertically in the field (this is what we call the *ophiolite log*), and we go on with the spatial and cartographic organization of the ophiolite, particularly along the strike of the nappe, parallel to the paleo-ridge axis (this is what we call the *axial segmentation*).

6.4.1 The synthetic Oman ophiolite log

In spite of local complications, the general description of the Oman ophiolite can be organized around the synthetic log of Figure 6.4A: it comprises a *crustal section*, several km thick everywhere, with an average of about 5 km, lying over a *mantle section*, mainly made of harzburgites. The transition zone between crust and mantle constitutes the 'petrological Moho'.

The crustal sequence

The volcanic unit and associated sediments (Figure 6.7)

The uppermost units of the Oman ophiolite are volcanic flows, organized in several units. At the base, unit V1 (400–1600 m thick, named *Geotimes* by the Open University team), grades downward into the sheeted dike complex. Fe–Mn-rich metalliferous sediments, composed of mixed Radiolarian cherts and hydrothermal components, and called 'umbers' because of their dark color, are locally interbedded as sedimentary lenses between the pillow flows. They were dated as Cenomanian by their Radiolarian microfaunas. V1 lavas are overlain by V2 units (the English *Lasail, Alley* and *clinopyroxene units*), themselves overlain by 50 m of micritic limestones and Radiolarian cherts, which gave a Campanian age. These sediments are overlain in places by huge, massive V3 flows (*Salahi unit*), showing spectacular columnar jointings.

V1 and V2 units, with their umbers and their massive sulphide deposits, are believed to represent the volcanic activity at an oceanic spreading center.

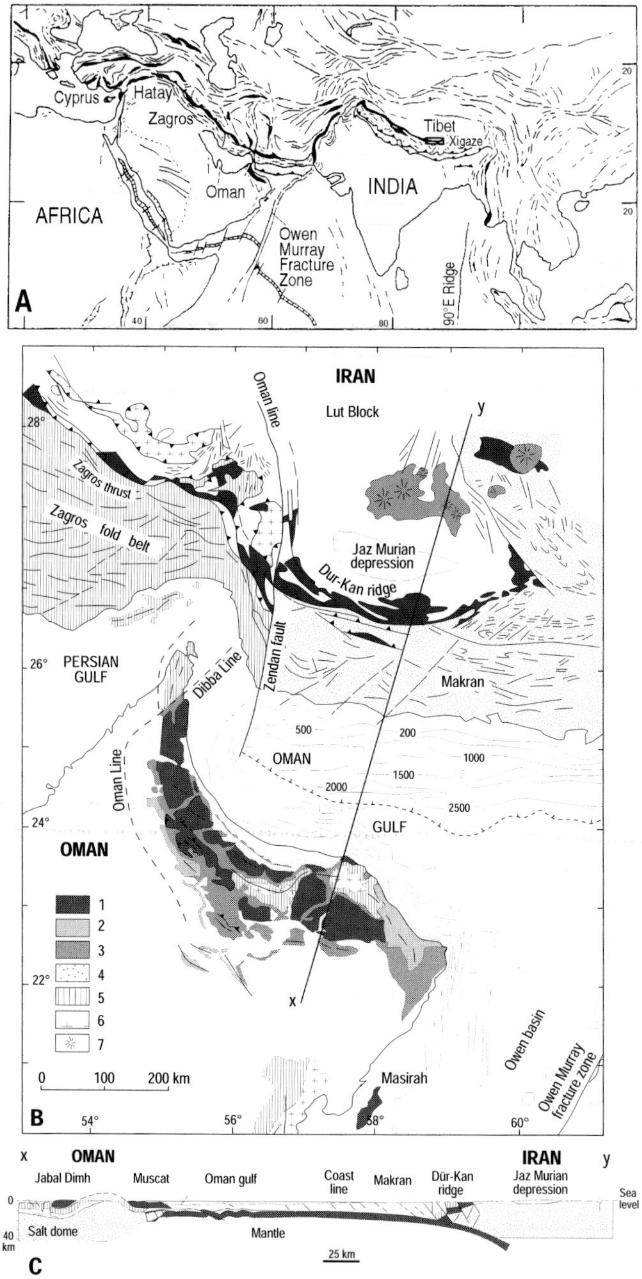

Fig. 6.6. (A) Emplacement of the Oman ophiolite in the whole Tethyan ophiolite belts of Asia and the Middle-East (after Gansser, 1966 and Coleman, 1981). (B) Geological and tectonic map of the Oman sultanate and southern Iran, showing the Oman and Iran ophiolites in their present-day context (after Coleman, 1981). (C) Interpretative geological cross-section through the Gulf of Oman, following the AB line on map B (after Coleman, 1981). 1. Ophiolites. 2. Shelf limestones. 3. Deep marine sediments. 4. Flysch deposits. 5. Shelf deposits. 6. Continental basement. 7. Arc volcanism (Iran).

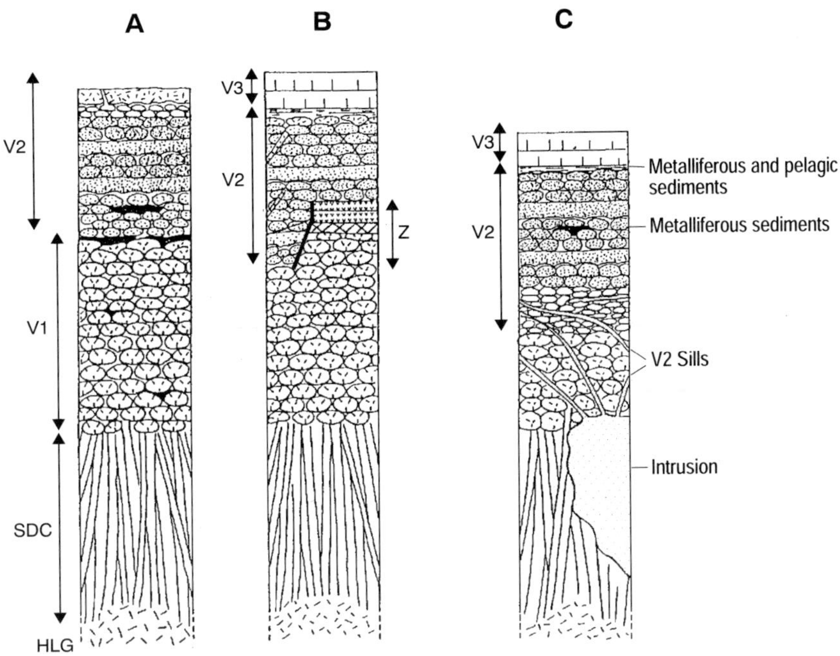

Fig. 6.7. Synthetic section of the volcanic units of the Oman ophiolite, in three different sections of the Salahi block (northern part of the ophiolite). GS: high-level gabbros. SDC: sheeted dike complex. V1 (or Geotimes unit): lower volcanic unit (MORB-type basalts, mainly pillow lavas). V2 (Alley, Lasail and Cpx units): upper volcanic unit (calc-alkaline basalts, picrites, boninites, cpx- and plagioclase-bearing basalts, pillows, massive flows, sills and dikes). V3 (or Salahi unit): episodic uppermost volcanic unit (mainly around wadi Salahi, transitional to alkaline basalts, mainly thick massive flows). In black: lens-shaped, Fe–Mn-rich metalliferous sediments (or 'umbers'), mostly at the V1/V2 interface. Z: Zuha oxidized polymetallic massive sulphide deposit (gossan), a fossil 'black smoker' field developed on the margin of a small tectonic graben (after Pflumio, 1988).

The V1 volcanic unit is mainly composed of pillow-lava flows, whose spectacular outcrops along Wadi Jizi were admired by the first observers (and made the cover of the *Geotimes* journal, hence the former name of the V1 unit): splendid tubular pillow lavas can be observed along steep natural volcanic slopes, forming cliffs more than 100 m high. The most delicate pillow structures are perfectly preserved. Massive flows and sills are frequent in the V1 unit. The V1 unit geochemistry, very close to that of modern MORBs, confirms their genetic links with the sheeted dike diabases and the plutonic gabbro sequence, and their belonging to the main oceanic accretion stage.

V2 lavas form localized volcanic edifices, and could represent submarine seamounts, emplaced on the seafloor immediately after the V1 episode (absence of sedimentary deposits between both units). These lavas are more varied (picritic basalts, plagioclase- and cpx-phyric basalts, andesites, dacites), and their geochemistry is closer to that of modern arc or back-arc lavas (Alabaster *et al.*, 1982). They are genetically linked to wherlitic intrusions of the plutonic sequence (see below).

The V3 volcanism, also very localized, has clear alkaline affinities, and was emplaced about 15–20 Myr after the end of the accretion stage, during the obduction process.

The sheeted dike complex (1000–1700 m thick)

A splendid sheeted dike complex, 1500 m thick on average, appears continuously all along the Oman ophiolite over 500 km in length, beneath the V1 volcanic unit. The contact between the lower V1 lavas and the dikes of the sheeted dike complex is rather sharp, along several tens of metres at most. These dikes are 0.5 to 2.5 m wide (1.0 m on average), and intrude into each other, showing one or two chilled margins. On the outcrops, they appear as 'packs' of 10 to 20 dikes intruded together. They are all parallel to each other, or intersect each other at low angles (<20°). They may split or point upwards with bevelled edges. In specific areas, two directions of dikes are visible, generally where evidence of axial discontinuities is demonstrated. These dikes are tholeiitic and deeply recrystallized in the greenschist facies, due to hydrothermal circulations. Late picritic, plagioclase-phyric or andesitic dikes linked to the V2 volcanic episode are observed.

High level 'isotropic' and foliated gabbros

At the base of the sheeted dike complex, 'screens' of cpx-amphibole-bearing gabbros appear between the diabase dikes.The transition zone coincides more or less with the horizon of intense hydrated recrystallization of foliated/layered gabbros into isotropic gabbros. These show large variations of grain size, probably linked to the circulation of hydrothermal fluids (Figure 6.8A), with development of secondary green amphiboles in more dioritic facies, or to local hydrated anatexis processes, responsible for the injection of hydrated, plagiogranitic magmas.

Isotropic gabbros and associated plagiogranites exhibit mutually intrusive relationships with diabase dikes, although they most often intrude the base of the sheeted dike complex. Magmatic breccias are common at that level.

The high-level gabbros have a more homogeneous grain size. These are cpx-bearing, locally noritic, 'dry' gabbros, or amphibole-bearing gabbros. Black amphibole, crystallized during the magmatic flow, is commonly foliated and lineated, whereas a later prismatic green and brilliant amphibole is more 'isotropically' oriented. Olivine-rich layers are scarce.

Close to the transition with the sheeted dike complex, the layering and the magmatic foliation tend to disappear, whereas lower in the section they are narrowly associated. The rotation of the magmatic foliation (and of the associated lineation) parallel to the dikes occurs at about 500 to 1000 m below the base of the sheeted dike complex.

The layered gabbros

This several-km-thick unit is characterized by its spectacular magmatic layering, which sometimes on the field mimics the layering of stratified sedimentary rocks

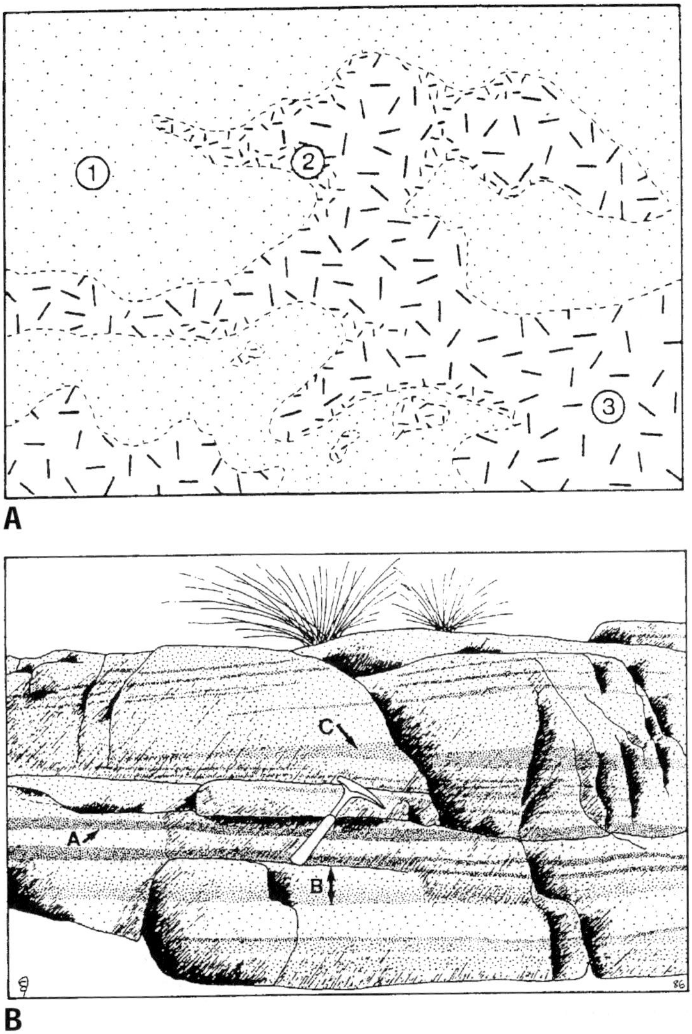

Fig. 6.8. (A) Aspect of the high-level gabbros on the outcrop, some tens of metres below the base of the sheeted dike complex. They are characterized by a very irregular grain ize, with numerous diffuse 'pockets' of pegmatitic gabbro. 1. Fine-grained gabbro. 2. Coarse-grained, amphibole-bearing gabbro. 3. Amphibole (epidote, sulphide)-bearing pegmatitic gabbro. Drawing from Ernewein (1987). (B) Aspect of the layered gabbros on the outcrop (wadi Hilti), showing isomodal layers, either with a constant thickness (A) or lens-shaped (C), and modal graded layers (B, positive sequence). Drawing from Ernewein (1987).

(Figure 6.8B). This layering appears on the outcrops as a combination of planar layers. Each layer is an elementary entity, characterized by its internal features (mineral composition, texture, grain size and modal variations, etc.), its thickness, its geometry and extension, and by the nature of its contacts with the adjacent layers.

The thickness of the layers is variable, ranging from several cm to several metres. The lateral extension of the layers is weak. The layers are commonly lens-shaped, showing lobated or bevel-edged endings. A kilometric extension of huge, massive gabbro layers remains hypothetic.

In Oman, as in most HOT ophiolite complexes, it is not possible to correlate the stratigraphic columns in the layered gabbros, even between closely spaced serial sections. This is a main difference with the Precambrian continental stratiform complexes: in the Bushweld complex for instance (South Africa), the famous 'Merensky Reef' horizon has a proven extension of more than 300 km. On the outcrop, the contacts between layers are marked either by modal proportion variations of the mineral phases (*modal contact*, the most common case), or by the appearance or disappearance of a major phase (*phase contact*), or else by a grain-size variation (*morphological contact*, relatively rare). The contact may be sharp or gradual.

An *isomodal layer* presents an homogeneous internal structure, without modal variations from bottom to top (Figure 6.8B, layer A). On the contrary, a gradual modal layer exhibits a progressive and continuous variation in the modal proportions between major mineral phases, commonly with the dense phases accumulated at the bottom (chrome spinel, olivine, clinopyroxene), and the less dense phases concentrated at the top of the layer (plagioclase). In this case it is a *positive sequence* (Figure 6.8B, layer B).

In the Oman ophiolite, the gradually modal layers are frequent in the lower part of the layered cumulates, associated with features of dynamic gravity settling (oblique stratifications, channeling, load marks, etc.). Isomodal layers are preponderant in the upper part of the layered-gabbro sequence, linked to viscous flow features (magmatic folds). See Chapter 8, § 3.4.

At the base of the layered-gabbro sequence, the layering is always parallel to the 'petrological Moho'; that is, to the contact between the gabbros and the residual mantle peridotites, and with the internal metamorphic foliation expressed in these peridotites. It is generally considered that this contact was initially horizontal (Nicolas, 1989). The internal magmatic layering of the layered-gabbro unit remains parallel to the 'petrological Moho' through most of the unit.

The transition to the high-level gabbros is marked by a progressive disappearance of the magmatic layering, and by a steepening of the magmatic foliation, which becomes finally parallel to the dikes of the sheeted dike complex (Lippard *et al.*, 1986; Nicolas, 1989). This observation is essential to modelize the geometry of the accretion magma chambers (cf. Chapter 8).

Kelemen *et al.* (1997) have recently emphasized the geochemical contrast between the REE-poor (rare-earth elements) cumulate-layered gabbros, and the high-level gabbros, which have REE abundance and slope similar to those of the sheeted dikes and lavas, and represent true liquid compositions.

Magmatic intrusions

The wehrlitic intrusive series and its differentiations. The crustal sequence of the

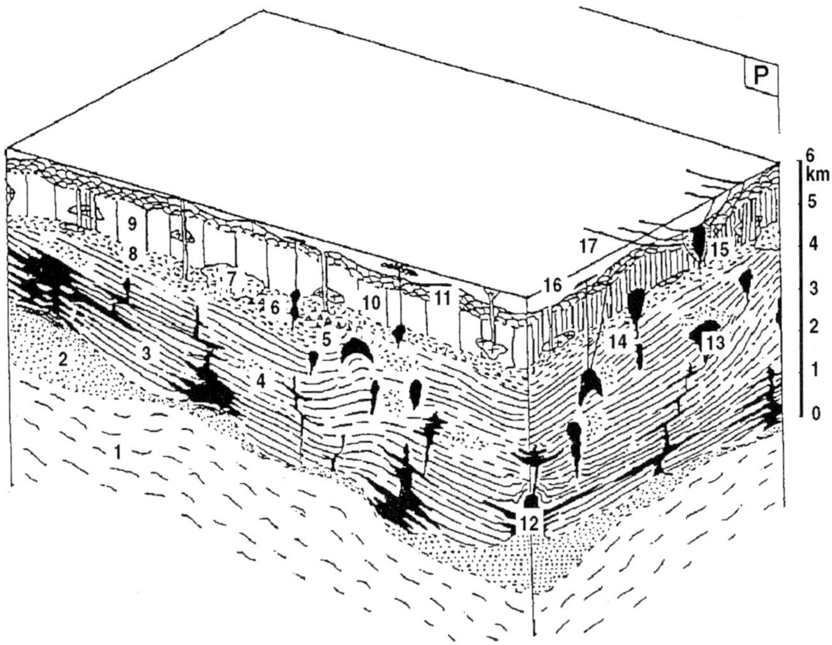

Fig. 6.9. Crustal structure and duality of magmatism in the Oman ophiolite. Block-diagram summarising the crustal structure of the Oman ophiolite and the relationships between the wherlitic intrusions and the gabbroic units. P: axial plane of the palaeo-ridge. 1. Foliated mantle harzburgites. 2. Residual dunites and dunitic cumulates. 3. Layered gabbros. 4. Recrystallized isotropic gabbros. 5. Laminated gabbronorites. 6. Magmatic breccias. 7. Intrusive isotropic gabbros. 8. High-level gabbros. 9. Sheeted dike complex (100% diabases). 10. Plagiogranites. 11. V1 volcanic unit. 12. Sills and laccolites of intrusive dunites-wehrlites. 13. Intrusive wehrlitic bodies. 14. Picritic dikes. 15. Differentiated intrusive bodies (wehrlites-gabbros-plagiogranites). 16. V2 volcanism. 17. Picritic basalts (after Juteau *et al.*, 1988a).

Oman ophiolite is intruded by a great number of *wehrlitic ultramafic bodies* (Juteau *et al.*, 1988a,b; Ernewein *et al.*, 1988; Benn *et al.*, 1988). Rooted downward in the residual dunites at the mantle-crust transition, these intrusions rise up to the dike complex (Figure 6.9). Dikes issued from these wehrlitic intrusions have obviously fed a part of the V2 volcanism (picritic basalts). In the upper levels of the plutonic crust, the wehrlites may differentiate and evolve to gabbros, diorites and plagiogranites (*late intrusive series* of the English authors, Lippard *et al.*, 1986).

The viscous deformations of the gabbros on contact with the intrusive wehrlites, and the systematic lobate contact between both rock types, without any chilled margin (Figure 6.10), indicate that the wehrlites and their differentiates were emplaced in the still-hot and not totally consolidated gabbroic crust. The source of these intrusions is located in the dunitic transition zone, just below the Moho, and have never been observed down in the harzburgitic sequence.

Most authors think that the ultramafic magma was intruded inside the accretion magma chamber or in its immediate vicinity. Some of them think that wehrlites represent a normal by-product of fast-spreading ridge activity (Benn *et al.*, 1988),

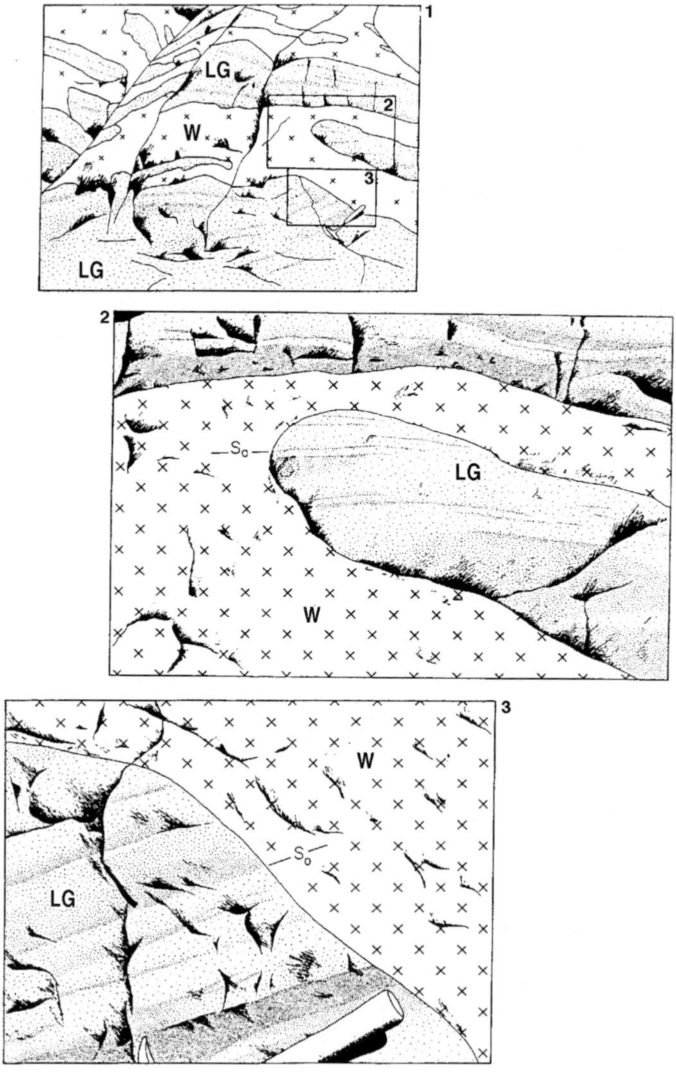

Fig. 6.10. 1. Aspect of wehrlitic intrusions on the outcrop. W: wehrlites. LG: layered gabbros. 2. Detail showing the characteristic lobate contacts between both rock types. 3. Detail showing the posteriority of the wehrlites (after Ernewein, 1987).

whereas others see in these intrusions the effects of the first stages of intra-oceanic slicings, with expulsion of the crystal mush accumulated at the mantle-crust transition (Ernewein *et al.*, 1988).

Such wehrlitic intrusions were described later on in a number of ophiolites. Therefore these intrusions are not a simple peculiarity of the Oman ophiolite, but seem to be an important feature of the HOT ophiolites.

The plagiogranite series. This intrusive suite is mainly located in the upper crust. It consists in veins, dikes and segregations invading the high-level gabbros and the base of the sheeted dike complex, and in small intrusions and stocks (hecto- to kilometric in size) in these same units, and up to the lava units (Lippard *et al.*, 1986; Nicolas, 1989). They are often found as magmatic breccias (numerous diabase and high-level gabbro inclusions in a plagiogranite matrix). It is established today that this group has several possible origins. A number of very different processes converge effectively toward the production of these highly differentiated magmas: (1) formation of residual liquids in the gabbroic magma chamber, or in some wehrlitic intrusions; (2) hydrated anatexis of high-level gabbros and diorites; (3) assimilation of hydrated crust at the roof and walls of the gabbroic magma chamber; (4) hydrated melting of amphibolites in the infra-ophiolitic metamorphic soles.

The 'petrological Moho'

Below the lower layered gabbros, the transition to the harzburgitic mantle is generally marked by a dunitic zone showing ample evidence of high-temperature plastic deformations. The thickness of this dunitic transition zone is highly variable, ranging from several metres to several hundreds of metres, with an average around 50 m.

The contact between the lowermost layered gabbro sequence and the dunites of the transition zone is sharp, and constitutes what is called the 'petrological Moho', that is, the mantle–crust transition zone. The expression is not very judicious, since the Mohorovicic discontinuity, or 'Moho'; is by definition a purely geophysical notion based on seismic velocities, which does not coincide always with the mantle–crust boundary established on petrological and structural criteria.

In detail, the transitional dunites include gabbroic sills, dikes and pockets, as well as numerous diffuse zones of 'magmatic impregnation', composed of intergranular plagioclase and diopside crystals, often poikilitic. Many of these impregnations are plastically deformed with the host dunite, but may be locally devoid of any deformation. Stratiform sills of layered gabbro, typically 2–3 m thick, are fed by gabbro dikes, a number of which escape the plastic deformation. It seems therefore that, at that level, magmatic activity and feeding lasted some time after the end of plastic deformation processes in the upper mantle (Nicolas, 1989).

The residual mantle

Below the transitional dunites, the residual mantle peridotites, visible up to 10 km thick, are petrographically very homogeneous. These harzburgites are depleted in orthopyroxene at the top, and are progressively more banded downward, parallel to the metamorphic foliation. Swarms of diabase, gabbro, websterite and dunite dikes (with or without chromite concentrations) intersect these rocks. Gabbro dikes, abundant in the first 2 km below the Moho, grade to cpx-rich olivine websterites in depth. Interstitial plagioclase definitely disappears in these dykes at about 4 km below the Moho.

Approaching the basal tectonic contact of the ophiolite, a 'cooler' lithospheric deformation (about 800°C), often mylonitic, overprints the high-temperature asthenospheric deformation. A new foliation/lineation is developed, parallel to the basal tectonic contact, and to the foliation of the metamorphic rocks of the infra-ophiolitic sole. This deformation is thought to be contemporaneous with the intra-oceanic slicings and with the formation of the metamorphic sole.

The infra-ophiolitic metamorphic sole

Like many other HOT ophiolites, the Oman ophiolite rests over a sole of metamorphic rocks, several hundreds of metres thick. These rocks, strongly tectonized and folded, are made of amphibolites, quartzites and schists. They are interpreted as representing fragments of upper oceanic crust (basalts) and associated sediments (cherts, clays, pelagic limestones, etc.), metamorphosed and sliced during the early intra-oceanic slicing events. On contact with foliated peridotites, garnet amphibolites attest to the high temperatures reached (750–860°C), confirmed by local evidence of hydrated anatexis, producing small quantities of plagiogranitic melts. They grade downward to epidotite amphibolites, then to greenschists, and finally to weakly metamorphic radiolarites and shales from the Hawasina formations. This inverse thermal gradient recorded in the metamorphic sole attests to progressive cooling of the ophiolite nappe while advancing toward the Arabic continental margin. The study of the statistical orientations of the metamorphic lineations in the sole allowed the directions of transport of the ophiolite nappe during this movement to be constructed.

6.4.2 Axial segmentation of the Oman palaeo-ridge: mantle diapirs and crustal segmentation

The description of the ophiolite log does not say much about the spatial organization of the main lithological units. It is necessary to have a look at the cartographic distribution of these units and of their internal structures (foliations, lineations), to have a better understanding of the 3D structure of the palaeo-ridge. Structural maps of the whole Semail nappe by A. Nicolas and his team have brought about a fundamental set of data, particularly:

(1) The existence of asthenospheric *mantle diapirs*, now frozen in the lithospheric mantle sequence. These structures are characterized by clusters of steep flow lines (metamorphic lineations) following the dip of the flow planes (foliations), surrounded by flat, diverging flow lines. These diapiric structures were mapped in at least three areas of the ophiolite, but the most beautiful diapiric structure is without doubt that of Maqsad (Figure 6.11), with a well defined elliptical zone with steep flow planes and lines (8 km in the direction perpendicular to the local sheeted dike complex, and 10 km parallel to the dikes). One part of the structure is hidden beneath the gabbroic blanket, so that the longer axis of the diapir structure is about 20 km long. At about 300 m below the contact with the layered gabbros, the steep flow lines vanish in harzburgites and transitional dunites rich

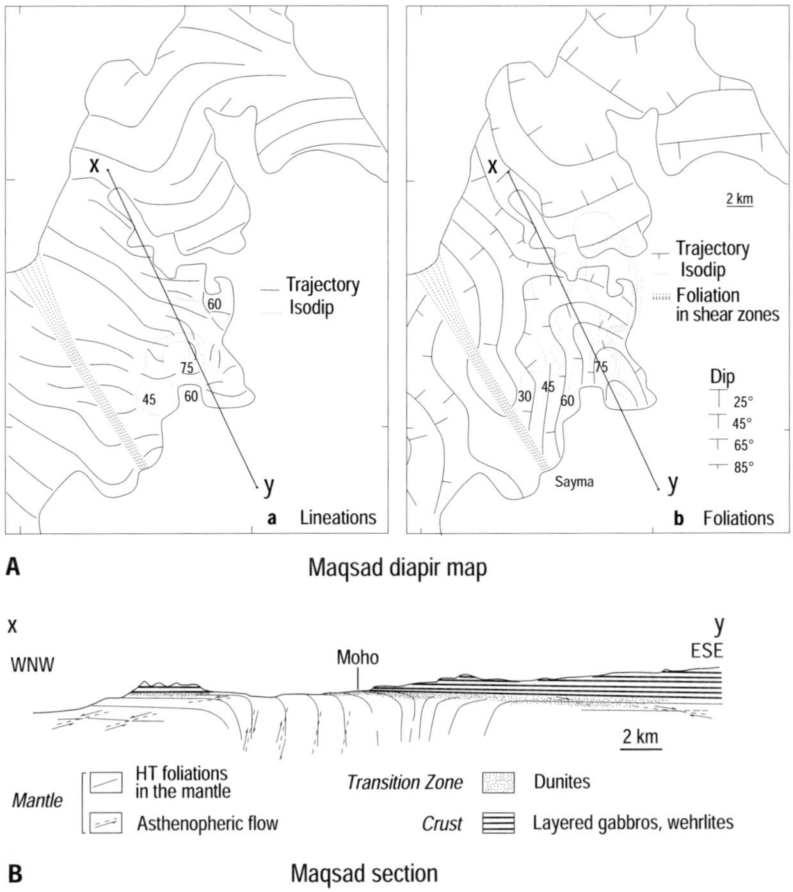

Fig. 6.11. Structure of the Maqsad mantle diapir, in the Samail massif (Oman ophiolite), partially outcropping below the overlying lower layered gabbros (GL). (A) Map of the mineral foliations and associated lineations in the harzburgites, showing characteristic concentric buckles with steep dips for the foliations. (B) Interpretative cross-section following the XY line, showing the rotation of foliations and flow planes at the top of the mantle diapir (after Ceuleneer *et al.*, 1988).

in podiform chromite bodies. Beyond a distance of about 30 km from the centre of the diapiric structure, the flow lines reappear with a flat attitude and a radial disposition with respect to the vertical axis of the diapir.

(2) The ubiquitous presence of a very homogeneous asthenospheric flow, oriented more or less perpendicular to the axial plane of the palaeo-ridge (about 70% of the measurements made in all the ophiolite nappe).
(3) The existence of asthenospheric flows parallel to the axis of the palaeo-ridge (about 15% of the measurements), particularly starting from the mantle diapirs.
(4) The existence of a strong structural continuity on both sides of the petrological Moho, between the high-temperature metamorphic foliations/lineations in the upper harzburgites (plastic deformations and tectonite fabrics), and the mag-

matic foliations/lineations in the overlying layered gabbros (viscous deformations, magmatic fabrics).

In the crustal sequence, T. Juteau's team mapped several areas of crustal discontinuities, interpreted as evidence of axial segmentation of the Oman palaeo-ridge, particularly in the Haylayn massif (Juteau *et al.*, 1988b), and in the Fizh massif (Reuber, 1988). These areas are characterized by the existence of two or several generations of dikes in the sheeted dike complex, differing in their orientation, and by the development of small, ephemeral magma chambers (because they were located at the tip of accretion segments), evolving in closed-fractionation systems, with important development of noritic gabbros and highly evolved ferrogabbros. These areas mark axial discontinuities of the fossil ridge, and hence should alternate along-strike with the mantle diapir areas, supposed to indicate the central part of the accretion segments.

A systematic mapping of these various markers of the Oman palaeo-ridge segmentation is far from being achieved, but one can hope that more detailed mapping efforts will allow a precise spatial representation of the lithological structure of a fast (or intermediate) spreading ridge to be obtained.

6.5 OPHIOLITES WITH A THIN, DISCONTINUOUS CRUST, AND A LHERZOLITIC MANTLE: THE LIGURIAN OPHIOLITES

The Upper Jurassic ophiolites of the Ligurian–Piedmont Basin crop out in the French–Italian Alps, in Corsica and in the northern Apennines in Italy, and constitute an excellent example to represent the other end-member of the various ophiolitic types. In the whole Alpine domain, from eastern Alps to Apennines, the Middle Jurassic period corresponds to the appearance of a Tethyan oceanic floor, consequence of the separation of the European and Apulian continental blocks. These events are contemporaneous with the opening of the central Atlantic Ocean.

Ophiolites of the inner zones of the western Alps belong to two types distributed in two distinct belts (Figure 6.12B): an 'internal' belt, metamorphosed in the eclogitic facies (Mount Viso-Rocciavre-Lanzo-Zermatt), and an 'external' belt, metamorphosed in the blueschist and greenschist facies (Queyras-Maurienne-Combin).

The lithostratigraphy of the blueschist belt is remarkably similar to that of non-metamorphic ophiolites of Liguria, in the Apennine Range (Figure 6.12D). In these ophiolites, pelagic sediments and mafic/ultramafic breccias lay directly over serpentinized mantle peridotites, or over gabbros. These gabbros do not constitute a continuous lithological unit, but rather hectometric/kilometric-sized pockets intrusive inside the serpentinized peridotites. Discontinuous pillow-lava flows, with either MORB or alkaline affinities, are separated in many places from the ultramafic/gabbroic basement by sedimentary ophiolitic breccias, representing ancient rubble talus.

Patient field studies on the least metamorphic massifs, particularly at Mount Geneva, in the Queyras region and in Liguria, in Corsica and in the northern

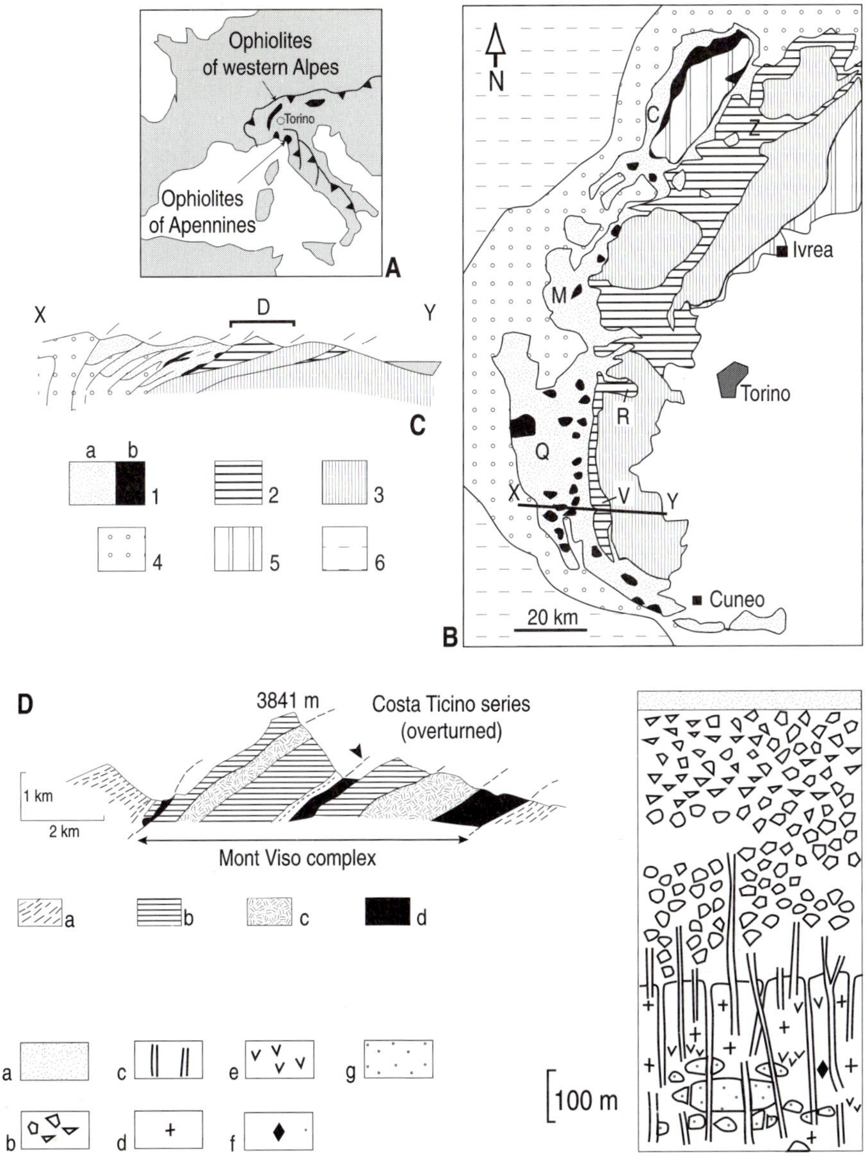

Fig. 6.12. Ophiolites of the western Alps and northern Apennines. (A) General location of the ophiolite massifs. (B) The two ophiolite belts of the inner Alps. 1. The blue schists–green schists-bearing ophiolite belt (1a, metasediments; 1b, ophiolites; Q, Queyras; M, Maurienne;C, Combin). 2. The eclogitic ophiolite belt (V, Mont Viso; R, Rocciavre; Z, Zermatt). 3. Internal crystalline massifs. 4. Briançon units (pre-Alpine continental basement and mesozoic cover). 5. Southern Alps units (continental basement). 6. External units (mesozoic to tertiary sediments). (C) Schematic transverse section following the x–y line on map B. (C) To the left: ophiolites of the internal western Alps (a, calc-schists and metasediments; b, metabasalts; c, metagabbros; d, serpentinites). To the right: ophiolites of the northern Apennines (Costa Ticino). a, Sediments; b, pillow lavas; c, basaltic dikes; d, gabbros; e, Ti-ferrograbbros; f, troctolites; g, serpentinized lherzolites (after Lagabrielle *et al.*, 1989 and Lagabrielle and Cannat, 1990).

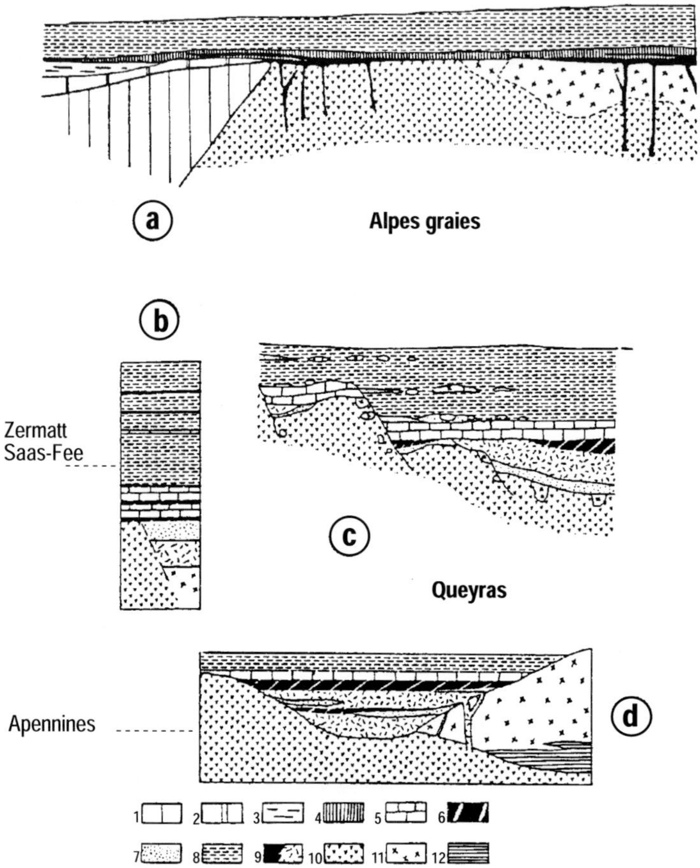

Fig. 6.13. Inferred structure of the oceanic crust in various areas of the Ligurian and Piedmont domain (Alps and Apennines), before the Alpine foldings, according to various authors: (A) according to Elter (1971); (B) according to Bearth, in Laubscher and Bernouilli (1980); (C) according to Lemoine (1980); (D) according to Decandia and Elter (1972); 1, Grand Paradis basement; 2, Triassic; 3, 'Liassic' glassy schist; 4, lower part of true glassy schists; 5, Calpionella-bearing marbles and limestones; 6, quartzites; radiolarites and metaradiolarites; 7, ophiolitic breccias; 8, calc-schists (Lower Cretaceous?), and palombini-bearing clays; 9, diabases, prasinites, pillow lava; 10, serpentinites; 11, gabbros; 12, cumulates (redrawn by Lagabrielle *et al.*, 1984).

Apennines, allowed to reconstruct the heterogeneous structure of the original oceanic floor to be reconstructed (Figure 6.13). Three main stages were distinguished in its pre-alpine oceanic evolution (Lagabrielle, 1987): (a) the constitution of an ultrabasic and gabbroic ocean floor; (b) the tectonization of this basement, with production of detrital sediments, followed or accompanied by the emplacement of tholeiitic basaltic flows; (c) the deposition of radiolarite and limestones of Malm-Berriasian age.

6.5.1 Constitution of an ultrabasic and gabbroic ocean floor

The Alpine geologists did not observe any field ophiolitic section resembling the 'standard' section corresponding to the definition of the Penrose Conference. Instead, they found everywhere a heterogeneous and discontinuous crustal structure, characterized by the absence of a clear sheeted dike complex in the upper crust, and of layered cumulates in the lower crust. Moreover, the basaltic pillow lavas often appear in direct contact with the serpentinized mantle peridotites, or with the products of their submarine dismantling on the ocean floor (Figure 6.14). Detailed studies in the field of the contact between ophiolitic rocks and overlying pelagic sediments indicate that, in most cases, the absence of a crustal plutonic sequence and of a well developed volcanic unit is primary, and not due to Alpine tectonics.

It appears therefore that the ocean floor of the Ligurian Basin was made of a serpentinized peridotite and gabbro basement, these deep rocks being directly recovered by a first sedimentary blanket of serpentine, and gabbro breccias.

The mantle sequence is everywhere composed of plagioclase lherzolite. It is intruded, as is the crustal sequence, by numerous isolated diabase dikes with chilled margins, and by pegmatitic gabbro dikes, and contains numerous small gabbroic intrusions.

Gabbros are everywhere intrusive in the serpentinized peridotites. Besides numerous dikes, they form intrusive massifs, with cumulate textures indicating ancient magma chambers (Figure 6.15). These coarse-grained gabbros (sometimes called '*euphotides*') are issued from the crystallization of a tholeilitic magma, gathered in small magma pockets. In detail, these gabbros, sometimes layered, exhibit a varied mineralogy, with a range of more or less evolved facies, including euphotides, ultramafic cumulates, troctolites and differentiated Fe–Ti-rich gabbros,

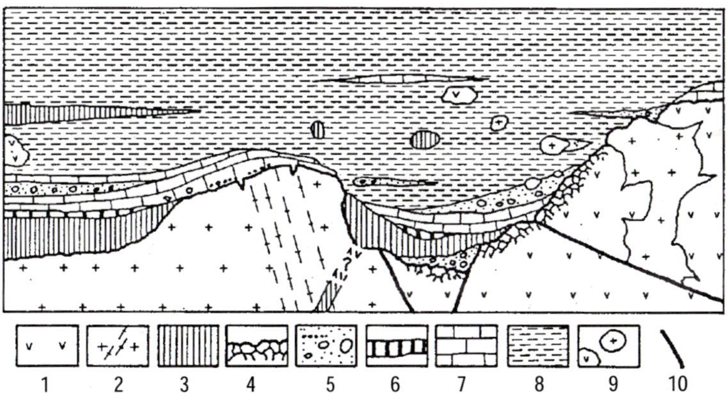

Fig. 6.14. Synthetic schema showing the relationships between the oceanic substratum and the sediments in the Ligurian and Piedmont domain, before the Alpine tectogenesis. 1. Serpentinites. 2. Gabbros and flaser-gabbros. 3. Basalts. 4. Ophicalcites. 5. Ophiolitic breccias. 6. Radiolarites and cherts (future quartzites). 7. Pelagic limestones (future marbles). 8. Limestones and clays (future calc-schists). 9. Ophiolitic blocks and boulders. 10. Intra-oceanic faults (after Lagabrielle *et al.*, 1984).

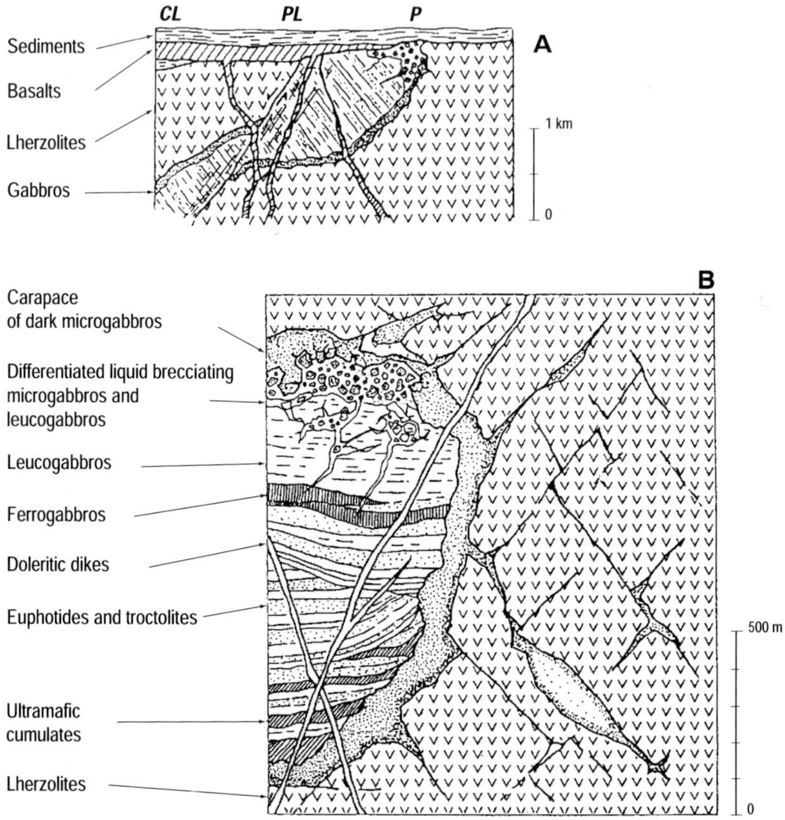

Fig. 6.15. Schemes illustrating the intra-mantellic magmatic pockets, characteristic of the Ligurian and Piedmont ophiolites. (A) General reconstruction, after the observations made in various massifs of the Queyras region, in the French Alps (PL: Pic des Lauzes-Roche Ecroulée; P: Peyroun; CL: Crête des Lauzes). (B) Detail of one of these magma pockets, about 1 km thick (after Lagabrielle, 1987).

fractured in places by residual silicic liquids with trondhjemitic affinities (magmatic breccias). Another characteristic of these massifs is the importance of extensional tectonics, expressed either by normal faults, or by ductile shear zones transforming the gabbros into foliated amphibolite corridors. On Mont Geneva, these foliated gabbros intersected by non deformed diabase dikes can be observed (Mével *et al.*, 1978). It has been demonstrated that these deformations, much earlier than the Alpine folding and metamorphism, are contemporaneous with the oceanic accretion stage which created these ophiolites.

Finally, detailed sedimentological, mineralogical and tectonic studies have shown that some serpentine breccias, called *ophicalcites*, were formed either by sedimentary filling of open fissures in the serpentinites of the ocean floor, or by sedimentary reworking of this serpentinite floor (Lagabrielle *et al.*, 1984).

6.5.2 The first supra-ophiolitic sediments and the emplacement of basalt flows

Immediately after their outcropping on the ocean floor, the ultrabasic rocks and gabbros were submitted to intense submarine erosion. The upper parts of these units were fractured and disaggregated *in situ*, sometimes in the presence of hydrothermal fluids (formation of ophicalcites). The products of this first episode of submarine dismantling, reworked by mass wasting, were accumulated in numerous topographic lows of the rough, tectonically active oceanic floor, forming sedimentary serpentine and gabbro breccias, ophicalcites and varied arenites. Simultaneously, pelagic limestone sediments were deposited directly over the ocean floor.

The extrusion of basaltic flows occurred after the deposition of the sediments described above. They are discordant and irregularly cover an already complex ocean floor, formed by ultramafic rocks, local gabbros, ophiolite breccias and pelagic sediments. Their emplacement corresponds to the activity of small volcanic centres distributed in the basin. Their thickness does not exceed a few hundreds of metres. The volcanic activity was far from being continuous, as many lava flows are separated by sedimentary horizons (Pic des Lauzes).

6.5.3 End of the volcanic activity, and deposition of radiolarites and limestones (Malm-Berriasian)

The red radiolarites that mold the top of the pillow-lava flows in Queyras and in Liguria gave an Oxfordian to lower Kimmeridgian age, marking the end of the volcanic activity. These lens-shaped and discontinuous radiolarites are followed by the deposition of carbonate oozes forming a continuous blanket (the future '*schistes lustrés*').

All these studies indicate that an important part of the floor of the Ligurian basin was made of mantle serpentinized peridotites and serpentinites, and not of basaltic upper crust. This small basin, probably occupied by a north–south axial rift offset by numerous east–west trending transform faults, was characterized by a thin, discontinuous crust (2–3 km at most), easily dismembered by vigorous extensional tectonic movements, favouring the tectonic unroofing and outcropping of the mantle.

The analogy with what is observed presently on slow-spreading oceanic ridges is evident, and was outlined in a number of recent papers (Juteau *et al.*, 1990; Lagabrielle and Cannat, 1990).

6.6 OPHIOLITES OF INTERMEDIATE TYPE

6.6.1 The Xigatse ophiolite

The Xigatse ophiolite in Tibet belongs to the Indus-Yarlung Zangbo suture, which separates the Indian plate to the south from the Lhasa block to the north. This suture marks the emplacement of a north-dipping palaeo-subduction zone, which

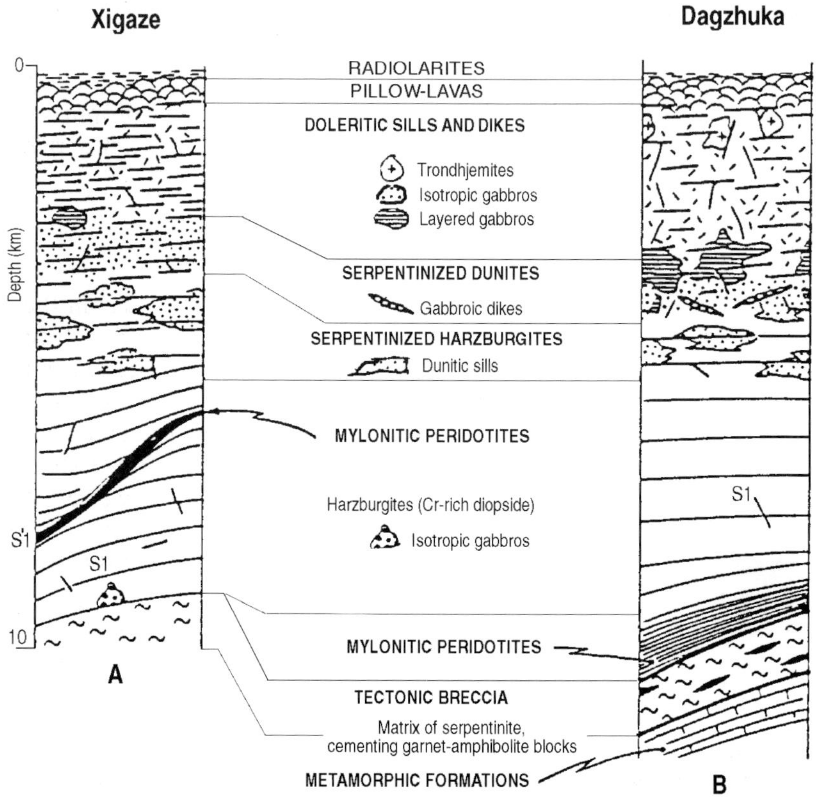

Fig. 6.16. Schematic logs of the Xigatse (A) and Dagzhuka (B) ophiolitic massifs (Tibet). S1 is a 'hot' (>1000°C), asthenospheric foliation. S′1 is a 'cold' (800°C), mylonitic and lithospheric foliation (after Nicolas *et al.*, 1981 and Girardeau *et al.*, 1985).

resorbed more than 2000 km of Tethyan ocean floors, before the complete collision of India with Eurasia at the end of the Eocene period.

The studies performed by French–Chinese teams on these ophiolites can be summarized by two characteristic logs, one across the Xigatse massif itself, the other one across the neighbouring massif of Dagzhuca (Figure 6.16). These sections, 8 to 13 km thick, exhibit a crustal sequence of hardly 3 km in thickness, lying over a mantle sequence (5 to 10 km), formed by alternating harzburgites and dunites.

Pillow lavas and massive basaltic flows form the upper part of the ophiolite (several hundreds of metres), and are covered by pelagic sediments and cherts of upper Albian–lower Cenomanian age. They grade downward to a diabase sill and dike complex, becoming rapidly a *sill complex* (diabase sills mutually intrusive in each other), intersected here and there by isolated diabase dikes, perpendicular to the sills. Scarce isotropic gabbro or plagiogranite 'screens' may appear between the sills. Small, irregular pockets of layered gabbros and ultramafic rocks are observed at the base of the 'sill complex', generally intrusive in troctolites or plagioclase-impregnated dunites. In total, the gabbros barely make up a few hundreds of metres in thickness.

The mantle part comprises residual harzburgites and dunites at the top, cut by numerous diabase sills (up to 50% in volume) in the first km, the proportion of sills decreasing progressively with depth. Going downward, harzburgites rich in Cr-diopside and lherzolites are more and more abundant.

6.6.2 The Trinity ophiolite

The Ordovician Trinity ophiolite (470 Myr) crops out over 3500 km^2 in the eastern Klamath Mountains (California). A thin crustal sequence, ranging from a few hundreds of metres to 2–3 km in thickness, rests over a mantle sequence, typically made of plagioclase lherzolites with steep foliations, including important dunitic bodies in the upper part (Figure 6.17). The upper crust, where preserved, is made of several hundreds of metres of pillow lava lying over a 600-m-thick sheeted dike complex, itself rooted in a thin layer of isotropic gabbro. The lower crust, visible over 300 to 500 m in thickness, is mainly represented by pegmatitic and coarse-grained gabbros, containing relicts of layered to foliated gabbros. All these gabbros contain primary amphibole. The pegmatitic gabbros appear in the ultramafic transition zone, as numerous several-metres-thick and vertical gabbro dikes. This ophiolite is considered by Nicolas (1989) as typically representing the LOT type.

6.7 SUMMARY AND CONCLUSIONS ABOUT THE CRUSTAL STRUCTURE OF PRESENT-DAY AND FOSSIL OCEANS

After reviewing the oceanic and ophiolitic, seismic and geologic data concerning present-day and fossil ocean crust, four main conclusions arise:

(1) The geological study of the structure of the oceanic crust, in the present-day oceans or in dismembered fragments of the oceans of the past, leads to the distinction between *two main types of crustal organization*:
- —A type characterizing the *fast- to intermediate-spreading oceans* (East-Pacific Rise, or Oman ophiolite type), with a thick and continuous crust (5–6 km), a well-developed sheeted dike complex (1 to 1.5 km thick), and a robust unit of layered/foliated gabbros (several km thick), overlying a harzburgitic residual upper mantle.
- —A type characterizing the *slow-spreading oceans* (Mid-Atlantic Ridge), with a thin, discontinuous crust (1–3 km), absence of a sheeted dike complex, or replaced by a sill complex (Xigatse type), and a reduced plutonic sequence, devoid of layered gabbros, overlying a lherzolitic mantle sequence (plagioclase lherzolites). In its extreme expression, this type may lead to the negation of any 'crustal sequence', when localized lava flows directly cover an ocean floor made of ultramafic and gabbroic rocks, or their dismantling products (Ligurian type).

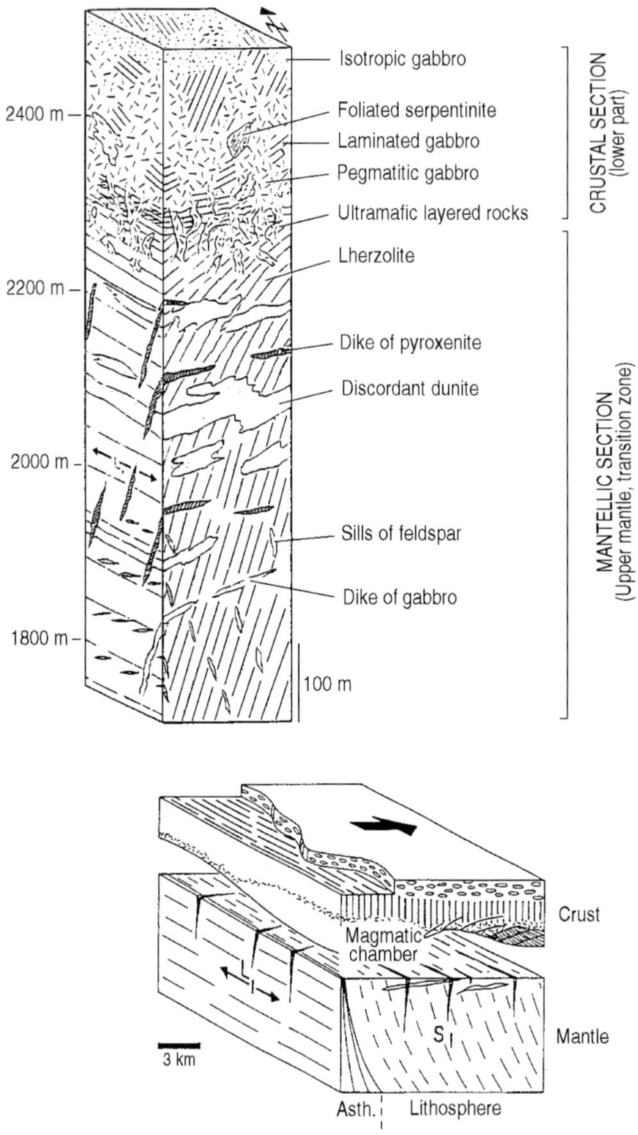

Fig. 6.17. Synthetic log of the Trinity ophiolite (South China Mountain, California), and interpretative structural block-diagram of the palaeo-accretion zone (after Boudier *et al.* (1989) and Le Sueur and Boudier (1986).

(2) The existence of ophiolites with intermediate characteristics between the Ligurian and the Oman types implies that intermediate-spreading ridges lead to such crustal structures.

(3) The paradox is that *seismic profiles across the crust of the present-day oceans do not reflect this duality at all*, and give, curiously, an identical image of the crustal

structure in all the ocean types, whatever the spreading rates. The depth of the seismic Moho below the ocean floor remains stable around 6–7 km below the Atlantic and the Pacific. How can we explain this paradox, and what is the significance of the seismic Moho below the Atlantic? It seems that there is a contradiction between the geological evidence, showing a thin and discontinuous magmatic crust, often reduced to zero at the very axis of the ridge (mantle outcropping in many places), and the seismic data, tracing invariably a regular seismic Moho at 5–6 km from the ocean floor. To resolve this contradiction, M. Cannat (1993, 1996) proposes an elegant solution (Figure 6.18): layer 3 in the Atlantic would consist of the serpentinized upper mantle, intruded by numerous isolated gabbroic pockets. The seismic Moho, in this case, would mark the lower limit of the gabbro pockets, and probably also that of the serpentinization of the host-peridotites.

(4) The comparison of ophiolites with the lithosphere of the present-day oceans must be conducted with an open mind, because:
   —Our knowledge of the present-day oceanic crust is still very fragmentary. layer 3 remains relatively unknown. Layered gabbros for instance, so frequent and so spectacular in many HOT ophiolites, have never been observed nor sampled up to now in the oceans. The transition zones layer 2/layer 3 and layer 3/mantle are poorly documented. Only the upper crust (layers 1 and 2) is reasonably studied in the various oceans, but many uncertainties remain about the composition and structure of the deep crust.
   —Ophiolites which exhibit on land complete sections across the crust and upper mantle of disappeared oceans, present some complications unknown

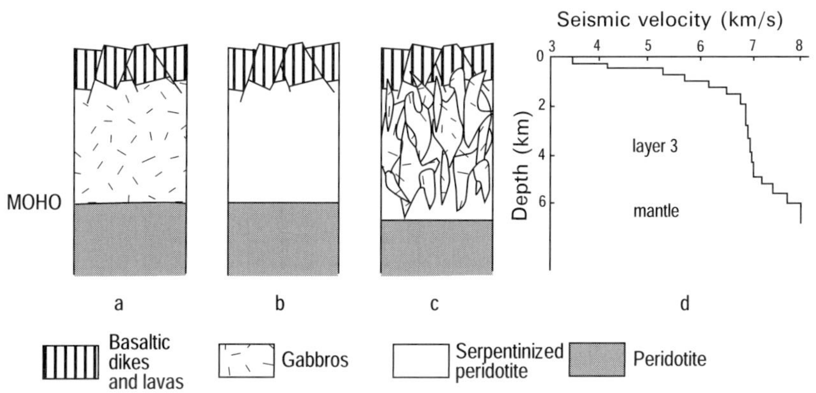

Fig. 6.18. Three possible lithological models for the oceanic layer 3. (A) A continuous gabbroic layer. This model can apply to fast-spreading ridges such as the east-Pacific rise, but certainly not to slow-spreading ridges. (B) A continuous serpentinite layer: this model has no ophiolitic equivalent, and meets serious geophysical objections (cf. Chapter 5, § 5.1.2). (C) A discontinuous gabbroic layer, forming isolated intrusive bodies in a serpentinized upper mantle. This model can apply to slow-spreading ridges such as the Mid-Atlantic Ridge (low magma budget, frequent outcropping of peridotites and gabbros), and remains compatible with the standard profile of the oceanic crust seismic velocities shown in (D) (after Cannat, 1993).

in the ocean domain, due to the fact that many of them were formed in marginal basins, or in back-arc, fore-arc or inter-arc basins, with complex volcanic and plutonic sequences.

Nevertheless, it remains that concerning the structures and accretion processes, ophiolites give up precious information: geometry, size and spacing of mantle diapirs; geometry, size and behaviour of accretion magma chambers; nature and petrology of the Moho; lithological transitions; hydrothermal circulations and associated ore deposits, etc.

7

Shallow processes: volcanic and tectonic activity of oceanic ridges

Beneath the sedimentary blanket covering the whole ocean floor, the hard basement constituting the top of the oceanic crust is everywhere made of basaltic flows. Hundreds of DSDP (Deep-sea Drilling Project) and ODP (Ocean Drilling Program) drill holes in all the oceans of the world have confirmed this fact.

This thick basalt layer, several hundred metres, was emplaced all along the world oceanic ridge system. It presently covers about $360 \times 10^6\,km^2$, a surface area far greater than that of the continents.

With a length of more than 60 000 km and a production rate of the order of $3 \times 10^6\,km^3$ of basalt per Myr, the oceanic ridge system is by far the largest volcanic belt in the world. This giant chain of volcanoes is hidden below several kilometres of seawater, except in rare places where the conjunction of the ridge with a hot spot results in its emergence, as in Iceland for instance.

In this chapter, we examine the volcanic and tectonic processes occurring along the axial zone of the oceanic spreading centres. Since the very first dives of the FAMOUS (French–American Mid-Oceanic Undersea Survey) program in the rift valley of the Mid-Atlantic Ridge (summer 1973), 25 years of detailed observations of the axial zone of the oceanic ridges by submersibles, deep-tow cameras and by acoustic imagery (side-scan and multibeam sonar systems, etc.), have permitted analysis of various expressions of deep-sea basaltic volcanism and better understanding of the volcanic activity of the world oceanic ridge system.

This deep-sea volcanism, however, remains poorly known. No deep-sea volcanic eruption has been observed up to now, even though very recent sheet flows, still hot and only a few days old have been observed by submersible on the East-Pacific Rise (see § 7.2.2). Some famous images showing the formation of pillow lavas beneath the sea have been taken at shallow depth (several tens of m) and a short distance from the shores of intraplate shield volcanoes (Hawaii and Reunion islands), under physical conditions very different from those of the deep-ocean floor, at several km in depth.

7.1 THE AXIAL VOLCANIC ZONE OF THE RIDGES

Whatever the spreading rate, the active volcanic zone of the ocean ridge forms a narrow belt, generally barely 1 km in width, at the very axis of the ridge. This active zone is characterized by the black and shining lustre of the basaltic glass, by the abundance of fragile, glassy buds decorating the base of many pillow lavas, and by the total absence of sediment and fauna. Except localized areas where hot springs create oases of animal life, this zone is a true biological desert, with scarce benthic fauna and isolated animals fixed on the bare lava.

The aspect, structure and morphology of the active neo-volcanic zone depends strongly on the spreading rate, which itself reflects the thermal state of the ridge (Figure 7.1). 'Fast' spreading ridges, having the highest spreading rates and the hottest thermal state, show a continuous and linear axial volcano, which is mainly made of fluid lava flows and lava lakes. 'Slow' spreading ridges, having the lowest spreading rates and the coldest thermal state, show an irregular and discontinuous axial volcano, made of less fluid lavas, mainly pillow-lava flows, building steep volcanoes.

7.1.1 Fast-spreading ridges: the axial volcanic zone of the East-Pacific Rise

The axial volcanism of the East-Pacific Rise is confined to a very narrow strip, about 500 m in width, and even narrower in many places (50 to 100 m), along a linear axial graben, often well delineated. The fissural and extremely linear nature of the volcanism is striking.

These lavas are distributed following a transverse zonation, from the axis to the flanks of the ridge (Figure 7.2):

- —The axial graben is mostly occupied by collapsed lava lakes and associated lobate lava flows.
- —The walls of the graben are generally made of truncated pillow-lava flows, interbedded sometimes with massive lava flows showing some columnar jointing.
- —On the top of the walls of the axial graben and on the flanks of the axial zone, the piling up of pillow-lava flows form elongated volcanic ridges, which can be observed sometimes on the inner floor of the axial graben.

7.1.2 Slow-spreading ridges: the axial volcanic zone of the Mid-Atlantic Ridge

Recent studies have shown that the structure of the upper crust of slow-spreading ridges like the Mid-Atlantic Ridge is very different from that of fast-spreading ridges. Effectively, no permanent magma chamber could be detected: instead, small magmatic pockets rise up in the crust to feed locally axial volcanoes distributed over the whole width of the inner floor of the axial valley. These volcanoes may be isolated or form linear ridges. In this case, they pile up and coalesce to form more important volcanic ridges.

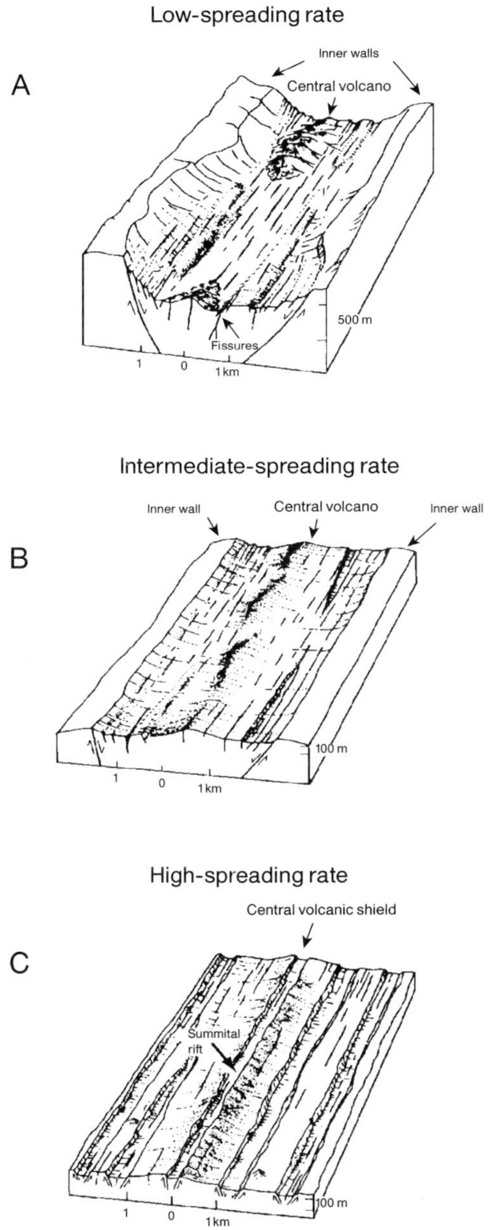

Fig. 7.1. Schematic illustration of the axial neovolcanic zone of oceanic ridges for various spreading rates. (A) Slow-spreading ridges. The axial volcano, made of pillow lavas over 200 to 300 m in height, is very discontinuous. (B) Intermediate-spreading ridges. The axial volcano, discontinuous and mainly made of pillow lavas (50 to 80 m in height), consists in several-km-long segments often disposed 'enchelon'. (C) Fast-spreading ridges. The axial features are extremely linear over tens of km. The axial volcano, essentially made of sheet flows, resembles a shield volcano of the Hawaiian type, with a summital rift (linear axial graben) (after Macdonald, 1982).

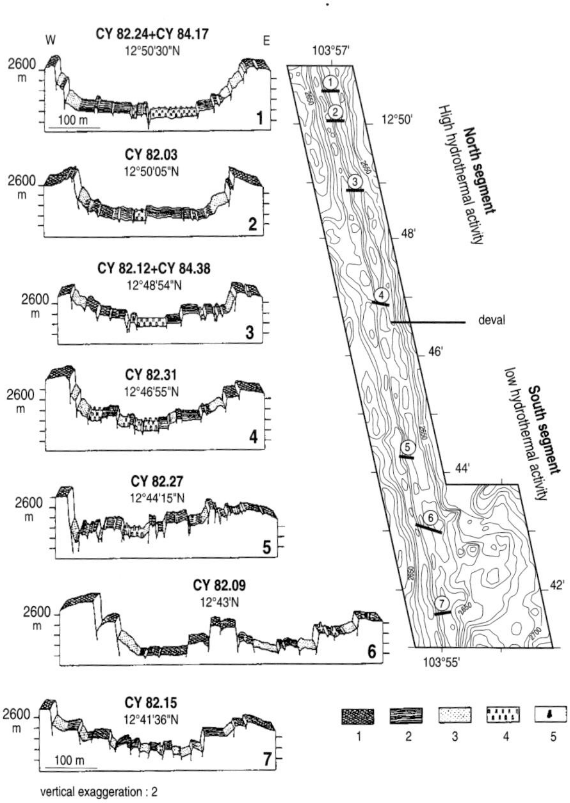

Fig. 7.2. Variations in morphology and in tectonic and volcanic activities along two segments of the East-Pacific Rise between 12°41′ and 12°51′N. These data were obtained by submersible observations, using the French submersible *Cyana*. 1. Pillow lavas. 2. Lobate lavas. 3. Rubble. 4. Lava lakes. 5. Hydrothermal site. The axial graben is typically 200–300 m wide and 50 m deep. The northern segment shows an axial graben with a flat bottom, occupied by lava lakes and sheet flows with lobate surfaces (very recent volcanic activity). The southern segment exhibits a much rougher morphology, due to the number of tectonic features (tectonic activity more recent than volcanic activity). Both segments are separated by a small axial discontinuity or 'deval' (for deviation to axial linearity) (after Gente,1987 and Gente *et al.*, 1986, simplified).

Let us come back to the Mid-Atlantic Ridge between 24° and 30°N in the northern Atlantic, between the Kane and Atlantis fracture zones. As we have seen in Chapter 3, 18 accretion segments were identified along these 800 km of mid-ocean ridge. Mapping by multibeam swath systems and the acoustic images obtained by the TOBI (Towed Ocean Bottom Instrument) British side-scan sonar system allowed a detailed morphological study of the volcanic reliefs (Smith and Cann, 1990, 1993):

Volcanic ridges. Most of the segments carry with them pre-eminent axial volcanic ridges on the inner floor of the valley, located either at the axis itself, or off-axis and close to the valley walls. These ridges are generally linear, and eventually sinuous. Their strike may be oblique to the strike of the walls of the rift valley, and sometimes

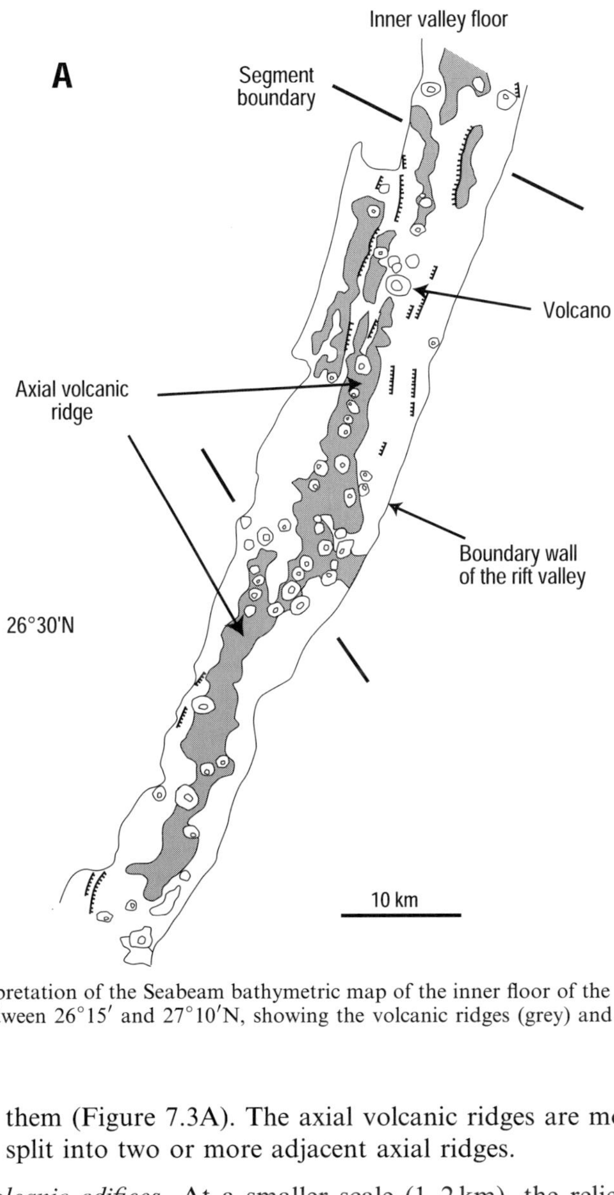

Fig. 7.3. (A) Interpretation of the Seabeam bathymetric map of the inner floor of the Mid-Atlantic ridge axial rift valley between 26°15′ and 27°10′N, showing the volcanic ridges (grey) and individual circular volcanic edifices.

end up against them (Figure 7.3A). The axial volcanic ridges are mostly discontinuous, and may split into two or more adjacent axial ridges.

Individual volcanic edifices. At a smaller scale (1–2 km), the reliefs of the inner floor of the rift are dominated by individual volcanic edifices (Figure 7.3B). Smith and Cann have numbered, from the Seabeam map, more than 450 circular volcanic edifices, ranging from 50 to 200 m in height along the Kane–Atlantis zone. Small volcanoes are distributed on the floor of the rift in all the mapped segments. They can be isolated or coalescent, forming the main axial volcanic ridges. Alignments of circular volcanoes can thus occupy the top of the volcanic ridges, and isolated volcanoes can be distributed on the top and flanks of the ridges.

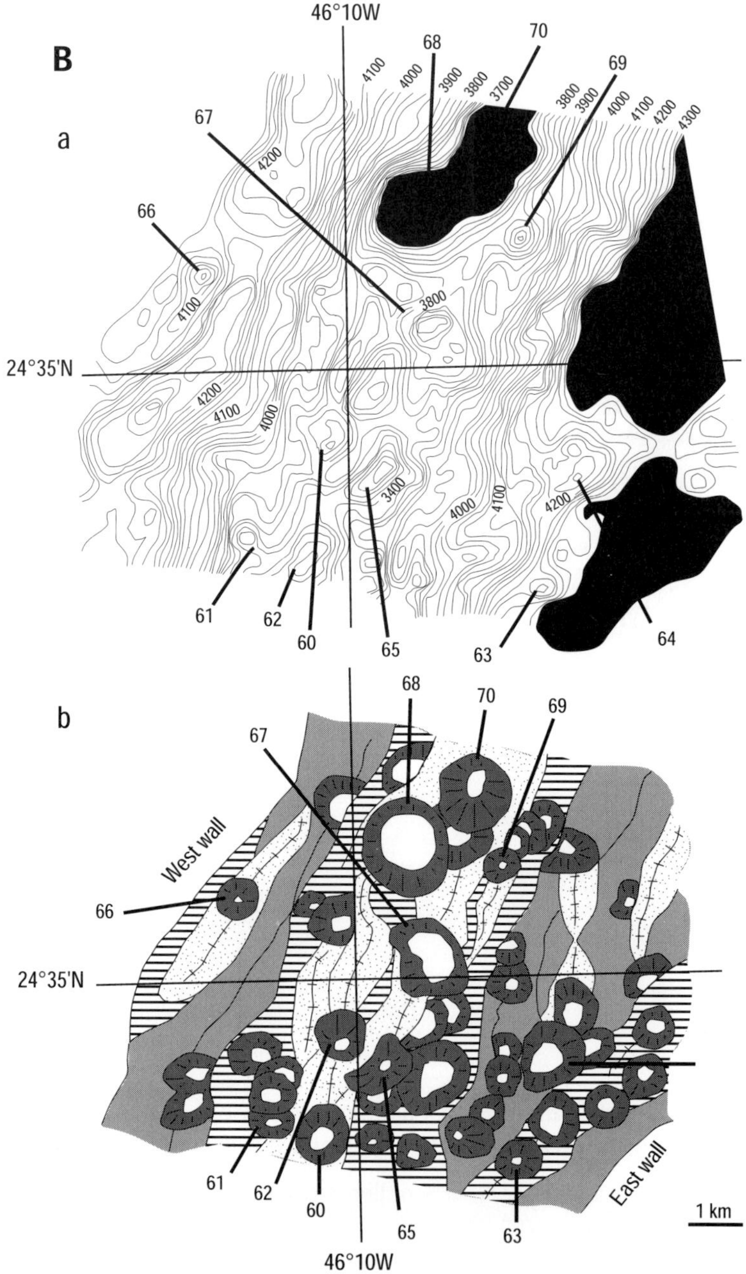

Fig. 7.3 (*cont.*). (B) Detail of the distribution of individual circular volcanoes and of a number of lava flows issued from these volcanoes on the inner floor of the rift valley at 24°35′N. These hummocky lava flows overlap each other (lighter stipple for the younger flows on schema b), and thus contribute to the edification of the axial volcanic ridge. (a) Bathymetric map. (b) Geological interpretation (each individual volcano has been given a number) (after Smith and Cann, 1993).

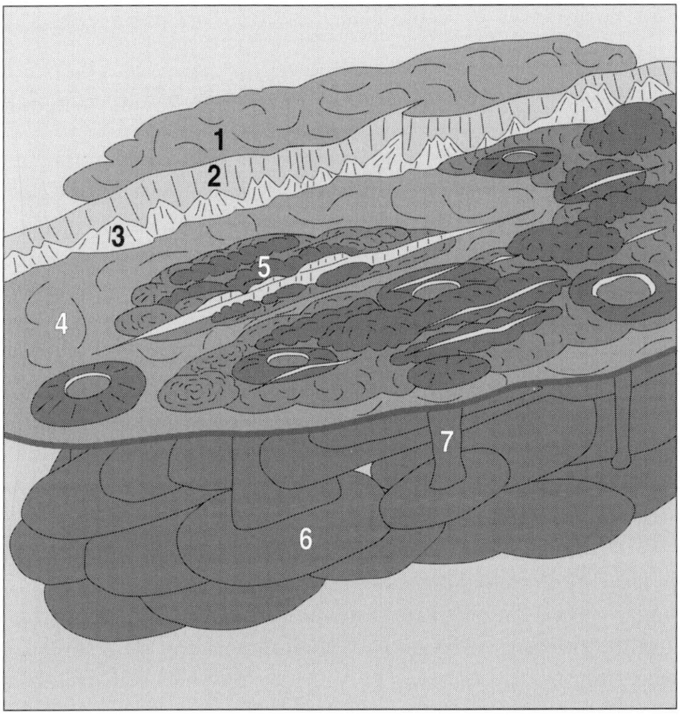

Fig. 7.4. Interpretative section and perspective view of a portion of an accretion segment from the Mid-Atlantic Ridge (after Smith and Cann, 1993). The figure shows—at the seafloor level: 1. The hills of the rift. 2. The major scarp bounding the rift (normal fault). 3. The rubble cones at the foot of the major scarp. 4. The inner floor of the rift. 5. The axial volcanic ridge presently active. The figure also shows—below the surface: 6. Small subsurface magma pockets. 7. The pipes and dikes with varied geometries issued from these magma pockets, and feeding the axial volcanic ridge.

The circular volcanoes present two main types of morphology: the majority are seamounts which pile up to form cones and domes. Others are edifices with a flat and smooth top, with or without a central crater. The acoustic images do not allow to recognize the morphology of individual flows, but do at least distinguish both smooth and hummock surfaces. The latter form mounds less than 50 m in height, which coalesce to form several-km-long ridges.

In summary, the coalescent circular volcanoes, the hummock mounts and flows and the smooth surface flows combine together in a complex 'patchwork' to form the reliefs of the inner floor of the rift, including the volcanic ridges and the topographic lows flanking them. It also appears that eruptions are distributed along the whole width of the rift floor, even though may of them contributes to building up the axial volcanic ridges. In the most recent interpretations, volcanoes or groups of volcanoes are fed by dykes or 'pipes' of variable geometry, issued from small, flattened magmatic bodies, mutually intrusive and making up most of layer 3 (Figure 7.4).

Table 7.1. Deep basaltic flow types: classification and main characteristics (after Bonatti and Harrison, 1988).

		Thermal constraints	Eruption temperature	Crystal/liquid ratio	Viscosity	Accretion rates
Quiet eruptions	Lava lakes	Weak	Elevated	Weak	Weak	Elevated
	Lava flows					
	Pillow lava					
Reactive eruptions	Paved lava					
	Deep hyaloclastites	Strong	Weak	Elevated	Strong	Weak (seamounts)

7.2 THE VARIOUS TYPES OF BASALTIC ERUPTIONS IN A SUBMARINE ENVIRONMENT

Among the physical or chemical factors that control the dynamism of the submarine basaltic volcanism, the temperature and viscosity of the erupting magma are certainly the most important (the viscosity itself depending on temperature, chemical composition and the volatile content of the magma), as well as the instantaneous discharge rate (which depends partly on viscosity).

Table 7.1 summarizes the volcanic styles and the products of the submarine basaltic volcanism presently known. Two main styles of eruptions must be distinguished: (a) the shallow eruptions (under less than 500 to 700 m of seawater), where hydro-explosions are possible; (b) deep-sea eruptions (more than 500 to 700 m), where gas expansion and volcanic explosions are prevented by the seawater hydrostatic pressure.

7.2.1 Shallow submarine eruptions

The shallow explosive submarine eruptions and their typical ‘cypress-shaped’ sheaves, mainly due to the expansion of magmatic gases and to the instantaneous vaporization of seawater on contact with magma, have been very well described in the Azores during the Capelinhos eruption (Fayal Island, Tazieff, 1958; Rittmann, 1963), in Iceland during the emergence of Surtsey (Thorarinsson, 1964), and also in Indonesia during the formation of the Anak Krakatau volcano (Decker and Hadikusumo, 1961). The lava, intensely fragmented by the explosions due to contact between magmatic gases and seawater vapor, falls back below water, forming stratified layers of explosion hyaloclastites. Since McBirney’s calculations (1963), we know that such eruptions are unlikely to occur beyond a water depth of 500 m (700 m for other authors).

For axial zones of the oceanic ridges located at an average depth of 2500–3000 m, this style of eruption is observed only in very specific areas of the ridge, in the immediate neighbourhood of oceanic islands built on the ridge (Iceland, the Azores, etc.). On the contrary, shallow submarine explosive eruptions constitute an obligatory stage during the edification of the numerous intra-oceanic volcanic islands.

7.2.2 Deep-sea submarine eruptions

Beyond 700 m in depth, the strong external pressure maintains the gas phase internally within the magma. Contrarily to what occurs in the case of aerial volcanism on land, the separation between liquid and gas phases does not occur at the time of the eruption: hence there are no explosions, nor volcanic projections. Deep-sea eruptions occurring at the ridge axis are 'quiet eruptions'.

The products of deep-sea eruptions were observed at fast and slow-spreading ridge segments, and also on various submarine seamounts. These observations allowed 'quiet eruptions' to be recognized, where the physical interaction between the erupting magma and seawater is very limited, and 'hyaloclastic eruptions' (also called 'reactive' by Bonatti and Harrison, 1988), during which an intense thermal fragmentation of the basalt occurs as it spreads on the ocean floor.

During the last few years, intensive exploration of ridge segments with variable spreading rates have shown the omnipresence of two main types of 'quiet-eruption' products: the pillow-lava flows on one hand, and the sheet flows and associated lava lakes on the other.

Paved lavas and deep hyaloclastites, rarely observed at oceanic ridges, were mainly described on intra-oceanic seamounts.

7.3 THE VOLCANIC DYNAMISM OF OCEANIC RIDGES: DESCRIPTION OF THE VARIOUS TYPES OF BASALTIC FLOWS AND INTRUSIONS

At great depths, the basalt magma emitted at 1200°C into a glacial seawater at 2°C suffers severe hardening. The effect of this thermal hardening on the morphology of the flows depends mainly on the volume of lava delivered:

- —The delivery of moderate volumes of lava results invariably in the formation of pillow lavas, constituting effusive bodies with variable morphology, more often tubular, and never planar.
- —The delivery of very important volumes of lava result in the formation of mainly planar effusive bodies: lava lakes, sheet flows and massive flows.

7.3.1 Pillow-lava flows

Fissural basaltic eruptions at low delivery rates typically produce pillow-lava flows. This type of lava flow has been known on land for a long time, in fact since the first half of the 19th century: the very first descriptions of 'globular', spheroidal or ellipsoidal lavas were given by Delesse (1846), and their interpretation as sub-aquatic flows was proposed at the end of the last century (Teall, 1894). It rapidly became clear that the pillow structure characterizes the emission of basic to intermediate lavas in various aquatic environments, for instance submarine (ophiolites), sub-lacustral or even sub-glacial (Iceland) environments. At the beginning of the present century, several authors suggested an emplacement into water-saturated sediments, for instance in the Devonian sedimentary formations of Cornwall in England. Lewis (1914) proposed the idea that pillow lavas propagate by successive bursts of the flow surface and growth of protuberances under the internal pressure of the lava, producing a chain of interconnected pillows.

The observation of ocean ridges from submersibles, and the numerous deep-sea drillings of the DSDP and ODP programs on reaching the basaltic layer, indicate that basaltic pillow lavas constitute by far the most abundant volcanic form in volume in the oceans, and consequently at the earth surface, taking into account that they also constitute the main part of the immerged parts of the intraplate shield volcanoes such as the Hawaiian or Reunion islands, and a significant part of island-arc volcanoes.

Mechanism of formation

The mechanism of formation and evolution of pillow-lava flows is paradoxically often misunderstood, and merits some detailed explanation.

The term 'pillow lava' itself is somewhat misleading. For a long time, pillow lavas were supposed to represent the piling up of isolated, more or less ellipsoidal lava 'bags', moulded one over the other, and evoking pillows or bolsters. This erroneous vision was based on limited observations, either along terrestrial geological sections providing a two-dimensional view, or in shallow submarine formations in which the pillows are effectively isolated in an abundant matrix of explosive hyaloclastites, as at Acicastello or Vizzini in Sicily (Rittmann, 1963; MacDonald, 1972).

The first pillow lavas observed directly from a submersible at great depth and at an ocean ridge were described during the French–American operation FAMOUS during the summers of 1973 and 1974, on the Mid-Atlantic Ridge (ARCYANA,* 1977, 1978; Ballard and Moore, 1977). The tubular and interconnected morphology of pillow lavas, on the scale of entire submarine volcanoes, then came into view.

In this case where the volumes of emitted lava are moderate, the effect of drastic thermal chilling undergone by the magma reaches a maximum: the surface of basaltic lava exiting at 1200°C in freezing seawater (2°C at 3000 m in depth) is instantaneously vitrified, and this several-cm-thick glassy margin prevents the flow from spreading at the surface, forcing it to flow as fingertips. The lava forms tubes which overlap one another and progress downslope by successively bursting the frontal glassy crust (Figure 7.5A). The internal pressure of the molten lava in the

* Collective name for the seven scientists of the French group, based on the names of the two French submersibles *Archimède* and *Cyana*.

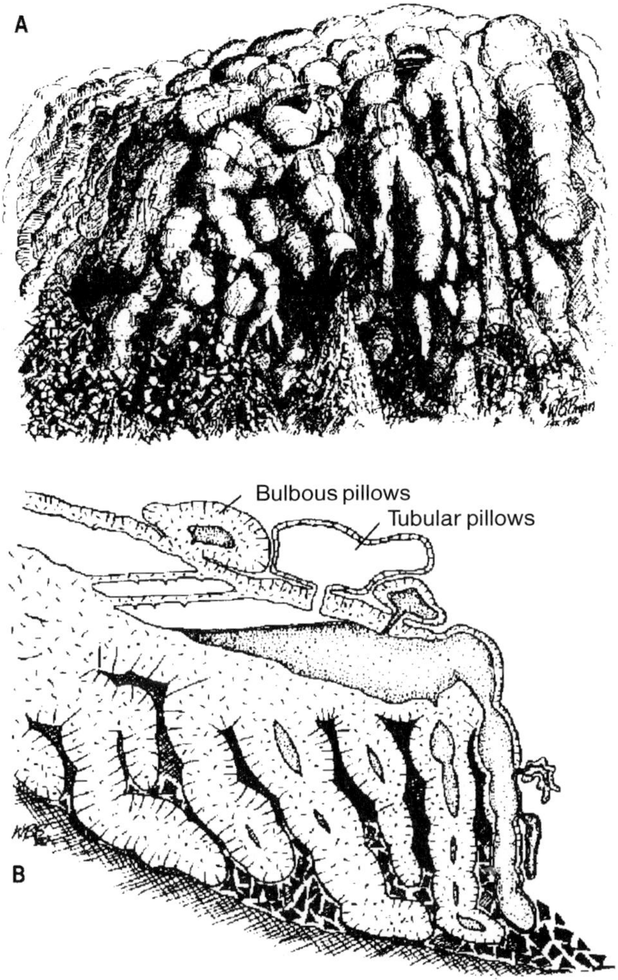

Fig. 7.5. (A) Drawing representing a pillow-lava flow front. The lava tubes are elongate along the steep slope, progressing downward by successive bursts of their glassy crust (drawing by Bryan, 1992). (B) Cross-section perpendicular to the flow front, showing the development of tubular pillows, starting from a submarine lava flow moving downward on an irregular slope. The standard structure of a pillow-lava flow includes various types of pillow morphology. The top of the flow has a gentle slope and is made of huge bulbous, quite spherical pillows, a number of which are hollow and collapsed, revealing internal voids. These hollow pillows, whose lava has been drained back toward their feeding conduit, form alignments revealing the hidden setting of the feeding conduits. The flow front has a steep slope and is draped with elongated, tubular pillows, whose diameter is much shorter than that of the bulbous pillows (after Bryan, 1992).

tubes provokes cracking of the glassy crust in many places, resulting in the formation of numerous glassy buds and small secondary flows.

So the term *pillow* is used to describe any submarine lava body not presenting a planar morphology.

Structure of a pillow flow

Pillow flows can spread over several hundreds of metres at most, and are often wider than long when they issue from a fissure. Downward they present a flow front made up, by finger production and budding, of many tubular pillow tongues, piling up one on top of the other over the flow front (Figure 7.5B). The flow fronts are very steep (30° to 60° or more), with an average height of 5 to 20 m, sometimes reaching 50 m. Along the steepest slopes, the pillow tubes are extremely elongated, with a round tip ('phallus' morphology). The spaces between fingers presents various shapes, including 'elephant trunks', 'tripe', etc.

The main pillow flow is constantly fed through by means of huge feeder tubes reaching 1 m in diameter, and continues its progression as long as the tubes are not obstructed by consolidated and cooled lava. The surface of the flow presents a much smoother slope, about 5° on average; it is occupied by huge, more or less spherical bulbous pillows, whose diameter can reach 4 m, and is rooted into one or several feeder tubes (ARCYANA, 1978).

Figure 7.6 shows the development of a pile of tubular pillows, starting from a single submarine lava flow moving downward over an irregular slope, assuming a constant and regular feeding of fresh lava from the top. When the lava flow meets an obstacle on the slope, the flow front inflates, then bursts under internal pressure of hot lava, producing an elongated pillow tube along the slope (stage A). This pillow goes on growing and spreading downward, as long as the lava pressure is sufficient to stretch and break its glassy crust. This latter tends on one hand to thicken and strengthen by drastic chilling on contact with the freezing seawater, and on the other to thin by stretching during expansion and spreading of the pillow tube. Sooner or later the thickening will prevail, and the resistance of the thickened crust prevents any further increase in volume of the pillow tube. Stage B shows that this first pillow constitutes a barrier for the feeder lava lobe which inflates, raises its roof and breaks the glassy crust at a point located above the first pillow, where the resistance is less. Simultaneously, the internal pressure in the feeding tube causes fracturing of its roof, and feeds the growth of bulbous pillows, which will form a true layer over the feeding tube. Stage C shows that elongated pillows go on forming at the flow front, dyking up the main feeder tube, and forcing it to inflate and burst above the pillows of the flow front, and to override them with new tubular conduits. These also split laterally, and may fragment into pieces accumulating at the foot of the flow front.

Several-metres-thick rubble talus form typically ahead of the flow fronts. They are made of angular pillow fragments ('camembert slices'), broken during the progression of the lava flow, and present a very steep talus angle (close to 45°). Some hollow pillows, which fell when they were still filled with liquid lava, can be observed sometimes: these are the spectacular 'egg shell' structures (Figure 7.7A). These rubble talus are progressively buried beneath the downward progress of the flow fronts and frequently constitute, in geological sections, breccia horizons below the pillow-lava flows.

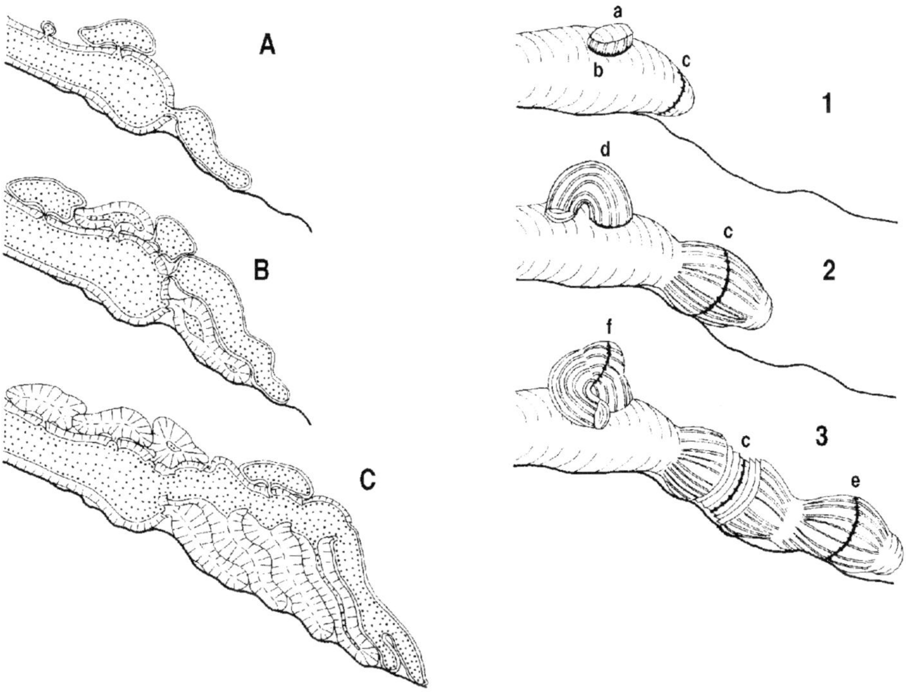

Fig. 7.6. Scheme showing the principle of formation of a tubular pillow-lava pile, analogous to that of Figure 7.5B. To the left, a pillow-lava tube progressing downward on an irregular slope by successive bursts of its glassy crust: secondary tubes are formed ahead (A) and mould one on top of the other, forming a steep flow front (much steeper than the topographic slope), whereas bulbous pillows develop at the roof of the feeder conduit (B, C). To the right, progression of an individual tube from stage 1 to stage 3, by successive bursts of its frontal lobes (bold lines 'c' or 'f'). Double lines along the tubes are corrugations, printed in the plastic lava by irregularities of the circular cracks. On the back of the flow, a circular piece of crust, raised by the internal flow pressure through a circular crack 'b' (trapdoor pillow 'a'), evolves as a 'toothpaste' tube ('d') (after Ballard and Moore, 1977).

Morphological nomenclature of pillow lavas

The morphological aspect of pillow lavas includes many variations. The most common forms observed are the following:

Bulbous pillows. (Figure 7.5) Convex, quite spherical and huge (currently one m in diameter, and up to 3–4 m), they usually develop over the smooth slope of the pillow flows, fed by the much larger feeding conduits. They develop upwards, forming an isolating carapace over the conduits.

Flattened pillows. (Figure 7.7B) Equi-dimensional in plan view, flabby and flattened, they occur near the summit of the main volcanic centres. Produced by rapid growth, then by partial drainage of bulbous-type pillows, they clearly indicate the sites of eruptive vents, where lava delivery rates are the highest. Their thin and fragile glassy crust, result of very fast growth, cannot withstand the spreading of

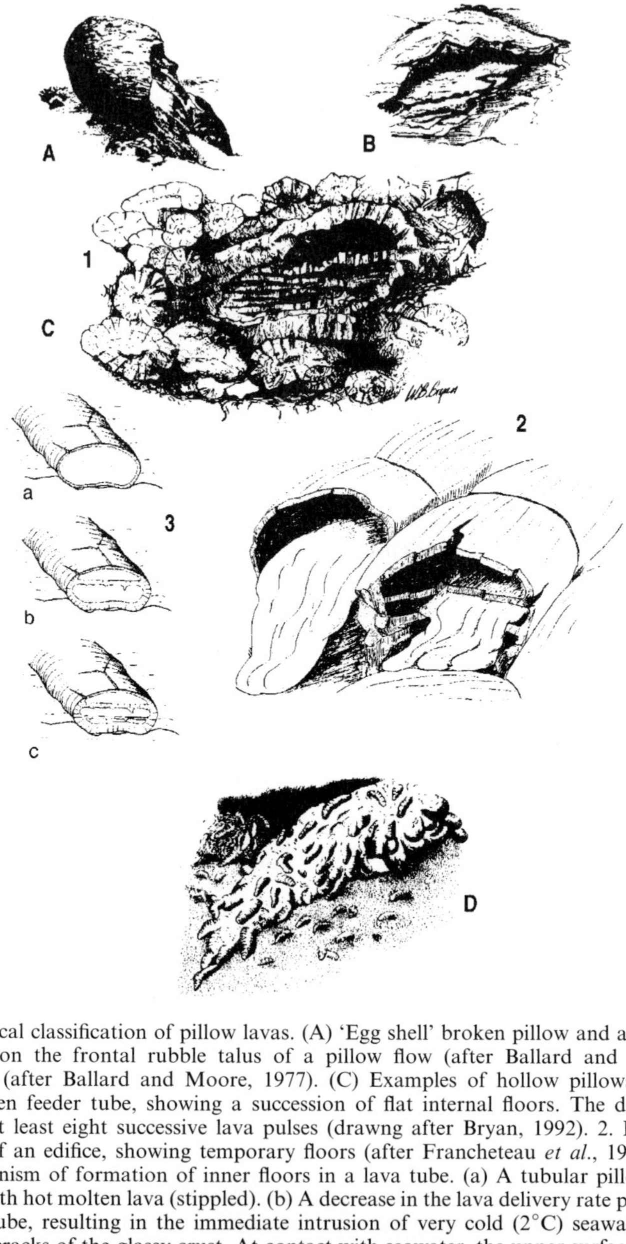

Fig. 7.7. Morphological classification of pillow lavas. (A) 'Egg shell' broken pillow and associated mini-lava flow, observed on the frontal rubble talus of a pillow flow (after Ballard and Moore, 1977). (B) Flattened pillow (after Ballard and Moore, 1977). (C) Examples of hollow pillows with internal floors. 1. Huge broken feeder tube, showing a succession of flat internal floors. The drainage of this tube occurred over at least eight successive lava pulses (drawng after Bryan, 1992). 2. Hollow broken tubes from the top of an edifice, showing temporary floors (after Francheteau *et al.*, 1980). 3. Scheme explaining the mechanism of formation of inner floors in a lava tube. (a) A tubular pillow with a thin glassy crust is filled with hot molten lava (stippled). (b) A decrease in the lava delivery rate provokes a drop in lava level in the tube, resulting in the immediate intrusion of very cold (2°C) seawater in the void created, through the cracks of the glassy crust. At contact with seawater, the upper surface of the mobile hot lava is chilled and solidifies, forming a glassy crust (first inner floor). (c) A second decrease in the lava delivery results in the formation of a second inner floor by the same mechanism. Finally, the lava supply is stopped, and the rest of the hot fluid lava escapes downward in the tube, leaving a void at the base of the tube. Glassy septa hang from the lower face of the temporary floors (good 'polarity' criterium), and the external glassy crust thickens considerably at the base of the tube. Note also that the first floor is thicker than the second one (after Ballard and Moore, 1977). (D) Pillow-lava surface decorated with numerous glass buds (drawing after Ballard and Moore, 1977).

the lava under the action of gravity. Some of them show a marked central depression.

Hollow pillows (Figure 7.5). They are recognized only when they are broken or collapsed. With a glassy crust 1 to 15 cm thick (5 cm on average), they are common in most other morphological types. They form when the lava feeding a lobe is suddenly cut from its source, and continues its course downward inside the tube. In the absence of a gas phase at these depths, it is seawater which immediately fills the void thus formed, infiltrating through the cracks of the glassy crust. As for flattened pillows, hollow pillows may also result from lava draining back towards the nearby feeding tubes, forming alignments marking out the path of buried feeder tubes.

Hollow pillows often exhibit *horizontal internal floors* (Figure 7.7C), which mark temporary stops of fluid lava movement through a partially filled tube. The delivery of fresh lava weakens by jerks, and the level of the lava in the tube decreases periodically. At every temporary stop of the flow, a new floor forms, while seawater penetrates through the cracks of the crust in the upper part of the pillow. Up to six or eight parallel *septa* may thus develop inside a hollow pillow. These septa are quite horizontal, generally slightly inclined downslope, but much less than the pillow tube in which they occur. The lateral walls of a hollow tube thicken progressively downwards, where solidification time is longer.

The lower surface of the roof of a hollow pillow, as well as of each temporary floor, may be decorated by thin septa or hanging 'stalactites'. The septa form curtains about 1 cm thick, and glassy basaltic stalactites may be 1 cm in diameter at their connection with the roof, and several cm in length. These vertical structures, as well as the almost horizontal floors, constitute excellent *polarity criteria*, and may help in finding the original orientation of a fragment of hollow pillow.

Pillows with glass buds. These elongated or bulbous pillows are full of small basaltic buds, 2 to 20 cm in diameter and 4 to 40 cm long, with round tips (Figure 7.7D). They are generally much more abundant in the lower half of pillows. When their diameter is less than 5 cm, they are entirely glassy. These glass buds are extremely fragile, and progressively disappear from the pillow surfaces with ageing of the crust. Broken by micro-earthquakes, and by tectonic, sedimentary or biological activities, they contribute to the formation of the hyaloclastites which in turn form the matrix of pillow lavas (see further). These buds are never observed in the ancient ophiolitic series, and their abundance in the axial zone of the ocean ridges has been a real surprise.

Trapdoor pillows. These pillows show a variety of protuberances, developed from a circular crack cutting through the roof of a huge pillow, and resulting in the extrusion of a viscous, squat lava panel over a brief time interval (Figure 7.6). The glassy crust of the support-pillow covers the surface of the trapdoor, whose flanks are intensely striated by corrugations parallel to the movement of the surrected part. The expansion may be symmetrical or asymmetrical. A trapdoor may evolve into a curved and hanging 'toothpaste' pillow, when the protrusion, still too fluid, cannot avoid the effect of gravity (Figure 7.6).

Huge irregular pillows, and mega-pillows. Basaltic bodies with an ellipsoidal section 5 to 15 m in diameter, often spectacularly radially prismatized, are sometimes observed in ophiolitic pillow series. These are probably mega-tunnels of lava, analogous to the tunnels known in intra-oceanic shield volcanoes (Hawaii). Typical examples have been described for instance in Cyprus along the Akaki river, in the northern Troodos ophiolitic massif (Schmincke and Bednarz, 1990). Moreover, lava channels and tunnels reaching several km in length have been mapped by side-scan sonar in the axial zone of the East-Pacific Rise.

Description at the pillow level

Now let us examine these flows at the pillow-lava level. On this scale, the elements to be distinguished are: the *core* of the pillow, its *margin*, its *glassy crust*, its *matrix* and the *cement* (Figure 7.8A).

Core and margin. The core and margin of pillow lava differ essentially in their content of gas bubbles. This differentiation is directly controlled by the degree of degassing of the lava (Juteau, 1975). If the lava is poor in gas bubbles, the scarce and small bubbles trapped inside the pillow will be adsorbed at the inner surface of the glassy crust already formed, and the whole core of the pillow will crystallize as a compact lava, devoid of vesicles: in this case we get a pillow with peripheral vesicles (often elongated radially), and a compact core. On the contrary, if the lava is rich in fluids forming gas bubbles during crystallization, the vesicles will coalesce and grow in size, being repelled progressively by the crystallization front towards the core of the pillow, until the formation of a central geode at the pillow core: this is the case of a pillow with a compact margin and a central geode, often filled later on with pelagic sediment. In some cases intermediate stages can be observed, where the lava is just rich enough in gas bubbles for a regular adsorption of these bubbles along the crystallization front at each stop: in this case we observe a pillow with vesicles distributed along concentric surfaces, with often a small central geode (Figure 7.8B).

Glassy crust and surface features of the pillows. All pillows have a 2–3-cm-thick glassy crust, which represents its first chilled and frozen envelope on contact with cold seawater. This crust is black, shiny and finely cracked. From a physical-chemical point of view, this volcanic glass may be considered as a supercooled liquid with an excessively high viscosity.

The surface of young pillows exhibits a number of cracks, ridges, striations, grooves and buds varying in size and shape (Figure 7.9). The direct observation of growing basaltic pillow lavas in shallow waters off the Hawaiian coast has shown that most of these surface structures are linked to the development of new glassy surfaces at expansion cracks (Moore, 1975; Tepley and Moore, 1974). These cracks form when the internal pressure of the molten lava inside the pillow provokes a burst in the external crust. The famous footage from *Fire under the Sea* from Tepley and Moore shows that when one of these cracks develops and widens, red lava is extruded through it and is immediately vitrified to a thickness over more than 1 cm on contact with cold seawater. Corrugations are formed when the opening of an

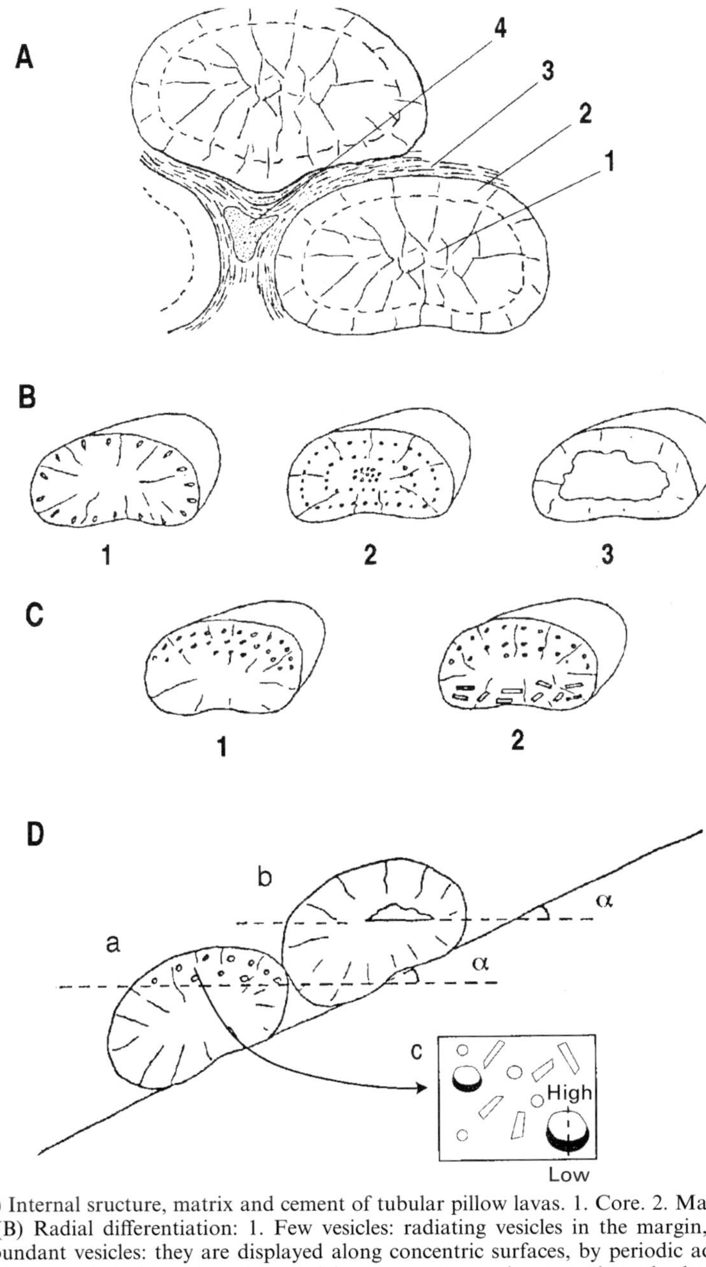

Fig. 7.8. (A) Internal sructure, matrix and cement of tubular pillow lavas. 1. Core. 2. Margin. 3. Matrix. 4. Cement. (B) Radial differentiation: 1. Few vesicles: radiating vesicles in the margin, compact core. 2. Rather abundant vesicles: they are displayed along concentric surfaces, by periodic adsorption along the solidification front. 3. Very abundant vesicles: compact margin, central geode, by coalescence of vesicles. (C) Vertical differentiation: 1. Aphyric and vesicular lava, with concentration of vesicles at the top. 2. Phyric and vesicular lava, with phenocrysts settling down and vesicles migrating upward. (D) Fossilization of the horizontal plane during immobilization of a pillow. (a) by the vesicles; (b) by a flat geode, with possibility of ascertaining slope *a* of the flow (after Juteau, 1975); (c) by the 'segregation vesicles', gas bubbles partially infiltrated by residual magma, forming a meniscus at the base of the bubble (after Bideau *et al.*, 1977).

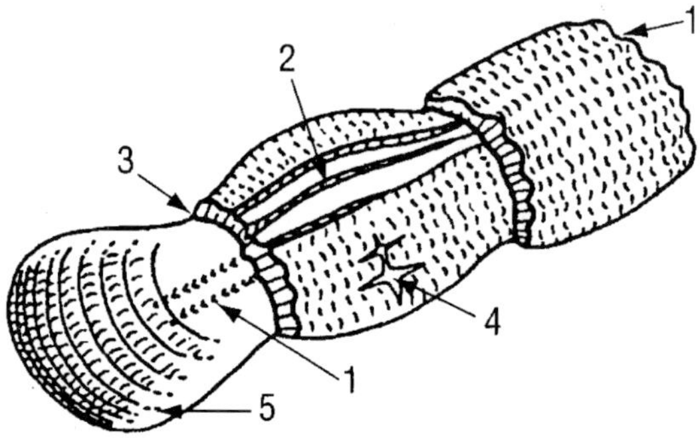

Fig. 7.9. Scheme showing the main features visible on pillow-lava surface. 1. Corrugations. 2. Longitudinal extension crack. 3. Transverse extension crack. 4. Tension groove. 5. Ropy surface (after Yamagishi, 1985).

expansion crack is fast enough (5 cm/s). These parallel ridges develop perpendicularly to the crack, with a wavelength of 1–10 cm, and a height of 1–3 cm.

The final growth of some pillows occurs through a 'bread crust' crack network, with cracks anastomosing and disposed radially from high-expansion centres. The opening of these cracks sealed with fresh lava is limited (a few cm).

The matrix. When pillow lava stands completely immobile, just after consolidation of its glassy crust, the matrix remains intact. On the contrary, when the movement of the flow goes on, this crust is broken and exfoliated by internal deformation, or by friction with neighbour pillows, forming the pillow matrix. The exfoliated glassy fragments surrounding the pillow are rapidly altered to chloritic and palagonitic products by reaction with superheated seawater trapped between the pillows. Several glass crusts may be successively exfoliated at the margin of the same pillow, which may explain the thickness of the matrix, which often exceeds the thickness of a single glassy crust. During the aging of the oceanic crust, a layer of manganese oxides/hydroxides usually grows slowly over the palagonite layer. Measurements of the thickness of the palagonite and/or manganese layers have been used sometimes with some success to estimate the relative age of various pillow volcanoes in the same area (Hekinian and Hoffert, 1975; Juteau *et al.*, 1980), calibrated sometimes with the help of the fission tracks method (Storzer and Selo, 1976).

The cement. Finally, the inter-pillow cement is the material filling the void spaces between them. This interstitial material must be examined carefully in order to demonstrate its sedimentary or hydrothermal origin: in the first case, the cement is nothing but pelagic sedimentary ooze slowly settling over the pillow flow, and infiltrating through all inter-pillow spaces. This is the case of numerous limestone cements rich in pelagic microfauna (radiolarians, diatoms, etc.). In the second case, the cement is formed from post-magmatic hydrothermal circulations. The case of radiating zeolitic cements is one of the less ambiguous. In other cases, the distinction may be more difficult: recrystallized limestone cements, silicic cements, etc. Core samples from the DSDP and ODP have given splendid examples of these various cases, and allowed them to be studied in great detail.

Polarity criteria and internal differentiation of pillow lavas

Morphological criteria. In the ophiolitic series, the morphological criteria of polarity are represented by the 'bean' shapes, the peduncles, etc. (Figure 7.8A). The 'standard pillow' is often represented as grossly ellipsoidal in shape, 1 to 4 m long, 0.5 to 3 m wide, and 0.5 to 1 m high. Pillow lavas, however, are most often elongated and sinuous lava tubes, sometimes several tens of metres long, and not ellipsoids. A two-dimensional view along geological sections has often masked this fact, but well-exposed outcrops in three dimensions (Semail ophiolite in Oman, Troodos ophiolite in Cyprus, etc.) show very well the fundamental *tubular aspect* of pillow lavas, also evident in the axial zone of present-day oceanic ridges.

Vertical differentiations. Gravitative vertical differentiations may occur inside a pillow (Figure 7.8C). This indicates that when a pillow stops its course, the magma below the glassy crust is still fluid enough to allow gravity settling of the phenocrysts, and ascent of the gas bubbles. This process is frequently observed: when the lava is aphyric, gas bubbles cluster in the upper part of the pillow; when it is phyric, the phenocrysts settle down, and the bubbles rise up.

Besides the information they give about magma viscosity when the lava flow comes to rest, these differentiations are valuable because: (a) they constitute excellent *polarity criteria*, in the case of folded palaeo-volcanic series for instance; (b) they precisely fossilize the horizontal plane, allowing calculation of the slope of the lava flow (Figure 7.8D). The floor of the intra-pillow cavities and geodes may also fossilize the horizontal plane, and glass 'stalactites' hanging from their roof may indicate the vertical plane (Figure 7.7C3c). Other criteria, for instance the observation of 'segregation vesicles' (gas bubbles partially infiltrated by residual magma, Figure 7.8Dc), have been used with success. Dredged samples could thus be reoriented in their original position with the help of these criteria, allowing study of their *in situ* magnetic properties (Bideau *et al.*, 1977).

Radial structures of pillows. Pillow lavas also show numerous radial structures (Figure 7.8B), such as:

—the radial orientation of peripheral gas bubbles;
—radial jointing. The presence of these typical cracks, developed during thermal contraction upon fast cooling of the pillow, is one of the best criteria to distinguish pillow lavas from any kind of spherical erosion of massive basic rocks;
—radial columnar jointing, spectacular phenomenon developing in huge, decametric pillow lavas.

Fig. 7.10. Sketch of a 'haystack' or pillowed cone marking a volcanic vent, typically made of radially draped elongate pillows, without summit depression (after Ballard and Moore, 1977).

Pillow volcanoes and volcanic vents

Pillow-lava flows build up volcanic edifices and ridges: they generally pond above the feeding fissures, and build *fissural volcanoes* characterized by very steep slopes (currently 60°), about 50 to 80 m high and several hundreds of metres wide, elongated parallel to the ridge or circular. Pillow-lava volcanoes may reach a height of 300 m on slow-spreading ridges, where they represent more than 80% of the volcanic products.

The *volcanic vents* are not easy to observe. They are sometimes marked by pillowed cones made of radially draped elongate pillow tubes, 5 to 15 m high, and 10 to 50 m in diameter (Figure 7.10). These are the '*haystacks*' described in the FAMOUS area (Mid-Atlantic Ridge, ARCYANA, 1978). In some cases, circular summit depressions, 10 to 20 m in diameter, or small central calderas were observed (Red Sea axial graben, Juteau *et al.*, 1983; Mid-Atlantic Ridge, Smith and Cann, 1990, 1993).

7.3.2 Lava lakes and fluid flows

Lava lakes

Bulky, voluminous fissural eruptions produce lava lakes. These structures were discovered at intermediate to fast ridge axes: first along the Galapagos rift with the help of deep-tow cameras (Lonsdale, 1977), then during the CYAMEX (for Cyana-

Fig. 7.11. Lava lakes. (A) Fossil glassy surfaces of lava lakes from the East-Pacific Rise, showing draperies, ropy and whirling structures (after Bryan, 1992).

Mexico) French–American cruise in 1978 in the RITA (from RIvera and TAmayo fracture zones) area (East-Pacific Rise at 21°N: Francheteau *et al.*, 1979, 1980; Ballard *et al.*, 1981). The chilling effect of cold seawater is negligible on a very large volume of hot lava issued from a wide fissure and rushing downslope: such an incandescent mass of lava will invade all the topographic lows located on its way, forming a real temporary lava lake whose planar surface over various square kilometres contrasts sharply with the surrounding pillow lavas.

Early drainage, occurring long before the cooling and complete consolidation of this lava lake, results generally in the collapse of its central part, leaving a field of residual, vertical basaltic pillars, 5 to 15 m high. Along the lateral margins of the lake, its former glassy surface, often lobated, hangs between the pillars, joining their summits. This surface sometimes shows whirling structures, reflecting the existence of turbulent convective currents in the lake before its consolidation (Figure 7.11A). These ruiniform landscapes, evoking swallowed-up acropoles, make up about 50% of the flows in the RITA area (Figure 7.11B).

The pillars are hollow: according to Francheteau *et al.* (1979), they result from the vertical percolation of hot seawater, trapped below the flow, through the still liquid lava lakes. Each geyser of hot water (about 400°C) has a cooling effect on neighbouring basaltic lava, and becomes surrounded by a vertical cylinder of early solidified basalt, which will remain standing during lava-lake draining (Figure 7.11B). Horizontal glassy rings, very regularly spaced, hang on the external

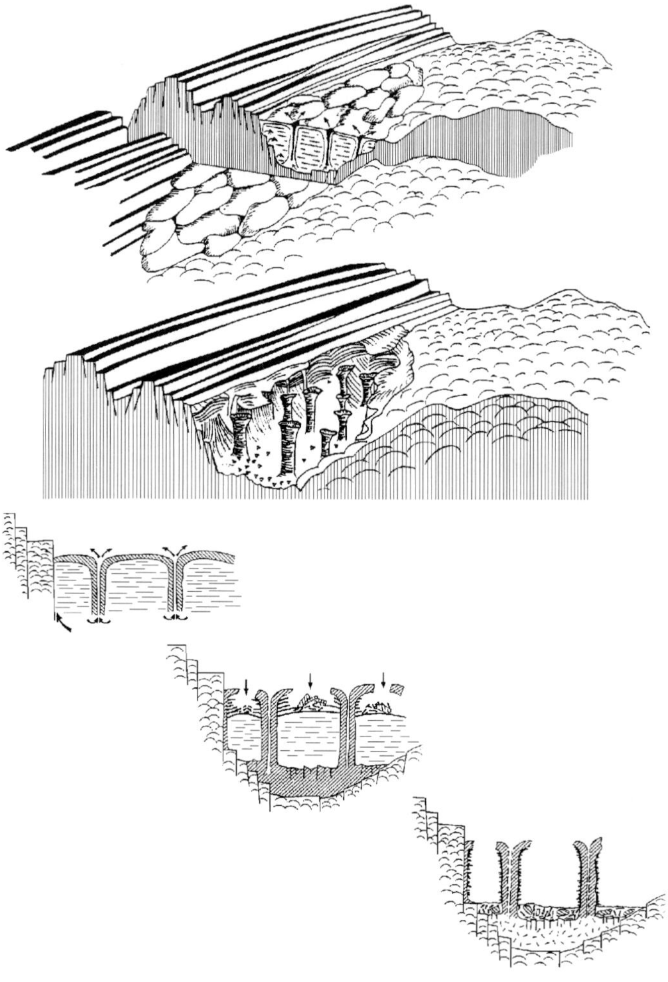

Fig. 7.11 (*cont.*). (B) Principle of evolution of a lava lake, with formation of standing pillars. A huge volume of fluid basaltic lava irrupts in a topographic low. The seawater trapped below the lava is heated and forms powerful hot-water geysers passing through the hot-lava lake, provoking crystallization of a basaltic sheath around each geyser (formation of hollow pillars, by relative thermal chilling: the lava is at 1200°C, the hot water at 400°C). Fast draining of the lake provokes the collapse of the central parts, leaving the pillars standing up (after Francheteau *et al.*, 1980).

wall of the pillars, as well as on the lateral chilled margins of the lava lakes, marking the progressive decrease in the level of lava in the lava lake, during its natural draining.

Sheet flows

The less thick of these lava lakes (1–3 m thick) generally do not suffer any drainage, and consolidate as massive basalts: their planar surface is often folded, like the *pahoehoe-type* lava flows of Hawaii. These sheet flows correspond to intermediate volumes of emission, between those of lava lakes and those of pillow lavas.

These sheet flows may exhibit nice columnar jointings when observed along fault scarps, the thermal contraction cracks being developed during the fast cooling of the flow. Sheet flows may also drape the more or less steep volcanic slopes, when they freeze before reaching a topographic low.

7.3.3 Breccia flows and deep hyaloclastites

'Reactive' flows are fluid flows which are auto-brecciated during their extrusion, producing hyaloclastic flows. This type of flow is difficult to observe directly on the ocean floor, but has been crossed many times by the DSDP and ODP drilling cores, interbedded with other kinds of lava flows. Excellent descriptions can be found for instance in the Initial Reports of Legs DSDP 46 and 49.

The mechanisms of formation of these deep hyaloclastites are very different from those invoked for the formation of hyalotuffs in shallow waters. The main mechanism seems to be due to the development of important stresses during the fast consolidation of phyric flows rich in phenocrysts, hence with a particularly high viscosity with respect to the average viscosity of MORB-type basalts. Developed on the scale of the whole flow, these stresses result in its complete brecciation as glassy basaltic clasts with angular morphology and straight contours (Figure 7.12).

7.3.4 Sills and dykes

Basaltic dykes and sometimes intrusive sills cut through the volcanic pile. They can be observed along fault scarps, and also along the flanks of transform valleys. In this latter case, the submarine erosion often leaves the dykes bare, standing as vertical walls (up to 30 m high), detached from the main cliff. DSDP/ODP holes in the upper oceanic crust have also met a number of basaltic sills and dykes, intruded in the volcanic pile.

7.4 RELATIONSHIPS BETWEEN PILLOW LAVAS AND LAVA LAKES

In the axial zone of intermediate to fast-spreading ridges, lava lakes, sheet flows and pillow lavas follow each other in time and in space. Lava lakes are generally set in tectonic grabens affecting pillow lavas previously emplaced, and may be themselves covered with younger pillow lava flows. The variation in the delivery rate of the feeding fissures control these alternations:

—The tips of sheet flows are often made of pillow lavas when the lava, at the end of its course, is channelized in tubes with moderate delivery rates (Van Andel and Ballard, 1979).

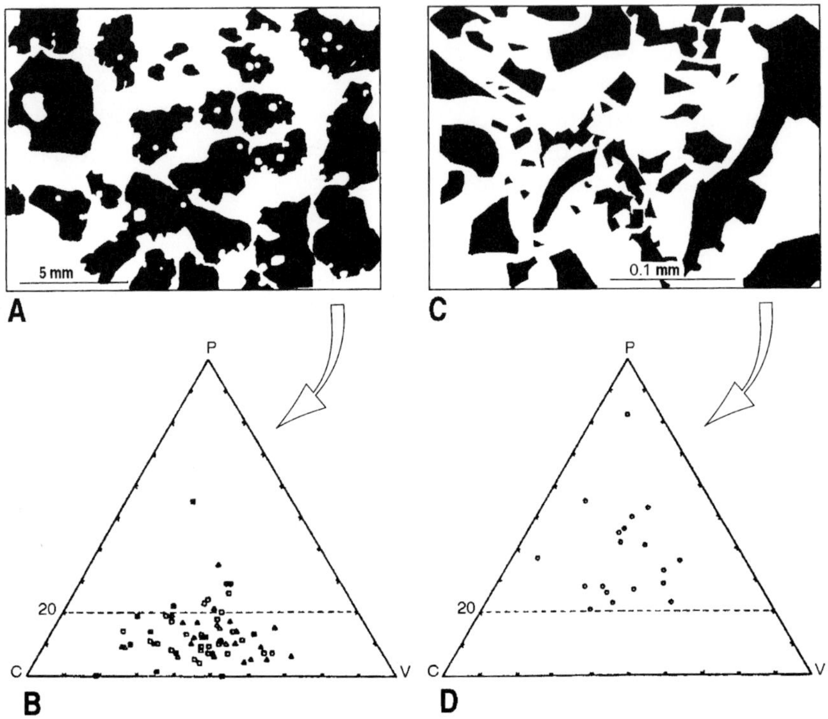

Fig. 7.12. Thin-section aspects and morphometric parameters used to distinguish hyalotuffs (formed in shallow waters) from deep hyaloclastites. (A) Very vesicular hyalotuffs from Surtsey, with typical ragged, curvilinear morphology of the fragments. (B) Corresponding P–V–C ternary diagram (P: planarity; V: convexity; C: concavity, in % of grain outline length). (C) Deep hyaloclastites from a Pacific seamount, with angular and linear morphology of the fragments. (D) Corresponding P–V–C ternary diagram (after Honnorez and Kirst, 1975). Dashed line: empirical boundary between both types.

—The walls of the graben allow pillow-lava edifices laying over sheet flows to be observed: every time the magmatic delivery rate slows down, the sheet flows or lava lakes disappear and give way to pillow lava. Figure 7.13 shows how variations in delivery rate, linked to variations in tectonic spreading rates in the axial domain, control the construction of the volcanic layer.

7.4.1 Influence of the spreading rate on the volcanic dynamism of the oceanic ridges

On the scale of one ridge segment, the ratio between the surfaces covered respectively by lava lakes (including sheet flows) and by pillow lava is correlated directly with the spreading rate (Figure 7.14): it is maximum and close to 100% on the fastest segments of the East-Pacific Rise (Renard *et al.*, 1985), and minimum and less than 10% on the slowest ridges, for instance on the Mid-Atlantic Ridge (Ballard and

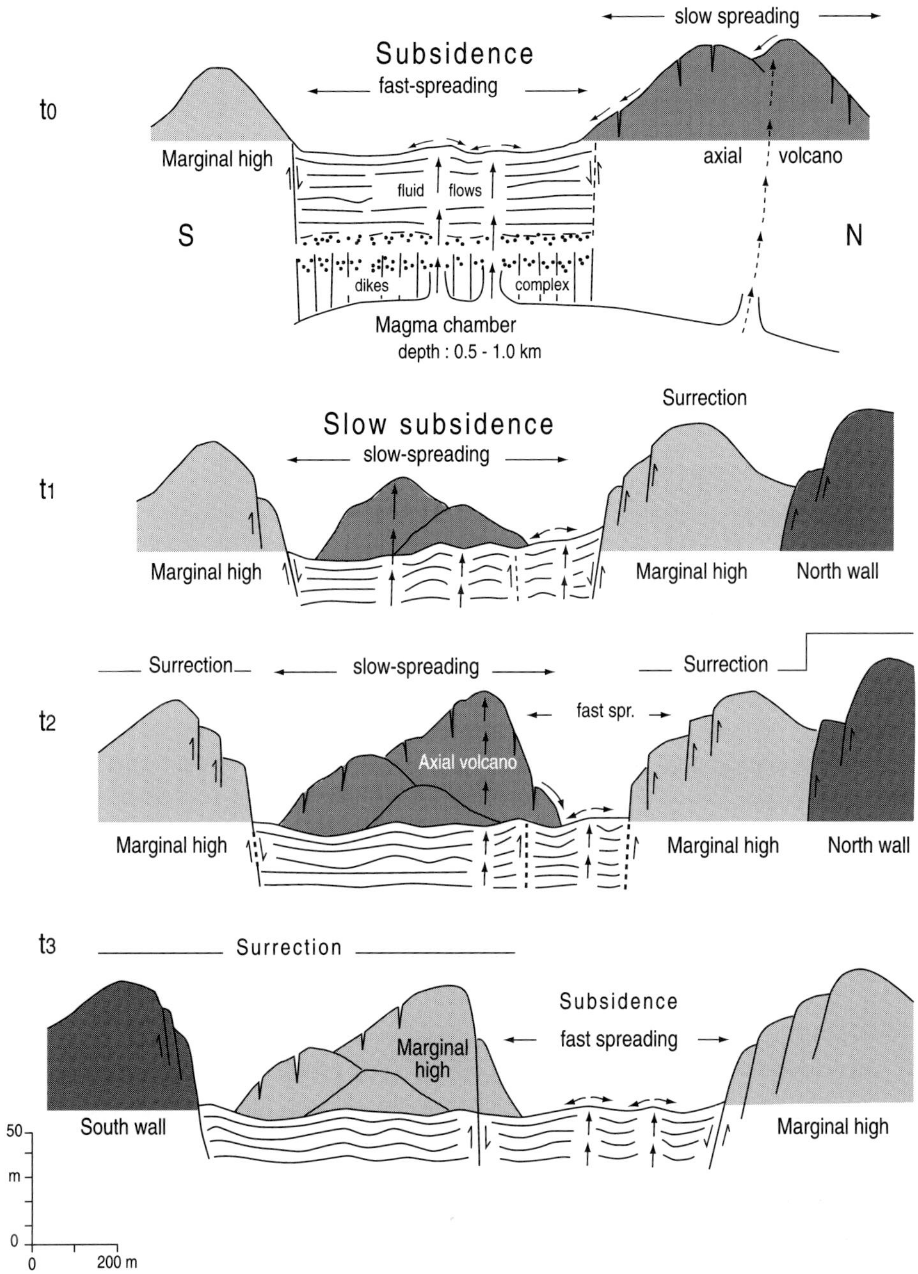

Fig. 7.13. Scheme illustrating in four stages the building of the axial zone of an intermediate-spreading ridge, during a 'volcanic period' of about 10 000 years. During the periods of fast extension linked to high-delivery rates of fluid sheet flows (horizontal lines), these lavas are ponded in topographic lows (t_0, t_3). They are succeeded by pillow-lava flows (stippled) as soon as the delivery rate decreases (periods of slower extension t_1, t_2). Pillow lavas build up volcanic edifices with steep slopes. Example based on a detailed study of the Galapagos rift at 86°W (Cocos–Nazca ridge) (after Van Andel and Ballard, 1979).

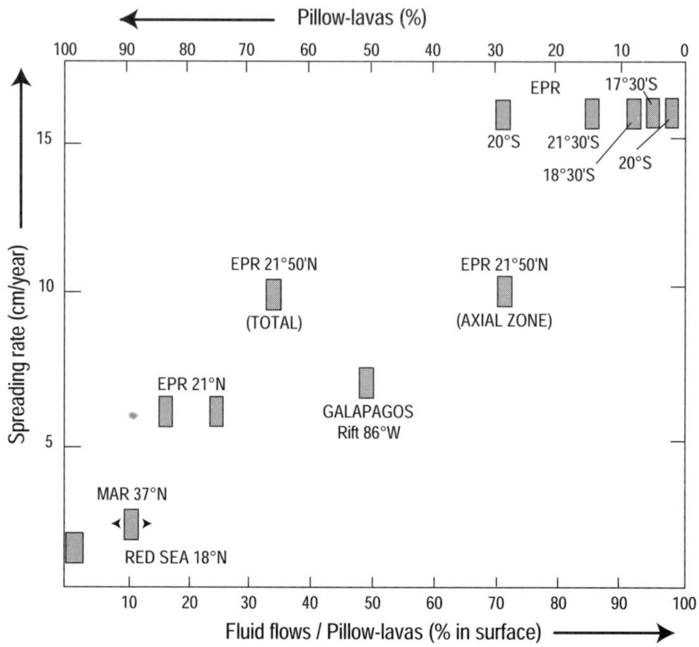

Fig. 7.14. Diagram showing the correlation between the sheet flow (including lava lakes)/pillow flow surface ratio and the spreading rate (after Bonatti and Harrison, 1988).

Moore, 1977; Crane and Ballard, 1981), or in the Red Sea axial rift, where the ratio of sheet flows/pillow flows is close to zero (Juteau *et al.*, 1983).

Detailed petrological studies have shown that the basalts from sheet flows and lava lakes are quite aphyric (devoid of phenocrysts) and show high chilled temperatures, whereas the basalts from pillow lavas always contain a certain charge in phenocrysts, and exhibit lower cooling temperatures (Juteau *et al.*, 1980; Thompson *et al.*, 1985): both observations clearly reflect the lower viscosity of sheet flow basalts with respect to that of pillow lavas.

Moreover, it has been shown that the least phyric basalts from the East-Pacific Rise correspond to the central and shallowest parts of the accretion segments, while their content in phenocrysts increases in the vicinity of the segment tips, near the transform faults for instance, where the depth of the ocean floor increases (Eissen *et al.*, 1981; Thompson *et al.*, 1985). Bryan (1983) has shown that the crystal/liquid ratio in basalts from the Mid-Atlantic Ridge is almost double that measured in the basalts of the East-Pacific Rise.

7.4.2 Relations between crystal charge of the basalts, viscosity and eruptive dynamism

The data presently available on basalts from oceanic ridges show that:

—On the same ridge segment, the chilling temperatures of basalts are the highest, and its content in phenocrysts as well as fractionation degree the lowest in the shallow, central areas, far from the segment tips. When approaching the extremities of the segment, the lava becomes more evolved, richer in phenocrysts, chilling temperatures are lower, and the lava lakes/pillows ratio decreases. These various parameters attest to a higher viscosity for basalts when approaching the tips of the ridge segments.

—The crystal/liquid ratio increases and chilling temperatures decrease from the lava lakes to the sheet flows, and then to the pillows. This indicates that viscosities at the moment of extrusion are increasing from the lava lakes to the pillow lavas.

—The pillows/lava lakes ratio and the phenocryst content of the basalts are a lot higher on the Mid-Atlantic Ridge (except in the vicinity of hot spots such as Iceland or the Azores) than on the East-Pacific Rise. The spreading rate seems to play an important part in the expression of volcanic dynamism, because it controls a number of important parameters: thickness of the crust in the axial zone, width of the feeding conduits, frequency of volcanic eruptions, volumes of delivered lavas, etc.

—On fast-spreading ridges (East-Pacific Rise), the central part of the accretion segments is shallower, the magma chambers very close to the ocean floor (1.5 to 3 km), the feeding fissures wide. All this contributes to slowing down the transit of basaltic magma between the magma chamber and the ocean floor. The basalt is extruded at high temperature, in surfusion, at low viscosity, without crystals and with a high delivery rate: internal stresses within the lava are minimized, hence sheet flows and lava lakes will be prominent. When approaching an axial discontinuity (fracture zone, etc.), the roof of the magma chambers is more distant from the ocean floor, the magmas have begun crystallising, and the time of transit is longer. The basalt is emplaced at lower temperatures, with a more important crystal load, the viscosity is higher and the volumes delivered are lower: the pillows/lava lakes ratio increases consequently (Francheteau and Ballard, 1983).

—On slow-spreading ridges (Mid-Atlantic Ridge), the magma chambers are small, ephemeral, more distant from the seafloor. The lavas delivered are often phyric, hence more viscous, and the volumes of lava delivered are moderate: pillow lavas are thus prominent, and even omnipresent.

7.5 CRYSTALLIZATION OF MORB

7.5.1 Mineralogical compositions and textures

Pillow lavas emplaced on the ocean floor are among the *most rapidly cooled igneous bodies* on Earth. This is due to the fact that seawater, despite its freezing temperatures at great depth (2° to 3°C on the ocean floor), has higher thermal conductivity and heat capacity than the atmosphere. These specific conditions have important consequences on crystal morphology and textural variation in the oceanic-ridge basalts.

The most common minerals present a strange and puzzling morphology for the student: 'hopper', 'lantern' or 'swallow tail' olivine crystals, 'buckle belt' plagioclase crystals, 'cockscomb' pyroxene crystals, etc.

Textural variations

At the beginning of the 1970s, before the DSDP drillings in the upper oceanic crust, Bryan (1972) distinguished three main textural zones in dredged, 'camembert slice' pillow samples: (1) an external, mainly glassy 'vitrophyric' margin, containing only scattered phenocrysts and microphenocrysts; (2) a variolitic zone, characterized by skeletal mineral growths in a glassy matrix; (3) an internal holocrystalline zone showing an intergranular, microlitic texture.

Kirkpatrick (1978) was the first to describe in great detail and to interpret the textural and mineralogical variations of pillow-lava margins, based on both laboratory experiments on lunar basalts, and on the nucleation/crystallization theory for silicate melts. After selecting various complete pillow sections in the 205 m of basalts drilled during Leg DSDP 46 in a 13-Myr-old oceanic crust (Mid-Atlantic Ridge), he defined *six concentric zones* parallel to the pillow margin. In the olivine-rich pillows, these zones are (Figure 7.15):

—Zone 1 (3 mm): glassy margin (100% glass).
—Zone 2 (4 mm): glass with isolated olivine dendrites and plagioclase spherolites.
—Zone 3 (3 mm): coalescent olivine dendrites and partially coalescent plagioclase spherolites, with interstitial glass.
—Zone 4 (6 mm): fully coalescent plagioclase spherolites, with clear boundaries between spherolites.
—Zone 5 (20 mm): fully coalescent plagioclase spherolites with characteristic 'bow-tie' morphology and fuzzy boundaries.
—Zone 6 (up to the pillow core): plagioclase microlites, with clinopyroxene dendrites between the microlites.

In pillows without olivine in the groundmass, the six previous zones are all recognizable, but their thicknesses and crystal sizes are considerably less.

Quantitatively, Kirkpatrick also observed that the spacings between the arms of skeletal plagioclases or dendritic clinopyroxenes increase systematically inward, in agreement with theoretical predictions (Figure 7.16).

Mineralogical compositions

Mid-ocean ridge basalts generally show at least three crystal generations:

—Idiomorphic, huge *early phenocrysts* (2 to 10 mm), which have crystallized slowly in the magma chamber.

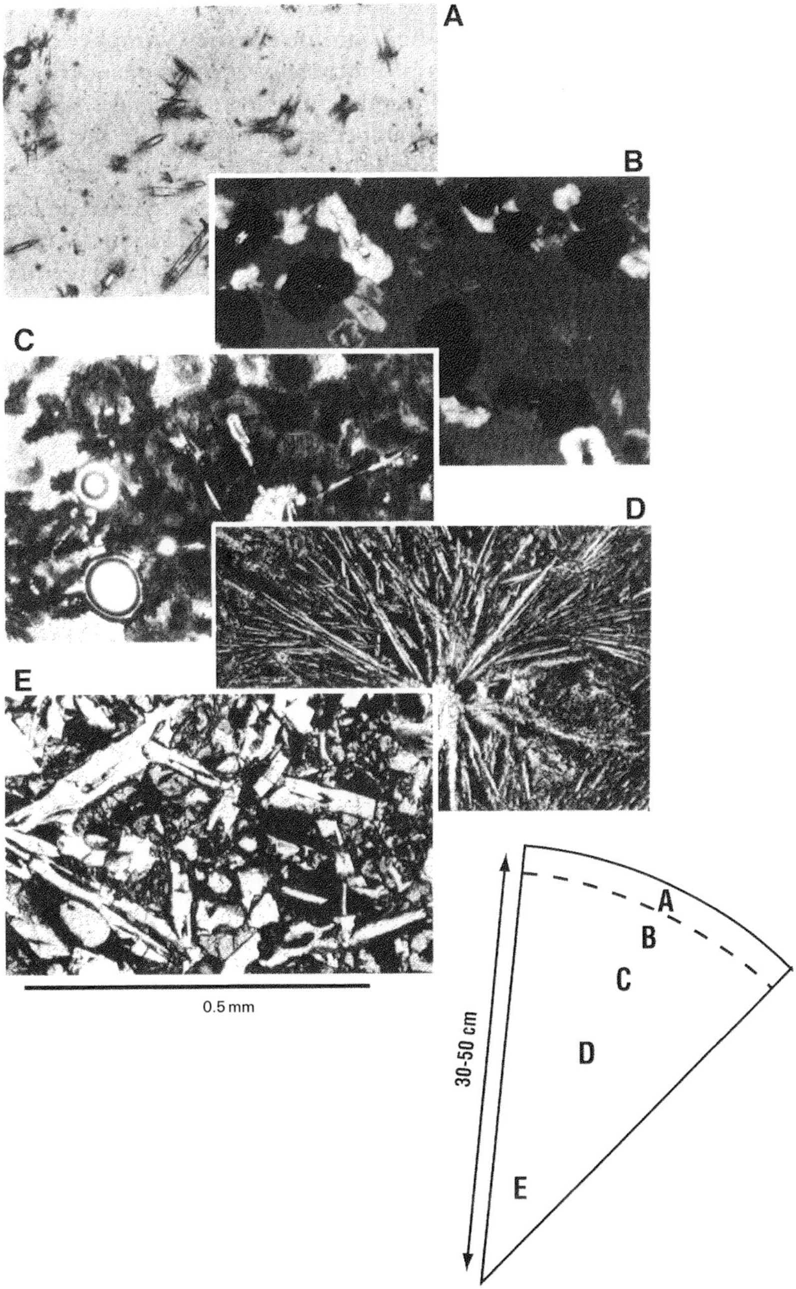

Fig. 7.15. Textural evolution of pillow lava from Leg DSDP 45 in the north Atlantic, from margin to core, showing the main zones defined by Naitland (after thin-section micro-photographs published by Naitland, 1978).

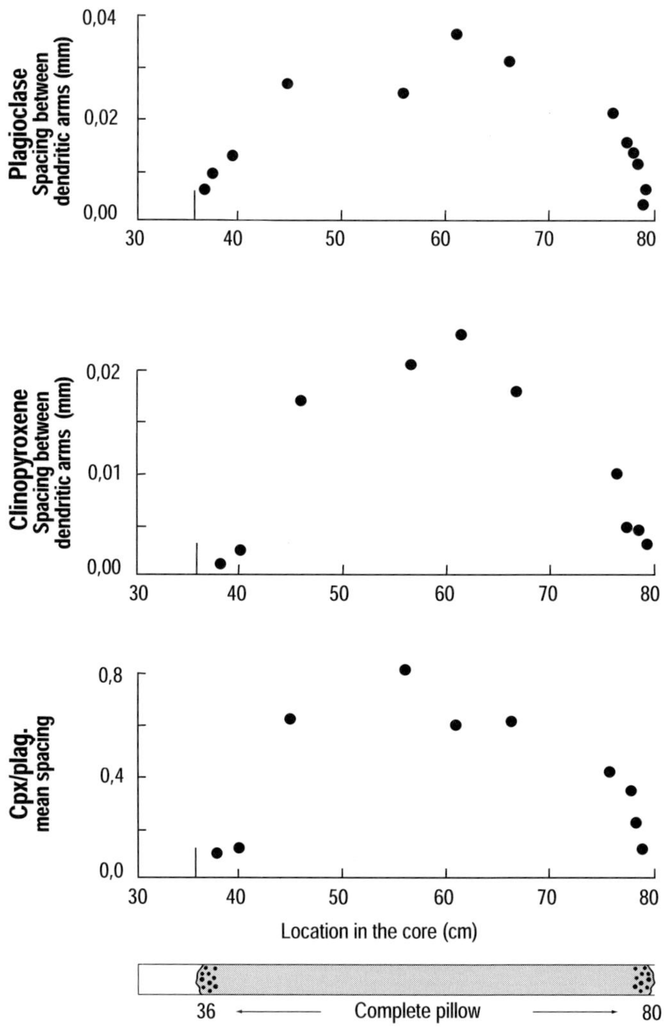

Fig. 7.16. Maximum spacing measured between the dendrites of skeletal plagioclase and pyroxene microlites, versus their location within the core of a complete pillow-lava core, from margin to margin (cm 36 to 80). This parameter increases from the pillow margins to the core, in agreement with theoretical predictions (after Kirkpatrick, 1978).

—Smaller, more elongated and skeletal *microphenocrysts* (0.5 to 2 mm), which probably crystallized during the ascent of the magma in the shallow conduits, just before eruption.
—Hastily crystallized *microlites* (<0.5 mm) with characteristic chilled textures, which crystallized in the lava flows, after extrusion on the seafloor.

To these three main crystal generations, one must add the more or less *resorbed xenocrysts* (rounded, or with 'corrosion gulfs'), which appear to be in clear disequilibrium with the basalt magma which transported them up to the surface. The origin of these xenocrysts may be related to various processes: early comagmatic phases crystallized under high pressures, mixing of magmas in the magma chamber, residual phases from the mantle or from the Layer 3, etc.

The various crystal generations in a basalt can be identified on both *morphological criteria* (sizes and shapes, Figure 7.17), and *chemical composition criteria*, the phenocrysts having more 'primitive' compositions, and the microphenocrysts and microlites having more and more 'evolved' compositions respectively (Figure 7.18).

Phenocrysts. Oceanic-ridge basalts generally contain plagioclase and olivine phenocrysts and microphenocrysts, in very variable modal proportions. Plagioclase is, however, the most common phenocryst phase, with rectangular shapes, sometimes rounded at the corners, an unzoned calcic core (An 75 to An 85, % anorthite) and a finely zoned margin, markedly richer in Na (up to An 50 on the external margin). The other common phase is a forsteritic olivine (Fo 88 to 90, % forsterite), as huge idiomorphic or rounded crystals, with an unzoned core here as well and a more fayalitic, finely zoned margin. Some samples exhibit only olivine phenocrysts, whereas others show only plagioclase phenocrysts. Many ocean-ridge basalts are totally aphyric, especially those from fast-spreading ridges (East-Pacific Rise). Small, translucent brown chrome-spinels, often octahedral, are commonly observed in the glassy margins, or as inclusions in the olivine or plagioclase phenocrysts or microphenocrysts, in holocrystalline samples.

Clinopyroxene is a much rarer phenocryst phase, often as corroded, rounded xenocrysts, in clear disequilibrium with the host liquid. The basic magmas in which these pyroxenes were formed were probably mixed with more evolved magmas in the magma chamber, resulting in corrosion of the primitive clinopyroxenes. Dungan *et al.* (1978a, 1978b) have studied these processes in detail by analysing the glass inclusions and the often complex zoning in the phenocrysts, and have demonstrated the importance of *magma mixing processes* in the oceanic accretion magma chambers.

Microphenocrysts. In this second generation, plagioclase is the most abundant phase, represented as elongated, rectangular crystals, often showing skeletal growth aureoles, and typically labradoritic compositions in equilibrium with the glassy margins of the pillow lavas. Olivine is also a common phase in this generation, appearing as small idiomorphic to skeletal crystals whose composition is also here in equilibrium with the glass. It is clear that olivine in most of the ridge basalts was at liquidus at the moment of extrusion from the seafloor, even if, later on, a reactional relation develops with the residual liquid and results in the disappearance of olivine in the groundmass (Cann, 1981). Augite is rare in this generation. When it appears, the clinopyroxene often cements olivine-plagioclase clusters, forming gabbroic glomerocrysts.

Microlites. Microlitic plagioclase is always less abundant in the groundmass of holocrystalline samples (pillow cores for instance), even if it is not present as phenocrysts. On the other hand, olivine is abundant in the groundmass only if it

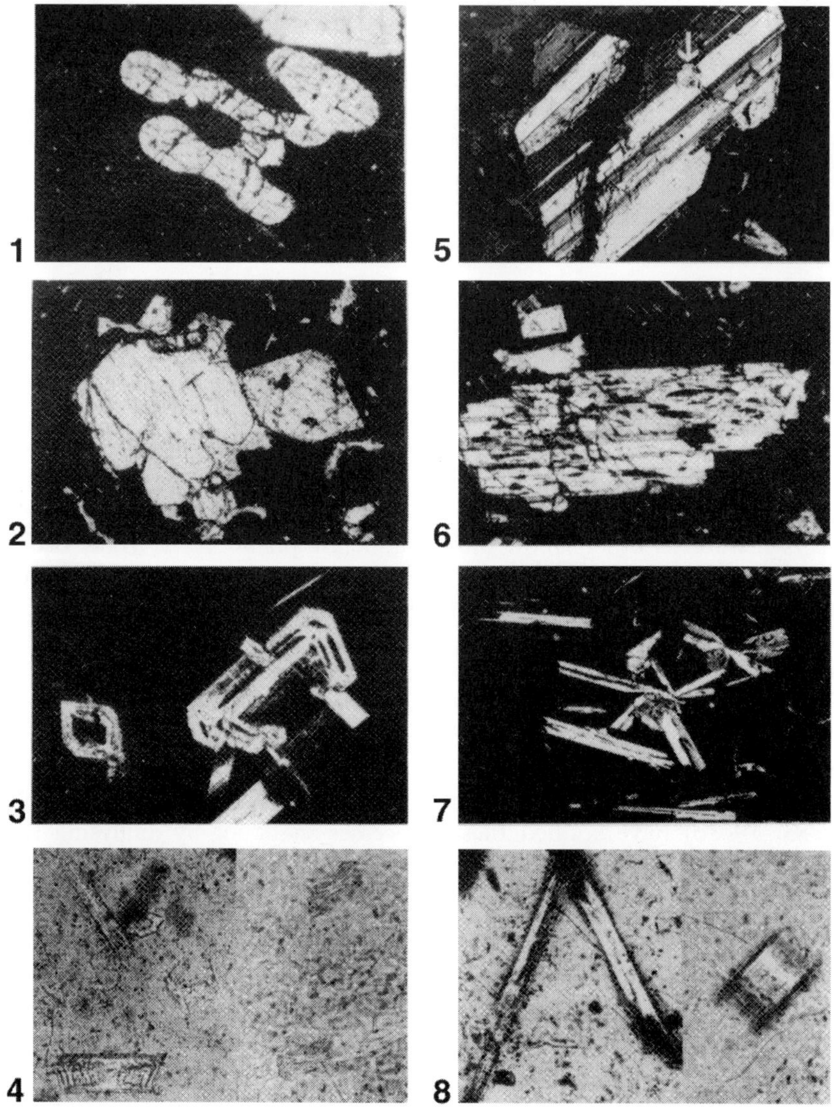

Fig. 7.17. Morphological identification of four crystal generations in a basalt set collected by submersible at the East-Pacific Rise axis (21°N, Cyamex cruise). **Olivine** (left side): 1. Olivine phenocryst, deeply resorbed (first generation, ×40). 2. Idiomorphic olivine phenocryst (second generation, ×40). 3. Skeletal, sub-idiomorphic olivine microphenocryst (third generation, ×80). 4. Olivine microlites and crystallites, quenched in the fresh glass, showing the typical '*ladder*' or '*swallow tail*' morphology (fourth generation, ×200). **Plagioclase** (right side): 5. The arrow shows the relict of a resorbed and rounded plagioclase phenocryst (first generation, ×40), included in an idiomorphic phenocryst of the second generation. 6. Idiomorphic plagioclase phenocrysts, stippled with numerous glass inclusions (third generation, ×40). 7. Idiomorphic plagioclase microphenocrysts, showing skeletal tips and '*buckle belt*' morphology (third generation, ×40), associated with olivine microphenocrysts of the same generation. 8. Plagioclase microlites, quenched in the fresh glass (fourth generation, ×200) (after Juteau *et al.* (1980).

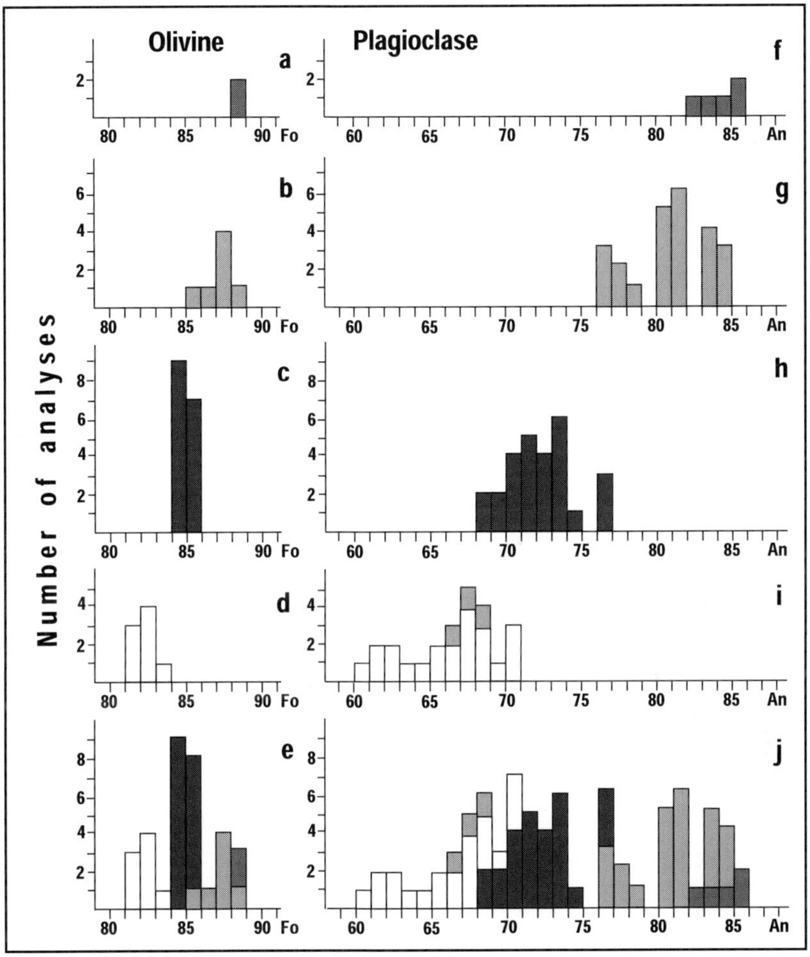

Fig. 7.18. Chemical identification of four crystal generations in the basalts of Figure 7.17. Frequency diagrams for olivine (Fo %) and plagioclase (An %) crystal compositions. Olivines: (a) First generation, resorbed phenocrysts. (b) Second generation, idiomorphic phenocrysts. (c) Third generation, skeletal microphenocrysts. (d) Fourth generation, quenched microlites. (e) Cumulated frequency diagram for all olivine crystals. Plagioclases: (f) First generation, resorbed phenocrysts. (g) Second generation, subidiomorphic phenocrysts. (h) Third generation, elongated and skeletal microphenocrysts. (i) Fourth generation, quenched microlites. (j) Cumulated frequency diagram for all plagioclases (after Juteau *et al.* (1980).

also appears as phenocrysts. Clinopyroxene may be abundant in the groundmass, generally poorly crystallized. Brown spinel, when appearing as isolated grains in the groundmass, is rimmed by secondary, black and opaque magnetite.

Magmatic sulphides. Spherical sulphide globules, 2 to 300 μm in diameter, are commonly scattered in basalt. They are generally composed of a mixture of pentlandite (Fe–Ni-rich sulphide) and chalcopyrite (Fe–Cu-rich), and result from

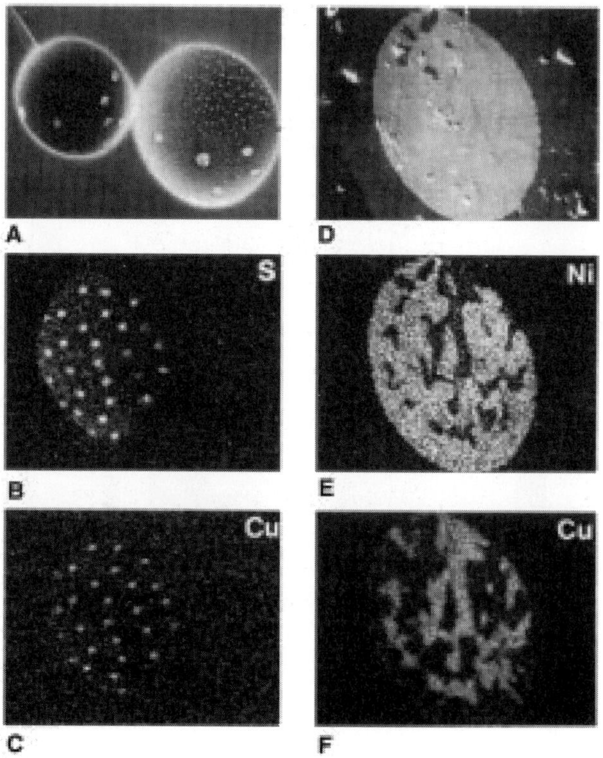

Fig. 7.19. Decorated vesicles and sulphide globules in MORB glasses. (A) Scanning electron microscope (SEM) picture of two decorated vesicles. The vesicle to the right is 30 m in diameter. The inner wall shows a regular arrangement of tiny sulphide globules. (B) Image of the sulphur distribution in a decorated vesicle by scanning. (C) Image of the copper distribution in the same vesicle. (D) SEM picture of a polished section showing a massive sulphide globule (180 m in diameter), in the glassy margin of pillow lava. (E) Image of the nickel distribution in the same globule by scanning. (F) Image of the copper distribution in the same globule (after Juteau *et al.*, 1980, Cyamex cruise).

the crystallization of an immiscible sulphide liquid phase, separated late from the silicate melt. Moreover, the walls of the gas bubbles (CO_2), always present though discreate at these depths (20–30 μm in diameter and <1% in volume), are delicately decorated with sulphide microglobules (<5 μm) of almost identical composition, as shown by scanning-electron-microscope studies (Figure 7.19).

Crystal morphologies

The specific crystal morphologies of oceanic basalts were described by Bryan (1972), and experimentally reproduced by Lofgren (1971, 1974, 1980) for plagioclase, and by Donaldson (1976) for olivine.

Olivine. This phase shows the greatest variety in crystalline morphology. It appears as a chilled phase in the external part of zone 2, where it can form skeletal

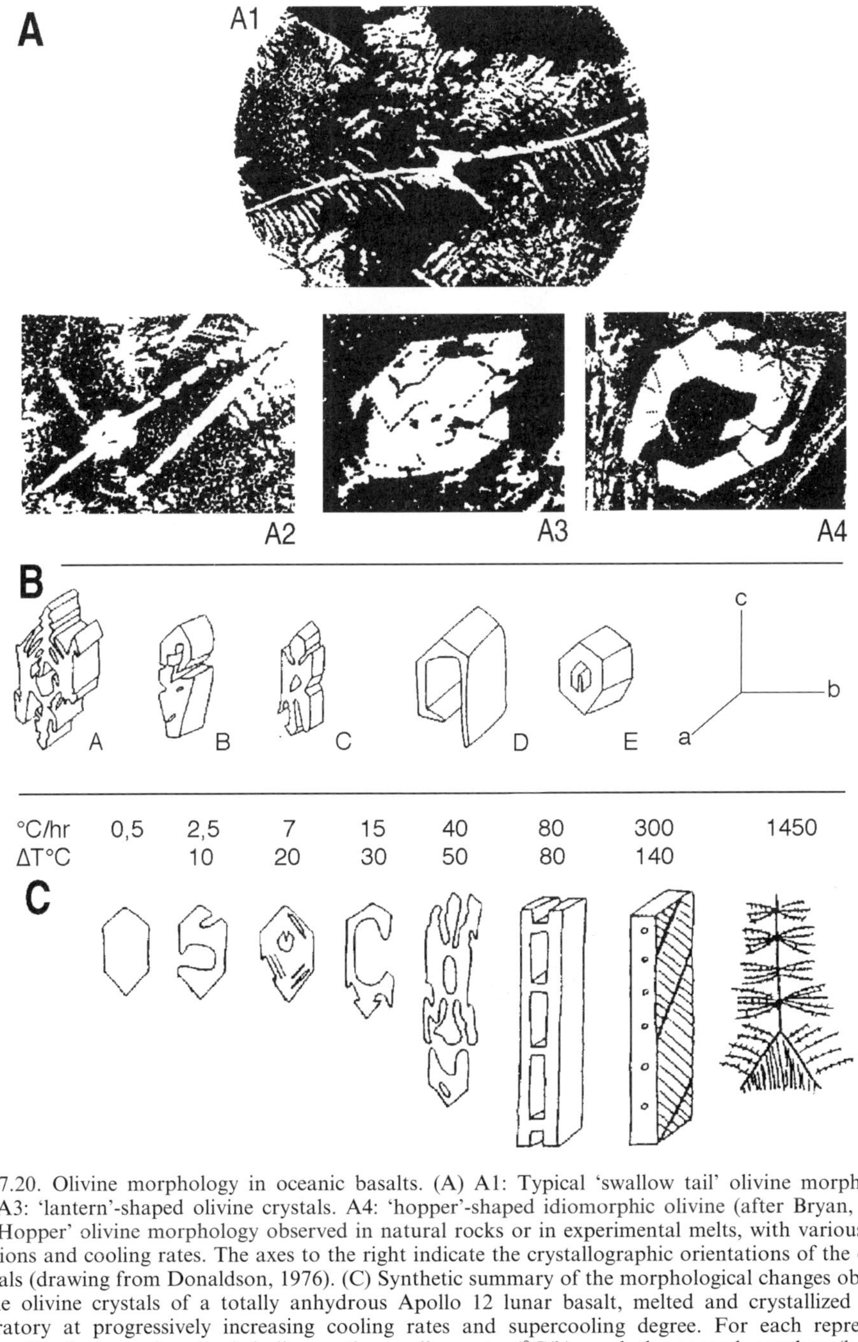

Fig. 7.20. Olivine morphology in oceanic basalts. (A) A1: Typical 'swallow tail' olivine morphology. A2, A3: 'lantern'-shaped olivine crystals. A4: 'hopper'-shaped idiomorphic olivine (after Bryan, 1972). (B) 'Hopper' olivine morphology observed in natural rocks or in experimental melts, with various compositions and cooling rates. The axes to the right indicate the crystallographic orientations of the olivine crystals (drawing from Donaldson, 1976). (C) Synthetic summary of the morphological changes observed in the olivine crystals of a totally anhydrous Apollo 12 lunar basalt, melted and crystallized in the laboratory at progressively increasing cooling rates and supercooling degree. For each represented shape, the first number (top) indicates the cooling rate (°C/h), and the second number (bottom) indicates the supercooling degree (Δt°C, gap with the liquidus temperature) (drawings from Donaldson, 1976). All crystals are drawn with *c* axis vertical and *a* axis perpendicular to the page, toward the reader.

growth networks on pre-existing olivine phenocrysts, or develop from fibrous nuclei as a finely reticulated and optically continuous structure. Towards the inner margin of zone 2, 'decorated chains' elongated following the 'c' axis of olivine are observed (Figure 7.20A). The central part of the chain may present the shape of a flattened 'Maltese cross', or of a 'decorated lantern', and some crystals have the appearance of a lantern hanging from a chain. When observed parallel to a direction perpendicular to the *a* axis, these same crystals show the characteristic 'double swallow tail' morphology, with more or less 'decorations'. Crystals cut parallel to the 'a' axis still show the double swallow tail morphology, but without decorations.

These various morphologies were all reproduced experimentally in the laboratory by Donaldson (1976), who identified 10 basic morphologies of olivine, and has shown that the crystal morphologies evolve systematically according to the cooling rate and the supercooling degree (Figure 7.20C). For increasing values of both parameters, the olivine crystals first idiomorphic and well crystallized evolve towards 'hopper' shapes (with external idiomorphic contours, but with numerous internal voids), and become more and more elongated following the *c* axis; these hollow and elongated crystals split and become chains, reticulated hoppers, plates and finally feathers and swallow tails. Each of these olivine forms is stable in a certain

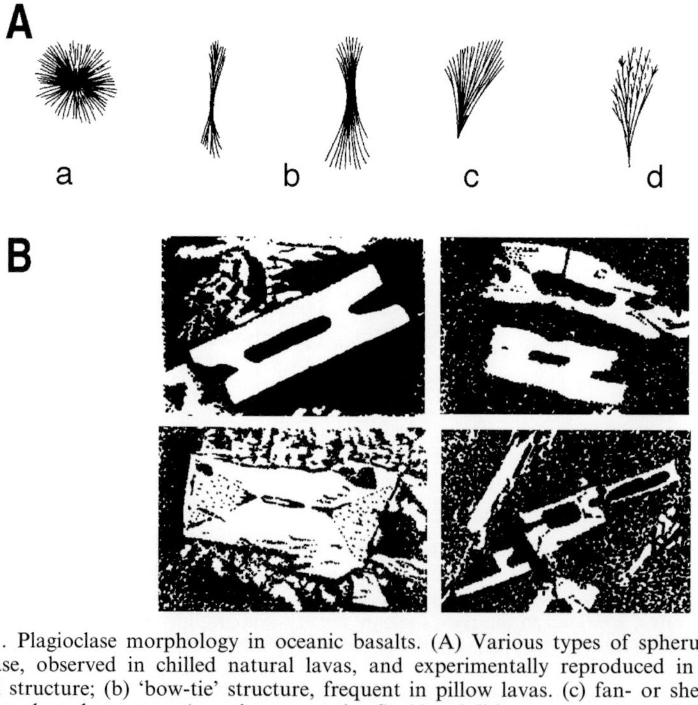

Fig. 7.21. Plagioclase morphology in oceanic basalts. (A) Various types of spherulitic morphology of plagioclase, observed in chilled natural lavas, and experimentally reproduced in the laboratory. (a) spherical structure; (b) 'bow-tie' structure, frequent in pillow lavas. (c) fan- or sheaf-shaped structure; (d) feather-shaped structure (or arborescent sheaf); (e) axiolitic structure: the spherulitic fibres develop along a line (drawings from Lofgren, 1974). (B) Typical 'buckle belt' morphology of microlitic plagioclase crystals in pillow lava, with frequent development of sector zonings, such as the common 'Martini glass' structures (lower left) (after Bryan, 1972).

Fig. 7.22. Typical 'cockscomb' clinopyroxene microlites in the glassy matrix of a basaltic pillow-lava.

temperature range, depending on magma composition: the richer the magma is in normative olivine, the lower the transition temperatures become between these various forms.

Plagioclase. Chilled plagioclase microlites also appear in the external part of zone 2, as very elongated acicular crystals following the *a* axis, with hollow cores parallel to the *a* axis filled with glass or with cryptocrystalline groundmass. These crystals have a tendency to cluster in radiating sheaves (Figure 7.21A). Cryptocrystalline intergrowth of plagioclase and pyroxene can develop with 'feather' morphologies when the plagioclase is not at liquidus. In this case, optical identification of these intergrowths is facilitated by observation under reflected light on polished sections. The plagioclase crystals observed on sections perpendicular to the *a* axis show typical rectangular and hollow 'buckle belt' morphologies (Figure 7.21B), frequent in the internal part of zone 2 and in zone 3. These rectangular buckles may cluster in the direction of the *c* axis. Zonings are common in these crystals, with the more calcic parts on the internal margin of the rectangles, close to the hollow cores. Sector zonings also appear as 'Martini glasses'. The albite polysynthetic twins develop normally and in perfect continuity on both sides of the hollow cores.

These chilling morphologies of plagioclase have also been reproduced experimentally in the laboratory (Lofgren, 1974).

Pyroxene. The optical identification of skeletal or intergranular pyroxene in pillow lavas is difficult. The chilled morphologies occur in zones 2 and 3 at the following aspects: (1) radiating intergrowths as feathers with the plagioclase; (2) intergrowths as reticulated networks with plagioclase; (3) clusters of curved crystals, like 'cockscombs' (Figure 7.22), often nucleated on a plagioclase lath or on a spinel; (4) xenomorphic interstitial grains between plagioclase and olivine crystals.

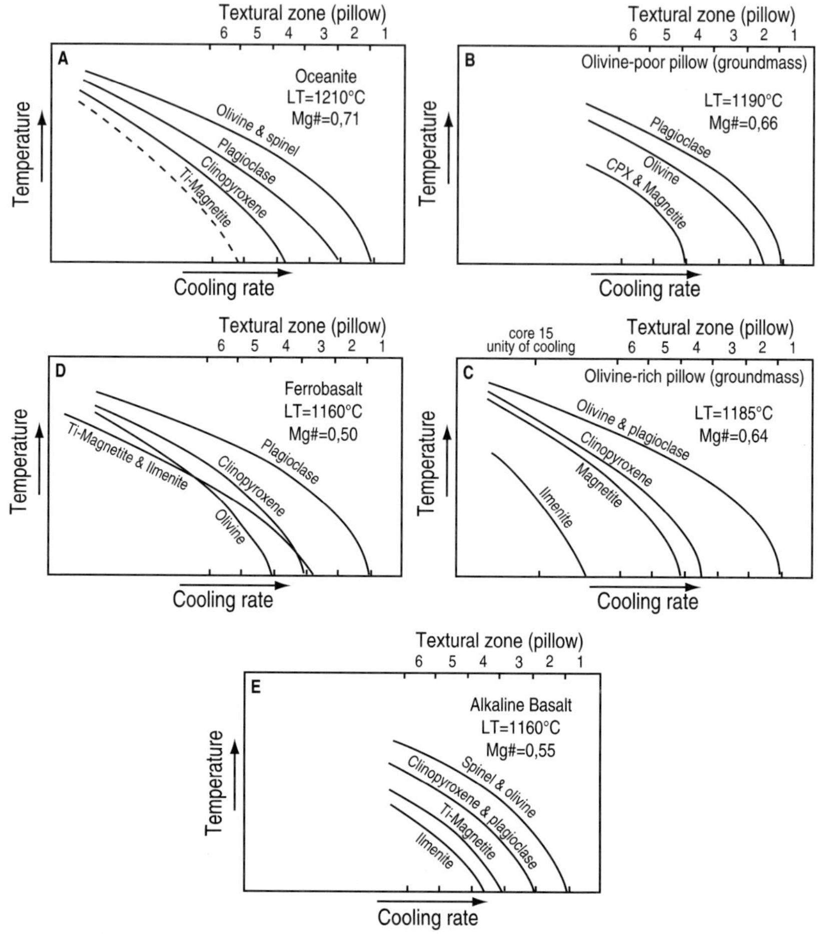

Fig. 7.23. These temperature/cooling rate diagrams, completing and extending Kirkpatrick's (1978) diagrams, predict the order of crystallization in five major oceanic basalt types, in the six textural zones defined by Kirkpatrick. (A)–(D) are abyssal tholeiites ranging from oceanites (Mg = 0.71) to ferrobasalts (Mg = 0.50), and (E) is an alkaline basalt. Liquidus temperatures (LT) were estimated from FeO and MgO content in the glass (after Natland, 1991).

Magnetite. Magnetite does not appear in zone 1, but begins to crystallize in zone 2 as tiny crystals in the groundmass. In zone 3, it is observed as interstitial grains or subidiomorphic crystals. Contrarily to chrome spinel which crystallizes early and appears as inclusions in olivine and plagioclase phenocrysts, magnetite crystallizes much later, when most of olivine and plagioclase have crystallized in the basalt.

7.5.2 Crystallization sequences in oceanic basalts

Kirkpatrick (1978) has reconstituted the crystallization sequences in the two varieties of basalts he has studied, plotted on temperature/cooling rate diagrams, well-known today. Natland (1991) has generalized these results by adding three diagrams to the first two, thus covering the whole mineralogic range of oceanic basalts (Figure 7.23).

It is clear that if distinct magma types can be identified today on oceanic ridges, the rocks cannot be reduced to simple classifications, nor to simplistic petrogenetic interpretations. The major problem resides in the fact that very few submarine basalts have crystallized under conditions close to equilibrium, and there are two reasons for this: (1) most of the MORB were extruded between 2 and 4 km on the seafloor, in freezing seawater near 0°C. The margins of the flows are hence chilled, as we have seen, as metastable glass and even the core of the thickest flows has suffered very high cooling rates, unknown in aerial environments; (2) a number of these basalts are phyric. The olivine and plagioclase (and sometimes also pyroxene or cpx-spinel) phenocrysts, megacrysts and glomerocrysts, may be abundant in MORB. It was first thought that these minerals could reflect zonings and crystallizations in magma chambers thermally stratified (Bryan and Moore, 1977), but more detailed studies have revealed the effects of magma mixing that cannot all be linked to simple differentiation processes at depth. Many of these minerals are probably the products of polybaric crystallization during magmatic evolution beginning in the mantle, and hence are indicative of these early magmatic stages, close to the initial melting events.

In summary, the idea that *very few phenocrysts, in oceanic basalts, are in equilibrium with the composition of the host glass should be retained*: the mid-oceanic-ridge basalts contain mineral assemblages in disequilibrium, whose compositions are mainly determined by the combined effects of magma mixing and of crystallization kinetics.

7.6 TECTONIC PROCESSES AT THE AXIS, AND THE VOLCANIC–TECTONIC CYCLE

7.6.1 Axial tectonic structures

Once created in the axial zone of the ridges, the young oceanic crust, just consolidated, is immediately submitted to the extension stresses of plate separation, and moves laterally away from the axial domain. A great number of ways to fracture develop in the fragile crust, which dissect it into a series of faulted blocks (Figure 7.24).

Open fissures are the most common superficial feature reflecting tectonic break up of the axial zone. By definition, they do not show any vertical displacement between both lips. Their width varies from several cm to several metres, and their length from several tens to several hundreds of metres. The walls are vertical and cut through the volcanic flows. Isolated pillow lavas may remain hanging at the surface, delicately 'perched' on one lip of the fissure, indicating the absence of significant earthquakes during the opening. In all the areas studied in detail by submersibles, hundreds of these open fissures have been described, all parallel to the ridge axis, and corresponding to a simple stress field.

Fig. 7.24. The axial graben of a very fast-spreading ridge is created by collapse of the axial zone along several steps, each of them several metres high. The basaltic bottom is extremely fissured, each open fissure being 1 to several metres in width. These fissures parallel to the axis are displayed in relay and 'enchelon', or are linked together by transverse, slightly oblique fissures. Active rubble talus are distributed at the feet of the major scarps. Thin, fragile standing walls cut through the basalts by active extension may collapse *in situ*. Lava lakes emplaced during a volcanic episode immediately preceding the extension tectonics are dissected by the fissure network (drawing from Lagabrielle).

Subvertical *normal fault* scarps, parallel to the ridge axis, represent the other type of common tectonic feature in the axial domain. Their vertical offset ranges from several tens of cm to several tens of metres. On average, the scarp increases with distance to the axis. On both sides of the active volcanic zone, these faults cut out the young oceanic crust in a series of elevated (horsts) or collapsed (graben) blocks, parallel to the axis. On the East-Pacific Rise, tectonic graben are the preferred site for ponding of fluid sheet flow and for the formation of submarine lava lakes.

When the submersible moves along a fault scarp, the observer can look at a natural section of the oceanic crust. He can see that pillow-lava flows are cut out neatly by the fault plane, showing an ellipsoidal section and a radial jointing (thermal contraction joints) of the pillows. The foot of the fault scarps is often masked by a *rubble talus*, made of pillow-lava fragments detached from the fault scarp. The sediment cover of these rubbles gives a valuable indication about the present or recent activity of the fault: if the rubble talus is devoid of any sediment, this signifies that the fault is currently active. On the other hand, a rubble talus buried beneath a thick sedimentary cover indicates a long-time inactive fault.

Low-angle *detachment faults* have been described in particular on slow-spreading ridges such as the Mid-Atlantic Ridge. These faults, similar to Vernicke's detachment faults, may explain a certain tectonic and volcanic asymmetry on both flanks of

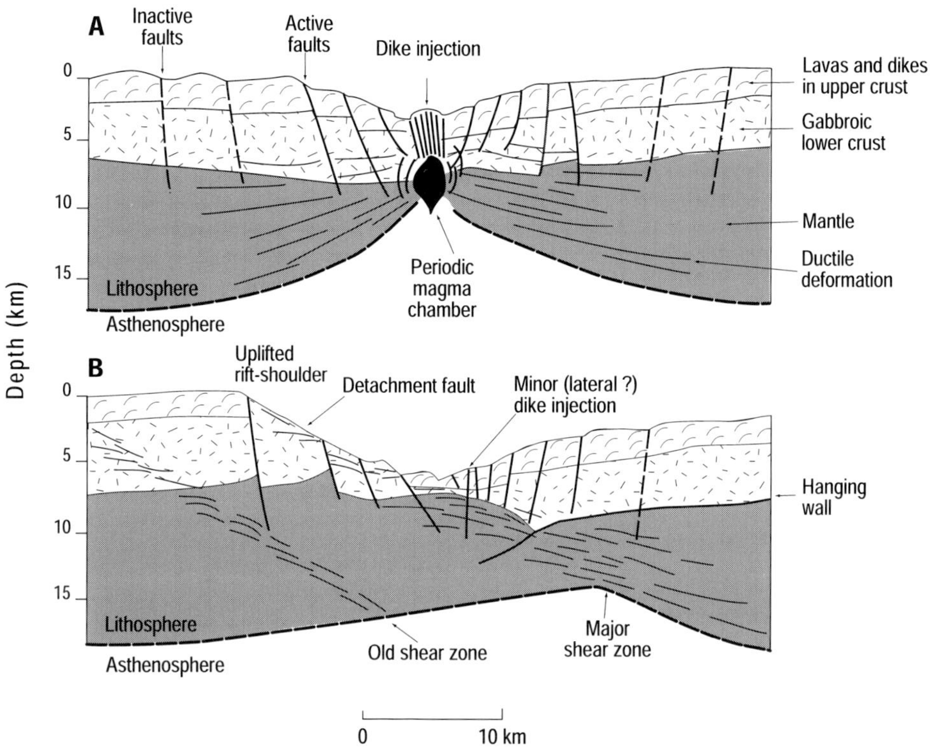

Fig. 7.25. The two major tectonic models explaining the morphology of slow-spreading oceanic ridges. (A) Pure shear extension model, with symmetrical stretching and creation of normal faults on both sides of the axial valley. (B) Simple shear extension model, with asymmetrical stretching along a flat fault (detachment fault), passing through the whole lithosphere and outcropping on one of the walls of the axial valley (after Karson, 1992).

the ridge, and the outcropping of deep rocks (mantle peridotites, gabbros) in the axial zone (Figure 7.25). Not all authors agree on the importance, and even on the presence of these detachment faults.

On rather fast- to fast-spreading ridges (6 to 16 cm/yr), a narrow and rectilinear axial graben cuts the axial zone, sometimes over very long distances. Volcanic and hydrothermal activity concentrate mainly in this axial graben.

7.6.2 The volcanic–tectonic cycle

Recent studies have shown that each ridge segment evolves with time following a fundamental volcanic–tectonic cycle, during which the magmatic, tectonic and hydrothermal processes interact together continuously. The necessity of a volcanic–tectonic cycle results from the fact that the deep submarine eruptions of the ocean-ridge system are intermittent, like those of aerial volcanism, whereas the accretion process is continuous. Each volcanic eruption is preceded and followed by

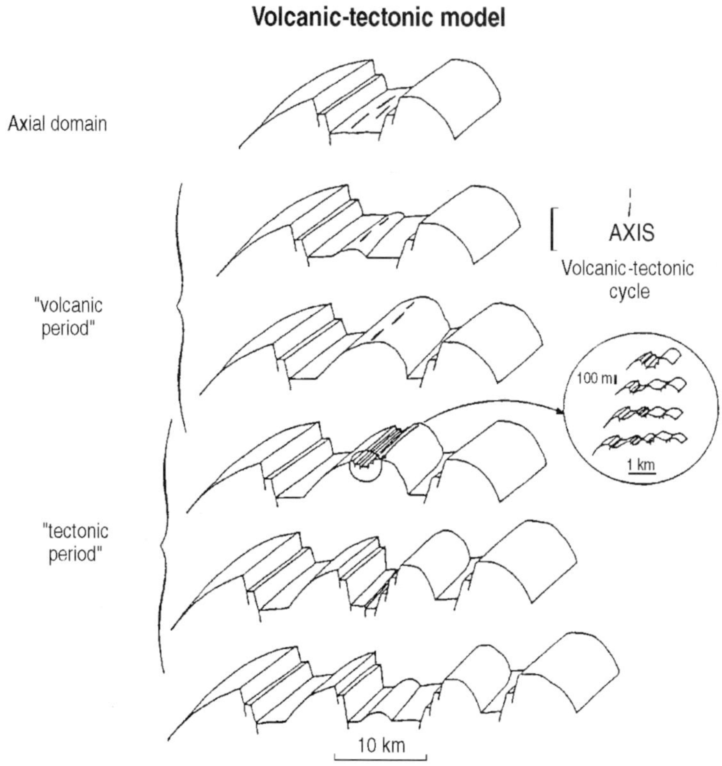

Fig. 7.26. The concept of a volcanic–tectonic cycle (after Gente, 1987). The succession of elementary cycles, first dominantly volcanic ('volcanic period'), then dominantly tectonic ('tectonic period') builds up the morphology of the oceanic ridge axial domain.

a more or less long stage of quiescence. It is then useful to distinguish the *magmatic accretion stages*, relatively brief in time, corresponding to volcanic eruptions on the seafloor, and the *tectonic accretion stages*, corresponding to pauses in volcanic activity, during which the young oceanic crust is submitted to purely tectonic extensional stresses. The first stages are characterized by an axial bathymetric profile with a *dome shape* (the magma chambers are full up and inflated), whereas the second ones present a profile with an intensely faulted *axial graben*.

On the East-Pacific Rise, it has been shown that building of axial reliefs is cyclic (Lonsdale, 1977; Gente *et al.*, 1986; Gente, 1987): the demonstration can be done either by studying the morphology of the flanks of the ridge when moving away from the axis, or by doing a number of transverse serial sections along the ridge axis, on several accretion segments, in order to deal with various stages of the volcanic–tectonic cycle.

On other ridges with a slower spreading rate, Gente (1987) has shown that the same kind of morphological variation can be observed in the axial zone, ranging from dome to graben structures.

Figure 7.26 summarizes the concept of the volcanic–tectonic cycle, whose various stages are responsible for variation of the axial morphology of the ocean ridges. The difference in size of the axial structures can be explained by the more or less important duration of activity:

—Volcanic episodes occurring at high frequency create an important axial relief with a dome shape. In this case, the succession of elementary volcanic–tectonic cycles is dominated by the volcanic activity.
—On the other hand, a succession of volcanic–tectonic cycles dominated by tectonic activity will create a graben.

Gente proposes calling such a succession of discrete events 'volcanic period' or 'tectonic period' respectively.

Using magnetic anomalies as chronological markers, it can be calculated that a complete cycle gathering two successive periods (formation of a dome, followed by formation of a graben) lasts about 0.7 to 1 Myr on average on the Mid-Atlantic Ridge at 14°N (spreading rate 2 cm/yr), 70 000 years on the East-Pacific Rise at 21°N (spreading rate 7.3 cm/yr), and 50 000 years on this same ridge at 13°N (10.3 cm/yr). The duration of the volcanic–tectonic cycle is thus linked to the spreading rate, and the offsets of the reliefs on the ridges increase with duration of the cycle. On fast - spreading ridges, the duration of the volcanic periods is often equal to that of the tectonic periods, resulting in a rather flat morphology of the axial domain (for example, the East-Pacific Rise at 21°N), or longer, resulting in a dome structure (idem, 13°N). On slow-spreading ridges, tectonic periods are longer (0.5 to 0.75 Myr) than the volcanic periods (0.25 to 0.5 Myr), resulting in a graben-dominated morphology in the axial zone.

Finally, Gente shows that the volcanic–tectonic cycle is 'desynchronized' along the ridges, evolving independently along elementary accretion segments of about 20 km on all the ridges, whatever their spreading rate. However, this observation remains to be confirmed by study of a significant number of ridge segments, before being generalized to the whole oceanic ridge system.

7.6.3 Tectonics, submarine erosion and mass-wasting products

Submarine erosion due to the action of gravity affects all submarine reliefs on the ocean floor, particularly those of tectonic origin: horsts and graben of the axial zone of the ridges, boundary fault scarps of the rift valley on slow-spreading ridges, transform valleys of the fracture zones, etc.

With respect to aerial erosion, submarine erosion is characterized by the omnipresence of seawater, which percolates in all the pores and voids of the oceanic crust, lubricating and softening the oceanic materials.

The products of this submarine erosion can be observed at various scales, from huge landslides of several km^3, to tiny grain flows, passing by blocky debris flows, avalanches, rubble talus, etc. (Figure 7.27).

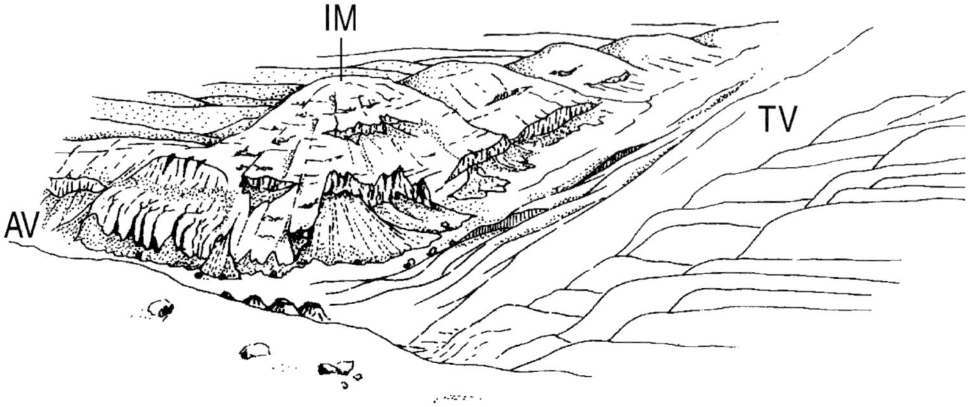

Fig. 7.27. Drawing showing the intersection of the Mid-Atlantic Ridge with a major fracture zone. The intersection–massif (IM) is affected by abundant mass wasting, gravity slidings and active rubble talus formation, accumulating at the feet of the reliefs. The denuded slopes from which mass wastings started directly show rocks of the deep crust and upper mantle (serpentinites, foliated gabbros, etc.), making up most of these reliefs. In the transform valley (TV), the accumulated debris is often reworked in recent or present-day shearing deformations. The flanks of the ridge axial valley may also be affected by mass wasting processes (drawing by Lagabrielle).

7.7 DETECTION OF DEEP SUBMARINE ERUPTIONS

On the geological time scale, the oceanic-ridge system continuously produces oceanic lithosphere all along its length. On the time scale of a human life, or even of several centuries, things are quite different: like aerial volcanoes, submarine volcanoes erupt during brief periods of crisis, separated by long periods of rest.

Calculations show that the frequency of volcanic eruptions at any point of a slow-spreading ridge like the Mid-Atlantic Ridge is of the order of 10 000 years, and of 500 to 2 000 years for a fast-spreading ridge lke the East-Pacific Rise. It then becomes clear that the probability of observing 'by chance' a deep volcanic eruption at a ridge axis is practically nil. However, there is no doubt that thousands of basaltic flows are emplaced every year all along the 60 000 km of the oceanic ridge system. But how can we detect and locate these eruptions?

The problem is that nothing indicates a deep submarine eruption at sea level: neither projections, nor gas bubbles, glowings, rumblings or temperature rises, etc. Even microseisms, linked to the magma ascent or to its propagation along the axis, generally have amplitudes that are too weak to be registered by seismic stations on land.

Up to now, *nobody has witnessed a deep submarine eruption*, but indications of present-day or very recent activity are accumulating in some specific areas. The seismicity linked to a ridge eruption may be registered in some cases. In 1990, Icelandic seismologists alerted the international community after detecting an important seismic crisis on the Reykjanes Ridge, south of Iceland. The activity lasted several days, during which unfortunately no oceanographic ship could be rerouted to the zone.

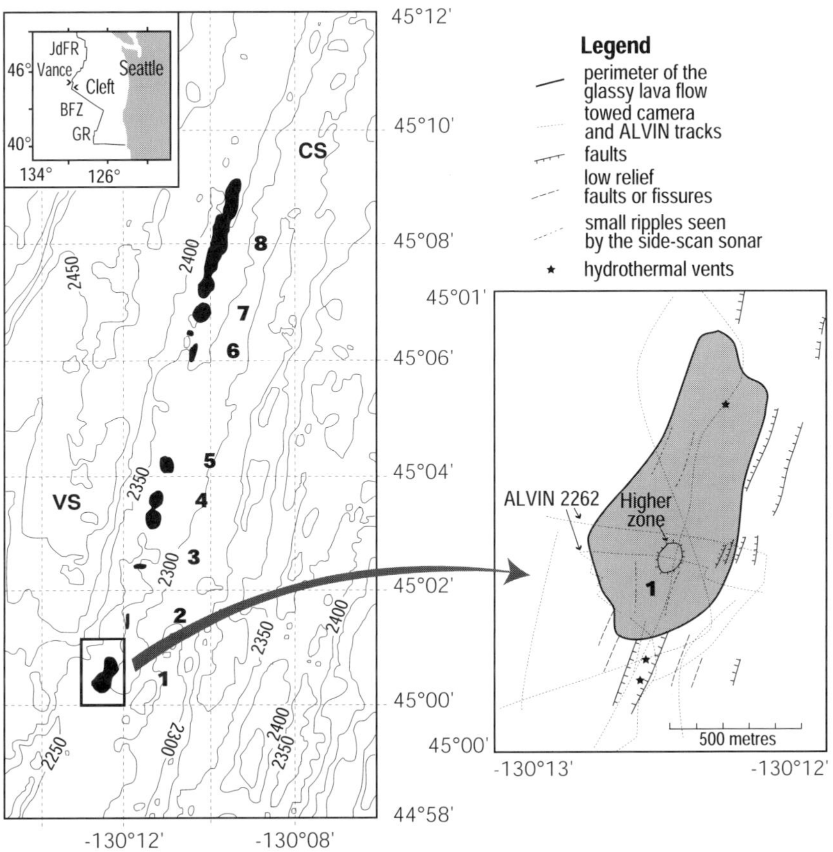

Fig. 7.28. Seabeam bathymetric map of the overlapping area between Cleft and Vance segments, southern tip of the Juan de Fuca Ridge. Bathymetric lines every 50 m. CS: northern tip of the Cleft Segment. VS: southern tip of the Vance Segment. The eight areas in black represent the newly emplaced pillow-lava volcanic hills (after Chadwick *et al.*, 1991).

Periodic visits in some study areas on some fast-spreading segments of the East Pacific have shown important modifications of the 'axial landscape', due to very recent eruptions, as in the two following cases:

7.7.1 Growth of a neo-volcanic ridge on the Juan de Fuca Ridge during the 1980s

During a routine cruise in 1989, three American scientists took a series of deep-tow camera pictures of the axial seafloor on the Cleft segment, southern part of the Juan de Fuca Ridge, off the coast of Washington state. The pictures revealed the existence of a nice, 25-m-high volcano, made of brilliant, glassy pillow lava, devoid of any sedimentary cover. To their immense surprise, there was no trace of such an edifice on the detailed Seabeam bathymetric map produced by themselves several years ago (in 1981): at its place, the map indicated only a gentle slope dipping eastward

(Chadwick *et al.*, 1991). Later complementary bathymetric surveys confirmed the existence of this new volcanic edifice, and acoustic images obtained by side-scan sonar showed a totally 'fresh' volcano, non-fissured, laying like a pustule over older volcanic terranes already abundantly fissured.

Moreover, a detailed comparison between the 1987–1990 and the 1981 surveys allowed seven other volcanic hills that were absent on the 1981 document (Figure 7.28) to be discovered. In total, it is a new 16-km-long segment that was emplaced at the axis and scattered with pillow-lava volcanoes, probably fed in depth by the same, sole basaltic dyke, according to the authors. Following their calculations, $0.05\,km^3$ of fresh lava was added to the surface, fed by a dyke at depth with a volume of at least $0.5\,km^3$. By comparison, the Kilauea volcano in Hawaii emitted $0.04\,km^3$ of lava in 1977, during an eruption which lasted 18 days.

Huge hydrothermal plumes, called 'megaplumes', may be associated with this volcanic activity. Independent of previous work, sampling in the water column at various depths done in 1986–1987 on the Cleft segment have revealed very high anomalies in some chemical elements tracing hydrothermal activity of the ocean floor in the seawater (manganese, methane, etc.).

This remarkable coincidence in time between a volcanic event and a major hydrothermal episode (a 'megaplume') seems to imply that the intrusion of the 'feeding dyke' caused a major fracture of the crust, greatly increased the local permeability around the existing hydrothermal site, and allowed the brutal irruption of a huge quantity of hydrothermal fluids in the local seawater column. 'Megaplumes' may thus be indicators of volcanic building episodes on the seafloor at ridge axes.

7.7.2 Carnage on the East-Pacific Rise, or the 'tubeworm barbecue' site

The 23 April 1991 on the East-Pacific Rise at 9°50′06″N, during the 'Adventure' diving programme with the submersible *Alvin*, the scientists noticed that a catastrophic event had decimated the 'oases' of life. Thousands of tubeworm and clam bodies lay strewn on the seafloor, and the active chimneys observed 2 years before had disappeared. In their place, fluid and brilliant sheet flows covered the ancient hydrothermal mounds. A recent volcanic eruption, since 1989, had buried the active hydrothermal sites, bringing death and desolation to the 'oasis'. The event was so recent that thousands of bodies exposed in the lava had not yet attracted the predators of the neighbourhood, particularly the Galathean crabs. Biologists in the team estimated that the lava flow occurred just a few days earlier.

Later on in the laboratory, datings performed on the basalts and based on the method of very short-lived radioactive disequilibria, confirmed this diagnostic: the sheet flow was emitted between 26 March and 6 April, 1991 (Haymon *et al.*, 1991)!

One month later, at the end of May, new dives at the same site showed that the crabs had invaded the 'BBQ Site' as the divers called it.

This time a deep volcanic eruption had been missed, by just two or three weeks! In both cases described above, it was not 'by chance' that the scientists discovered evidence of very recent volcanic activity at the ridge axis: only periodic visits to these ridge segments chosen along with others to constitute 'natural laboratories' can lead to the discovery of such events, either by high-resolution mapping, or by analyses of the seawater column, or else by direct observation with submersibles or deep-tow cameras.

7.7.3 'InterRidge', an international network of exploration and survey of the oceanic-ridge system

The international scientific community is now well organized, and aims at developing new methods and tools to detect, to localize and to react to such events. If the whole 60 000-km-long ocean-ridge system is to be watched, then a very important international effort is necessary.

This effort does exist. It was put in place, at the beginning of the 1990s, by setting up the 'InterRidge' program, gathering 25 countries actively involved in oceanographic research. The main objectives are to coordinate exploration and complete mapping of the ocean-ridge system, to watch its activity, and to organize fast-response plans in case of detection of a deep volcanic eruption.

The task of detecting eruptions is difficult. Three kinds of events are potentially observable: earthquakes, volcanic eruptions and hydrothermal plumes.

Earthquakes provoked by ocean-ridge activity are the only 'signals' lkely to be recorded at distance. Seismic and acoustic methods are therefore the most promising. We know that there are two kinds of seisms linked to an oceanic spreading episode:

—'Tectonic' earthquakes, due to fracturing and fault movements during the extension of the newly formed crust;
—'Volcanic' tremors directly linked to the movement of magma in the crust: inflation and deflation of the magma chambers, intrusion and extrusion of the magmas.

Unfortunately, the amplitude of seismic waves from volcanic seisms is much lower than that of tectonic seisms, and generally below the detection threshold of terrestrial seismic networks. In fact, it seems that most of the events recorded on the oceanic ridge by these networks are due to tectonic movement rather than to magma movement.

However, it appears that some acoustic frequencies can propagate over thousands of km through the oceans, practically without attenuation, thanks to the existence of a real 'conductive layer' called the 'SOFAR channel' (used for a long time by cetaceans). Specialists in the US Navy have installed such a listening network in the north-eastern Pacific, and have recently agreed to making these confidential data accessible to researchers.

The first studies have shown that hydrophones displayed as a network on the Pacific seafloor increase the sensitivity to natural seismicity of the oceanic ridge by a factor of 30, and so seismic events of volcanic origin should be routinely detectable.

In summary, scientists are no longer unable to detect deep volcanic eruptions. By refining seismic and acoustic methods, by multiplying bathymetric surveys on carefully selected ridge segments (notion of 'natural laboratory'), and by routinely analysing the seawater column (detection of 'megaplumes'), it should be possible in the near future to locate and study in real time one or several deep volcanic eruptions on the ocean-ridge system, and to determine their effects on hydrothermal activity and on the associated fauna.

8

Deep processes: mantle diapirs, magma extraction and models of accretion magma chambers

Active or passive, the oceanic accretion process implies the upwelling of the asthenospheric mantle, and its partial melting by decompression (Chapter 5). What are the sources of the MORB (Mid-Ocean Ridge Basalts)? How do the partial melting liquids appear in the ascending mantle diapirs, how are they extracted from the upper mantle, and how do they rise to the surface? Do these magmas en route to the surface stop temporarily in an accretion magma chamber? In this case, what are the dimensions, the geometry and the parameters controlling the evolution of these chambers?

The answers to these fundamental questions are examined in the present chapter.

8.1 THE SOURCES OF THE MORB

We have seen (Chapter 2) that the oceanic ridge basalts represent a very homogeneous group with respect to the other groups of oceanic basalts (Figure 2.2). The frequency diagrams of the strontium, neodymium, lead and helium isotopic ratios for MORB, OIB (Ocean Island Basalts) and IAB (Island Arc Basalts) confirm the homogeneity of the MORB group (Figure 8.1).

However, it appears that this group is made of an array of compositions, from the N-MORB for the most depleted, to the T- and E-MORB for the most enriched.

In order to discuss MORB genesis, it is useful to examine how these various types are distributed in space and time.

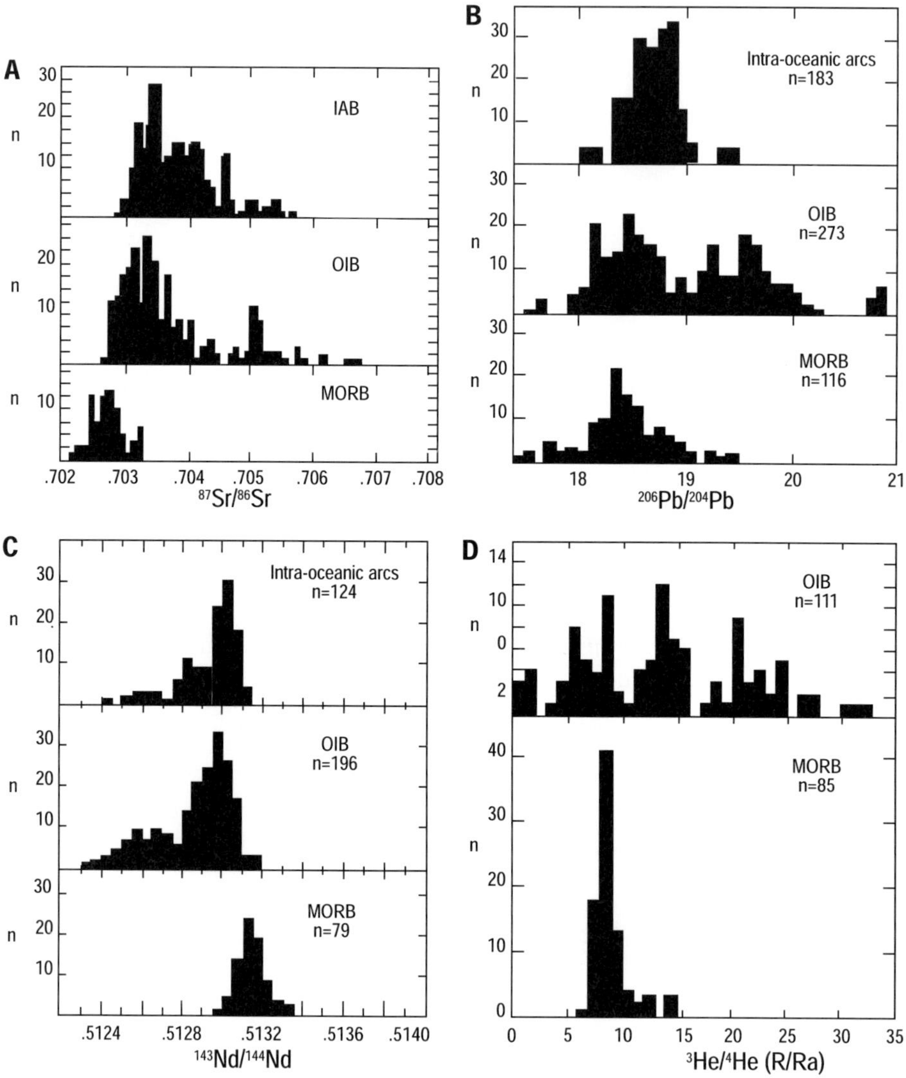

Fig. 8.1. Frequency histograms of the $^{87}Sr/^{86}Sr$, $^{143}Nd/^{144}Nd$, $^{206}Pb/^{204}Pb$, and $^{3}He/^{4}He$ isotopic ratios in MORB-, OIB- and IAB-type oceanic basalts, showing the geochemical homogeneity of the MORB group (after Zindler and Hart, 1986).

8.1.1 Vertical variations

The DSDP (Deep-sea Drilling Project)/ODP (Ocean Drilling Program) oceanic drillings have gained valuable information about the vertical geochemical variations of the basaltic pile. In the Atlantic Ocean in particular, a number of drill holes have reached a penetration of 250 to 550 m in the upper oceanic crust, on both the Mid-Atlantic Ridge and the old Upper Cretaceous crust as well.

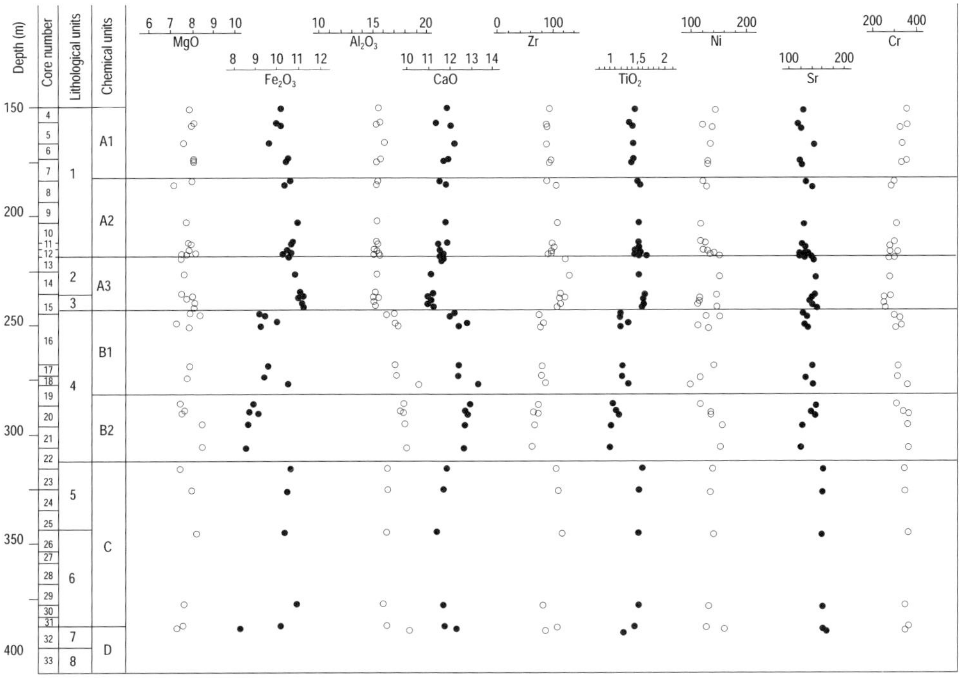

Fig. 8.2. Stratigraphic column of DSDP Hole 396B basalts (eastern flank of the Mid-Atlantic Ridge, 22°N), with plots of some major and trace element content versus depth (whole rock analyses, realized on board). The chemical discontinuities observed, along with other criteria, were used to define the 'cooling units'. Leg DSDP 46, Site Reports, 1979.

Definition of the cooling units

With the help of relatively simple geochemical parameters, combined with other criteria (presence of chilled margins and of interlayered volcanic breccias, variations of modal phenocryst contents, etc.), the individual cooling units (flow and sill intrusions) making up the volcanic pile can be identified with a good degree of precision. The boundaries between *cooling units* (especially of pillow lava) are frequently marked by more or less clear geochemical discontinuities (Figure 8.2).

Source heterogeneities on the same vertical

In a number of cases the superposed cooling units show, along the same vertical profile, variable multi-element spectra, ranging from N-MORB ('Normal') to E-MORB ('Enriched'), and even to T-MORB ('Transitional'). Figure 8.3A illustrates these variations, which constitute a decisive argument against the existence of huge steady-state magma chambers beneath the axial zone of the Mid-Atlantic Ridge. It is clear that the existence of such magma chambers would result in homogenization of the source heterogeneities. The observed variations are in favour of the existence of multiple, small and temporary magma chambers, functioning more or less as closed

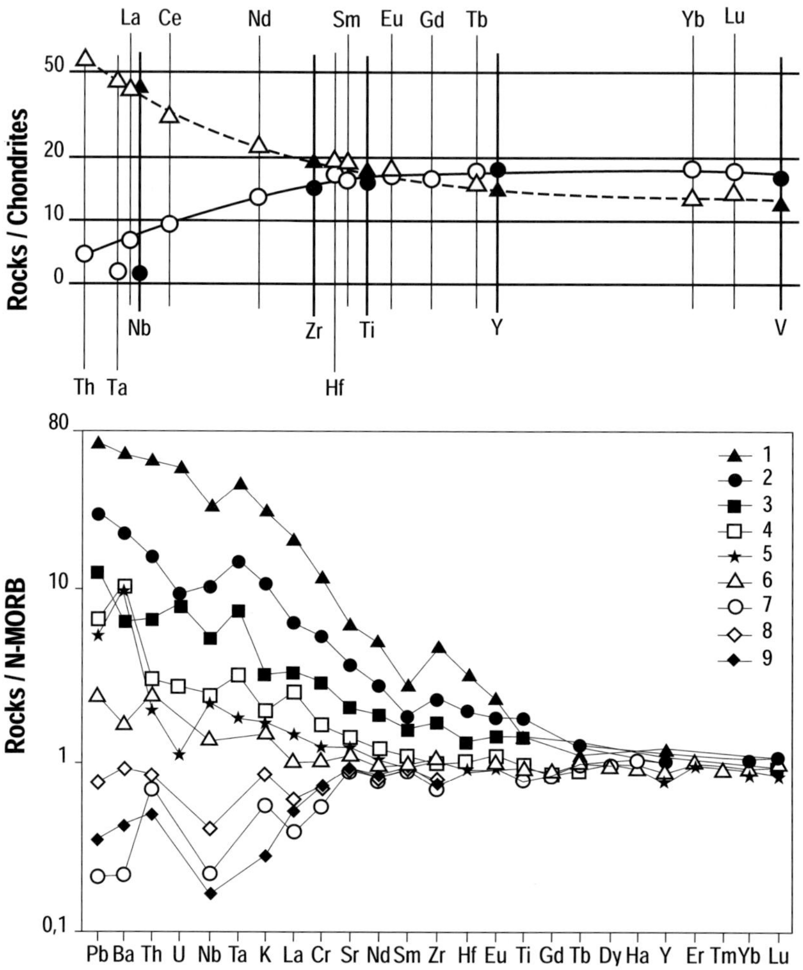

Fig. 8.3. Variations of incompatible element spectra at small scale. (A) Along the same vertical. Site 561 basalts, Leg DSDP 82 (Bougault *et al.*, 1985). (B) Along a single ridge segment. 'West Valley segment', Juan de Fuca Ridge, about 50 km long (after Cousens *et al.*, 1995).

fractionation systems, each of them characterized by a particular mantle signature. Such variations enhance the idea of a strong heterogeneity of the upper mantle at small scale.

8.1.2 Local-source heterogeneities

In a given segment of the oceanic ridge, the spectrum of the MORB geochemical and isotopic variations can be quite extended. In the north-east Pacific for instance, one of the segments of the Juan de Fuca Ridge, about 50 km in length (called the 'West Valley segment'), has given a collection of dredged basaltic samples showing extreme diversity of composition, ranging from N-MORB completely depressed in incompatible elements, to very enriched E- and T-MORB compositions (Figure 8.3B). The La/Sm_N ratios vary from 0.3 to 1.8, the isotopic ratios $^{87}Sr/^{86}Sr$ from 0.702 35 to 0.702 60, and $^{206}Pb/^{204}Pb$ from 18.22 to 19.15 (Cousens *et al.*, 1995). These characteristics are best explained by mixing models between two magmatic sources, confirmed by a lot of evidence of chemical and isotopic disequilibrium between phenocrysts and glasses. The underlying mantle is modelized by the authors as a *plum pudding* in which *plums* of enriched and amphibole-bearing peridotite float in a matrix of depleted mantle. Low melting rates preferentially melt the *plums*, producing E- to T-MORB, whereas more important melting rates affect both the *plums* and the depleted matrix, producing N-MORB.

8.1.3 Horizontal variations along oceanic-ridge axes, and effects of segmentation

The bathymetric, morphologic and structural segmentation of the oceanic ridges, revealed by modern mapping and geophysical methods (see Chapter 4), is outlined spectacularly by the geochemical variation of fresh basaltic glasses collected along the Ridge axis (Schilling *et al.*, 1983).

Along the Mid-Atlantic Ridge between 10° and 70°N for instance, the $(La/Sm)_N$, $(Nb/Zr)_N$ and $^{87}Sr/^{86}Sr$ ratios faithfully reproduce, at least qualitatively, large-wavelength bathymetric undulations linked to the hot spots. Looking closer, it appears that only the strontium isotopic ratios reveal the Azores 'superstructure' (23–50°N), whereas well-chosen trace element ratios outline the second-order bathymetric anomalies (Figure 8.4).

At shorter wavelength, the second or third order of segmentation is accompanied also by more subtile geochemical variations.

8.1.4 Source heterogeneities at large scale

The MORB sources of the Indian Ocean, central Pacific and of the seas lying between these two main domains (southern China Sea, Filippinas Sea, etc.), are less depleted than the sources of the northern Atlantic for instance, and exhibit lower $^{143}Nd/^{144}Nd$ and $^{206}Pb/^{204}Pb$ isotopic ratios (Figure 2.2), and higher $^{208}Pb/^{204}Pb$ and $^{207}Pb/^{204}Pb$ ratios. This particularity has been called the *DUPAL anomaly*, in hommage to Dupré and Allègre (1983) who were the first to describe it.

The distribution of the DUPAL anomaly is shown in Figure 8.5. It concerns not only the MORB, but also the OIB and IAB of the same large province. The DUPAL anomaly in fact reflects the presence of enriched components (of the EM type) in the source of the magmas. Its wide distribution reflects the existence of large-scale convective currents in the mantle.

8.1.5 Genesis of MORB

Large numbers of oceanic-ridge basalts, collected far from any hot-spot volcanism, present depleted isotopic and geochemical characteristics (multi-element spectra),

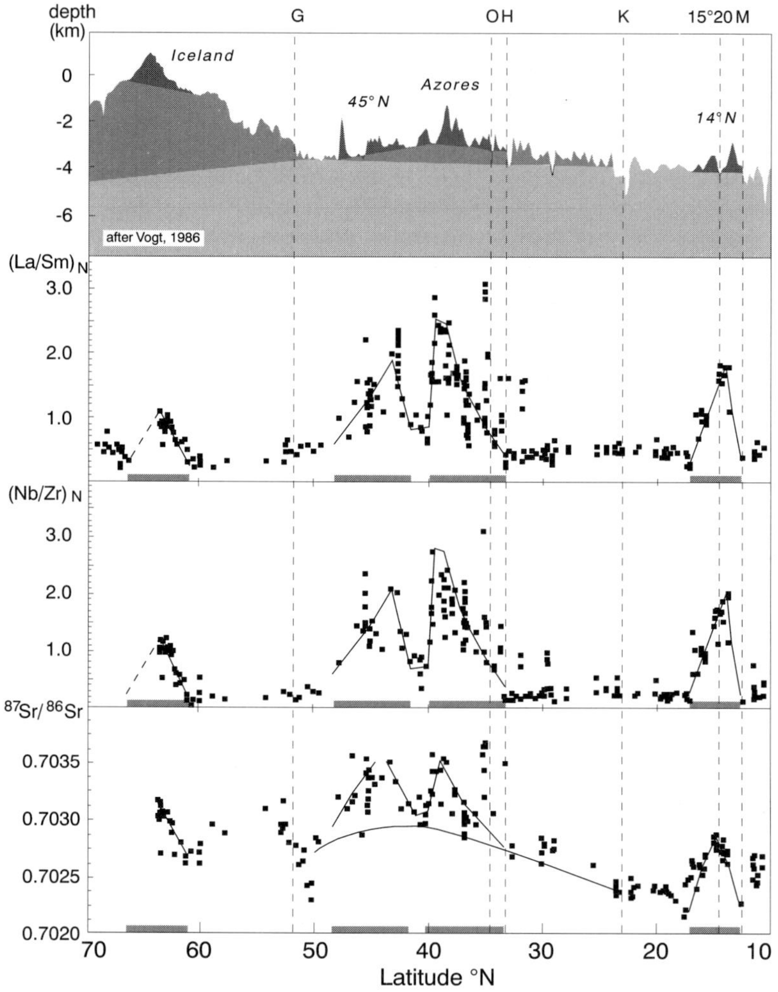

Fig. 8.4. Variations of trace element ratios (La/Sm and Nb/Zr), and of strontium isotopic ratios in the basaltic glasses of the Mid-Atlantic Ridge axial zone, between 10°N and 70°N (northern Atlantic), and comparison with bathymetric variations (top profile). The Iceland and Azores topographic 'superstructures' are indicated in intermediate grey. Four topographic anomalies of second order appear in dark grey (Iceland, 45°N, Azores and 14°N). Only strontium isotope ratios reveal the Azores superstructure, whereas these same ratios plus the trace element ratios clearly outline the second-order bathymetric anomalies. Fracture zones: Kane (K), Hayes (H), Oceanographer (O) and Gibbs (G) (after Dosso *et al.*, 1993).

unknown in all the other oceanic or continental volcanic domains. This fundamental observation is a strong argument in favour of the existence of a depleted mantle domain, source of the N-MORB, which should constitute volumetrically the main part of the source-reservoir of the MORB. This depleted domain is commonly designed under the term 'DMM' (for *Depleted MORB Mantle*, Zindler and Hart, 1986). Located beneath the world oceanic ridge, it lies at the level of the lithospheric mantle and of the underlying asthenosphere (Figure 8.6A).

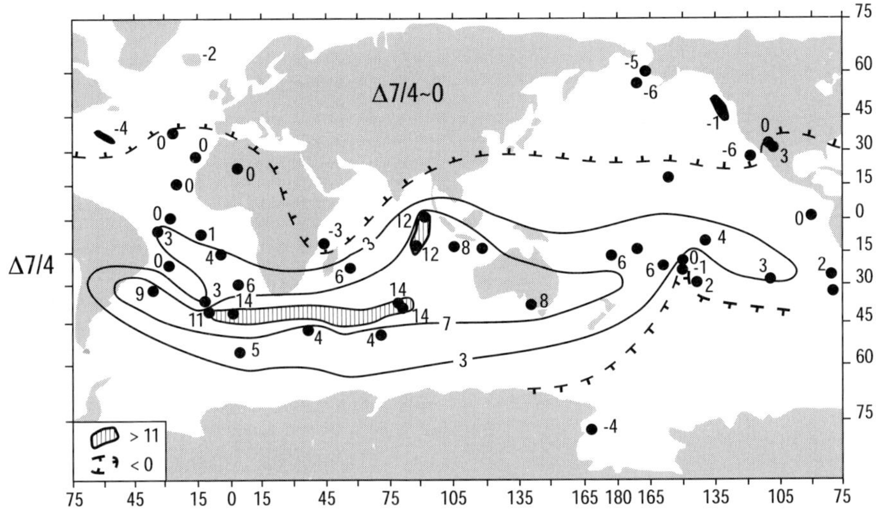

Fig. 8.5. World distribution of the DUPAL anomaly, using the Δ7/4 parameter (gap of the $^{207}Pb/^{204}Pb$ ratio for oceanic basalts to the NHRL, Northern Hemisphere Reference Line) (after Zindler and Hart, 1986).

Starting from there, any oceanic-ridge basalt presenting a more or less enriched isotopic signature would result from the melting of an heterogeneous source comprising enriched components set in a DMM-type matrix. It is conceivable that these enriched mantle components come from small ascending diapirs, issued from much deeper levels in the mantle, and potential sources for the intra-oceanic OIB volcanism (hot spots). Mixing processes between the DMM depleted matrix and the enriched components would adequately explain the negative correlation observed between the Nd and Sr isotopic ratios in the MORB group (Figure 2.3), as well as the positive correlation between the Sr and Pb isotopic ratios.

Following this concept, the enriched mantle diapirs would be more abundant beneath slow-spreading ridges, resulting in clearly visible bathymetric and geochemical anomalies (Iceland, the Azores, etc.) (Figure 8.6B1). Beneath fast-spreading ridges, the enriched mantle diapirs would be smaller and more numerous, with more 'diluted' bathymetric and geochemical effects (Allègre *et al.*, 1984) (Figure 8.6B2)

8.2 THE UPRISE OF THE ASTHENOSPHERIC MANTLE: BIRTH AND EXTRACTION OF MAGMAS

Beneath the oceanic ridges, the asthenospheric mantle is uprising in the form of diapirs. The initial movement of uprise is due either to advective movements of the asthenosphere, or, in the case of hot spots, to a local density contrast with respect to the surrounding mantle. The study of mantle xenoliths in alkali basalts has shown that the ascending peridotite is generally a fertile lherzolite.

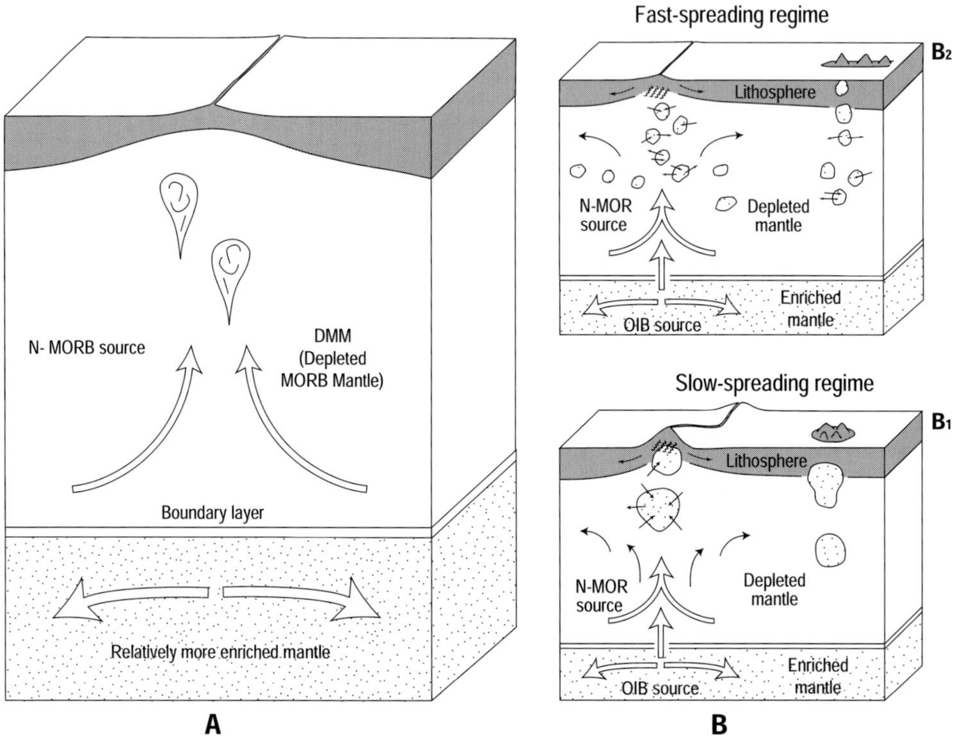

Fig. 8.6. Schematic models explaining the various mantle components producing MORB-type oceanic basalts (after Allègre *et al.*, 1984). (A) Away from hot spots, the oceanic ridges produce N-MORB-type basalts, issued from the partial melting of a mantle source depleted in incompatible elements, and homogeneous isotopically: this is the *DMM (Depleted MORB Mantle)*, constituting the upper mantle beneath the oceanic ridge system and the oceanic lithosphere. (B) E- to T-MORB-type basalts are issued from the partial melting of an upper mantle composed of enriched-mantle diapirs, moving up in a DMM matrix. These enriched-mantle diapirs, issued from a deeper mantle reservoir (the OIB source), would be larger beneath slow-spreading ridges (B1 case), and smaller and more numerous beneath fast-spreading ridges (B2 case).

Partial melting in the ascending lherzolite diapir begins when the adiabatic uprise trajectory intersects the solidus of the lherzolite (supposed to be anhydrous). Laboratory experiments have shown that this happens at about 75 km in depth, in the field of garnet lherzolites (Figure 8.7). Starting from the moment when partial melting begins, the density contrast between the mantle diapir and the surrounding asthenosphere will progressively increase, favouring upward acceleration toward the surface.

8.2.1 The beginning of melting

Under conditions of high pressure and temperature, that is those of incipient melting, any polycrystalline material is constituted of polyhedral minerals. It is the

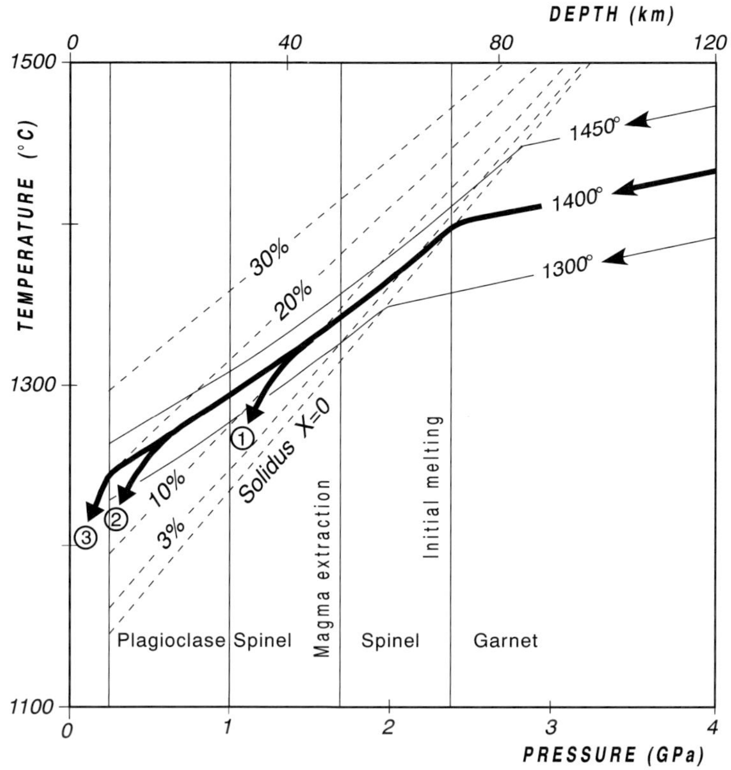

Fig. 8.7. P-T trajectory of an ascending mantle diapir (bold line), in McKenzie's partial melting diagram (1984). Dashed lines indicate the partial melting percentage (0% at the solidus, 100% at the liquidus, not visible on that part of the diagram and never reached). 1. Trajectory in the field of spinel lherzolites. 2. Trajectory in the field of plagioclase lherzolites. 3. Trajectory in the field of harzburgites. Garnet: field of garnet lherzolites. Spinel: field of spinel lherzolites. Plagioclase: field of plagioclase lherzolites (after Nicolas, 1989), slightly modified.

case for natural mantle lherzolites, whose textures have been studied in detail (Mercier and Nicolas, 1975). In the most common case, natural lherzolites under the microscope show a nice polygonal mosaic texture, with many triple junctions (eventually quadruple) between the mineral grains. The angles between the crystal faces are usually close to 120°, particularly between identical species (Figure 8.8).

8.2.2 A simplified phase diagram

Let us first examine a phase diagram, for instance the system *forsterite–diopside–pyrope* at 40 kbar (Figure 8.9). This ternary system can be considered a simplified representation of a natural garnet lherzolite. Point E in this diagram can be considered as a ternary eutectic: it corresponds to the composition and temperature (1670°C) of the first partial-melting liquid from any solid mixture of the three phases, whatever their relative proportions. If L represents the composition of a natural

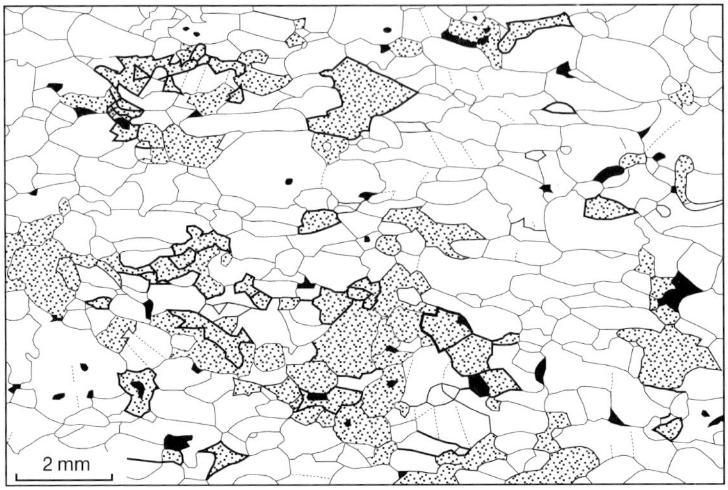

Fig. 8.8. Typical mosaic texture of a spinel lherzolite. The mineral phases have polygonal outlines, with numerous triple junctions with angles close to 120°. Quadruple junctions may occur also. Olivine in white, orthopyroxene stippled in bold with thin outlines, clinopyroxene stippled in bold with bold outlines, chrome spinel in black (drawing after Mercier and Nicolas, 1975).

lherzolite, the first partial melt liquid appears in *E*. Supposing that the melting process is fractionated, that is that the liquid is extracted continuously during its formation thanks to good permeability of the medium (see further), the points R_1, R_2, etc., represent the successive residual compositions of the peridotite, aligned on the straight line E–L–R_N. If the composition of the residual solid reaches R_N, this means that the diopside phase has disappeared from the solid assemblage. We enter then in the olivine–garnet binary system, whose eutectic point B is at 1770°C, 100°C above the temperature of point E. The melting process will hence stop at point R_N, except in the improbable case of a strong increase in temperature.

The reality of course is more complex than in this simple phase diagram. First, all the mineral phases present extended solid solutions (particularly between Mg and Fe end members), which complicates their behaviour. Besides, the evolution of an ascending mantle diapir must be polybaric and quite adiabatic: melting occurs by decompression, and not by heating at constant pressure, as in the phase diagram presented. Lastly, the polybaric evolution results in the destabilization of the garnet, in favour of a spinel phase. The garnet transforms into a mixture of orthopyroxene, Cr-spinel and olivine, and the R-type rocks become *residual spinel harzburgites*. Finally, in the real 'multi-component mantle system', there are no mantle eutectics. Mantle melting produces a range of compositions, and melts form and segregate over a range of pressures (Langmuir *et al.*, 1992).

Despite these remarks, the phase diagram of Figure 8.9A remains very helpful in understanding the global evolution of the system. The schemes of Figure 8.9B indicate that during all the partial melting process, the solid residue deforms and recrystallizes continuously.

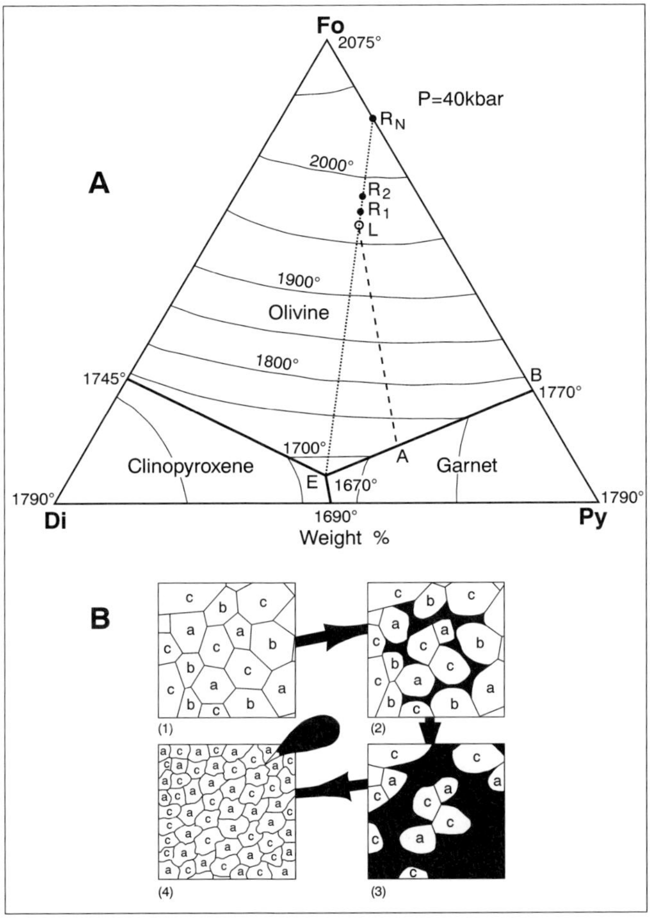

Fig. 8.9. (A) Ternary phase diagram forsterite (Fo)–diopside (Di) and pyrope (Py) at P = 40 kbar (after Davis and Schairer, 1965). Point L represents the composition of a fertile lherzolite before melting. (B) Schematic aspect of lherzolite under the microscope, at different stages of melting, here supposed to be fractionated (this means that a good overall permeability of the medium allows an efficient extraction of the liquid, as and when it is produced). (1) Aspect of the polyhedral mineral assemblage before melting (temperature lower than that of ternary eutectic E, subsolidus conditions). (2) The partial melting process has begun. The liquid (in black) has composition E. It appears preferentially at the triple junctions Fo–Di–Py. The composition of the refractory solid residue evolves towards R_1, R_2, etc., and progressively approaches point R_N, on the E–L–R_N straight line. (3) At a much more advanced stage of melting, the liquid still has composition E, when the diopside phase is entirely used up (resorbed) and disappears from the solid residue (residual composition R_N). (4) The liquid has been drained off, and the solid residue (olivine plus garnet) has recrystallized with plastic deformation. Chemically, the rock is a residual harzburgite. In the case of a melting process *at equilibrium* (and not fractionated), a poor permeability of the medium maintains the liquid *in situ*, in contact with the residual crystals. When the diopside phase disappears, the liquid of composition E evolves towards point A (on the univariant curve diopside–garnet). At point A (in straight line with Fo and L), the garnet phase disappears, the solid residue becomes purely dunitic (forsterite alone), and if the melting process goes on, the liquid will follow the line from A to L. If the point representing the liquid reaches point L, then a theoretical complete melting is realized (100% liquid), but this is not likely to occur under natural conditions. Drawing after Wilson (1989).

8.2.3 Distribution of the first magma droplets

The partial-melting process must begin where the solidus temperature is minimum; that is, where a maximum of mineral phases have mutual contacts with the lherzolite. The first magma droplets should consequently appear at some triple or even quadruple junction between the polyhedral mineral grains, where olivine, clinopyroxene, orthopyroxene and garnet are in contact. These first droplets will first appear at the summits of the crystalline polyhedras, and then extend along the crystal edges, and finally along the crystal faces, with progressive extension of melting (Figure 8.10C).

This initial distribution of the magma will control the permeability of the mantle diapir, and can be detected in seismic profiles by some attenuation of the V_p and V_s seismic wave velocities. After partial-melting experiments done in the laboratory, it seems that the distribution of initial melt at the summits and edges of the mineral phases is most common for partially melted rocks. The distribution of the magma depends on two main factors: the '*wetability*' of the magma, and the excess of internal pressure of this liquid phase.

In a partially melted polycrystalline aggregate containing a fluid phase (CO_2, H_2O, magma, etc.), the *wetability* is a very important parameter, which characterizes the shape of the pores occupied by the fluid phase. This parameter is described by the dihedral angle θ, formed by two sidewalls of a pore, at the junction of two solid grains (Figure 8.10A). It has been demonstrated that:

$$\cos(\theta/2) = G_{ss}/G_{sf}$$

where G_{ss} and G_{sf} respectively represent the energy (per surface unit) of the interface solid–solid, and the energy of the interface solid–fluid.

The value of the dihedral angle θ determines the shape of the magma droplets (Figure 8.10B):

—If $120° < \theta < 180°$, the magma forms isolated and convex droplets at the summits of the crystals. For $\theta = 180°$, the magma droplets are perfectly spherical and remain isolated, even for important melting rates.

—If $60° < \theta < 120°$, the droplets are convex and the magma penetrates partially along the edges of the mineral grains. For $\theta = 60°$, the section of the droplets or of the micro-channels is an equilateral triangle.

—If $\theta < 60°$, the droplets are concave, extend easily along the edges of the mineral grains, and the magma penetrates partially along the faces of the grain. In this case the wetability of the magma with respect to the solid phases is very good, allowing a tridimensional permeability to develop rapidly (Figure 8.10C and D).

Figure 8.11 is the illustration of laboratory experiments (melting of natural peridotites), suggesting an average value lower than 50° for θ. The curvature of the magma droplets is negative (concave). The mantle is permeable for small partial-melting rates, and the magmatic flow in an ascending diapir will be mainly controlled by pressure gradients and permeability.

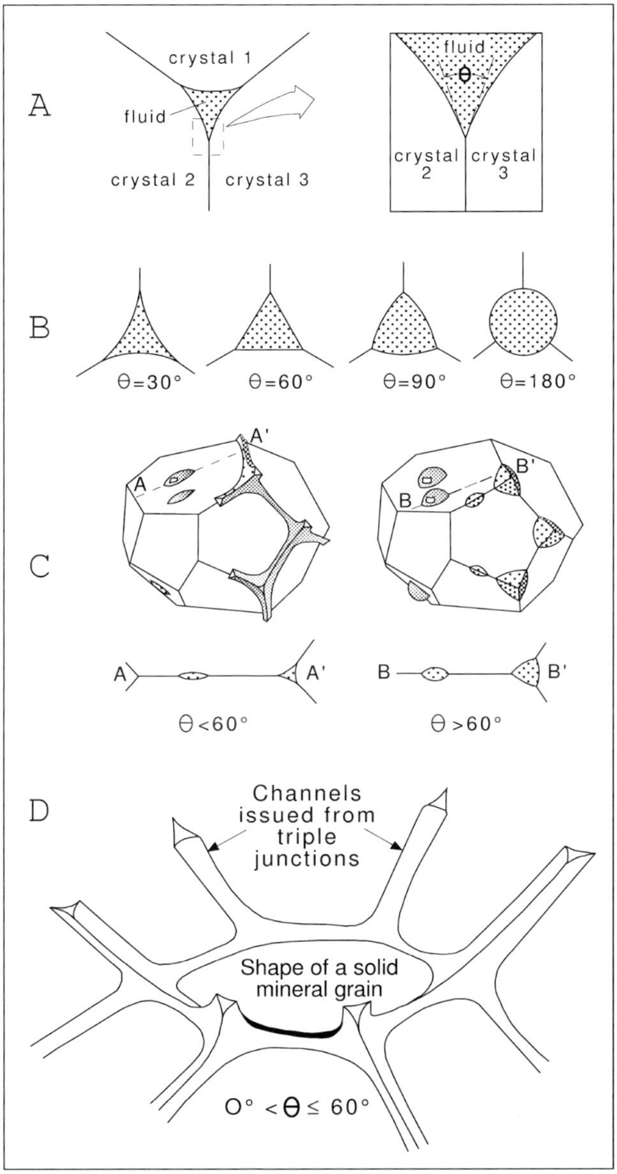

Fig. 8.10. (A) Transverse section of a micro-channel filled with a fluid at the junction of three mineral grains in a rock. The dihedral angle θ is defined in the close-up sketch to the right. (B) Morphology of the channel section for various values of dihedral angle θ. (C) Perspective view of the distribution of magma droplets around a crystal grain, in a rock beginning to melt, for θ values lower than 60° (to the left, *good wetting* of the magma), and higher than 60° (to the right, *poor wetting* of the magma). To the left, the magma droplets connect along the grain edges, and constitute a connected tridimensional network. This is not the case to the right, where the liquid pores remain isolated (nul permeability). Sections AA′ and BB′, perpendicular to the grain surface, distinguish the different morphologies of the magma droplets in both cases (after Watson and Brenan, 1987). (D) Aspect of the geometry of magmatic micro-channels around mineral grains in the case of good wetting ($\theta < 60°$).

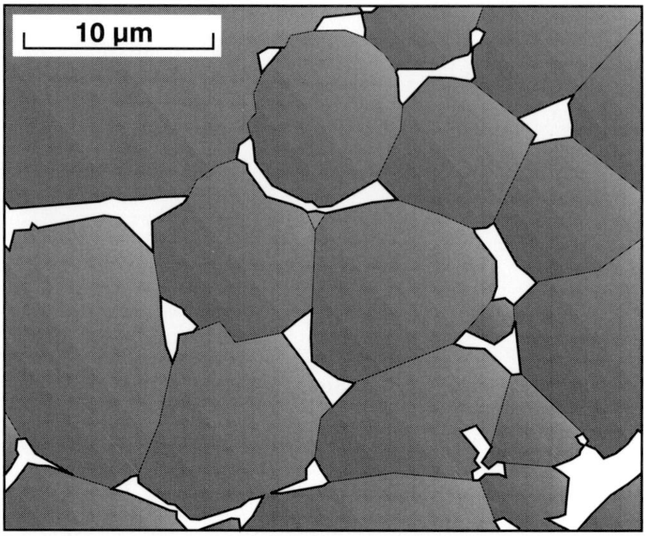

Fig. 8.11. In this laboratory experiment, a mechanical mixture of olivine powder and MORB-type basalt powder (about 7% in weight) has been compressed at 30 kb and 1300°C in order to melt completely the basalt and to observe its distribution. In this image obtained by SEM, the olivine grains appear in grey, and the basaltic pores in white. The liquid appears at all triple or quadruple junctions, indicating the formation of an interconnected network, with a measured dihedral angle of 38° (after Kohlstedt, 1992).

On the other hand, the internal pressure of the liquid phase increases rapidly with the partial-melting rate. Effectively, the melting process is accompanied by a volume increase estimated at about 16%. Taking into account the very low compressibility of silicate melts, the production of magma results in an over-pressure of the liquid phase which may exceed the lithostatic pressure, and provoke 'hydraulic' fracturing of the surrounding rocks.

8.2.4 The permeability threshold

It seems that the permeability threshold of a dunite is reached for about 2 to 3% of partial melting. For a lherzolite, the first magma droplets appear only at the contact points between the four major mineral phases of the lherzolite, and a *critical percentage of partial melting* is needed to observe an increase in size of the droplets, followed by their extension along the arêtes and faces, and finally their junction across the crystalline matrix.

Laboratory experiments have shown that the more fertile the lherzolite, hence the richer in pyroxenes, the higher the permeability threshold. In anhydrous conditions, at least 3 to 5% of partial melting are needed to reach this threshold. Of course, the presence of very weak quantities of water may lower this permeability threshold considerably, the presence of water having the effect of decreasing strongly the value of the dihedral angle θ of the pyroxenes.

8.2.5 Extraction of magmas

Once the permeability threshold is reached, the magma can percolate through the residual crystalline matrix, provided that a pressure gradient exists. The magma flux F passing through a solid permeable medium will be a function of the viscosity η of the liquid, of the permeability K of the rock, and of the pressure gradient dP/dx in the liquid. The equation linking these parameters is called Darcy's law, and is written:

$$F = -K/\eta \, dP/dx$$

The flux is therefore proportional to the permeability K, and inversely proportional to the viscosity η. The permeability is of the order of 10^{-7} to 10^{-10} cm^2. Two processes favour the migration of the liquid upwards:

—The compaction of the crystalline matrix, submitted to a compression: $dP = h(\rho r - \rho m)g$, with h representing the depth, ρr and ρm the densities of the host rock and of the magma respectively, and g the acceleration of gravity. The compaction first develops in the deeper parts of the mantle diapir, where the compression is the strongest. The plastic flow of the residual solid exerts a pressure on the interstitial magma, which begins to move upwards through interstitial channels towards the top of the column. The pressure gradient is equal to $(\rho r - \rho m)g$.

—The 'hydraulic' fracturing process, or rather *magmatic* fracturing (or *magma-fracting*, as proposed by Yoder, 1976), when the over-pressurized magma opens new cracks at its roof, the internal pressure of the liquid exceeding the brittle threshold of the peridotite. A crack appears into which magma is injected to form a dyke.

Finally, a mantle diapir ascending beneath a ridge would pass by the five following stages (Figure 8.12):

—At about 75 km in depth, beginning of the partial melting process, with local segregations of magma. At this early stage, no interconnected network exists (the permeability is null).

—At about 60 km in depth, creation of an interconnected network of magmatic veins, due to the continuous increase in the adiabatic melting process, during the diapir uprise. The veins are preferentially oriented following tension cracks. The formation of veins and cracks is strongly helped and oriented by the plastic flow of the residual solid at high temperature. A magmatic flow through a porous matrix develops actively at this stage, with compaction of the residual solid.

—At about 50 km in depth, the hydraulic fracturing of the peridotite roof develops efficiently and helps vertical extension of the network. A crack will propagate upwards, through which the magma is extracted from the system.

—If the fracture reaches the ocean floor, the magma filling the interconnected net around the crack, particularly at its base, is efficiently extracted and

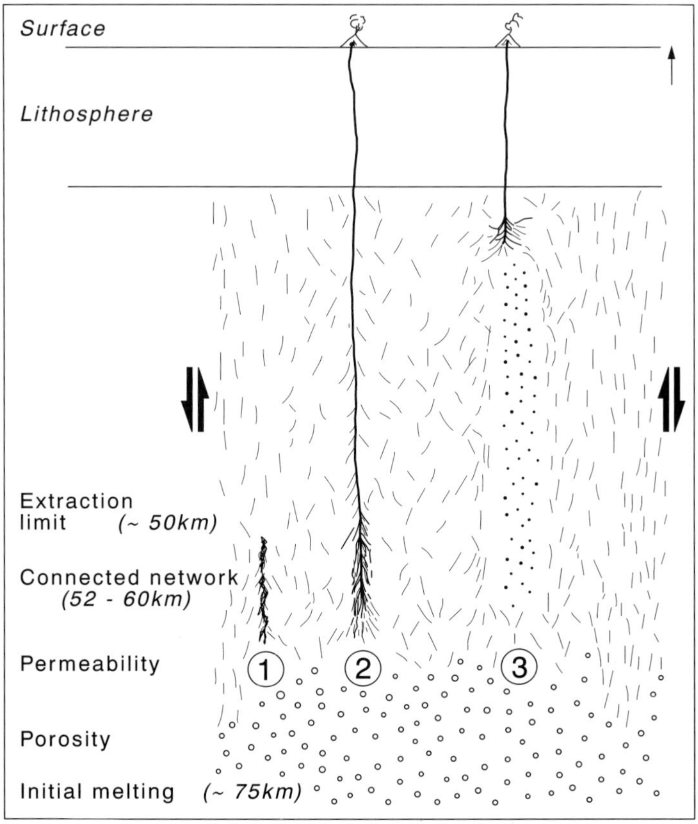

Fig. 8.12. Schema summarising the different stages of magma extraction from a mantle diapir. Empty circles represent isolated magma droplets. Dashed lines represent magmatic veins. Stippled areas represent the residual mantle after extraction of the liquids. The magmatic conduits (bold, continuous lines) follow S1–S2 trajectories and are not necessarily vertical (after Nicolas, 1989). 1. Creation of a net of magmatic veins, mainly interconnected in the vertical dimension. 2. Newly created magmatic conduit, draining up the magmas towards the surface. 3. Dying magmatic conduit, still draining shallow liquids, and letting in depth a zone of residual peridotites (dunites). Sequence 1–2–3 may be repeated many times in various parts of the mantle diapir.

expulsed from the mantle diapir, due to the strong depression created at the bottom of the column. The lithostatic pressure tends to close up the conduit.

—The initiation of a new hydraulic instability may occur in any other area in the ascending diapir, inducing a new extraction sequence, and so on.

8.3 THE ACCRETION-MAGMA CHAMBERS

The world oceanic-ridge system is a giant plant producing oceanic crust continuously. The central organ of this plant is the *accretion magma chamber*, whose

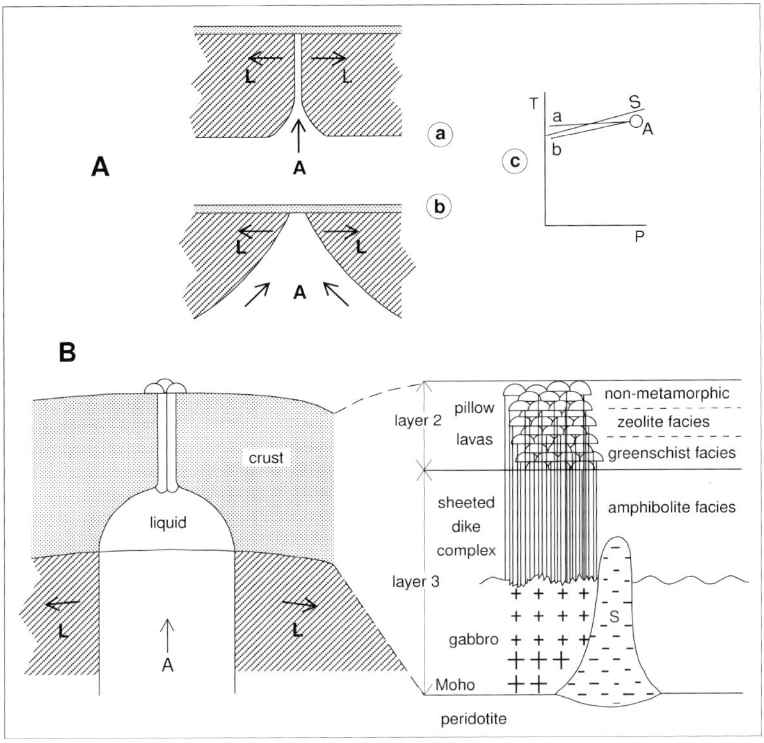

Fig. 8.13. One of the very first schemas of an oceanic ridge, and of the corresponding crustal structure (Cann, 1970). (A) Principle of the asthenosphere (A) upwelling, either narrowly (a) or widely (b) channelized. The asthenosphere is frozen later on into rigid lithosphere (L). In (c), the P-T diagram shows an asthenospheric mantle volume A moving up to the surface following two possible pathways: (a) in adiabatic conditions (fast upwelling), with crossing of the solidus S and partial melting; (b) in non-adiabatic conditions (very slow upwelling), without partial melting. (B) Crustal structure of the mid-ocean ridge. The magmas issued from the partial melting of asthenosphere A cluster in a magma chamber at the level of layer 3. This magma chamber crystallizes gabbros along its floor and walls. Its roof cracks periodically to feed submarine basalt flows on the ocean floor (pillow lava). The fissures filled with basalt cool as diabases and progressively constitute a sheeted dike complex. The basalt/seawater interaction develops an oceanic metamorphism becoming hotter with depth, because of the high thermal gradient (zeolite, greenschist and amphibolite facies develop in the crust from top to bottom). S: serpentinite diapir due to local hydration of the mantle rocks.

necessary existence beneath oceanic ridges was postulated at the time of the birth of the seafloor spreading concept.

At the present time, the 'ophiolite log' is still poorly known, and nobody has observed the axial domain of an oceanic ridge. Cann (1970) postulated the existence of a liquid pocket beneath the ridge, at the level of the lower crust, collecting the partial-melting liquids rising up from the decompressed asthenosphere (Figure 8.13). This chamber periodically fractures its roof and emits MORB-type submarine lava flows on the ocean floor, through vertical cracks which become basaltic dykes, once the magma is chilled. Simultaneously, coarse-grained gabbros crystallize slowly in depth at the lateral walls of the magma chamber, progressively constituting layer 3.

The necessity for a magma chamber stems from the fact that the great majority of MORB, sampled at the oceanic ridges, represent already evolved tholeiitic basalts, neither the glass nor the phenocrysts of which are in chemical equilibrium with the upper mantle. This can be easily explained if it is accepted that the magma has stayed some time in a reservoir intermediate between the mantle and the surface, where it may have cooled and evolved by crystal fractionation.

8.3.1 Bidimensional models

The 'infinite onion' model. During the 1970s, and after the 1972 Penrose Conference dedicated to the definition of ophiolites, the so-called 'infinite onion' model became extremely popular (Cann, 1974). It was a 2D model, a transversal section ignoring the along-strike dimension, considered as infinite.

The chamber (Figure 8.14) has the shape of an onion, and like it 'peels off' by crystallising on all its margins. At the roof and on the flanks, it crystallizes isotropic, non-layered gabbros. At the floor, magmatic cumulates pile up, mainly by gravitative settling, along with layered gabbros, lying over a variable thickness of ultramafic cumulates, themselves resting over the residual mantle rocks. Over the roof of the magma chamber, a layer of basaltic dykes is extruded and extends progressively. This sheeted dyke complex is overlain by the submarine volcanic layer, mainly made of pillow lava.

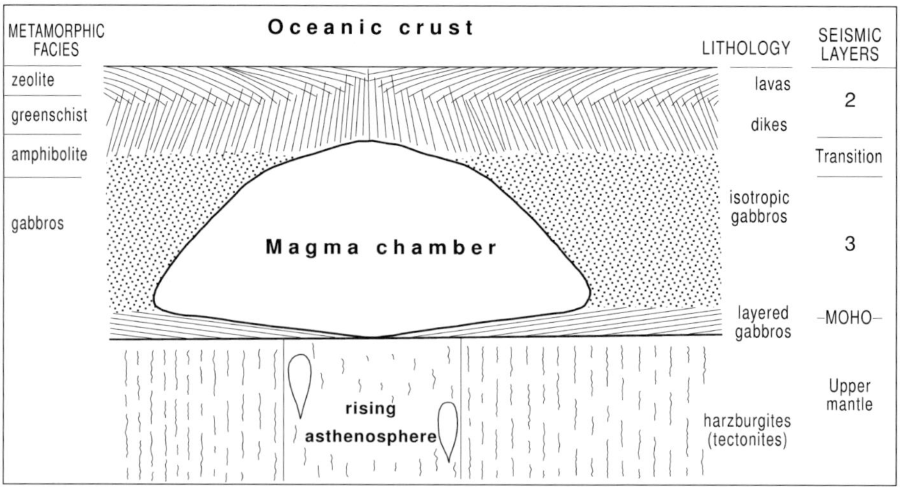

Fig. 8.14. So-called 'infinite onion' bidimensional model of an accretion magma chamber (for a 'non-rifted', rather fast-spreading ridge), and corresponding crustal structure (after Cann, 1974). This model has been extremely popular during the 1970s. The magma chamber is supposed to be steady state in volume, the extension due to the oceanic accretion process compensating for crystallization on the walls. This chamber was supposed to be huge (20–30 km wide, 4-5 km high), and entirely liquid. It was called 'infinite onion' because of the geometry of the chamber (bulbous-shaped, domed roof, flat floor slightly inclined toward the axis), and because it 'peels off' by crystallization along all its walls: layered gabbros accumulating on the floor (magmatic cumulates), and isotropic gabbros solidifying at the roof.

This model was mainly based on the petrology of oceanic rocks and on ophiolitic data, particularly from the Troodos ophiolite in Cyprus. It explained satisfactorily (at that time) the 'standard' ophiolite suite defined by the Penrose Conference. It was not constrained by any oceanic structural data, the geology of the oceanic ridges still being very poorly known, but seemed to be in agreement with the seismic stratification of the oceanic crust, whose seismic layers were assimilated to magmatic lithological units (we have seen in Chapter 5 that things are not so simple).

The infinite onion model recognizes a number of variations in its geometry ('Chinese hat', bowl-shaped, funnel-shaped magma chambers, etc.), or in its behaviour (subsiding floor, thermal and mineralogical zoning, etc.). All these variations have in common:

—The huge dimensions of the accretion magma chamber, either in width (20 to 30 km, or even more), or in height (several km). The reservoir occupies the whole thickness of layer 3 at the core of the ridge, and lies directly over the foliated harzburgites of the residual upper mantle.

—The liquid nature of the chamber, which is often conceived as being vigorously convecting, favouring crystal settling processes. The discovery of thick-layered gabbro series in the major Tethyan ophiolites of the Middle East gave support to this idea of huge, liquid-magma chambers, by analogy with the large continental mafic/ultramafic stratiform complexes, such as the Skaergaard or the Bushweld complexes.

—The permanent character of the accretion magma chamber, maintained indefinitely in a steady state and keeping a constant volume, the extension due to the accretion processes compensating the loss of volume due to crystallization on the margins.

Among the various versions of the 'infinite onion' model, several of them had an important impact on the scientific community; in particular: (a) the subsiding floor model (Figure 8.15) from Dewey and Kidd (1977), based both on ophiolitic observations (Bay of Islands ophiolite, Newfoundland) and recent oceanic data (FAMOUS diving project on the Mid-Atlantic Ridge). This is the first 'tent shaped model' (or 'Chinese hat'), taking into account the narrowness of the ridge's neo-volcanic axial zone, and in agreement with thermal modelling of a ridge axis; (b) the zoned magma chamber model (Figure 8.16) from Bryan and Moore (1977) for the FAMOUS area, postulating a huge magma chamber, thermally and mineralogically zoned, in order to explain the occurrence of olivine-rich basalts (Mt Venus picritic basalts) in the axial neo-volcanic zone (hotter and more primitive magmas), and of plagioclase-rich basalts on the flanks of the inner floor of the rift valley (colder and more evolved magmas); (c) the 'funnel-shaped model' (Figure 8.17) from Pallister and Hopson (1981), based on a detailed study of the Ibra massif, in the Oman ophiolite; d) the 'bowl-shaped model' (Figure 8.18) from Smewing (1981), based on the study of other massifs of the Oman ophiolite.

The 'infinite leek' model. Toward the end of the 1970s, a vigorous reaction developed against the concept of huge and liquid accretion magma chambers. This

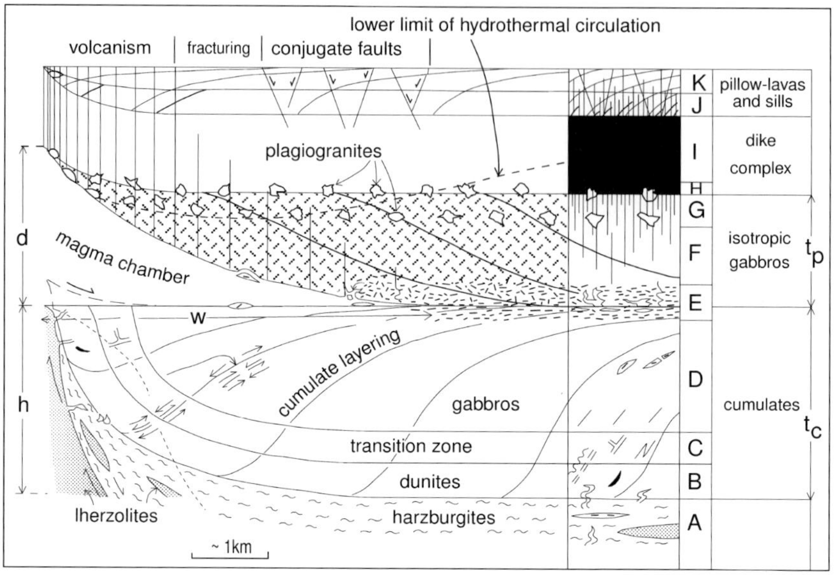

Fig. 8.15. Dewey and Kidd's model of the accretion magma chamber. This model is based: (1) on a detailed structural study of the Bay-of-Islands ophiolite (Newfoundland); (2) on the observations made at the axis of the Mid-Atlantic Ridge (FAMOUS cruises, 1973–74); (3) on thermal models of oceanic ridges available at that time. This is the first 'tent-shaped' model, taking into account both the narrowness of the neovolcanic zone of the oceanic ridges and thermal constraints. It proposes underplating and progressive thickening of the isotropic gabbros beneath the upper crust lid (dike complex and lavas). At the floor level, an active subsidence of layered cumulates compensates for mantle upwelling at the axis. Axial tectonics and hydrothermal circulation in the crust were also taken into consideration for the first time.

reaction came mainly from geophysicists, who had tried without success to detect huge magma chambers beneath the oceanic ridges. Besides, thermal modelling showed that it was dubious that a huge magma chamber may maintain a steady-state indefinitely. Lastly, Smewing had shown, by a detailed mapping in the Troodos ophiolite, that the plutonic sequence consists of a series of small magma pockets intersecting each other. In a now classical paper, Nisbet and Fowler (1978) refuted the infinite onion model, and proposed instead, especially for slow-spreading ridges, the 'infinite leek' model (Figure 8.19), in which small gabbroic ephemeral magma chambers crystallize completely beneath the axial zone and intersect each other, emitting a dense swarm of dykes towards the surface (hence the image of the leek).

8.3.2 The 3D models, and the contribution of seismic tomography

During the 1980s, ideas concerning accretion magma chambers evolved considerably thanks to the introduction of new mapping and geophysical tools (multibeam sonar systems, side-scan sonars). The data acquired in the ocean progressively overtook ophiolitic data, and eliminated the notion of huge liquid magma chambers. The

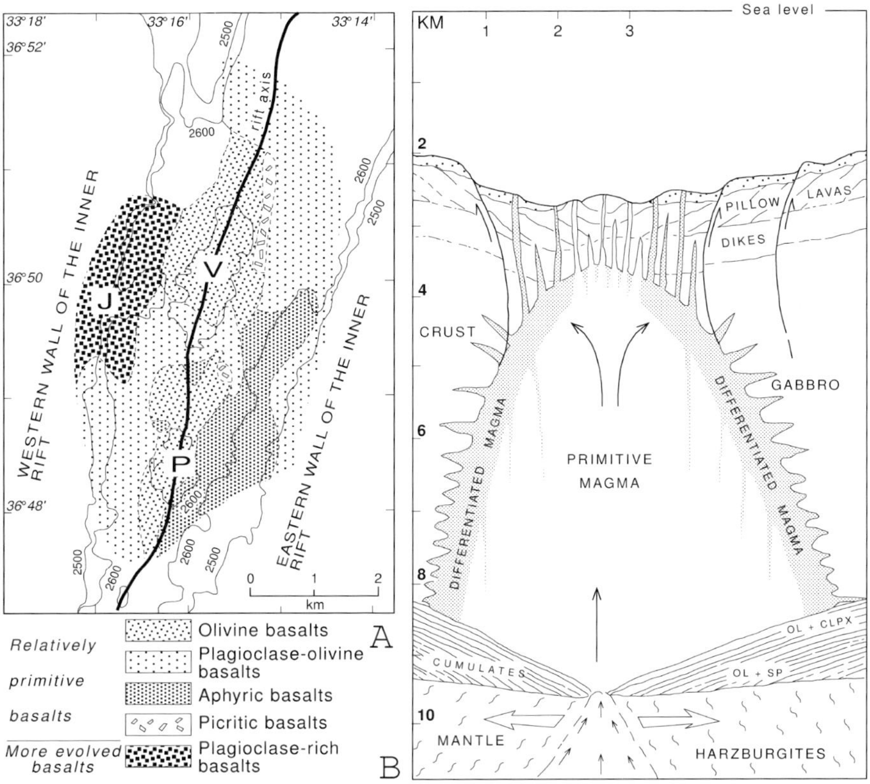

Fig. 8.16. Model of a chemically and mineralogically zoned accretion magma chamber, proposed for the FAMOUS area (Mid-Atlantic Ridge at 36°50′N). (A) Distribution map of the various basalt types in the rift valley (FAMOUS zone), after Hékinian (1982). Mounts Venus (V), Jupiter (J) and Pluton (P) are the volcanic hills presenting the most recent volcanic activity. (B) Interpretation of this distribution by Bryan and Moore (1977). A huge accretion magma chamber (6 km wide, 5 km high) is emplaced beneath the rift valley. It is thermally zoned, resulting in chemical and mineralogical zoning. The colder flanks of the chamber are occupied by a differentiated magma rich in plagioclase phenocrysts, whereas the hotter central part, directly filled with hot, primitive basaltic magma, is occupied by a magma richer in MgO, crystallising olivine phenocrysts.

more and more accurate seismic and gravimetric data provided drastic constraints concerning the size of magma chambers: small becomes beautiful!

East-Pacific Rise (EPR). Several geophysical cruises that took place at the end of the 1980s on the East-Pacific Rise between 9° and 13°N (average spreading rate: 11 cm/yr) have revealed severe constraints about the dimensions and geometry of accretion magma chambers. These experiments were conducted either with multichannel seismics using two oceanographic ships, or using an array of OBS (ocean bottom seismometers) laid out on the ocean floor, and disposed parallel and perpendicular to the ridge axis.

The main results of these experiments are (Figure 8.20):

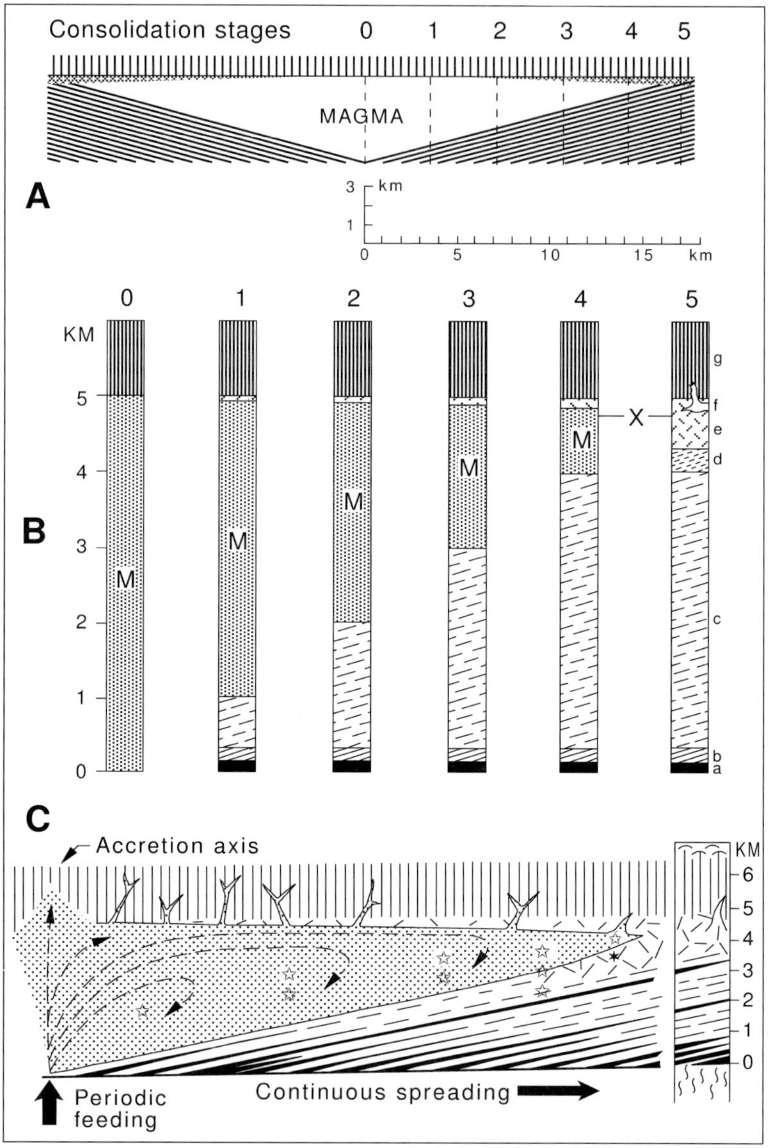

Fig. 8.17. Pallister and Hopson's model (1981) of a steady-state accretion magma chamber, based on the study of the Oman ophiolite (Ibra massif). This model has been very popular at the beginning of the 1980s. (A) Geometrically, this huge chamber (36 km wide, 4 km high) is the reverse of Dewey and Kidd's chamber: the roof is flat and the floor inclined towards the axis. (B) The oceanic crust is totally solidified when both crystallization fronts meet together, the first one thin and going downward (isotropic gabbros crystallising at the roof), the other one several km thick and going upward (layered cumulate gabbros). Transitional foliated gabbros and very differentiated magma pockets (quartz diorites to plagiogranites) are emplaced during the last stages of the crust solidification. (C) The existence of recurrent wehrlitic levels (indicated in black), interbedded within the layered gabbros, is explained by frequent and periodic inputs of primitive magma in the chamber. This interpretation is supported by the 'zigzag' cryptic evolution of the mineral phases in the gabbroic pile.

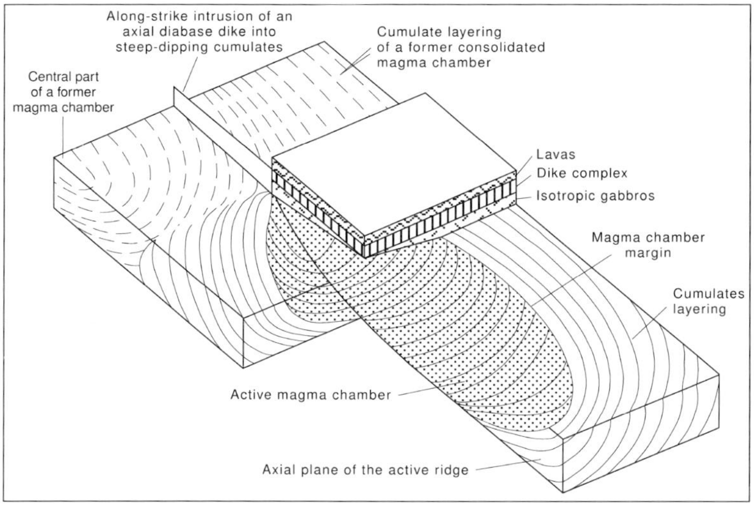

Fig. 8.18. Lippard *et al.*'s model (1986) of an accretion magma chamber, based on the study of the Oman ophiolite (central and northern massifs). The English Open University team has proposed a 'bowl-shaped' geometry, taking into account the progressive steepening of the magmatic layering and foliations in the gabbros going upsection. The evolution of the magma chamber, with periodic replenishment with primitive magma, is analogous to that of Pallister and Hopson's model.

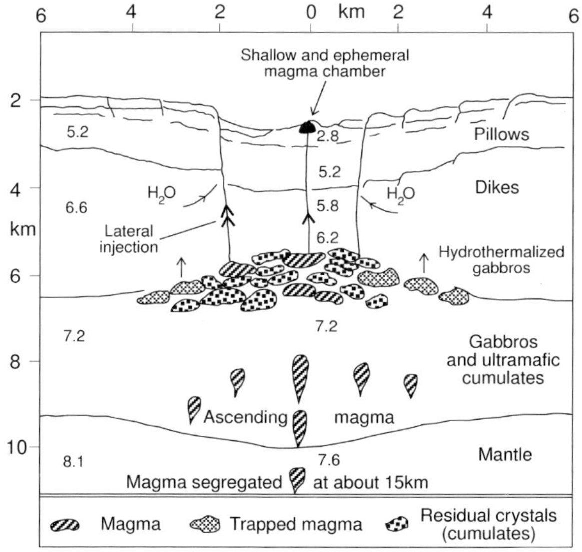

Fig. 8.19. The 'infinite leek' model, proposed by Nisbet and Fowler (1978) for slow-spreading ridges. The authors have re-interpreted the FAMOUS zone, using available seismic profiles. The ascending magmas are trapped in the 7.2 km/s layer, at the base of the elastic crust. From there, they may either migrate to the surface, leaving a solid cumulate residue, or crystallize completely *in situ*, forming small gabbroic pockets. The numbers correspond to V_p seismic velocities (km/s). Numerous swarms of dykes are supposed to escape upwards, forming a dense net of parallel vertical structures, hence the shape of the 'leek'. Compare with Bryan and Moore's huge zoned-magma chamber for the same FAMOUS area (Figure 8.16).

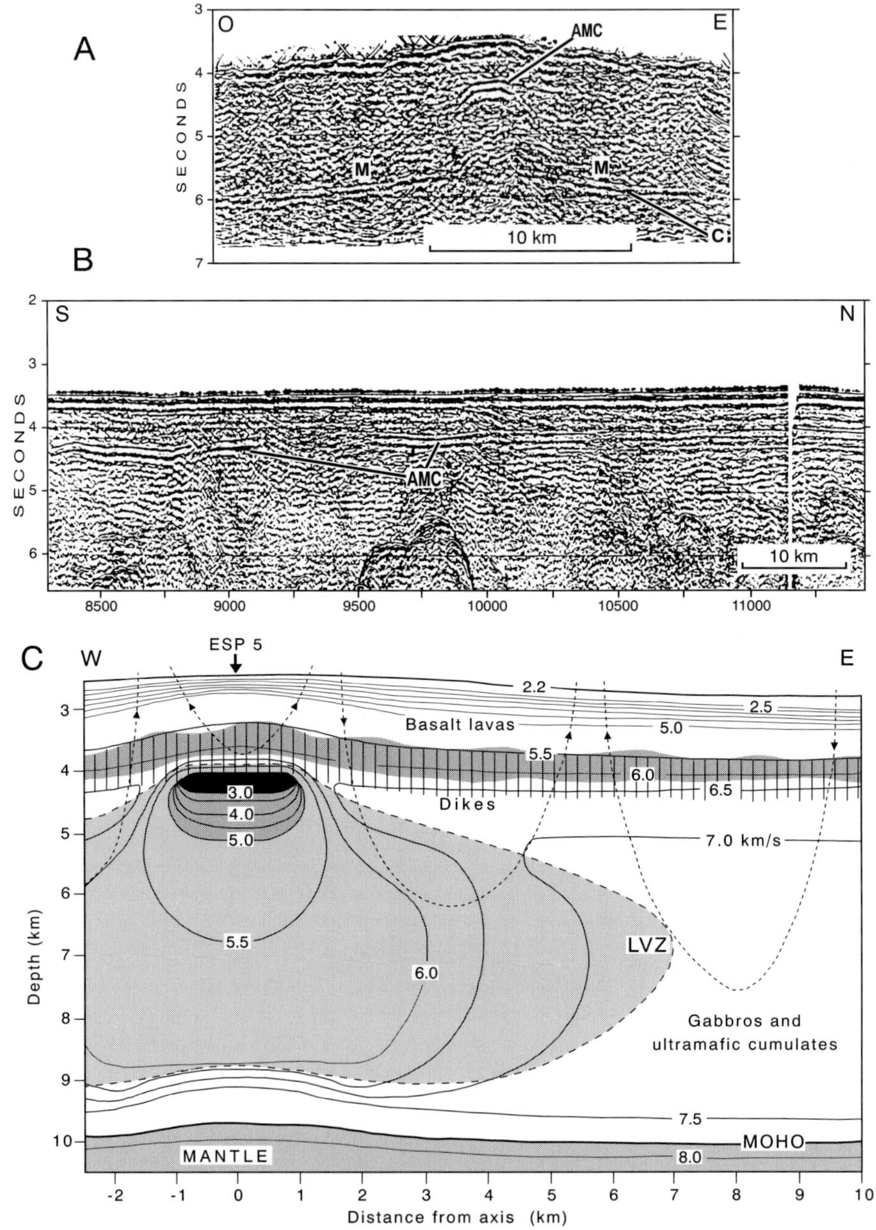

Fig. 8.20. Seismic images of the East-Pacific Rise. (A) One of the seismic lines, transverse to the ridge, shows a high- amplitude reflector at the level of the roof of the *axial magma chamber* (AMC), beneath the ridge axis, at about 9°30′N. The reflexions of the Moho (M) are also very visible below the flanks of the ridge, interrupted only by a narrow axial zone (after Vera *et al.*, 1990). (B) Along-strike axial profile, showing the remarkable continuity of the AMC reflector. (C) V_p velocity structure model for the East-Pacific Rise at 9°30′N, based on ESP solutions proposed by Vera *et al.* (1990). Oblique hashing outlines the low-velocity zone (LVZ).

—The existence of a *flat and brilliant reflecting horizon* at about 1.5 km beneath the ridge axis, barely extending over 2 km in width (across the ridge), and continuous over tens of km along strike, interrupted only at axial discontinuities recorded by bathymetry;

—The existence of a wide axial zone characterized by low seismic velocities, with respect to those of the normal crust. This 'low-velocity zone' extends from 1 to 2 km beneath the ocean floor to the base of the crust, has a flat top and widens with depth below the flanks of the ridge. At its base, the 'low-velocity zone' has a width of 10 to 12 km. The V_p negative anomalies are <1 km/s with respect to the velocities of normal layer 3. The lowest velocities ($V_p < 5$ km/s) are confined in a thin (less than 1 km thick) and narrow (<2 km in width) zone, just beneath the ridge axis. Along axis, tomographic images indicate that the 'low-velocity zone' is a continuous structure which follows the axis, even through small axial discontinuities.

These results exclude any possibility of a huge chamber filled with liquid magma beneath the explored segments of the EPR. Instead, the proposed interpretative model comprises a thin (several hundreds of metres thick) and narrow (1 to 2 km wide) liquid magma lens perched at the top of a wider zone made of a 'crystal mush', itself surrounded by a transition zone of more and more consolidated crust, containing isolated pockets of magma (Figure 8.21A).

The 'low velocity zone' is interpreted as a largely consolidated but still hot section of lower crust, which can contain several percent magma. The truly liquid magmatic body would be restricted to the zone of V_p anomalies >2–3 km/s, and located just beneath the axis. Its roof would correspond to the high-amplitude, flat and brilliant reflector, located at 1.5 km on average beneath the ocean floor. The lower limit of this liquid lens is loose, which seems to indicate a very progressive transition to the crystal mush. The gravimetric data confirm this interpretation.

Similar experiments conducted on 'ultrafast' segments of the EPR gave analogous results, with a brilliant reflector located at hardly 1 km beneath the axis. Similarly, intra-crustal reflectors were registered beneath the Juan de Fuca Ridge (6 cm/yr, 3 km in depth), and on the Valu Fa Ridge, in the Lau back-arc basin (7 cm/yr, 1 to 2.5 km in width, with V_p passing from 5.3 to 2.9 across the reflector).

It can be concluded then that the dimensions and the geometry of the crustal magma chamber detected beneath the East-Pacific Rise are typical for a wide range of ridges with intermediate to fast-spreading rates.

Mid-Atlantic Ridge. Until very recently, no proof of this kind was found along the Mid-Atlantic Ridge, in spite of numerous geophysical experiments using multichannel reflexion and refraction seismics. The natural micro-seismicity and the study of focal movements have shown that seisms can occur at depths over 8 km or more beneath the axial valley of the ridge, suggesting that the whole crustal section is cooled down to the brittle domain. The existence of huge, steady-state magma chambers seems to be excluded beneath the Mid-Atlantic Ridge (Figure 8.21B). However, modest-sized, more or less ephemeral magma bodies may be locally present.

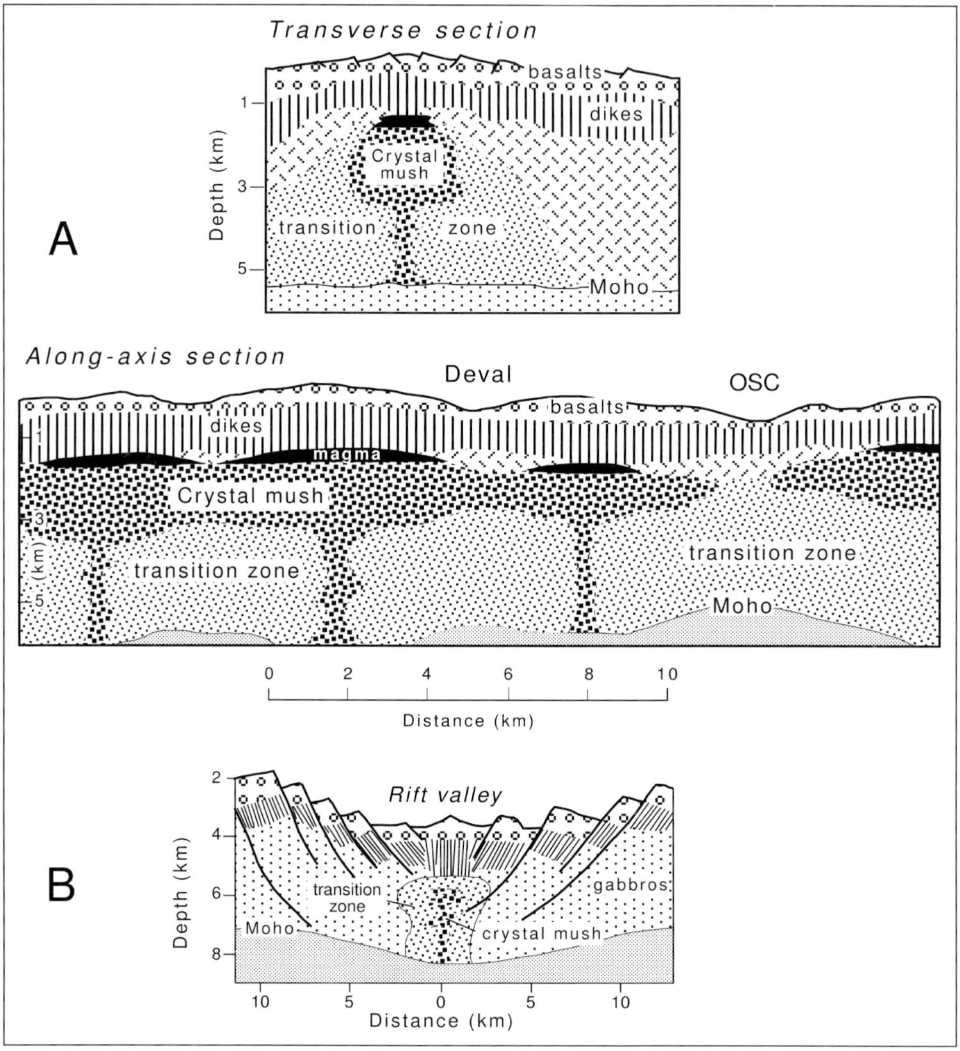

Fig. 8.21. Geological interpretation of the structure of fast- and slow-spreading ridges, based on available seismic data (after Sinton and Detrick, 1992). **(A) Fast-spreading ridges (high-magmatic budget):** interpretation of the East-Pacific Rise data, transverse and along-strike geological sections. The main features of this model are: (1) a lens-shaped liquid magmatic body, located at 1.5–2 km below the ridge axis, passing progressively downward to (2) a zone constituted of a partially solidified crystal mush, grading itself to (3) a transition zone toward the still hot and solidified crust. The lava emplaced on the seafloor is extracted from the liquid sill and is typically aphyric. The along-strike profile shows that this magmatic sill is probably interrupted at the same level as most axial discontinuities recorded by bathymetry. **(B) Slow-spreading ridges (poor-magmatic budget):** interpretation of the Mid-Atlantic Ridge data. These ridges are not likely to comprise steady-state and purely liquid magma chambers. Instead, a *dyke-shaped crystal mush zone* is proposed, which would emit sill-like intrusive bodies, crystallising progressively to constitute an heterogeneous oceanic crust. Extrusions of phyric lava should coincide in time with episodes of primitive magma inputs issued from the upper mantle. This model also outlines the *importance of extensional tectonics:* numerous major normal faults starting from the margins of the axial rift valley may deeply penetrate the hot crystal mush zone.

Below the rift valley, some segments show abnormal crustal thicknesses and seismic velocities, whereas others show a thin, anomalous crust devoid of a developed layer 3, with low mantle velocities. In the MARK (Mid-Atlantic Ridge Kane) area, Purdy and Detrick (1986) have found lower crustal velocities below some topographic highs, associated with negative gravimetric anomalies. As shown in Chapter 4, these topographic highs indicate the location of focalized basaltic magma upwellings, beneath the central part of ridge segments.

8.3.3 Towards a new concept of oceanic accretion-magma chambers

The geophysical approach

It appears clear that over the last few years, the concept of what is exactly an accretion magma chamber has evolved considerably. We know today that the the truly liquid (or molten) part of this reservoir can only occupy a restricted volume, and that the geometry of this liquid part approaches a lens or a sill shape, rather than a 'chamber'. Beneath this liquid-magma body, what was formerly called the 'magma chamber' consists in fact mainly of a crystal mush, containing variable proportions of interstitial magma: this is the 'low velocity zone' whose major part, with more than 60% crystals, behaves rheologically and seismically as a solid.

The rheological and dynamical processes accompanying the crystallization of crystal mushes therefore become crucial in understanding the geochemical evolution of MORB magmas and the structure of the lower crust.

The geological model recently proposed by Sinton and Detrick (1992) takes these new geophysical data into account. The model necessarily includes two versions, one for fast-spreading, EPR-type ridges, the other for slow-spreading, MAR-type ridges (Figure 8.21).

The ophiolitic approach

The detailed studies conducted parallelly on ophiolites, and especially on the Oman ophiolite, resulted in new propositions of magma-chamber models, more realistic about geometry and dimensions, and in agreement with the new oceanic geophysical constraints (Figure 8.22A). In this model, the main part of the layered-gabbro cumulate series crystallizes at the roof of the magma chamber, the layering resulting mainly from the deformation of crystallising gabbros by viscous flow. Another important characteristic of this model is the effect of mechanical coupling between the divergent asthenospheric mantle and the lower part of the cumulate pile: detailed structural measurements in the field have shown effectively, as we have seen (Chapter 5), a remarkable and systematic parallellism, on both sides of the 'petrological Moho', between the *metamorphic mineral lineations* in the residual dunites and harzburgites, and the *magmatic mineral lineations* in the lower gabbros.

The Oman accretion-magma chamber is presently conceived as having an inclined 'tent'-shaped roof (narrow apex) and a wide floor, in agreement with oceanographic, geophysical and thermal data and models obtained on fast-spreading ridges. The

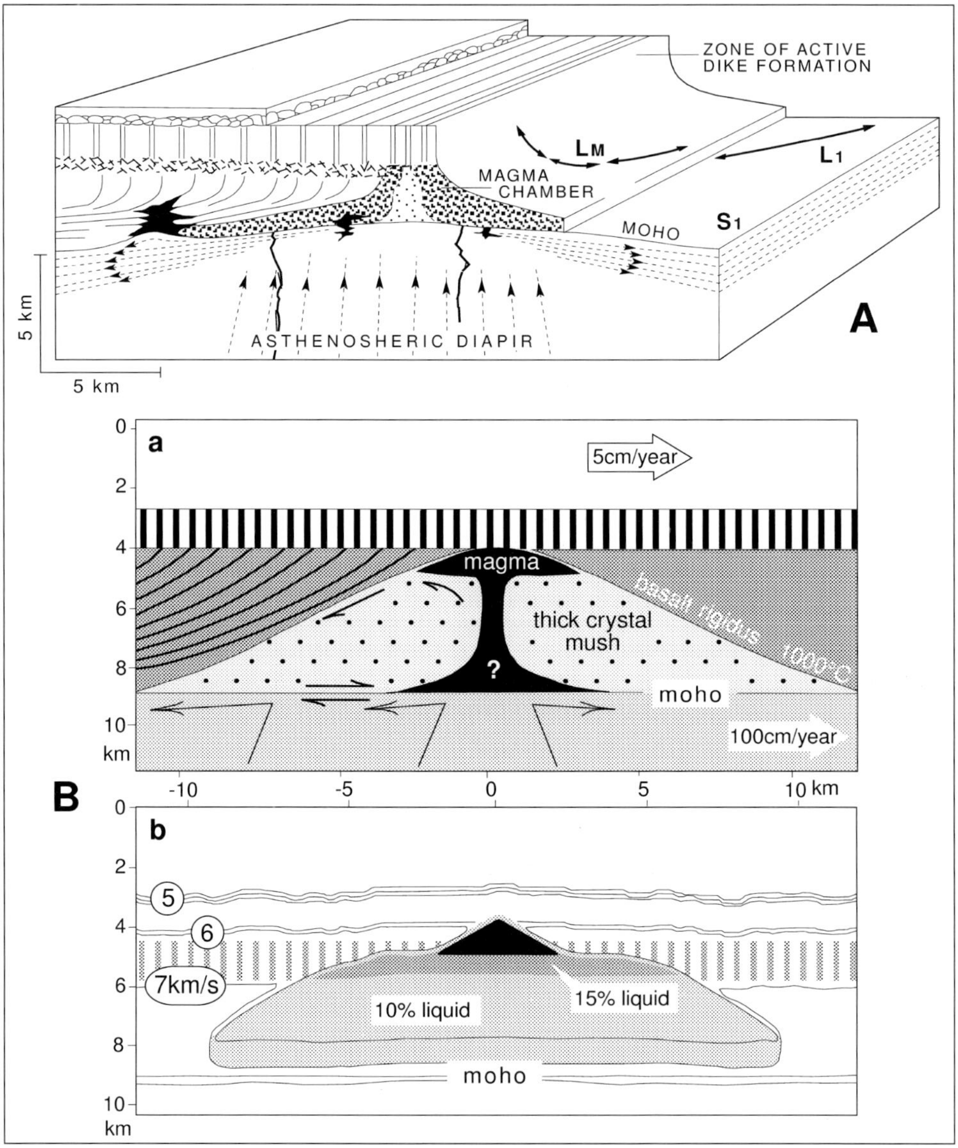

Fig. 8.22. Reconciliation between ophiolitic and oceanic models, and synthesis for both major oceanic ridge types. **Fast-spreading ridges:** (A) Geometrical model proposed by Nicolas *et al.* (1988) for the Oman ophiolite accretion magma chambers, based on the most recent mapping and structural studies. The chamber has a 'tent-shaped' profile (inclined roof, flat floor). The isotropic gabbros constitute a thin boundary-layer underplated beneath the sheeted dyke complex. Most of the cumulate sequence is formed along the inclined roof, the layering resulting mainly from viscous flow processes. This deformation is induced by the continuous drag exerted by the underlying asthenosphere (dashed lines with arrows). Lm magmatic lineations in the cumulates are effectively parallel to the L1 metamorphic mineral lineations in the tectonites (lineations observed on the S1 foliation planes). (B) Comparison of the Oman model ('a', simplified) with the East-Pacific Rise model ('b', also simplified). The interstitial liquid percentages are deduced from seismic velocity attenuations (after Nicolas, 1992).

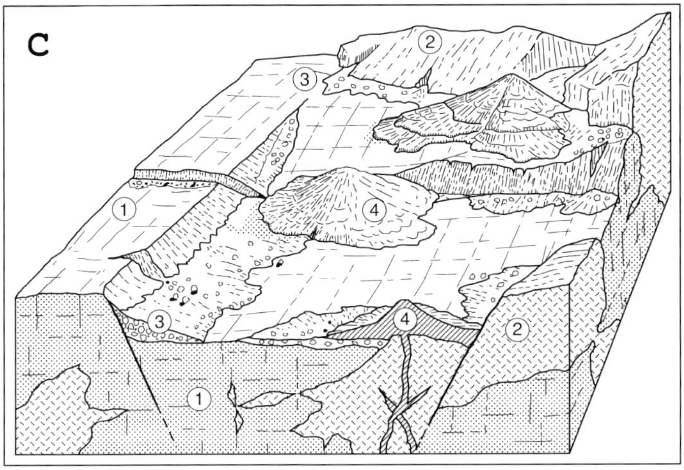

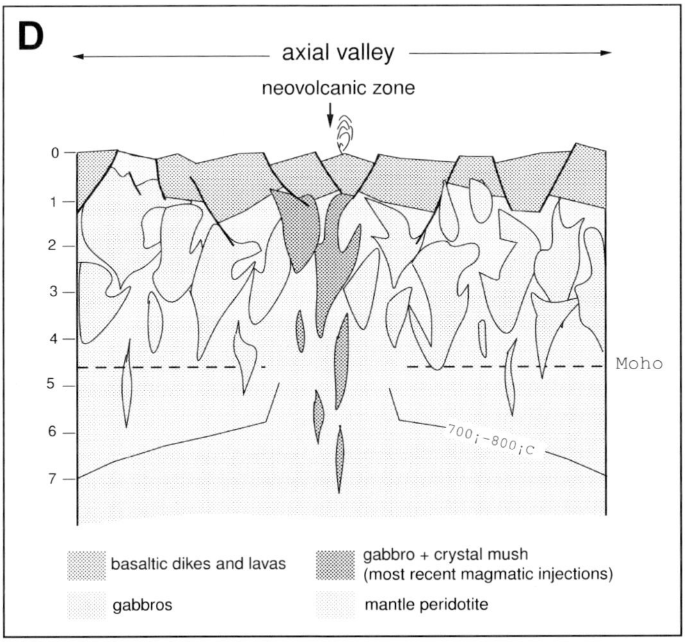

Slow-spreading ridges: (C) Schematic reconstitution of the Upper Jurassic Tethyan oceanic basement in the Ligurian and Piedmont basin, before Alpine foldings (French–Italian Alps, Apennines). 1. Serpentinized lherzolites. 2. Intrusive gabbros in the upper mantle. 3. Dismantling products of gabbros and serpentinites (talus, breccias). 4. Pillow-lava basaltic volcanoes and associated volcanic rubble that may rest directly over serpentinized peridotites or their rubble (drawing from Lagabrielle (in: Lagabrielle and Cannat, 1990)). (D) Model for the axial zone of a slow-spreading ridge, combining geological and geophysical observations in present-day oceans. Layer 3 is made of numerous individual gabbroic pockets, intrusive in the upper, serpentinized part of the mantle. The depth of the Moho and that of the brittle–ductile transition are constrained by seismic, gravimetric and microseismic data (MARK area). Tectonic accretion processes are dominant. Layer 2 is discontinuous, and rocks of the upper mantle frequently outcrop on the ocean floor (after Cannat, 1993).

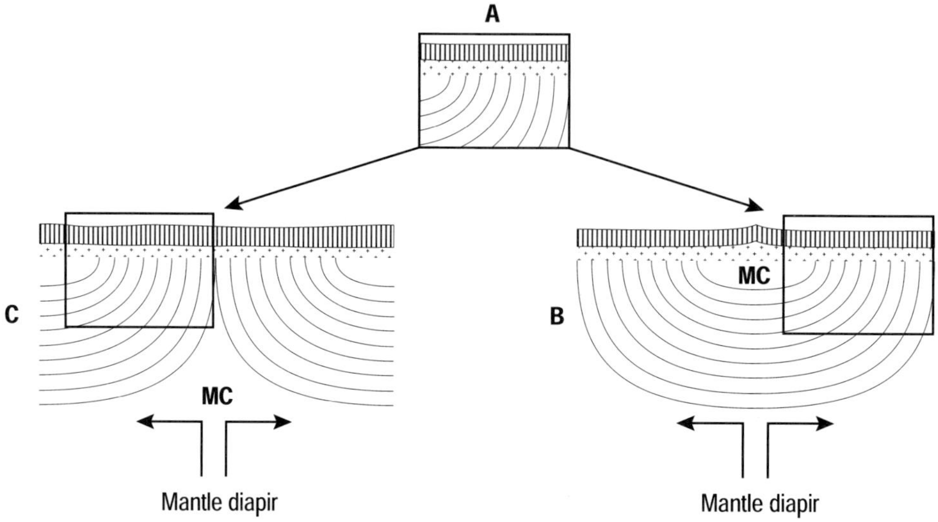

Fig. 8.23. Schemas showing two radically different interpretations of the geometry of the Oman ophiolite magma chamber, starting from identical mapping and structural data. (A) Schematic section of the crustal sequence in any massif of the Oman ophiolite, showing *progressive steepening* of magmatic layerings and foliations when going upsection. (B) 'Bowl-shaped' interpretation (to the right), implicitly showing that layered cumulates crystallize at the floor (Lippard *et al.*'s model, cf. Figure 8.18). (C) 'Tent-shaped' interpretation (to the left), implicitly showing that layered cumulates crystallize at the roof (Nicolas *et al.*'s model, cf. Figure 8.22A). Note that according to the model chosen, the piece of studied ophiolite is not issued from the same flank of the palaeo-ridge.

truly liquid part of the chamber is restricted to the upper, central zone, the rest of the reservoir being occupied by a crystal mush (Figure 8.22B).

The reader will perhaps be perplexed to see that very different and even contradictory geometrical models have been proposed over the last 20 years for the Oman ophiolite magma chambers. The quality of geological maps and surveys done by the different teams is not in dispute. It must be understood that in the case of steady-state, permanent accretion-magma chambers, the contacts of the chambers are not visible in the field: only the internal structures of the rocks issued from the solidification of this chamber are observable, and can be interpreted quite differently, according to the hypotheses proposed about the origin of these structures, as explained in Figure 8.23.

In conclusion, there is a clear convergence of the accretion-magma chamber models presently proposed for fast-spreading ridges (EPR) and for HOT-type (harzburgitic ophiolite type) ophiolites (Oman).

The same convergence also appears between the reconstructions proposed for the LOT-type (lherzolitic ophiolite type) ophiolitic paleo-ridges (Ligurian basin, Figure 8.22C) and the models published in recent years for the axial zones of slow-spreading ridges, characterized by a dominant tectonic accretion and by ephemeral magma pockets, crystallising inside the upper mantle (Figure 8.22D).

8.3.4 Functioning of accretion-magma chambers: significance of texture, structure and composition of the layered cumulates, and cryptic evolution

Texture of layered cumulates

We recall that Wager *et al.* (1960) were the first to propose replacing the term 'cumulative rocks' (Bowen, 1928) by the term 'cumulate', to describe 'any magmatic rock formed by an accumulation of crystals from a cooling magma'. Such a rock is typically made of two kinds of crystals:

—*cumulus crystals*, individual crystals successively separated from the magma and forming, by accumulation, a crystalline pile in whose pores the *intercumulus interstitial liquid* remains. In the original definition by Wager *et al.*, the cumulus crystals separate downwards under the effect of their own density. In more modern definitions, the accumulation of cumulus crystals can be due to various processes (including the action of gravity).

—*postcumulus material* is described as a 'crystalline material occupying the position of the intercumulus liquid'. In other words, it cements the cumulus crystals together.

According to their texture, cumulates were subdivided by Wager *et al.* (1960) into *orthocumulates*, *mesocumulates*, *adcumulates*, *hetradcumulates* and *crescumulates*. For these authors, each of these terms had a genetic significance, and corresponded to a precise mechanism (Figure 8.24).

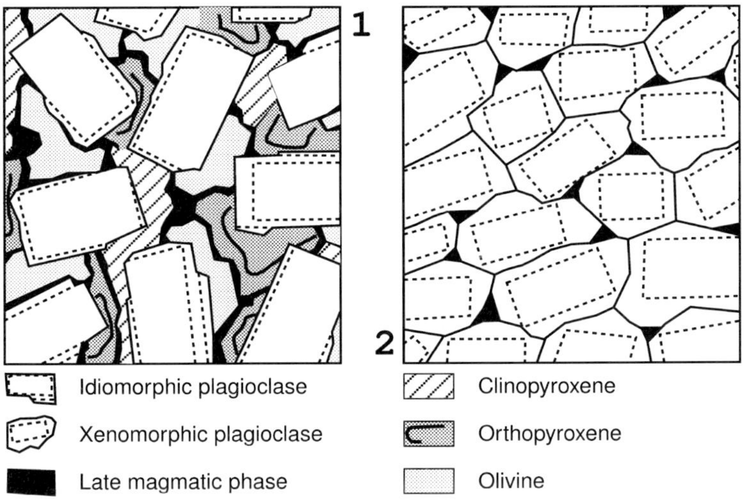

Fig. 8.24. Schematic representation of two common textures observed in ophiolitic cumulates (after Wager, Brown and Wadsworth, 1960 and Mehier, 1995). (A) Plagioclase orthocumulate. The interstial magma, trapped *in situ*, has enlarged the cumulus crystals (with zoning), then crystallized the other gabbroic phases during cooling (clinopyroxene, olivine, oxides, etc.). (B) Plagioclase adcumulate. The interstitial magma, renewed by advection and keeping a constant temperature, has enlarged the cumulus crystals (without any zoning), until they are welded together. Residual phases may crystallize in some remaining pores (several %).

A number of studies have shown that crystal settling due to the effect of gravity is not the main process of cumulate formation, and that other mechanisms are much more appropriate (density current deposits, *in situ* crystallizations, differentiation in viscous laminae, etc.). Irvine (1982) has redefined cumulate terminology, in order to eliminate any genetic connotation, and to remain in the purely descriptive domain.

For Irvine, a cumulate is 'a magmatic rock characterized by a network of crystals formed and concentrated in a first stage of crystal fractionation'. These crystals, generally more or less idiomorphic, are the *cumulus crystals*, whatever the fractionation and accumulation mechanism. They are cemented by a second generation of crystals, the *postcumulus crystals*, whose principal habits are the filling of interstitial spaces, their growth on cumulus crystal faces, or the reaction replacement of them.

The various types of cumulate are based on the postcumulus material volume ratio: 25 to 50% of postcumulus material in *orthocumulates*, 7 to 25% in *mesocumulates*, and 0 to 7% in *adcumulates* (Figure 8.25).

Moreover, Irvine proposes replacing the inelegant term 'heteradcumulate' by the more descriptive term *poikilitic ortho-*, *meso-* or *adcumulate*, the postcumulus material being made of poikilitic crystals (which may commonly englobe several tens of small cumulus crystals).

Structures that can be interpreted in terms of active or passive 'magmatic sedimentation' (gravity settling)

The vertical succession of cumulate layers showing important gradual modal variation, including positive rhythms (density-graded layers, with dense phases such as olivine and pyroxenes at the bottom, and plagioclase concentrated at the top of each individual layer), has been interpreted as the result of gravity differentiation, with crystal fractionation according to their density during their settling on the floor of the magma chamber (Figure 8.26). This is the older interpretation, proposed for the huge mafic/ultramafic layered complexes or for fossil volcanic reservoirs such as the Skaergaard intrusion in Greenland (Bowen, 1949; Wager and Brown, 1968).

A more modern interpretation supports the intervention of 'density currents': convective currents sweeping across the magma chamber are able to drag large numbers of crystals, and to deposit them on the floor of the chamber (Figure 8.27). Experimental studies and theoretical works (Irvine, 1982) have shown that crystal deposits issued from density currents, borne along the steep side-walls of a magma chamber, and flowing along a gentle slope of the floor, may accumulate modal graded layers. Data from analogous laboratory experiments indicate that these density currents flow at velocities of the order of 1 to 6 km/h; that is, one hundred times faster than the settling velocities of olivine or pyroxene cumulus crystals.

Other mechanisms have been proposed, often based on experimental observations of simple systems in the laboratory (salt or carbonate solutions): capture of the crystals of a convective cell by a viscous front, 'springs' of residual liquids expulsed after crystallization, downgoing columns charged with crystals suddenly collapsing from the reservoir roof and spreading across the floor, etc. (Figure 8.28).

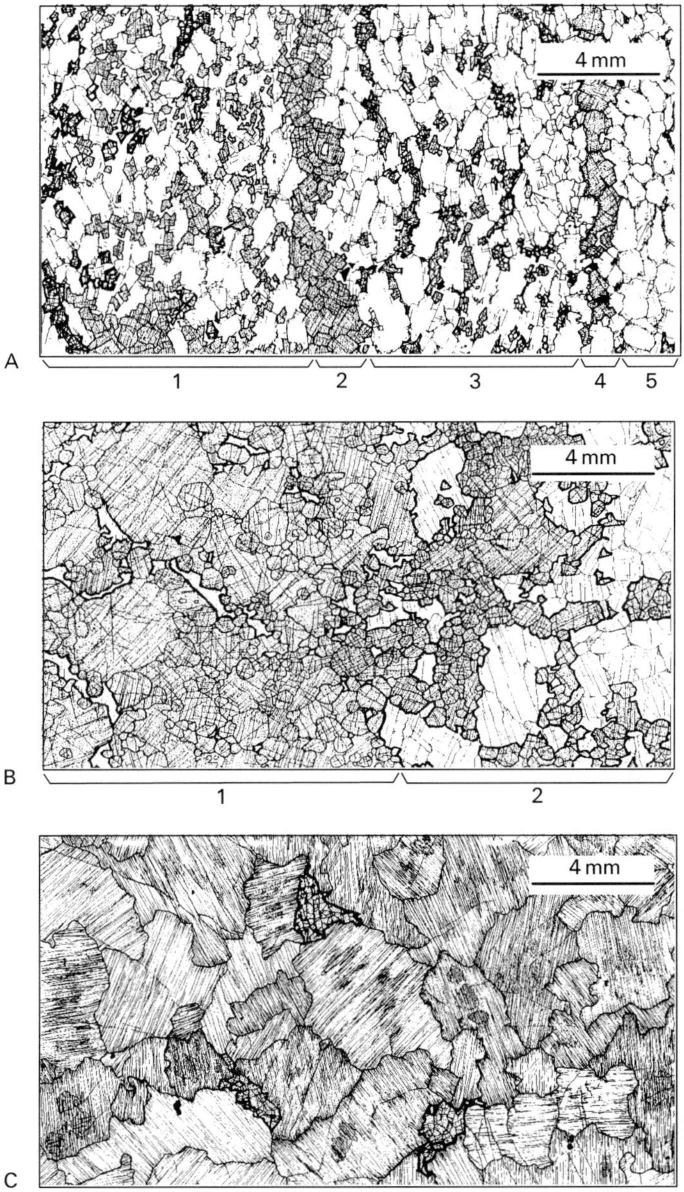

Fig. 8.25. Drawings showing entire thin sections below binocular, cut-in ophiolitic-layered cumulates (Antalya ophiolites, Turkey) (drawings from Bard, 1980 after pictures from Juteau, 1975). (A) Cumulative micro-rhythms in layered gabbros. 1. Plagioclase and clinopyroxene (cumulus phases) mesocumulate layer, with post-cumulus olivine. 2. Clinopyroxene adcumulate lamina. 3. Gabbro mesocumulate layer analogous to (1), richer in plagioclase. 4. Clinopyroxene adcumulate lamina. 5. Plagioclase adcumulate layer, with post-cumulus olivine. (B) Pyroxenite/gabbro contact. 1. Pyroxenic heteradcumulate. Huge clinopyroxene crystals englobe small, idiomorphic and globular orthopyroxene crystals, and are cemented by post-cumulus plagioclase. 2. Plagioclase and orthopyroxene adcumulate. (C) Orthopyroxenite: adcumulate with some post-cumulus olivine.

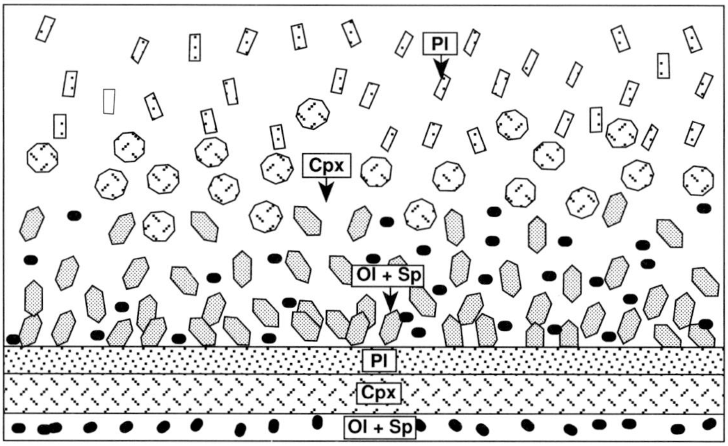

Fig. 8.26. Gravity settling model, resulting in a recurrent succession of Cr-spinel dunite, clinopyroxenite and gabbro, according to concepts of Bowen (1949) and Wager and Brown (1968) (drawing after Whitechurch, 1993).

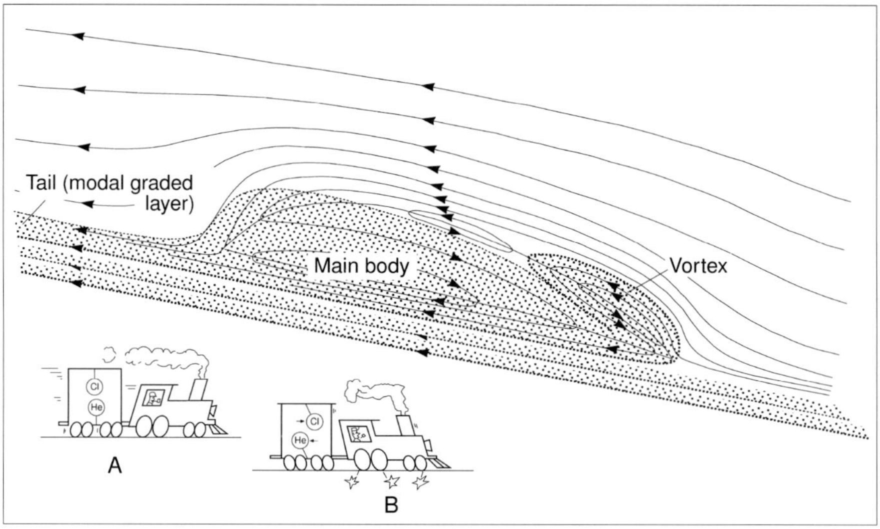

Fig. 8.27. Structure of a density current forming a modal graded layer, according to Irvine (1980). The flow lines describe the instantaneous movements of the fluid with respect to the frontal extremity of the current (point R). Small trains A and B drawn by J. Irvine schematically illustrate the concept of 'inertial sorting velocities' for particles with different momentums.

Structures and deformations linked to magmatic flow

All sections performed on the plutonic unit of the Oman ophiolite show a magmatic deformation, more or less imprinted in the layered gabbros, and marked by a magmatic foliation and associated lineation, resulting from a dynamic flow under a viscous regime. The magmatic foliation plane (or *lamination* plane) is defined by the

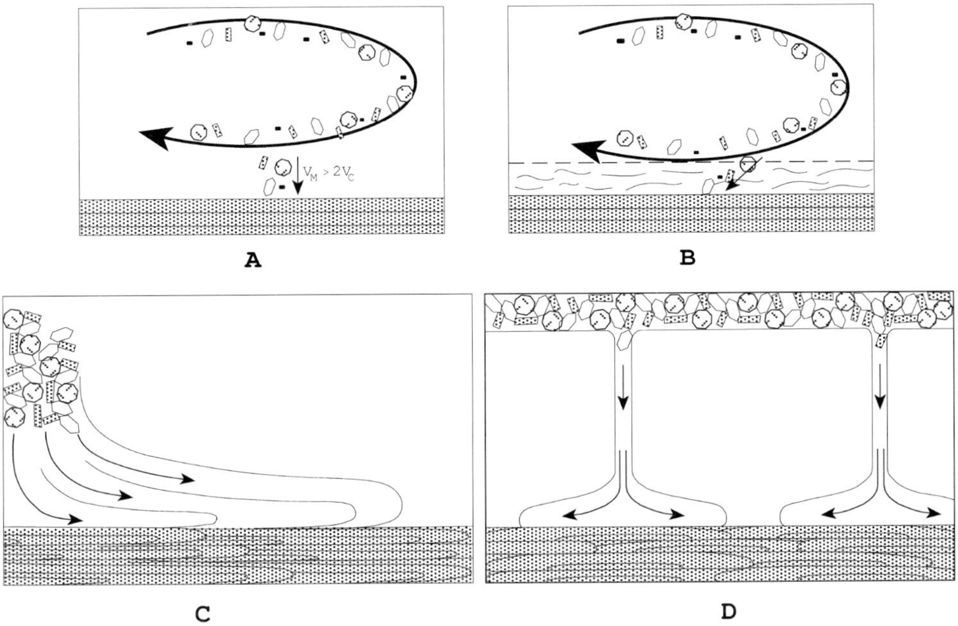

Fig. 8.28. Various mechanisms proposed to explain the formation of layers with very contrasted modal variations. (A) Gravity settling below a convective cell. The settling velocity of the crystals must be at least twice higher than that of the convection current. (B) Capture of minerals from a convective cell by a viscous front. (C) Viscous flow at the wall/floor transition, with formation of flow-folds at the floor. (D) Collapse and flow of crystal mushes issued from the chamber roof (drawings after Whitechurch, 1993).

preferential orientation of non-deformed tabular minerals. It is often accompanied by a magmatic lineation, frequently parallel to the axes of magmatic folds.

It is necessary to distinguish between *plastic deformations*, developed in the solid state during high-temperature deformations, and *viscous deformations*, acquired by dynamic crystal mushes during their flow movements, in the presence of an interstitial liquid magma (Figure 8.29).

Both types of deformation can be observed in the transition zone between mantle and crust, at the 'petrological Moho' level (Nicolas, 1989, 1992). At the base, in the upper parts of the residual mantle, decametric gabbro lenses may show important solid state deformations by high-temperature plastic flow. These deformations result in typical sub-structures for olivine, and the development of fine, sinuous and bevel-shaped deformation lamellae in plagioclase. The minerals themselves are bent and deformed with rolling extinctions under the microscope. Somewhat higher in the basal gabbros of the plutonic section, the gabbros have been deformed by viscous flow. The minerals are not bent, the plagioclases have wider linear-growth twinnings across the whole crystal.

For Nicolas (1989, 1992), most of the layering of the plutonic section would be due to such viscous flow processes, even in mineral graded layers. In the Oman

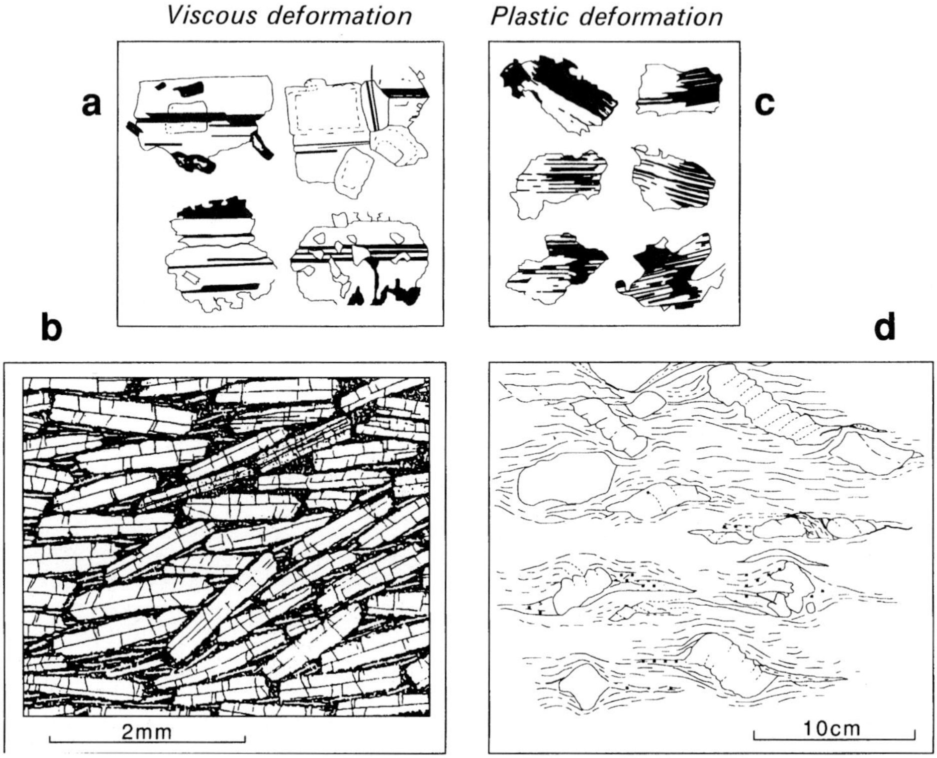

Fig. 8.29. Viscous deformation and plastic deformation (after Nicolas, 1989). *Viscous deformation:* (a) the plagioclase crystals in viscously deformed gabbros are not deformed, and exhibit linear-magmatic growth twin planes. (b) These sub-idiomorphic plagioclase crystals define a *magmatic foliation. Plastic deformation:* (c) the plagioclase crystals of solid-state plastically deformed gabbros are bent and tapering, and show narrow and sinuous deformation lamellae. (d) These plagioclase crystals form residual porphyroclasts in a finely recrystallized matrix, and define a *metamorphic foliation.*

ophiolite, gravitative differentiations often result in a spectacular layering in the lower layered gabbros, and are interpreted today as resulting from the combined action of gravity and viscous flow inside multiple intrusive sills of metric thickness (Figure 8.30) (Boudier *et al.*, 1996; Nicolas, 1997; Kelemen *et al.*, 1997). According to the shear rate applied to the crystal mush and to its state of solidification, the mush will deform with more or less intensity. The minerals tend then to orient themselves parallel to the shear planes. The tabular minerals, particularly, are at an angle of 10° to 40° with the magmatic foliation, and are piled up one upon another. A total transposition of the original layers may result in structures parallel to the shear plane, and mimic a layering, and oblique structures constitute a magmatic foliation. At the end of the solidification of the crystal mush, when viscosity has drastically increased, these rocks may deform by normal faults and asymmetric folds (Figure 8.31).

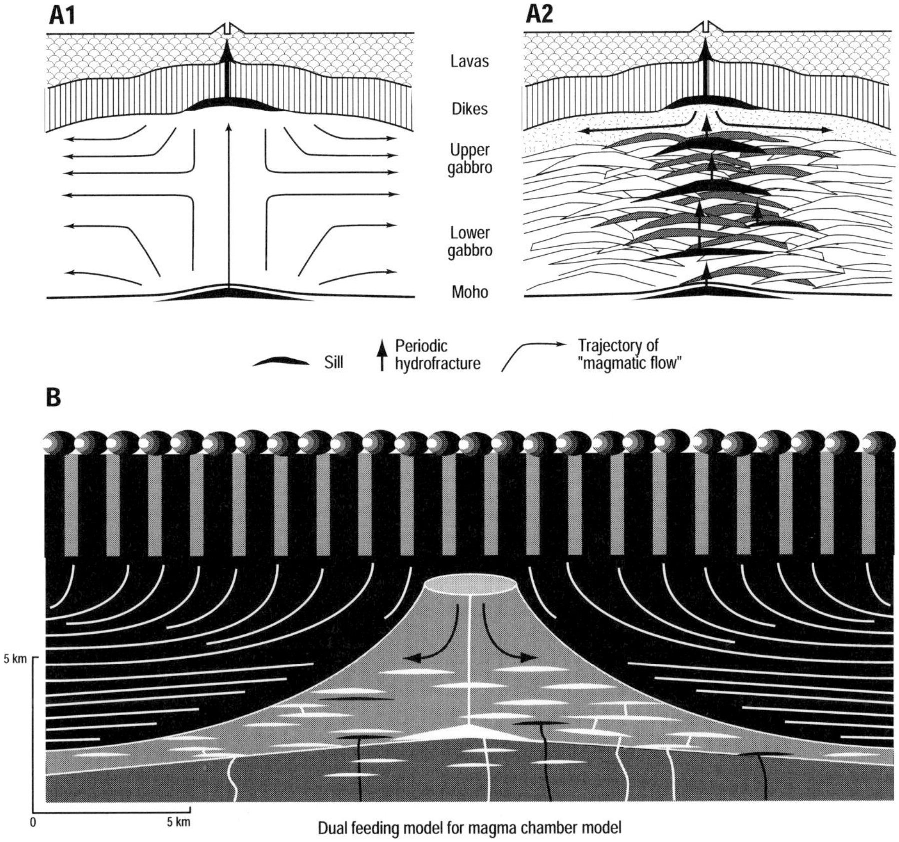

Fig. 8.30. Role of deep crustal sills in forming lower gabbros beneath oceanic spreading ridges. (A1) In this 'two sills' model, the high-level gabbros crystallize in a shallow magmatic sill located just below the base of the sheeted dike complex (similar to the 'perched' magmatic lens of the EPR), and the layered gabbros crystallize in a deeper sill located at the Moho Transition Zone. Flow of crystal mush from the two sills forms the gabbroic lower crust. Periodic hydrofracture carries evolved liquids out of the lower sill, leaving cumulate gabbros (after Kelemen *et al.* (1997). (A2) Alternatively, lower gabbros form in 'sheeted sills', that are emplaced and crystallize near their final depth within the crust. Upper gabbros crystallize as in (A1) (after Kelemen *et al.*, 1997). (B) 'Dual melt delivery' model. The foliated upper gabbros partly crystallize at the floor of the perched magma lens, then subside as a magmatic mush from that elevated position, and acquire their magmatic foliation during subsidence and subsequent motion. The lower layered gabbros are a composite, in which the mass of foliated gabbros is intruded by a variable number of sills (gabbros, wehrlites) issued from melt ponding at the Moho level. The contrasted layering is accentuated by tectonic transposition related to drag caused by the fast asthenospheric flow beneath (after Boudier *et al.*, 1996).

Cryptic evolution

The term *cryptic variations* (or 'hidden', because they are not visible under the microscope) is descriptive of the chemical variations of the plutonic-section mineral phases according to their stratigraphic position in the lower crust; that is, in the fossil magma chamber. These phases are few and may vary widely in composition between

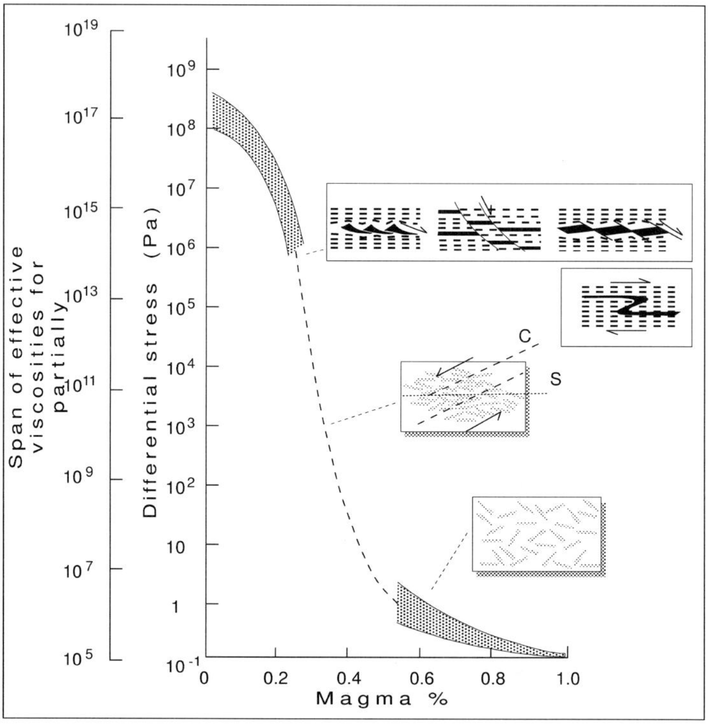

Fig. 8.31. Deformation of crystal mushes with different rheologies submitted to various shear conditions, according to liquid percentage and to viscosity (after Nicolas, 1992, slightly modified).

isomorphic end-members: olivine, plagioclase, clinopyroxene, orthopyroxene, spinel and hornblende.

(a) Fast-spreading ridges. A number of studies on the Oman ophiolite have shown a remarkable similarity between the cryptic variation profiles in the layered gabbros: these profiles have a typical 'zig-zag' pattern, without any general shift towards the differentiated end-members, so that the average composition of the mineral phases at the top of the cumulate pile is identical to the composition at the bottom.

This fact is commonly interpreted as being due to periodic and frequent injection of primitive magma into the chamber, preventing any significant evolution of the residual magma. Accepting, as do the most authors, that the Oman ophiolite represents a fast-spreading ridge, it can be imagined that the accretion-magma chambers of present-day fast-spreading ridges have their magmatic composition buffered by periodic feeding with juvenile magma. Unfortunately, as already mentioned, no complete gabbroic sequence from a fast-spreading ridge has been sampled up to now.

(b) Slow-spreading ridges. The Ligurian-type ophiolites, like the gabbroic

sequences studied on present-day slow-spreading ridges, clearly show that these small and ephemeral magma chambers, rapidly cooled, evolve on the contrary towards strongly fractionated terms (ferrogabbros, diorites, plagiogranites). This implies probably *closed fractionation systems*, without refilling. The oceanic gabbros studied so far all come from fracture zones (Hole ODP 735B), or from segmentation areas (MARK zone). The evolution of gabbroic series from the central parts of the segments remain unknown, and could well be notably different.

9

Basalt–seawater interaction: metamorphic and hydrothermal processes

Once consolidated in the axial zone of oceanic ridges, the young oceanic crust constituting the new ocean floor is immediately exposed to seawater, which infiltrates through all the open fissures, cavities and pores of the basaltic pile. The interaction between seawater and the rocks of the oceanic crust results in deep modifications of the physical, chemical and mineralogical properties of the oceanic crust. These chemical exchanges contribute to the balance of seawater chemistry and to keeping the composition of the oceans constant.

An important consequence of these processes is that the old, altered oceanic crust which returns to the mantle through the subduction zones no longer presents the pristine magmatic composition it had in its birth site, at the axis of an ocean ridge. To understand the chemical heterogeneities of the mantle, and their influence on the genesis of magmas, it is necessary to integrate these modifications in geochemical modelling.

The metallogenic consequences of the basalt–seawater interaction along oceanic ridges are also very important. The relatively recent discovery of hot hydrothermal vents precipitating iron-, copper- and zinc-rich polymetallic sulphides (the famous 'black smokers') on the East Pacific Rise (EPR), along with iron and manganese oxides, has been extended to all ocean-ridge systems, including back-arc basin ridges.

These genuine ore deposits, being formed before our eyes, represent an unique opportunity to study in real time metallogenic processes, and to better understand the fossil volcanogenic massive sulphide deposits (or VMS deposits), known and mined on land, especially in ophiolite terranes (Cyprus, Oman, Newfoundland, etc.). Moreover, the subduction of these metalliferous deposits from the oceanic crust, and their subsequent distillation in depth, during recrystallization (or even melting) of the subducted crust, should have important repercussions on the metallogeny of active continental margins and island arcs.

9.1 MODIFICATIONS OF THE PHYSICAL PROPERTIES OF THE CRUST

The water–rock interaction modifies practically all the physical properties of the oceanic crust (Thompson, 1991):

—*Acoustic properties*. The hydration of rocks lowers their density. The velocities of compressive seismic waves, which depend on the density of traversed materials, are hence modified. These modifications, registered according to the distance to the ridge axis, were used to calculate the durations and velocities of chemical reactions implicit in hydrothermal alteration (Hart, 1973);

—*Magnetic properties*. Fresh, non-altered oceanic basalts exhibit high values of magnetic susceptibility and thermoremanent magnetization. These values decrease drastically with aging of the crust, reducing along the ridge axis. This is due to the fact that seawater–basalt interactions at low temperature oxidize and destabilize the iron–titanium-rich oxides (mainly Ti-magnetites) bearing the magnetization. High-temperature hydrothermal reactions result in still more drastic modifications: recent studies have shown that active hydrothermal areas have a 'magnetic signature' characterized by very weak intensities of the remaining magnetization.

—*Porosity*. The original porosity of basaltic layer 2A, which is very high as we have seen, decreases progressively with aging of the crust. The fissures, pores and numerous natural cavities of the basaltic pile are gradually filled-in by precipitation of metamorphic or hydrothermal minerals, produced by the seawater–basalt interaction. This filling-in results, paradoxically, in an increase in the mean density of layer 2A, hence of the compressive seismic-wave velocities, although altered basalts are less dense than fresh, non-altered basalts.

—*Heat flow*. Terrestrial heat flow is directly affected by the circulation of seawater in the oceanic crust, which drains off the heat very efficiently. According to recent calculations, this circulation removes about one-third of total terrestrial heat flow measured along oceanic ridges. The observed deviation from the theoretical conductive model (Figure 9.1) is due to convective cooling of the crust by hydrothermal circulation. This gap with the conductive model allows estimation of the amount of heat removed by the circulation of seawater in the crust, and the volume and flux of the global oceanic hydrothermal system (Wolery and Sleep, 1976). In a given area, the distribution of heat-flow measurements allows the distinction between *recharge zones* (cold seawater moving down, hence negative heat-flow anomalies), and *discharge zones* (hot hydrothermal fluid moving up, hence positive heat-flow anomalies).

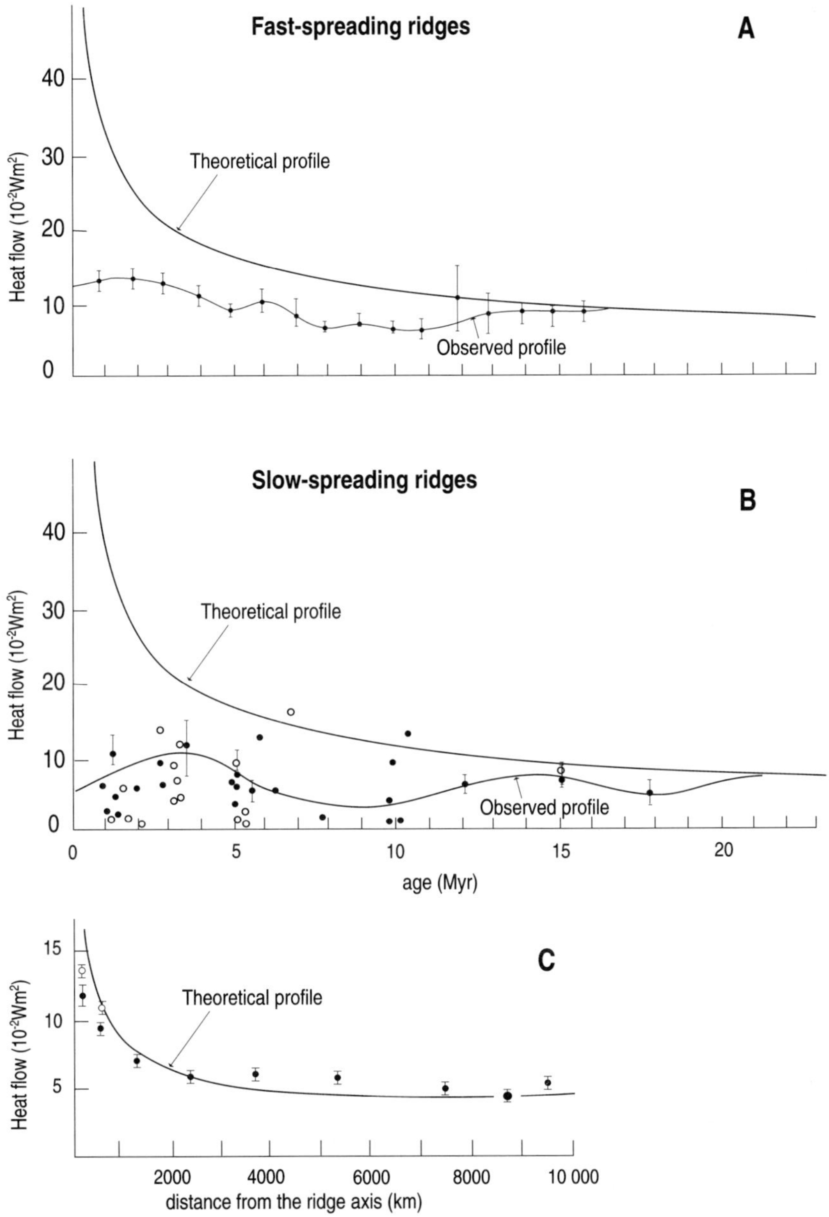

Fig. 9.1. Variation in oceanic-ridge heat flow, versus the age of the crust or the distance to the axis (A and B) (after Wolery and Sleep, 1976 and (C) Le Pichon *et al.*, 1973). Comparison of the theoretical heat flow (supposing that all the heat is evacuated by conduction), and of the heat flow effectively measured: (A) for a fast-spreading oceanic ridge (East-Pacific Rise); (B) for a slow-spreading ridge (Mid-Atlantic Ridge). In both cases, the important gap recorded between both curves in the axial zone of the ridge can only be explained by a convective cooling of the crust, due to the contribution of an active hydrothermal circulation, draining-off the heat efficiently from the axial zone; (C) shows that when looking at heat flow on the scale of an ocean, the gap between both curves becomes nil at a certain distance from the axis.

9.2 'ACTIVE' AND 'PASSIVE' CONVECTION

Lister (1982) proposed two types of convective regimes in the oceanic crust (Figure 9.2):

—*Active systems* are strictly located at the axial zone of oceanic ridges, either above the roof of magma chambers (the magmatic lens along the EPR, for instance), where heat is extracted through a boundary layer, or above still-hot (but cooling) gabbroic plutons, where a downward fracturing front allows penetration of fluids and extraction of heat (Mid-Atlantic Ridge (MAR) for instance). In such systems, temperatures are high ($T > 300°C$), circulation is vigorous and fast, and discharge zones are much more focused than recharge zones. High-temperature hydrothermal vents (250° to 400°C) are restricted to a narrow zone, about 100 m wide in the axial zone of oceanic ridges.

—*Passive systems* develop off-axis, heat being supplied simply by conductive cooling of the crust and lithosphere. The temperatures are much lower ($T < 200°C$), and the velocities of circulation much slower than in active systems. Whereas the active axial systems remain fixed, because they are linked to the heat removed from the axial magma chamber (itself fixed), the convective cells developed off-axis are moving with the oceanic crust, moving away from the ridge axis. Passive systems are much less spectacular than active systems, and hence much more difficult to detect and study. These systems, however, remove the greatest amount of heat, at least 10 times more than the active systems.

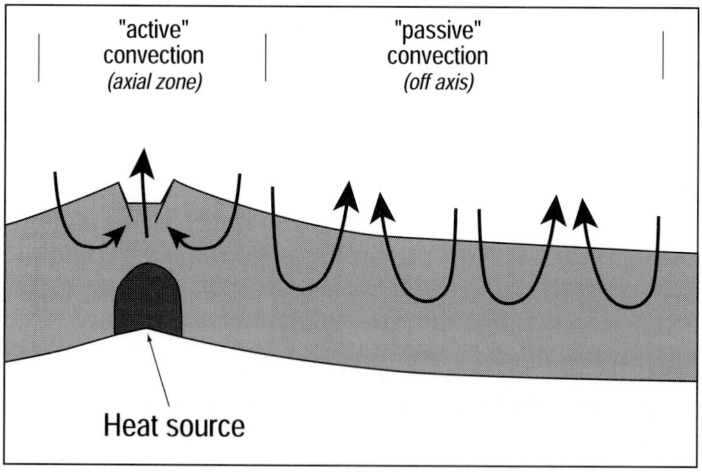

Fig. 9.2. Scheme illustrating the two main convective regimes in the oceanic crust (after Lister, 1982). In the axial cell, a high-temperature active circulation is maintained by the permanent presence of a strong heat source (magma chamber, or very hot plutonic rocks). In the off-axis domain, a lower temperature circulation ($T° < 200°C$) is maintained by passive thermal cooling of the crust and the whole lithosphere.

9.3 CHEMICAL AND MINERALOGICAL MODIFICATION OF THE OCEANIC CRUST

The effects of the seawater–oceanic crust interaction can be gathered under the general term of 'oceanic metamorphism', since the chemical and mineralogical composition of the crust is deeply modified, whatever the reaction temperature. Practically, however, this oceanic metamorphism is considered by the great majority of specialists as resulting from two distinct processes:

(1) Low-temperature reactions ($T < 70°C$). developed mainly at ambient seawater temperature; that is, just a few degrees Celsius. Penetrative and ubiquitous, these reactions are active over very long periods of time ($> 10^6$ Myr), and affect mainly the upper basaltic crust.

These low-temperature processes are described as *low-temperature seafloor weathering* by Laverne (1987), whose terms *halmyrolysis* (reserved for sediments), and *diagenesis* (concerning only clay minerals) must be discarded. For Honnorez (1981), low-temperature seafloor weathering includes all reactions, excluding the presence of clearly metamorphic minerals, such as lawsonite, wairakite and chlorite; that is, up to about 150°C in temperature. Cann (1979) proposed the term *brownstone facies* to describe the assemblages at very-low oxidizing temperatures of brownish-rusty haloes along cracks in the basalts, at temperatures ranging between 0° and 70°C.

The 70°C temperature corresponds to the well-known inversion of behaviour of alkali metals, especially potassium, experimentally observed during laboratory experiments. Seyfried and Bischoff (1979), for instance, caused seawater and powdered MORB-type basalts to interact for months in pressure-sealed autoclaves, under various temperatures and pressures of several hundred bars, corresponding to the seafloor conditions of the oceanic ridges (Figure 9.3). They observed that below 70°C, the alkali metals are extracted from seawater and taken up by the basalt, whereas above this temperature, they are leached and extracted from the basalt.

(2) High-temperature reactions (from 70° to 400°C or more). Strictly located to plate boundaries in spreading areas, or to active intraplate seamounts, where high thermal gradients are generated either by shallow magma chambers, or by still-hot plutonic rocks. The reaction times are much less (10^2 to 10^4 yr). These high-temperature processes, resulting from vigorous convective circulation of seawater across the basaltic crust, are given the term *hydrothermal alteration* (Laverne, 1987; Thompson, 1991).

Figure 9.4 describes the various parts of an active hydrothermal system at a spreading oceanic ridge. This system comprises many *recharge zones*, from where cold seawater penetrates the crust and moves down towards the roof of the magma chamber, reacting at increasing temperatures with the basaltic crust. In the *High-Temperature Reaction Zone* (HTRZ), seawater has transformed into a hot hydrothermal fluid and is charged with dissolved metals and reducted sulphur. This low-density fluid, close to the critical conditions, rises up towards the seafloor, and is channeled in focused or diffuse *discharge zones*.

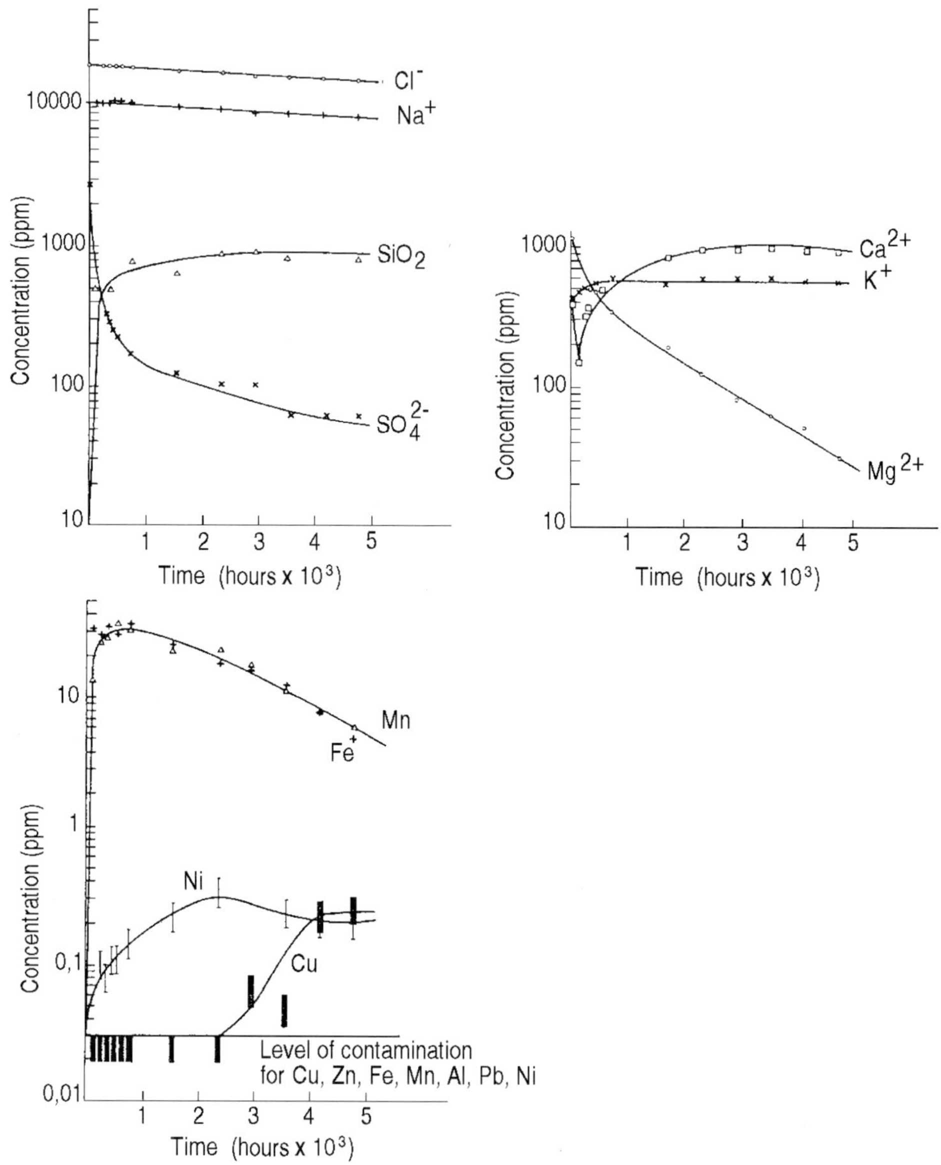

Fig. 9.3. Concentration variation in a number of dissolved ions in standard seawater, during its interaction with a powdered MORB-type basalt (after Bischoff and Dickson, 1975). The imposed conditions are 200°C and 500 bar. The concentrations are in ppm (parts per million in mass), and reported on a logarithmic scale. This laboratory experiment lasted 5000 h; that is, about 7 months. The contrasted behaviour of the various elements is clear: fast increase of Si, Fe and Mn extracted from the basalt (SiO_2 reaches a plateau, whereas Fe and Mn decrease slowly after a strong increase, because of the precipitation of insoluble Fe–Mn hydroxides); fast decrease of SO_4 and Ca (precipitation of anhydrite), then progressive recovery of Ca, extracted from the basalt); increase in K, extracted from the basalt (since $T^\circ > 70^\circ C$), and rapidly reaching its plateau (MORB-type basalts are very K-poor); concerning metals, Ni is rapidly extracted from the basalt (as soon as olivine is destabilized), and Cu takes much longer to appear in the solution, because of the later destabilization of clinopyroxene.

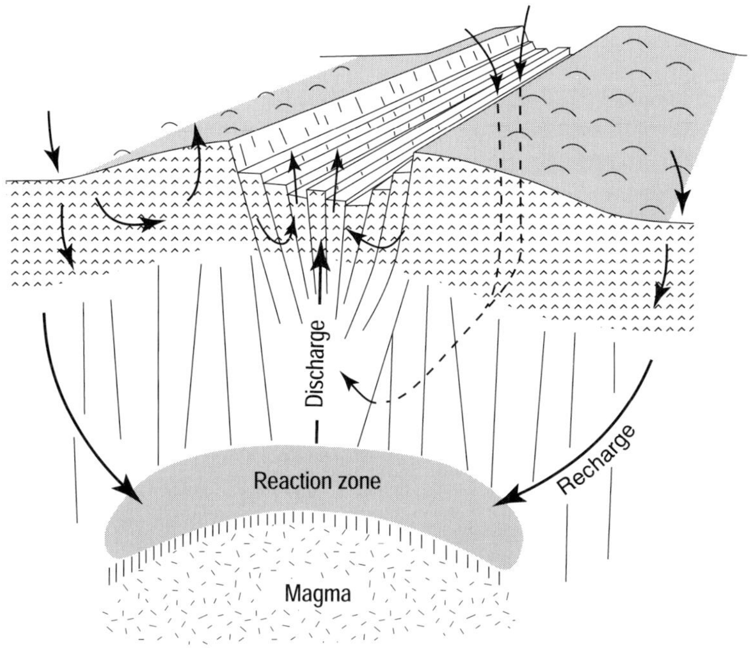

Fig. 9.4. Schema showing the various parts of a submarine hydrothermal system, developed at the oceanic spreading ridges (after Alt, 1995). The cold seawater penetrates down into the crust through large recharge zones, and reacts with the crust at growing temperature during its downgoing migration. High-temperature chemical reactions ($T^\circ > 350^\circ C$) develop in the HTRZ (High Temperature Reaction Zone), located just above the heat source (generally the magma chamber), from where the hot hydrothermal fluids rapidly ascend towards the seafloor, through localized or diffuse discharge zones.

We describe now in detail the main steps of this convective circulation of seawater through the oceanic crust based on the excellent synthesis by Alt (1995).

9.4 DESCRIPTION OF THE CONVECTIVE SYSTEM

9.4.1 Recharge zones

Huge volumes of seawater circulate through the porous, very permeable volcanic pile. Seawater infiltrates through inter-pillow spaces and natural cavities (hollow lava tubes, collapsed lava lakes, etc.), talus rubbles, or hyaloclastic breccias, and also through the countless open fissures, cracks and normal fault planes developed in the axial zone of the spreading oceanic ridge. These are the *recharge zones*. Only a small percentage of this cold seawater circulating in the basalts will penetrate in lower levels, down to the sheeted dike complex and the deeper, hotter parts of the active convective system (Alt, 1995). The rest of the circulating water constitutes the 'off-axis' circulation, by far the most important in volume.

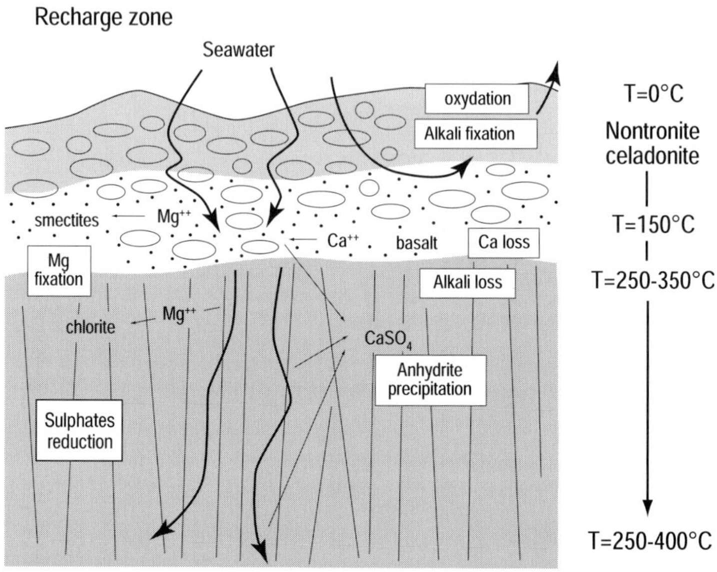

Fig. 9.5. Scheme illustrating the processes developing in the recharge zones (after Alt, 1995). Chemical reactions and fluid compositions are modified while seawater is heated and penetrates the crust, with progressively lower water/rock ratios. The upper volcanic flows are affected by a low-temperature open system alteration (light grey). The lower volcanic flows undergo a low-temperature alteration under more restricted circulation conditions (white zone on the schema). The underlying sheeted dyke complex suffers a higher-temperature hydrothermal alteration ($T > 250°C$, dark-grey area).

Figure 9.5 shows the general scheme of these recharge zones. The reactions developing during the initial penetration of seawater in the crust belong typically to the low-temperature seafloor weathering type, under open and oxidizing circulation conditions. They consist mainly in oxidation and fixation of alkalis in the upper basalts of the volcanic pile, with development of the well-known 'alteration haloes", about 1 to 20 mm wide along the cracks and exposed surfaces. At lower levels in the volcanic pile, at about 300 m in depth, the seawater becomes hotter and circulates less easily, conditions becoming more reductive.

Low-temperature, open and oxidizing circulation

Mineralogically, 'rusty' iron oxi-hydroxides (goethite, amorphous materials, rare hematite) replace olivine and primary sulphides, fill the pores and veins, and invade the cryptocristalline groundmass of the basalts (red–brown haloes). Smectites develop often as intergrowths with the oxi-hydroxides. Moreover, *celadonite* (an iron-rich sheet silicate) and *nontronite* (ferric smectite), often associated with inter-layer mixed crystalline structures, fix the alkali metals (K, Li, Rb, Cs), forming dark 'haloes' (Figures 9.6 and 7). Iron oxi-hydroxides-rich red–brown haloes frequently obliterate early celadonite–nontronite-rich dark haloes. In some cases of maximum oxidation, a *K-feldspar* replaces plagioclase at the top of the volcanic pile, increasing the general enrichment in alkalis (Figure 9.8). The primary Ti-magnetite is oxidized

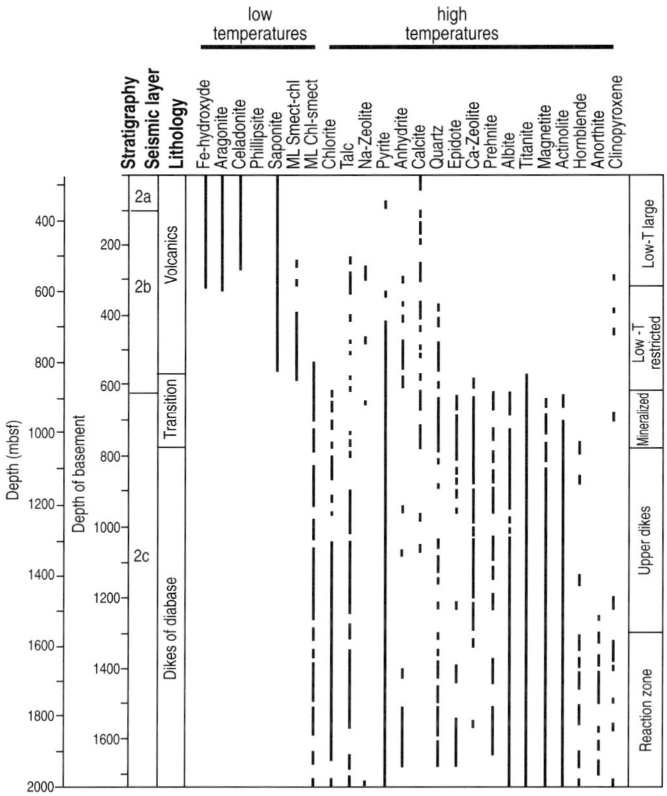

Fig. 9.6. Schema showing the evolution of secondary mineral parageneses with depth in Hole ODP 504B (after Alt, 1995). The depths are indicated to the left in metres below the seafloor (mbsf), and in metres below the sediment/basalt contact (depth in the crust). Seismic stratigraphy and lithology are also indicated to the left. The various regimes of hydrothermal alteration are indicated to the right. The main fact, generally observed in the oceanic crust and in ophiolites, is the sharp transition (in barely several tens of metres) between the low-temperature alteration zone developed in the volcanic pile, and the high-temperature hydrothermal alteration zone, in the 'transition zone' and the underlying sheeted diabase dike complex.

into *Ti-maghemite*. Only the clinopyroxene remains unaltered. Some late *zeolites* (natrolite, analcite), associated with *carbonates*, crystallize in the off-axis domain, fixing alkali metals here again, as well as CO_2.

Palagonitization

The basaltic glass of the pillow margins exhibits a characteristic alteration into *palagonite*, an assemblage of low-temperature hydrated mineral phases, comprising *zeolites* (phillipsite), *smectites* and *Fe–Mn oxi-hydroxides* (Honnorez, 1981). The complete palagonitization of a basaltic glass occurs slowly, between 5 to 10 Myr, with an evolution in three major steps (Honnorez, 1981): the glass is first hydrated and oxidized, with crystallization of sodic phillipsite ($Na > K$) and K–Mg-rich

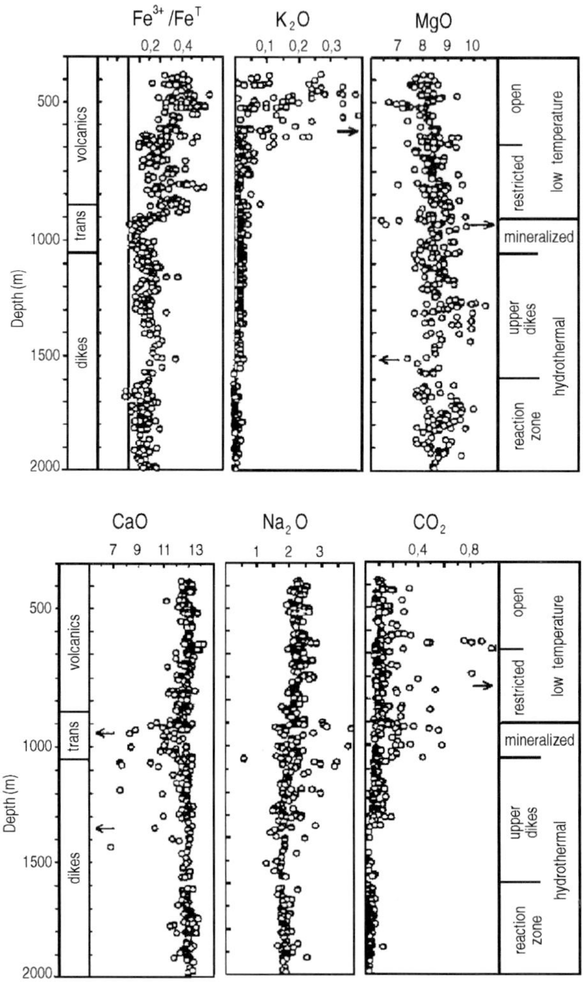

Fig. 9.7. Variation of some chemical parameters (total rocks) with depth in Hole ODP 504B (after Alt, 1995). The relative enrichment in K_2O and Fe^{3+}/Fe^T at the top of the volcanic pile reflects the fixing of K in the crust (smectites and celadonite) under oxidising conditions. The mineralized zone is chemically perturbed. The vertical line in the Fe^{3+}/Fe^T column indicates the normal value of this ratio in non-altered MORB glasses.

smectites; during this first step, the glass fixes K, Na, Mg and looses Ca. During the second step, a K-rich phillipsite ($K > Na$) develops with Fe–Mg- and K-rich saponites or nontronites. The third step is characterized by the crystallization of F–Mn oxi-hydroxides, as well as K-phillipsite and F–Mg smectites. Finally, the palagonitized basaltic glass has yielded Ca, Mg and Si to seawater, and taken up K, Mn and Na.

Chemically speaking, this low-temperature oxidising alteration is characterized by gains in H2O, P, K, Rb, Cs, Li and U for the crust, by losses in Mg, Si, S (sometimes

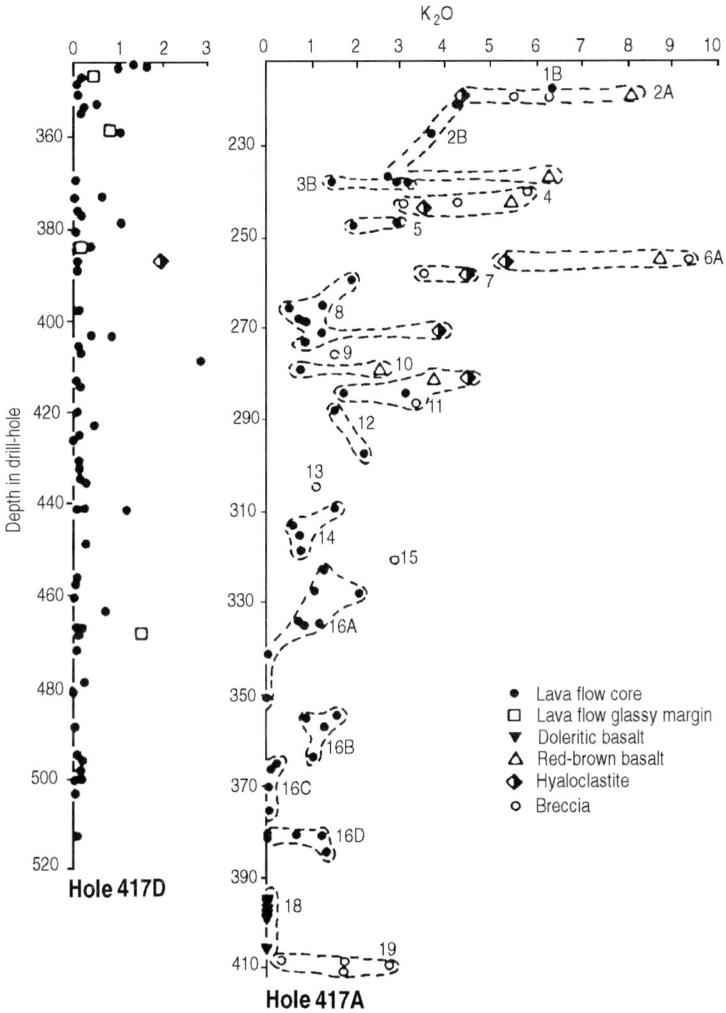

Fig. 9.8. Variations in K_2O content (recalculated on a water-free basis) with depth in the basalts of Holes DSDP 417A and 417D, drilled in the old crust of the Atlantic ocean, north of Puerto Rico (after Donnelly *et al.*, 1979). It immediately appears that the intensity of the alteration in the basaltic column is very different in both holes, although separated by only a few km. The age of the crust is the same, but Hole 417D was drilled in a small graben rapidly filled with carbonate sediments, whereas Hole 417A was drilled on a basement horst, exposed to seawater interaction for several million years. The high content in K_2O at the top of the volcanic pile in the latter hole is due to the development of adularia (a low temperature K-spar), smectites and celadonite.

in Ca, Co and Ni), and by an increase of the Fe^{3+}/Fe^{T}, $^{87}Sr/^{86}Sr$, and $^{18}O/^{16}O$ ratios. Some elements such as Fe, Mn, Na, Cu, Ba, Sr and light REE can show a variable mobility inside the same pillow lava, according to the local conditions. Finally the immobile elements, not affected by the low-temperature alteration, are Al, Ti, Zr, Ta, Hf, Nb and the heavy REE.

Restricted and reductive circulation (intermediate temperatures)

When seawater penetrates further into the crust (300–500 m), the 'open' and oxidising circulation gives way to a more restricted circulation and to a reducing alteration. The lower lavas drilled in Hole 504B for instance show a slight increase in alkalis and in the Fe^{3+}/Fe^{T} ratio, reflecting a decrease in seawater volume having penetrated at these depths, and a longer reaction time between seawater and basalts (depletion in alkalis and oxygen). At this level, *saponite* (Mg-smectite, containing small amounts of alkalis) invades the pores and cracks in the basalts, and replaces the olivine, the glass, and even sometimes the plagioclase, along with calcite. The clinopyroxene still remains unaltered. The high content of ferrous iron in the saponite, and its frequent association with secondary sulphides (pyrite, chalcopyrite, marcasite), indicates far greater reducing conditions than before. Here, zeolites (natrolite, analcite) and carbonates also form late phases, precipitated in the off-axis domain. Oxygen-isotopic ratios in saponite indicate temperatures of formation of about 15° to 170°C.

The recharge zones are also characterized by the fixation of the magnesium of seawater in the crust, by precipitation of anhydrite ($CaSO_4$), and by a loss of alkalis by the basalts at intermediate temperatures. The Mg^{2+} and $(OH)^-$ of seawater combine together to enter the network of a saponite-type smectite at low temperature ($T < 200°C$), or of a chlorite at higher temperature (Alt, 1995). This uptake of Mg by saponite develops in the fissures of the basalts, the matrix of the breccias and the palagonites replacing the glassy margins. It cannot be detected in basalt samples devoid of these structures. The result is a *drastic, quite complete decrease of magnesium in the percolating seawater.*

The anhydrite $CaSO_4$ precipitates simply by heating of seawater. Experimental data show that between 150° and 200°C, all the calcium contained in seawater, and two thirds of the sulphate are thus eliminated. Veins of anhydrite were effectively observed in the lower basalts of Hole 504B and at the top of the sheeted dike complex.

Finally, another important process in this zone is the *inversion of behaviour of the alkali elements at intermediate temperatures*. Experimental studies have shown that potassium, for instance, is fixed by basalts and extracted from seawater below 70°C, but is leached from basalts and returns to seawater at 150°C or more. For Li, the inversion temperature is higher.

In this intermediate temperature zone, the alteration minerals reflect a high thermal gradient increasing downwards, the temperatures ranging from 100–150°C in the lower lavas, to 250–350°C in the upper part of the diabase sheeted dike complex, with the typical appearance of the greenschist facies.

9.4.2 The high-temperature reaction zone (HTRZ)

This is the zone of water–rock interaction, where seawater becomes a hot hydrothermal fluid. It is generally thought that the hottest fluids (350°C and more) sampled at the exit of the 'black smokers' acquired their chemical signature in the HTRZ. This zone is located at about 1.5–2.0 km in depth, at the base of the sheeted dike complex and in the upper part of the gabbro sequence, at the roof of the magma chamber. Theoretical models predict a steep thermal gradient in this zone, the fluid temperatures inceasing from 200°C to 350–400°C and more. The chemical reactions are modified and their knetics strongly enhanced. The fluids approaching the critical conditions become much less dense, resulting in their swift ascent to the surface and discharge zones.

The mineralogical result of the reactions can be observed along some natural geological sections in the present-day oceans, either drilled (Hole 504B, Hole 735A) or outcropping (fracture zones), and also in the ophiolite massifs.

Oceanic data

Epidote is rare in dredged oceanic samples (only in scarce veins, associated with quartz, or replacing plagioclase in small patches), whereas it is abundant in the ophiolites, in agreement with experimental predictions.

Some hot springs show low Fe/Cu ratios, suggesting reducing conditions and stability of the pyrite–pyrrhotite–magnetite assemblage in the reaction zone, in agreement with observations made on oceanic diabases and gabbros, and on ophiolite samples. In other hot springs, however, high Fe/Cu ratios indicate more oxidising conditions and the stability of anhydrite in the reaction zone, as in Hole 504B (anhydrite replacing the plagioclase in the lower dykes).

The base of the sheeted dike complex in Hole 504B. The reaction zone comprises a Mg–hornblende-type amphibole (enriched in Al and Ti), coexisting with a calcic secondary plagioclase, implying equilibrium temperatures of about 400–500°C (Laverne *et al.*, 1995; Figure 9.6). The base of the dike complex is almost completely leached in Zn, Cu and S (Figure 9.9). In the basal part of the dikes, the rocks do not show any enrichment in Mg, which is interpreted as due to fluids strongly depressed in Mg because of formation of Mg-smectites and chlorites in shallower zones, during the recharge path. Moreover, variable enrichments in Na and impoverishments in Ca are observed, as well as enrichments in H_2O, losses in K and Rb, and variable gains and losses in P and Ti.

Alteration of gabbros in Hole ODP 735B. As already seen, this hole has provided a section of 500 m in the gabbros of the Indian Ocean. The gabbros are variably altered, according to local deformations and to permeability variations. Brittle faults at the ridge axis have penetrated below the brittle–ductile transition, allowing massive penetration of hydrothermal fluids into the hot gabbroic rocks, and provoking hydration reactions at high temperature. This early stage of plastic deformation and metamorphism has formed *foliated amphibolites* at temperatures ranging from 600° to 700°C. These effects are well developed in the uppermost 40 m, and are also locally met at deeper levels in the section. The subsequent cooling of the gabbros with time has provoked a brittle deformation with formation of cracks,

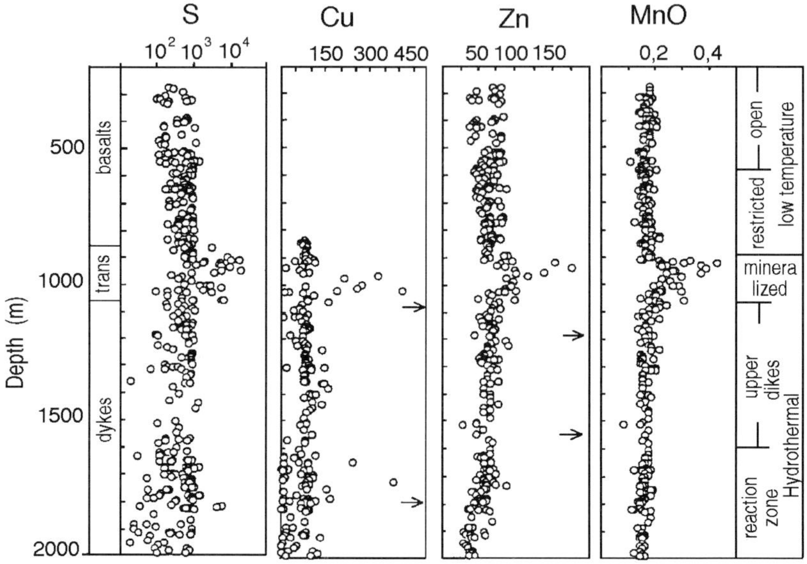

Fig. 9.9. Variation in content of sulphur and metals (Cu, Zn, Mn) with depth in the basalts of Hole ODP 504B (total rocks) (after Alt, 1995). The reaction zone (lowermost part of the sheeted dyke complex) is characterized by an important decrease in Cu, Zn and S, due to high-temperature hydrothermal fluids ($T^\circ > 350^\circ C$) leaching out of the zone. Higher in the column, these same hot and ascending fluids have met cold, downgoing seawater at the level of the 'transition zone', resulting in the precipitation of Cu, Zn and Fe sulphides, and of Fe–Mn oxi-hydroxides, and in the formation of a mineralixed stockwork.

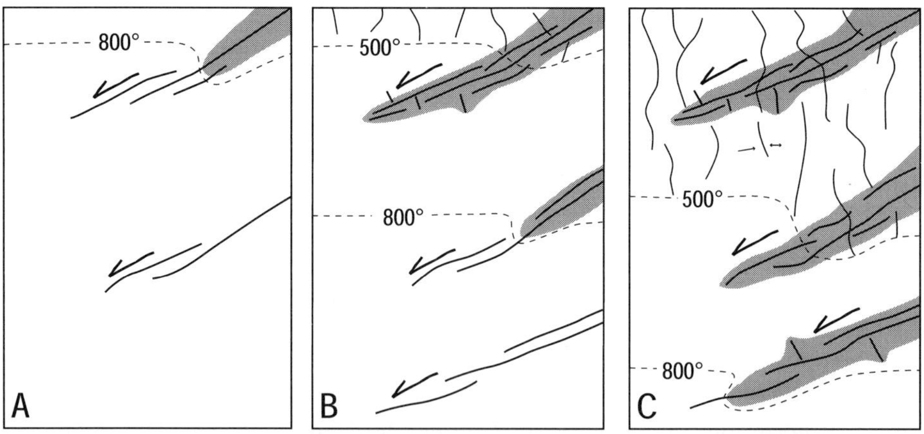

Fig. 9.10. Scheme showing the link inferred between tectonic deformation and metamorphic recrystallization in the gabbros of layer 3, at a slow-spreading ridge; Hole ODP 735B (after Mevel and Cannat, 1991). A dense net of normal faults developed in the axial zone of the ridge allows penetration of fluids in the hot gabbroic rocks, resulting in ductile deformations and high-temperature metamorphic recrystallizations 600–700°C, grey areas). The time sequence A–B–C illustrates the downward displacement of isotherms during the cooling of the crust, accompanied by a vertical fracturing front, developing below 500°C. This net of cracks allows further circulation of hydrothermal fluids, and the leaching of metals and sulphur.

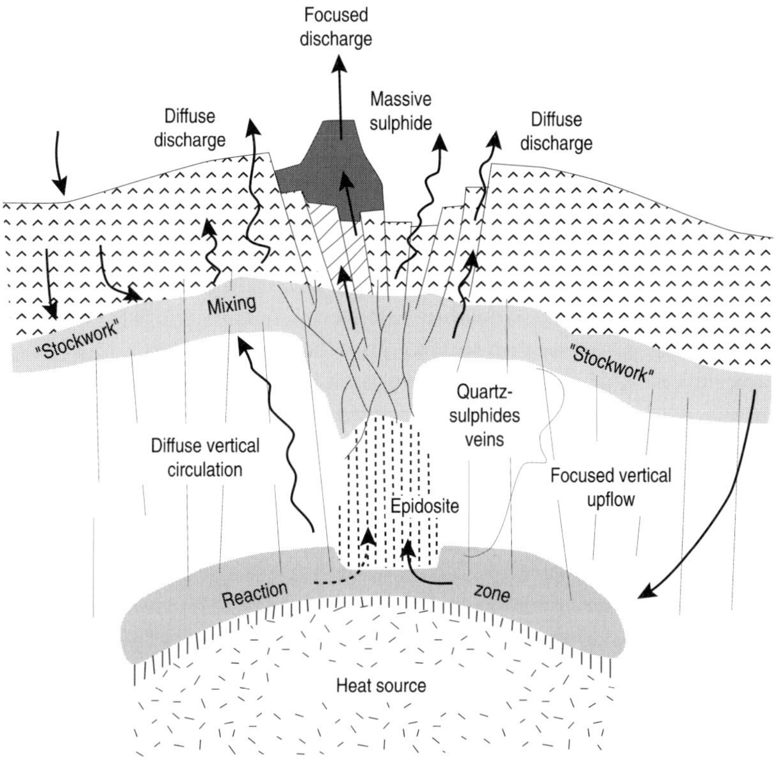

Fig. 9.11. Scheme of an hydrothermal discharge zone in the axial zone of an oceanic ridge, after detailed studies in several ophiolitic complexes (Cyprus, Oman, Newfoundland, etc.) (after Alt, 1995). Rooted in the HTRZ (High Temperature Reaction Zone), massive epidosites mark the pathways of focused ascending hot fluids, and give way to shallow mineralized 'stockwork', feeders of 'black smokers' depositing polymetallic sulphides on the seafloor.

followed by a new penetration of hydrothermal fluids, and the formation of steep hornblende and Na-plagioclase veins at 500–600°C (Figure 9.10).

Ophiolitic data

The study of fossil hydrothermal systems preserved in ophiolites has provided much information about the geometry of the whole hydrothermal system (Figure 9.11), about the composition and temperature of the fluids, and about the effects of water–rock interactions, particularly in the deeper parts of the system (reaction zone), which are difficult to reach in the oceans.

This reaction zone has been studied in detail in Oman, Cyprus, Newfoundland and Josephine (California) ophiolites. Located at the roof of the magma chambers, the reaction zone is characterized by mutually intrusive relationships between gabbros, plagiogranites and diabase dikes, and by a great abundance of altered diabase xenoliths in plagiogranitic and gabbroic intrusions. The direct interaction between the magma chamber and the hydrothermal system occurs at that level, with exsolutions of magmatic volatiles into the hydrothermal system, and incorporation of hydrothermal fluids into the magma chamber, by inclusion of altered and hydrated rocks of the roof.

The density of hydrothermal veins decreases considerably at this level, whereas the highest density of veins occurs typically inside the sheeted dike complex. The veins network is highly anisotropic, and exhibits a marked preferential orientation, parallel to the mean orientation of the sheeted dike complex (Nehlig and Juteau, 1988; Nehlig, 1989).

The study of fluid inclusions trapped inside the minerals of the reaction zone has shown that important phase separations occurred at high temperatures in this zone, with production of dense salt brines (Kelley *et al.*, 1992; Nehlig, 1989, 1993). These results have led several authors to propose a double-floor, double-diffusion convective model, with diffusion of heat and diffusion of salts at the boundary layer between both convective cells (Figure 9.12). This model explains rather well the observed salinity variation of fluids exiting from the black smokers on the seafloor (Bischoff and Rosenbauer, 1989; Kelley *et al.*, 1992; Nehlig, 1993).

9.4.3 Discharge zones

Alt (1995) has distinguished *focused upflow zones*, when hot hydrothermal fluids are channelized enough to exit directly and without modifications onto the seafloor, and *diffuse upflow zones*, when hot hydrothermal fluids mix with cold seawater before exiting on the seafloor, and form low-temperature diffuse vents.

Focused-discharge zones

These focused discharge zones are well known in ophiolites, where they are characterized by *epidosites* (Richardson *et al.*, 1987; Schiffman *et al.*, 1987; Nehlig *et al.*, 1994). These granular, yellowish-green rocks are exclusively formed of quartz, epidote and sphene, and occur as steep areas, 10 to 100 m wide, intersecting the grey–blue diabases (actinolite, albite, chlorite, sphene). The epidosite zones are located at the base of the sheeted dike complex, just below polymetallic massive sulphide deposits formed by 'black smokers' in the volcanic pile (Figure 9.11). Compared with diabases of the dike complex, the epidosites are enriched in Ca, Sr, ^{34}S, Fe^{3+}/Fe^{T} and $^{87}Sr/^{86}Sr$, and have lost Mg, Na, Zr, K, Cu, Zn and S (Alt, 1995). They are also poorer in ^{18}O. Fluid inclusions and the leaching of metals and sulphur indicate reaction temperatures of at least 350–440°C. The residual mineralogy indicates fluids devoid of Mg and enriched in Ca, analogous to those exiting from the 'black smokers, with high water–rock ratios (of the order of 500 to 1000).

The lack of epidosite samples from present-day ocean floors probably reflects a sampling bias, but could also reflect some fundamental differences between ocean crust and ophiolitic crust. In most studied cases, the ophiolitic epidosites are associated with late, differentiated intrusions, which may have formed off-axis.

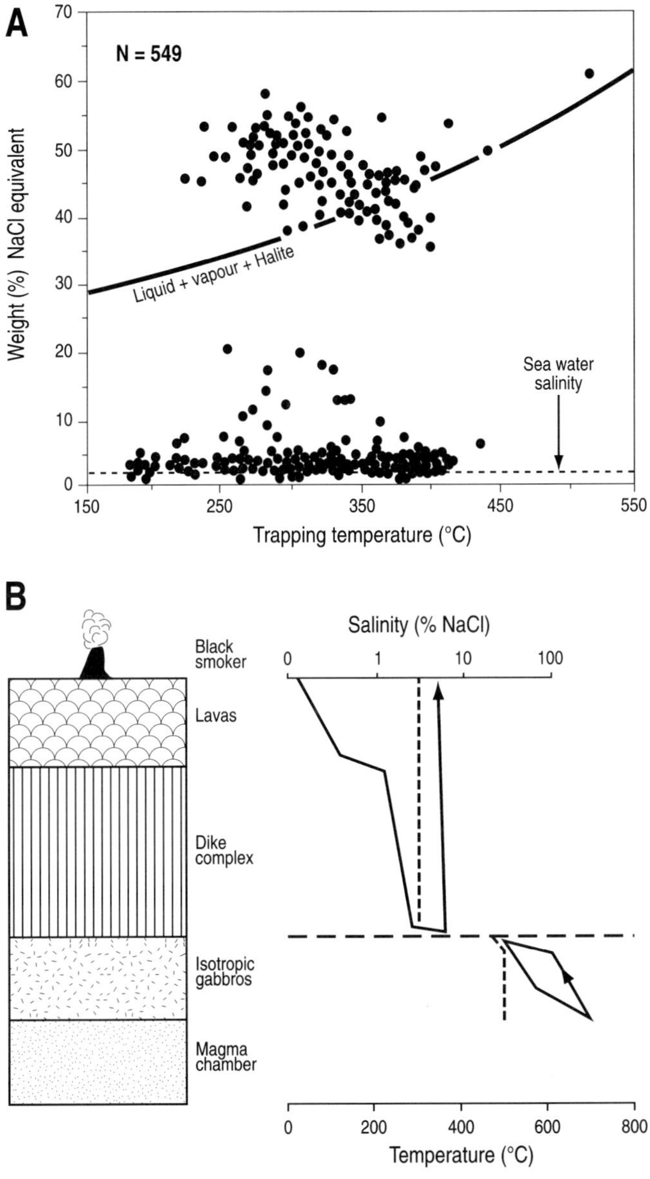

Fig. 9.12. Evidence of phase separations at high temperatures, at the base of oceanic hydrothermal convective systems. (A) Microthermometry of fluid inclusions from the HTRZ in the Troodos ophiolite (Cyprus): salinity versus trapping temperature. The diagram clearly shows two groups of inclusions. The first group has on average the salinity of sewater, the other group is constituted of hot, extremely salty brines (30 to 60% equivalent NaCl). N represents the number of analysed inclusions (after Kelley *et al.*, 1992). (B) Double diffusion convective model, proposed for the Oman ophiolite. A dense layer of hot, salty brine is sandwiched between the heat source (the magma chamber) and a diffusing thermohaline surface, separating brine from seawater convecting above. An hypothetical thermal cycle (full line) and salinity variation (dash line) are represented for both convective cells (after Nehlig, 1993).

Diffuse-discharge zones

When the ascending hydrothermal fluids are not channelled enough, particularly inside the volcanic pile, the fluids wander through the inter-pillow spaces and mix sooner or later with cold, downgoing seawater. The resulting tepid fluid, denser and weakly channelled, will exit as diffuse vents on the ocean floor. The hydrothermal vents of the Galapagos spreading centre are a good example of these diffuse uprises, with exiting temperatures of 3° to 13°C and very low contents in metals, indicating that the polymetallic sulphides have precipitated in depth, at the level of the meeting of hot fluids and cold seawater. *It is estimated that these diffuse discharge zones convey ten times more heat than the focused ones.* (Alt, 1995)

In Hole 504B, a mineralized stockwork has formed in the transition zone between pillows and dikes (Figure 9.9).The basalts are intensely altered into chlorite, albite and sphene. The sulphides (pyrite, chalcopyrite, sphalerite and rare galena) are abundant in the rocks and in a network of quartz veins. The mixing of high-temperature hydrothermal fluids (250–350°C), going up along the cracks of the dike complex, with huge volumes of cold seawater, circulating through the more permeable overlying volcanic pile, has provoked the precipitation of quartz and polymetallic sulphides. The corresponding transition zone is clearly enrichd in metals and sulphides.

In the Josephine, Semail or Troodos ophiolites, the transition zone between the sheeted dike complex and the overlying volcanic pile also exhibits rocks enriched in Cu, Zn and S, and abundant quartz and suphides veins. All these observations suggest that the effects of the diffuse uprising are common, and probably represent a typical feature of the oceanic crust. They incite to think that the volumes of polymetallic sulphide ores contained inside the oceanic crust are not restricted to the spectacular seafloor deposits around the 'black smokers', but are mainly precipitated in the depths of the crust.

9.5 METALLOGENIC ASPECTS OF OCEANIC HYDROTHERMALISM: THE POLYMETALLIC MASSIVE SULPHIDE DEPOSITS

The mineral resources of the deep ocean are the *massive sulphide deposits* in the accretion zones (oceanic ridges, back-arc basins, etc.) and seamounts, the *manganese nodules* of the abyssal plains, and the *manganese crusts* on the flanks of seamounts and volcanic islands. The potentially minable metals in these deposits are copper, zinc, lead and silver (sulphides), plus gold and barium (sulphates) in the polymetallic sulphide deposits; cobalt, nickel, copper and manganese (oxides–hydroxides) and eventually platinum-group elements (PGE) in the manganese nodules and crusts. Iron is abundant in all these deposits, but has no commercial value.

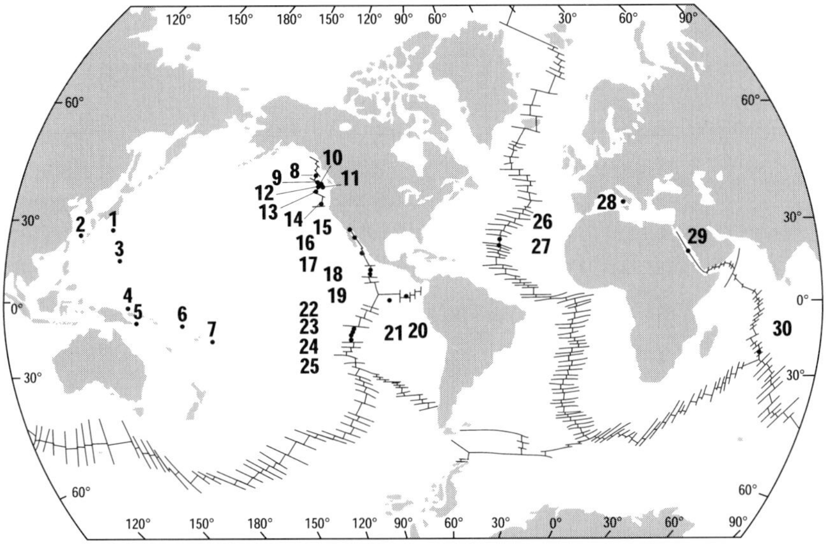

Fig. 9.13. Location of active or fossil polymetallic sulphide deposits presently known on the ocean floor of present-day oceans (after Rona and Scott, 1993). 1: Ogasarawa. 2: Okinawa Trough. 3: Marianas Trough. 4: Manus Basin. 5: Woodlark Basin. 6: North Fiji Basin. 7: Lau Basin. 8 to 13: Juan de Fuca Ridge. 14: Escabana Trough. 15: Salton Sea. 16: Guaymas Basin. 17: East-Pacific Rise (EPR), 21°N. 18: EPR, 13°N. 20: DSDP/ODP Hole 504B. 21: Galapagos Rift, 86°W. 22 to 25:EPR, 17 to 21°S. 26: Mid-Atlantic Ridge (MAR), TAG Site. 27: MAR, Snake Pit Site, Palinuro Seamount. 29: Red Sea, Atlantis II Deep. 30: Carsberg Ridge, Sonne Field.

9.5.1 Massive sulphide deposits on the seafloor: the discovery of the 'black smokers'

During the last 20 years of oceanic exploration, an increasing number of polymetallic sulphide deposits have been discovered on the deep ocean floor, in various geodynamic contexts. They have in common the proximity of a heat source, inducing hydrothermal circulation of seawater through fractured and faulted rocks, and the mixing of hot hydrothermal fluids with cold seawater, provoking the massive precipitation of metallic sulphides and other minerals on the seafloor or just below it.

These deposits contain high concentrations of Cu, Zn, Pb, Ba, Ag or Au. In sedimentary environments, some of them can reach a size of several million or tens of million tons, comparable with analogous deposits mined on land.

Although scarcely 1% of the deep ocean has been explored in detail, more than 100 massive sulphide deposits have already been recorded on the ocean floor, most of them located in the Pacific Ocean (Figure 9.13).

The very first discoveries of metallic sulphides occurred in the middle of the 1960s, in several deeps of the axial zone of the Red Sea. In 1977, tepid vents surrounded by spectacular hydrothermal faunas were discovered and sampled on the Galapagos rift. In 1978, dead sulphide chimneys were discovered and sampled at 21°N on the East-Pacific Rise, during the Cyamex French–American diving cruise. One year later in 1979, the Rise American–French cruise in the same area made the remarkable

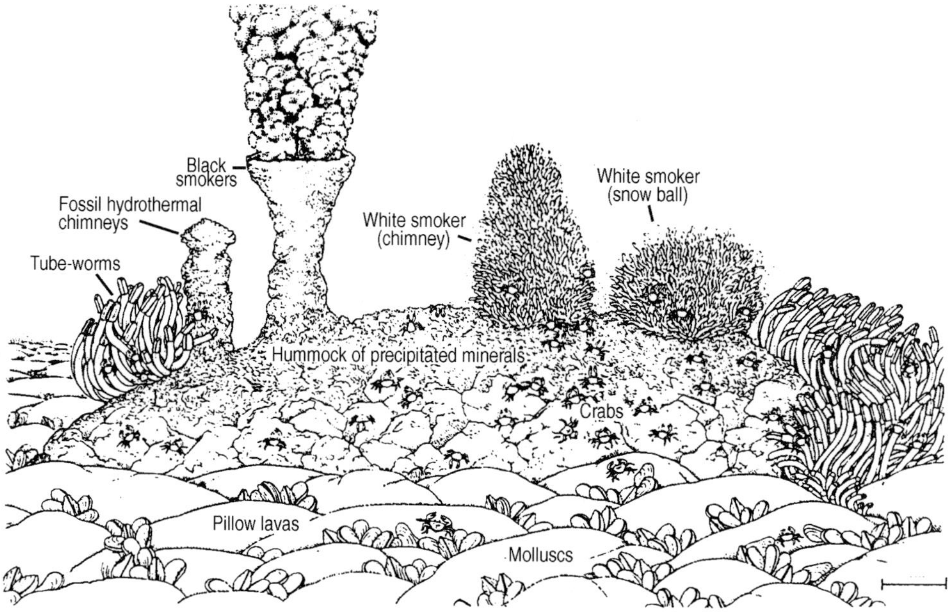

Fig. 9.14. Scheme of an hydrothermal field with active white and black smokers, and their typical hydrothermal faunas (after a drawing from Juteau and Rangin, 1981). An alignment of hydrothermal vents is visible at the top of a mound made of chimney fragments, mineral precipitates and organic debris. The 'white chimneys' are made of an accumulation of *Alvinella pompeiana* tube worms. The whitish fluids emitted by these chimneys exit at temperatures ranging from 20° to 200°C. A much hotter fluid exits from 'black smokers' (350°C). These chimneys, after collapsing, construct sulphide mounds over the basaltic seafloor. Numerous colonies of giant clams (Vesicomydaes) live in the low gaps between pillow lavas.

discovery of high-temperature hydrothermal activity, by observing sulphide chimneys pouring out a hot, black and whirling fluid at 350°C, the famous 'black smokers'. Surrounded by a spectacular hydrothermal fauna (Figure. 9.14), these 'black smokers' are mainly made of polymetallic sulphides, amorphous silica and calcium sulphate (anhydrite, with eventually some barium sulphate).

Similar active hydrothermal areas with black smokers were discovered later on in many sites along the world oceanic ridge, in a number of back-arc or marginal basins, and on the top of some seamounts.

9.5.2 The genetic model

In metallogeny, it is rare to observe *in real time* the formation of an ore deposit. The active black smokers provide this exciting opportunity. The hot hydrothermal fluids directly precipitate iron, copper and zinc sulphides on the inner wall of the chimneys. The chimneys themselves collapse, and their accumulated and imbricated fragments, cemented by lower-temperature minerals, construct little by little a *volcanogenic massive sulphide (or VMS) deposit* on the seafloor.

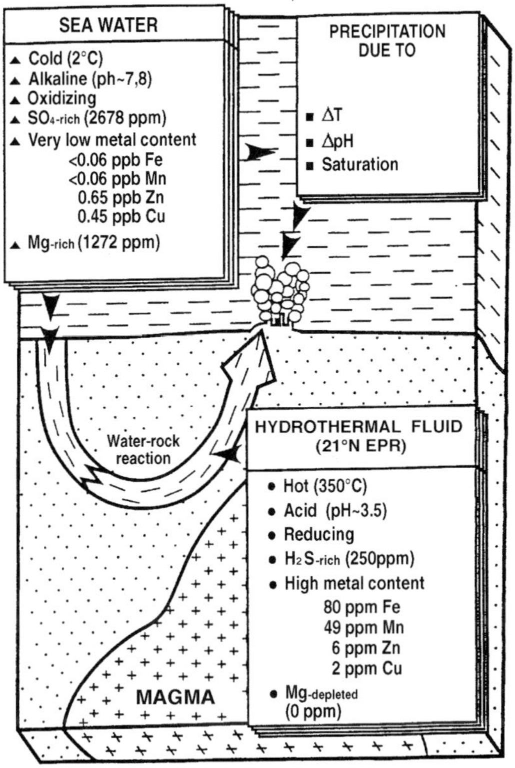

Fig. 9.15. Metallogenic model for hydrothermal deposits of the East-Pacific Rise, using the data from 21°N (after Scott, 1992). The oxidising and metal-poor, cold seawater results in a hot, reducing and metal-rich hydrothermal fluid, after interaction with the hot rocks of the oceanic crust located at several km below the seafloor. The hydrothermal fluid rapidly ascends and mixes with ambient seawater to precipitate the main part of its mineral charge. This precipitation is a consequence of fast cooling and of recovery of pH, by addition of seawater sulphate. Note: ppb stands for parts per billion (10^{-9}).

The genetic model is particularly clear here, at least in the main lines (Figure 9.15): the cold seawater percolates down inside the young oceanic crust through its countless open fissures and pores; this slightly alkaline water (pH close to 8) precipitates its sulphates and carbonates while going down. Its temperature is increasing progressively with depth and at the approach of the magma chambers. At about 2–3 km in depth, it is no longer seawater, but a hot hydrothermal fluid (about 350–400°C), acidic (pH < 4) and corrosive. This fluid, close to its critical point and submitted to a drastic thermal expansion, raises back towards the surface at the ridge axis, dissolving on its pathway the metals and sulphur contained as trace elements in the basaltic layer. When exiting in cold seawater (about 2°C), it massively precipitates its metallic suphides, along with anhydrite and amorphous silica. A fraction of these precipitates remains in the hydrothermal plume, which spreads several hundreds of metres above the active chimneys, and will slowly deposit its fine particles all around the site (mainly Fe–Mn oxi-hydroxides).

The 'black smokers' activity and the polymetallic massive sulphide deposits can thus be explained as a true convective circulation of seawater through the young and fractured oceanic crust. The three essential conditions for starting a convection are effectively ideally fulfilled at a ridge axis:

(1) The presence of a *convecting fluid*, here seawater, an inexhaustible reservoir;
(2) The existence of good *permeability* of the medium: this permeability, which allows seawater to circulate effectively through the crust, is created by the intense fracturing of the basaltic layer, in the extensive tectonic context of the ocean spreading processes;
(3) The existence of a permanent *thermal motor*, generating the steep thermal gradient necessary to start up the movement of water in the crust and the creation of convective cells, and to maintain strong temperature differences during long periods of time. Below the oceanic ridges, the magma chambers, with their basaltic magma close to 1200°C constitute an ideal thermal engine. Between the roof of the magma chambers, at about 1000°C and located at 2–3 km in depth, and the surface of the seafloor at 1–2°C, the thermal gradient is enormous: 350 to 500°C/km on average, that is, between 10 and 20 times the normal geothermal gradient on land.

Such a gradient will force a vigorous hydrothermal convection in the crust, with the development of multiple and diffuse *recharge zones*, and more localized and focused *discharge zones*, as we have seen.

It happens frequently that cold, downgoing seawater mixes with hot, ascending fluids below the seafloor. In this case, the diluted and chilled hydrothermal fluid will precipitate its metallic charge in depth, and will exit tepid and purified on the seafloor, forming 'white diffusors' whose tepid waters (20°C to 150°C) exit with a much more moderate flow.

9.5.3 Towards a better understanding of terrestrial ore deposits

We understand better now how a number of metallic ore deposits, presently observed on land, have formed in the oceans of the past. Beginning of course by the copper-rich polymetallic massive sulphide deposits in ophiolites, these fragments of oceanic lithosphere obducted on continental margins or island arcs. Many of them contain minable copper, iron and zinc sulphide deposits, with sometimes interesting contents in gold or silver.

These sulphides form massive lenses inside or at the surface of ancient submarine volcanic piles (mainly pillow lavas), exactly as at the axial zone of present-day oceanic ridges. Moreover, circular fragments of fossil hydrothermal chimneys and even, in some cases, fossil remains of hydrothermal faunas, have been found in these deposits (for instance the fossil tubes discovered in 1985 in the ore deposits of the Bayda mine, in the Upper Cretaceous Oman ophiolite, very similar to the *Alvinella* tube worms proliferating on and around the white and black smokers of the East-Pacific Rise).

The analogies are so conclusive that the general consensus is that the polymetallic massive sulphide deposits known and mined in Cyprus, Oman, Turkey, Newfoundland or the Philippines represent ancient hydrothermal deposits formed by black smokers on the seafloor of the oceans of the past.

9.6 BIOLOGICAL ASPECTS OF THE OCEANIC HYDROTHERMAL ACTIVITY: OASES ON THE DEEP SEAFLOOR

Only 23 years ago, marine biologists were convinced that they knew reasonably well the deep abyssal fauna living on the deep ocean floor, far from the continental margins.

These settlements they believed were poor—almost a biological desert—because of the poor nutritive supplies coming down from the surface. These supplies included tiny planktonic particles constituting the 'sedimentary snow' raining on the deep ocean floors, and also the carcasses of big animals (fishes, marine mammalians, etc.), living close to the surface and falling down after their depth. In both cases, shallow photsynthetic production is an important and indispensable step of the food chain.

Who could imagine, in 1975, the unthinkable for any reasonable marine biologist? That is, the existence of swarming oases of life on the deep ocean floor, and of biological communities totally independent of photosynthesis?

9.6.1 The discovery of deep hydrothermal faunas

In 1977, a cruise of the American submersible *Alvin*, helped by deep-tow pictures taken the year before by Peter Lonsdale (an American geophysicist), discovered the first 'oases of life' in the axial zone of the Galapagos Rift, settled around tepid water vents exiting at 15–20°C. The vents were of the 'white diffusor' type, but these first oases, named poetically 'Paradise Garden' or 'Rose Garden' by the divers, are settled by the same faunas as those discovered 2 years later around the hot 'black smokers' of the East-Pacific Rise. The biologists could not believe their eyes: they discovered tens of entirely new species.

The discoveries increased rapidly over the ensuing years:

- 1978 the dives of the French–American Cyamex cruise discovered and sampled the very first polymetallic sulphides on the East-Pacific Rise at 21°N. These sites were inactive, only fossil edifices and chimneys were observed, along with graveyards of half-dissolved giant clams (*Calyptogena magnifica*) and serpulids (small worms living in several-cm-long tubes).
- 1979 the American–French Rise cruise discovered the first active black smokers in the southern part of the same area, surrounded by swarming and varied hydrothermal faunas, analogous to those of the Galapagos Rift, in particular the typical giant worms Vestimentifera, species *Riftia pachyptila*.
- 1982 French teams discover a number of active hydrothermal sites between 11° and 13°N along the East-Pacific Rise.

1983 Canadian researchers discover original settlements on the Juan de Fuca ridge, off the coast of Oregon, and Americans discover an active hydrothermal site in the middle of the sedimented Guaymas basin, in the Gulf of California.

1985 a team from the American NOAA discovers the first hydrothermal active sites on the Mid-Atlantic Ridge at 26°N. The same year in December, the submarine camera of the *Joides Resolution* drilling ship discovers by chance the Snake Pit hydrothermal field at 23°N, just south of the Kane fracture zone. Slow-spreading ridges, thought to be sterile, are indeed hydrothermally active too!

1986–89 this is the period of the discovery of numerous active hydrothermal sites in the back-arc basins of the south-western Pacific: Manus Basin (1986), Marianas Basin (1987), Lau and North Fiji Basins (1989).

The conclusion is clear today: hydrothermal activity exists all along the 60 000 km of the world Mid-Ocean Ridge, on slow, intermediate and fast-spreading centres. Its existence is also demonstrated in many spreading ocean basins, such as back-arc basins and marginal seas, located above a subduction zone. Up to now, all the explored sites are settled with swarming animal communities, including fairly large species, most of them unknown (more than 400 new species recorded presently). The observation of graveyards of clams and worms close to dead vents demonstrates the direct and vital link between these communities and the hot hydrothermal fluid.

The concentrations of living matter are among the highest in the oceanic medium. The biomasses reach 50 to 100 kg per square metre around the hydrothermal vents, and this in the middle of an abyssal desert!

But what are these communities living on?

Hydrothermal fluid contains 1 to 8.5 millimoles of hydrogen sulphur (H_2S) by litre, and also some methane (CH_4). A totally unexpected discovery: *chemiotrophic bacteria* were found inside the cells of tissues of various invertebrates, who use the energy liberated by chemical transformation of H_2S or CH_4 to synthetize organic molecules from the CO_2 and nutritious salts.

The large Riftia tube worm for instance is totally devoid of a digestive tube. It lives from the organic matter produced by bacteria living inside the cells of its own 'troposome', a very vascularized tissue occupying two-thirds of the volume of its body. The metabolism of the bacteria is fed by the red blood of Riftia, which transports hydrogen sulphur, oxygen, carbon dioxide, nitrates and phosphates. Once dead, the bacteria are used as food by the worm.

Other invertebrates exhibit the same association: the giant Calyptogena clams, the Pompei worms, the Gasteropods, etc. This symbiosis is the base of the hydrothermal ecosystem. It produces the organic matter indispensable for the rest of the community: shellfish, fish, cephalopods, etc. One can watch, for instance, the fish and shellfish grazing on the red branchias of Riftia, or attacking the giant clams. These animal communities are totally autonomous with respect to supplies coming down from the surface of the sea, and are especially independent of photosynthesis.

9.6.2 Towards the origins of life?

The discovery of deep hydrothermal faunas has relaunched the debate on the fascinating problem of the origin of life. There is general agreement that life started in the primitive ocean, some four billion years ago, in much hotter and reducing conditions than today. This is the prebiotic 'primitive soup' concept, proposed for the first time by the Soviet biochemist Oparine in the 1920s.

The remarkable context constituted by the oceanic hydrothermal vents has probably existed since the creation of the primitive ocean, several billion years ago, and has probably not substantially altered since that time.

This context is obviously a good candidate for the first appearance of life: presence of strong physical and chemical gradients facilitating encounters between organic molecules; high contents in the hydrothermal fluid in dissolved gases or elements essential to life (H_2S, H_2, N_2, CH_4, CO_2, CO, etc.); abundance of support-minerals suitable for stabilizing and fixing organic molecules (pyrite FeS_2 and various sulphides, Fe–Mn oxides, Fe, Al-rich clays, etc.); protection against solar radiation, especially UV radiation, and against the incessant meteorite showers thanks to the thickness of the water column. All these factors were at that time extremely favorable to the appearance of life.

The discovery, at the end of the 1970s, of archaeobacteria in the hydrothermal fluids, true living fossils whose RNA has a totally different structure from that of other known bacteria, has given support to the theory of the origin of life starting at hydrothermal vents. These bacteria live and reproduce at high temperatures ranging from 85 to 115°C. They are strictly anaerobes, and perhaps very close to the bacteria of the very first ages of the Earth, at a time when seawater temperature was above 70°C, and when the atmosphere of the Earth was totally devoid of any oxygen.

The invertebrates who colonize the hydrothermal fields apparently invaded this ecological niche much later, in successive waves, but many question marks remain concerning the history of this colonization, and the strange association between bacteria and invertebretaes.

10

Intraplate oceanic magmatism

Between the stages of its formation at ocean ridges and of its disappearance and recycling in subduction zones, the oceanic lithosphere undergoes important modifications: its thickness increases with time (as a function of the square root of its age), marine sediments accumulate on its top, and additional changes occur when it passes over hot spots. The frequency of the latter event is highly variable, as well as its consequences. In some sectors from the Pacific and Indian Oceans, the crustal thickness is nearly doubled by emplacement of an intra-oceanic plateau over the original MORB-type (mid-ocean ridge basalts) crust. In the Atlantic, on the other hand, plume-related magmatic formations are less common, even though some bathymetric and geochemical variations observed along the ridges may be considered as resulting from deep plume/ridge interactions (Chapter 8).

10.1 FROM SEAMOUNT TO ATOLL: THE FRENCH POLYNESIA LINEAR CHAINS

Oceanic intraplate magmatism results in the formation of two major kinds of structures: oceanic plateaus and linear island chains. The evolution of the latter, from seamounts to shield volcanoes and finally atolls, is well illustrated in the Pacific by the Hawaii-Emperor Chain and the French Polynesia linear chains.

10.1.1 Location

French Polynesia is a part of the huge oceanic domain of the South Pacific. Together with the Cook and Pitcairn islands, it occupies a vast triangular area of 15 million km^2 (Figure 10.1) including less than $5000\,km^2$ of land. The Late Cretaceous to Oligocene ocean crust from this area is 250 to 750 m shallower than the Atlantic and Indian crusts of equivalent ages, which allowed definition of a 'South Pacific superswell' (McNutt and Fischer, 1987). Indeed, the oceanic substratum of most volcanic islands lies above the -4000 m isobath. The Polynesian islands form five

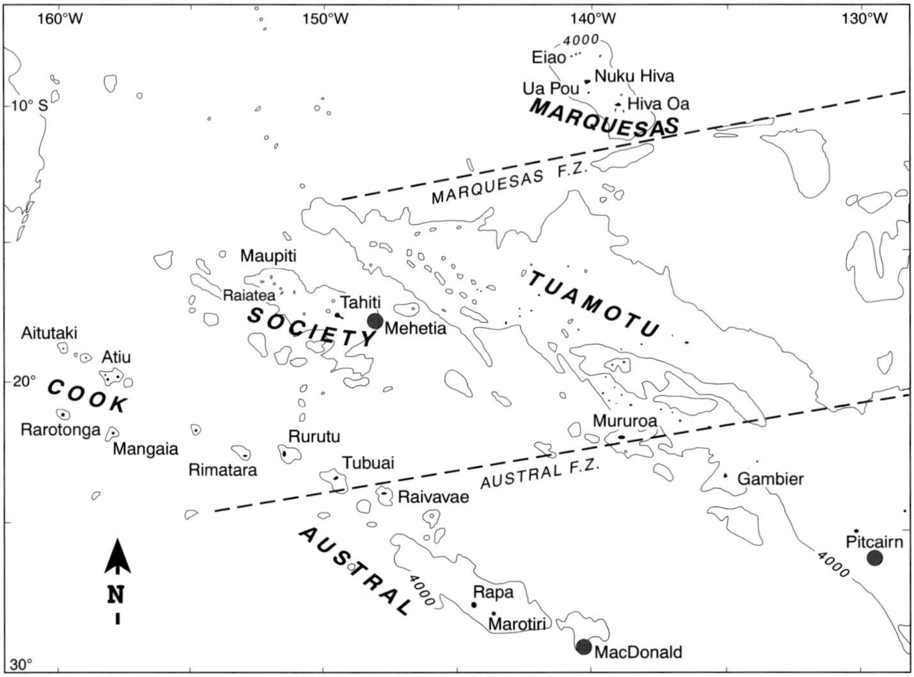

Fig. 10.1. The Polynesian oceanic domain. The size of the area is 3000 × 4000 km. The curves correspond to the −4000 m isobath. The positions of hot spots are shown as shaded circles.

linear chains, i.e. from south to north the Austral-Cook, Society, Pitcairn-Gambier (prolongated by the Mururoa and Duke of Gloucester atolls), Tuamotu and Marquesas archipelagoes. The lengths of these chains are variable: 2300 km for the Austral-Cook, 1700 km for the Pitcairn-Gambier including the Hereheretue atoll, 1000 km for the Tuamotu, 700 km for the Society from Mehetia to Motu One atoll, and finally 350 km only for the Marquesas. Their overall SE–NW trends (N110°–N120°E) fit very well with the absolute motion of the Pacific plate, with the exception of the Marquesas which trend N140°E (Figure 10.1).

10.1.2 Age patterns

Most of the Polynesian linear chains include 'high' islands exposing volcanic rocks. However, the Tuamotu linear chain is only composed of atolls, and the age of their volcanic substratum remains unknown. This archipelago is considered as much older than the four other chains.

The Austral-Cook, Society and Pitcairn-Gambier linear chains include active volcanoes at their southwest tip (Figure 10.1). The summit of the permanently active MacDonald seamount in the Australs is only 40 m below sea level. The Society active area includes a number of submarine volcanoes and one emerged cone, Mehetia islet. Two large seamounts and about 20 smaller volcanic edifices mark the present-day

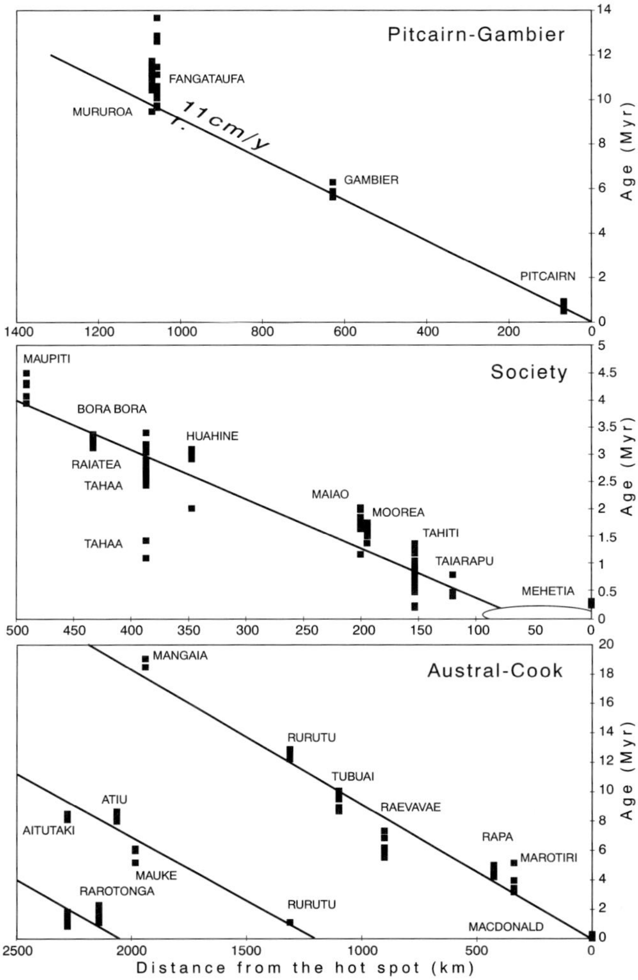

Fig. 10.2. Correlation between ages of the islands and their distances to the hot spot for the Pitcairn-Gambier, Society and Austral-Cook linear chains.

location of another active volcanic area 60 km southwest of Pitcairn. The positions of these active volcanoes reveal the sites of hot spots which can be considered as fixed with respect to the asthenospheric mantle. The Pacific plate migrates towards the northwest (N110°) over these hot spots with an average speed of 11 cm/yr.

In all the archipelagoes except the Tuamotu the island morphologies vary with distance from the hot spot, the primary volcanic landforms being progressively degraded. The coral reefs evolve from fringing reefs to barrier reefs which isolate progressively widening lagoons from the ocean. The radiometric ages of lavas (Figure 10.2) increase rather regularly with distance from the hot spot. Most ages are largely consistent with the 11 cm/yr motion of the Pacific lithospheric plate above a deep mantle plume/hot-spot system considered as having a fixed position with respect to the asthenospheric mantle.

The reliability of these age/distance correlations, however, is variable from one chain to another. The simplest case, similar to that of the Hawaii-Emperor chain, is observed for the Pitcairn-Gambier linear chain, with a very good fit of the age data with the 11 cm/yr plate motion. However, it is based only on four islands: the young volcanic edifice of Pitcairn, the partly submerged Gambier large shield volcano and the atolls of Mururoa and Fangataufa, the volcanic substratum of which is known from a number of deep drillings (Guille *et al.*, 1995). The K–Ar ages from the bottom of these drilled piles are older than predicted by the model, but they may have been modified by submarine alteration processes.

The Society island measured ages are generally well correlated with the predicted ages. However, several islands (Tahiti Nui, Maiao, Huahine, Tahaa) are characterized by late volcanic events which post-date by up to 1 Myr the end of the main volcanic episode (Huahine, Tahaa). The corresponding lavas are usually different from the previously erupted ones (picritic character; abnormal REE (Rare Earth Element) enrichment in the case of Tahaa).

The Austral-Cook chain presents distinctive features with respect to Gambier-Pitcairn and the Society. Besides its unusual length, it contains quite old high islands (Mangaia, 18 Myr), whereas in the other Polynesian linear chains those islands older than 6 Myr (Gambier) are converted to atolls. Most of the islands located far away from the MacDonald hot spot are characterized by Mio-Pliocene coral reefs uplifted to altitudes of 70 to 100 m (Rurutu, Mangaia). The most distinctive feature of the Austral-Cook chain lies in the occurrence of young volcanic events far away from the Macdonald active seamount. For example, in Rurutu, the Miocene volcanic units (12 Myr) and the uplifted coral reef are overlain by Quaternary basaltic flows (1.1 Myr) and associated strombolian and hydromagmatic breccias. Rarotonga is a well-preserved shield volcano with a central caldera, circled by a fringing reef. Its activity ended at 1.1 Myr as in Rurutu. Aitutaki was built in two volcanic episodes, during the Miocene (8 Myr) and the Quaternary. It is clear that no conventional hot-spot model can account for the complex age–distance relationships shown in Figure 10.2, except by postulating the existence of three hot spots 1000 km apart along the linear chain!

The Marquesas archipelago is also very different from any Hawaiian-type chain. Although the ages if its islands are similar to those of the Society, the length of the chain—which is not really linear—is much shorter (Figure 10.1) and its trend (N140°) is different from the present-day motion of the Pacific plate. However, it fits its Paleocene–Early Eocene trend (prior to 43 Myr), which corresponds to the age of its oceanic substratum. The Marquesas apparently lack active volcanoes, the youngest measured K–Ar age in the area (0.76 Myr) corresponding to a seamount located in the southwestern part of the chain, close to the Marquesas fracture zone. The K–Ar ages of the Marquesas islands tend to increase towards the northwest (Brousse *et al.*, 1990): 1 to 2.5 Myr for the southern group (Hiva Oa), 1.5 to 5 Myr in the central islands (Ua Pou, Nuku Hiva) and 4.5–6 Myr for the northern islets (Eiao), but they vary much less regularly than in the other chains.

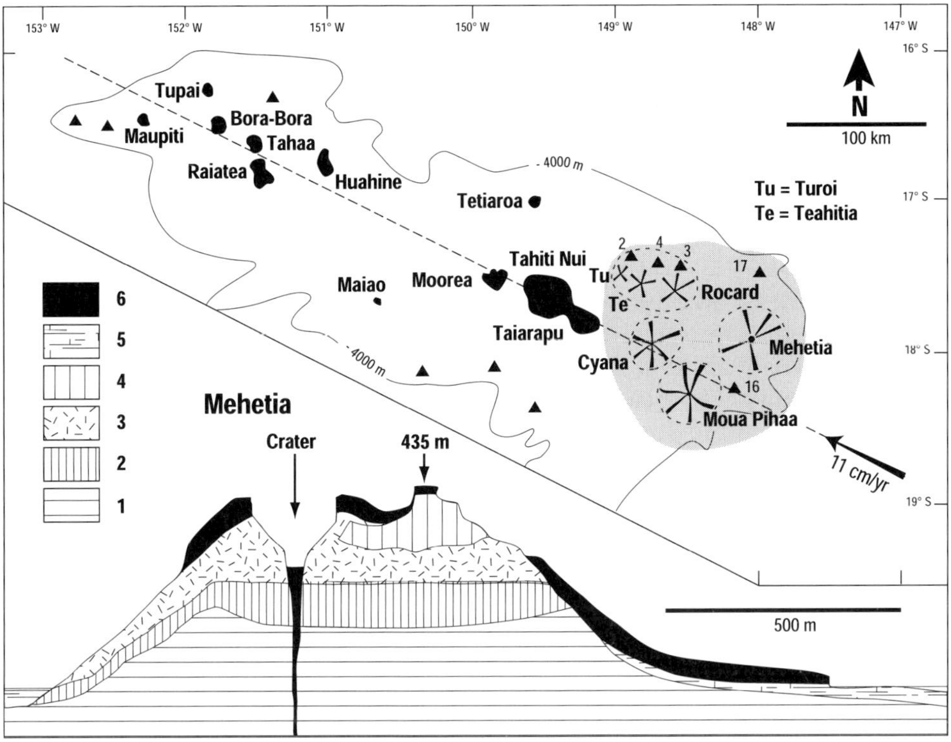

Fig. 10.3. The Society hot spot and archipelago (from Binard *et al.* 1991). The shaded area is the locus of present-day volcanic and/or seismic activity. The crests of the main seamounts, interpreted as rift zones, are shown. Bottom: schematic cross-section of Mehetia Islet, with: (1) Surtseyan-type hydromagmatic breccias; (2) basaltic flows (75 000–60 000 yr); (3) strombolian cone; (4) upper basaltic flows; (5) fringing coral reef; (6) recent lava flows (< 40 000 yr) and lava fountain deposits.

10.1.3 The initial stages: the Society hot spot

The Society linear chain corresponds to a N110° domain 700 km long and 200 km wide, delineated by the −4000 m isobath, which contains 11 volcanic islands and an equal number of seamounts (Figure 10.3). Each island corresponds to a shield volcano, except the largest one, Tahiti (1042 km^2) which is made of two shields connected by a narrow isthmus: Tahiti Nui (2241 m) and Taiarapu (1332 m). Southeast of Taiarapu, whose volcanic activity ended 400 000 years ago, the active Society hot-spot zone has a diameter of 150 km (Figure 10.3). It is 400 to 500 m shallower than the adjacent abyssal plains. This topographic high (bulge) results from a combination of factors including: (i) uplift of the Pacific oceanic crust above an ascending mantle plume, (ii) accumulation of glassy clasts from the very young submarine eruptions, and finally (iii) the reworking of volcanic materials from the slopes of Mehetia, Moua Pihaa seamount and mostly from Taiarapu by large submarine gravity landslides. There is no evidence for a thickening of the bulge crust by accumulation of lava flows, as commonly observed in oceanic plateaus (§ 10.3).

The Society hot-spot area contains a dozen young volcanic edifices of various sizes (Figure 10.3). The summit of the largest one, Mehetia (50 km in diameter, 4000 m high) forms an islet 1500 m wide and 435 m high, with a very well-preserved summital crater. The Moua Pihaa seamount, only slightly smaller than Mehetia, has its top at −143 m. The deeper volcanic edifices named Cyana, Rocard, Teahitia, Turoi and Volcan 16 are nevertheless 1000 to 2000 m high, and the smaller seamounts labelled by numbers culminate 400 to 500 m above the crust of the bulge. Although no eruption has been observed, there are numerous indications of activity in the area: hydrothermal springs at 30°C at the top of Teahitia, seismic foci below seamounts 2, 3 and 4, and finally the 1981 genuine volcanoseismic crisis below the southeast flank of Mehetia. 3450 seisms with magnitudes up to 4.3 on the Richter scale, corresponding to submarine eruptions at a depth of 1600 m, were recorded during this crisis (Binard *et al.*, 1993).

The geology of the volcanoes at the Society hot spot is known from SeaBeam bathymetric maps, dredgings, submersible dives and field data on Mehetia Islet. The deep parts of the volcanoes are characterized by slopes of 11° to 15°, similar to the low emerged slopes of large shield volcanoes. Their base is composed of flat fluid flows with ropy glassy surfaces similar to the fluid flows from the ridge axis (§ 7.3), but made up of alkali basalts. Above the −3000 m isobath they are progressively replaced by bulbous and tubular pillow flows, together with hyaloclastic breccias resulting from fragmentation of pillows. These flows were mostly emitted from submarine crests corresponding to radial rift zones which show a star-shaped pattern (Figure 10.3). Their trends are controlled either by the present regional field of constraints (N110°–N130° and N30°–N40° trends) or by old discontinuitities and weakness zones from the underlying oceanic crust (N80°, transform fault orientation and N150°–N180° corresponding to a Farallon Ridge trend). The relief of the rift zones is due to lava chimneys similar to hornitos and to intrusions (dykes, necks) which mark the position of submarine vents.

Above the −500 m isobath, pillow basaltic flows are progressively replaced by hydromagmatic breccias, then, above sea level, by strombolian cones and subaerial lava flows. For instance, Mehetia Islet (Figure 10.3) is made up of a Surtseyan-type hyaloclastic ring, formed by pulverization of basaltic magma within shallow sea water (§ 7.2). The crater of this ring was later filled by first subaerial lava flows 60 000–70 000 yr old, which in turn were overlain by a steep sloped (40°) strombolian cone made up of lapilli, scoriae and bombs interbedded with lava flows. The last volcanic events, younger than 30 000 yr, emplaced fluid alkali basalt flows over the fringing coral reef, as well as Hawaiian-type lava fountain ejecta covering the slopes of the former strombolian cone.

10.1.4 The shield-volcano stage: Nuku Hiva (Marquesas)

Although partly collapsed, Nuku Hiva island (Figure 10.4), located in the central part of the Marquesas, is representative of a large Polynesian shield volcano. Its

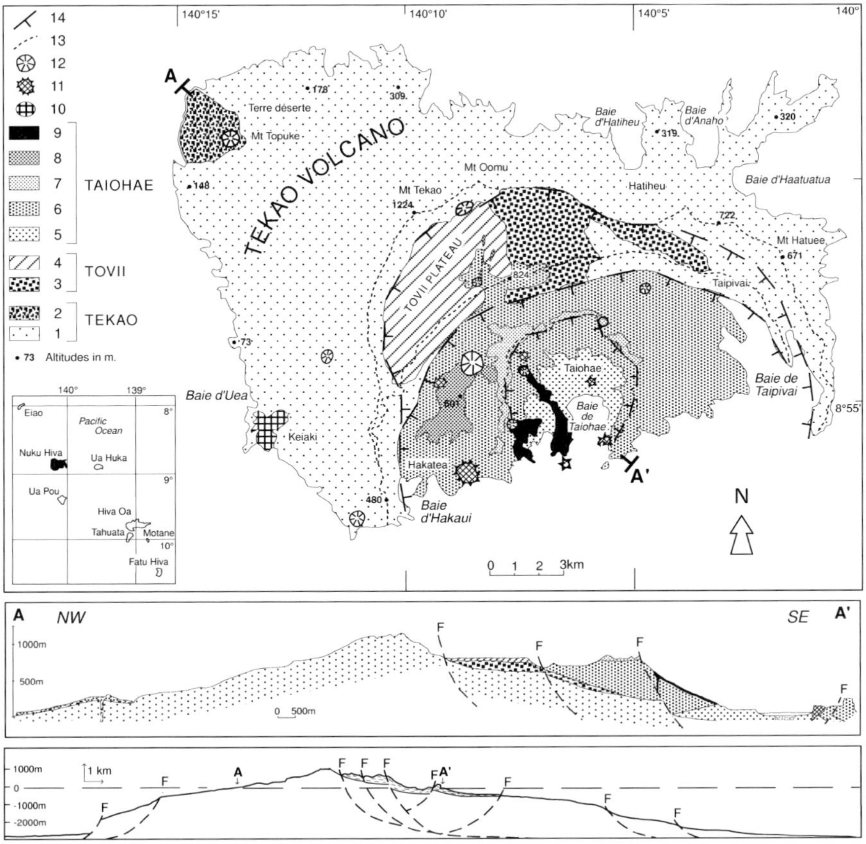

Fig. 10.4. Geology of Nuku Hiva, Marquesas (from Le Dez *et al.* 1996). Inset: position in the archipelago; bottom: cross-section of the island and of the bulk volcanic edifice (no vertical exaggeration). Tekao volcano: (1) tholeiitic basalt flows; (2) hawaiitic flows. Tovii Plateau: (3) alkali basaltic and hawaiitic flows; (4) pyroclastic breccias. Taiohae volcano: (5) basal unit (alkali basalts and hawaiites); (6) middle unit (hawaiites, mugearites, benmoreites); (7) upper unit (hawaiites, mugearites, benmoreites); (8) late products emplaced outside the caldera; (9) intracaldera late products (alkali basalts, hawaiites); 10 trachytic extrusions; (11) intrusions; (12) strombolian cones; (13) crest lines; (14) collapses.

emerged part (20 km in diameter and 1224 m high, the second largest French Polynesian island after Tahiti) represents only 1.4% of the total initial volume of the shield volcano, estimated at 13 000 km^3. For comparison, the total volume of the Mehetia shield volcano is about 1000 km^3, still more than that of the largest intracontinental Neogene stratovolcano from Europe, the Cantal, French Massif Central (750 km^3). On the other hand, the volume of Mauna Loa (Hawaii), the largest shield volcano on Earth, reaches up to 42 500 km^3.

The Nuku Hiva volcanic edifice is built over a 3000-m-deep substratum which does not correspond to the top of the Pacific crust but that of the archipelagic apron of the Marquesas, 2000 m thick in the vicinity of Nuku Hiva. This apron is made up of volcanic debris reworked from the slopes of Nuku Hiva and neighbouring islands by huge submarine landslides (Filmer *et al.*, 1994). It overlies a less than 100-m-thick pile of pelagic sediments, which mark the roof of the 40-Myr-old oceanic crust, at a depth of 5000 m. Allowing for subsidence of the oceanic crust linked to the overload, the total height of the Nuku Hiva volcanic stucture at the end of its edification can be estimated at 7500–8000 m (Le Dez *et al.*, 1996). The shape of the submarine part of the shield (Figure 10.3, bottom) is dominated by gentle slopes of about 10°, abruptly interrupted by cliffs interpreted as the starting zones of huge gravity collapse landslides which fed the surrounding parts of the archipelagic apron.

The island represents the northern half of a basaltic shield, the Tekao volcano, the southern part of which collapsed after the end of volcanic activity. This gentle-sloped (10°) shield corresponds to a 1000 m pile of fluid tholeiitic basalt flows, the individual thickness of which rarely exceeds 2 m. Its basal part is 4.8 Myr old, and its summital flows (which are no longer tholeiitic but made up of transitional or even alkali basalts) have been dated to 4.2–4.1 Myr. Immediately after being built, the Tekao shield was affected by a caldera collapse at 4.1–4.0 Myr. The collapsed sector, 15 km in diameter, was partly filled by alkali basalt and hawaiitic flows and hydromagmatic breccias (Tovii plateau). A second collapse occurred, followed by the construction of a new volcano, the Taiohae, on the floor of the second caldera, between 3.8 and 3.6 Myr. The Taiohae lava represents a typical alkali basalt–hawaiite–mugearite–benmoreite–trachyte series (Table 2.7 and Figure 2.10). The central part of this volcano collapsed again, forming the third nested caldera (Figure 10.4). Finally, late eruptions and intrusions sporadically occurred inside and outside the youngest caldera up to 3.1 Myr. The age of the collapse of the southern half of the island is unknown. It is probably due to listric faults (Figure 10.4, bottom) linked to huge landslides such as those documented in Réunion Island, where the growth of the lava pile together with repeated intrusions led to increasing imbalance on the seaward side of the volcano, which collapsed through giant landslides bounded by rift zones on its unbuttressed flank (Gillot *et al.*, 1994).

The Nuku Hiva example illustrates the difficulties of any reconstruction of the geological (and petrologic) histories of large shield volcanoes based on small samples of material (mainly from the late stages of activity). The volcanoes from Hawaii (Mauna Loa and Kilauea) are typical of a further stage of growth (Figure 10.5), when the added intraplate lava pile doubled the original thickness of the oceanic crust. The resulting overload leads to a 4-km subsidence of this crust. The magmas, first segregated in diapirs uprising from the 100–80-km-deep melting zone into the asthenospheric mantle, continue their ascent through a fracture network into the lithospheric mantle and finally the ocean crust. Before their emplacement, they are temporarily stored in magma reservoirs located at depths of a few kilometres, i.e. within the volcanic edifice. Geophysical methods determined with precision the shape of these reservoirs below Kilauea (Figure 10.6). At depths greater than 5 km, the magmas are stored within relatively narrow channels corresponding to an interconnected fracture network (the volcano's 'plumbing system'). The magma reservoir underlying the summit crater is small, and its shape is controlled by the caldera's concentric fracture system. Most of the magma is stored within the large permanent reservoir located below the Eastern Kilauea rift zone, along which are

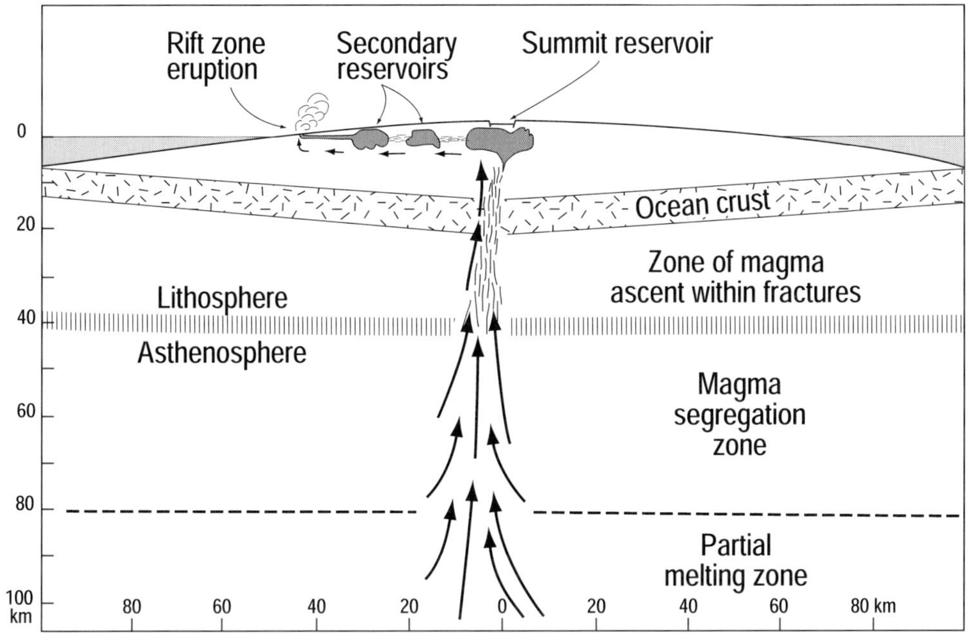

Fig. 10.5. Crustal and mantle structures beneath a large Hawaiian volcano (from Tilling and Dvorak, 1993). Based on seismic data obtained in Hawaii, this schematic cross-section shows the effects of overload exerted by the volcanic edifice on the ocean crust.

located the vents of the most recent eruptions. It corresponds to a coalescent open fracture system which extends over 12 km along the southeastern slope of the volcano.

The late stages of shield volcano building are well illustrated by the Nuku Hiva example. The rates of edification decrease from 10 to 1 mm/yr from the base of the Tekao shield to its top, and are correlated with decreasing partial melting degrees which lead from tholeiitic to transitional and finally alkali basalt magmas (§ 10.2.1). Concentric or sector collapse events lead to nested calderas. The post-shield (or post-caldera) alkali basalt volcanism which follows is quantitatively minor with respect to former tholeiitic activity (5% of the emerged part of Nuku Hiva, i.e. 0.07% of the total volume of the edifice). Late minor alkali basalt/basanite eruptions can occur well after the end of this stage (500 000 yr later in the case of Nuku Hiva), emplacing their products within an already degraded volcanic morphology (post-erosional stage).

10.1.5 The stage of subsidence: Tubuai (Australs)

The progressive subsidence of intraoceanic volcanic islands and their burial below reef limestones have been well known since Darwin (1842). The main cause of their subsidence is the increase in depth of the ocean floor due to cooling of the oceanic

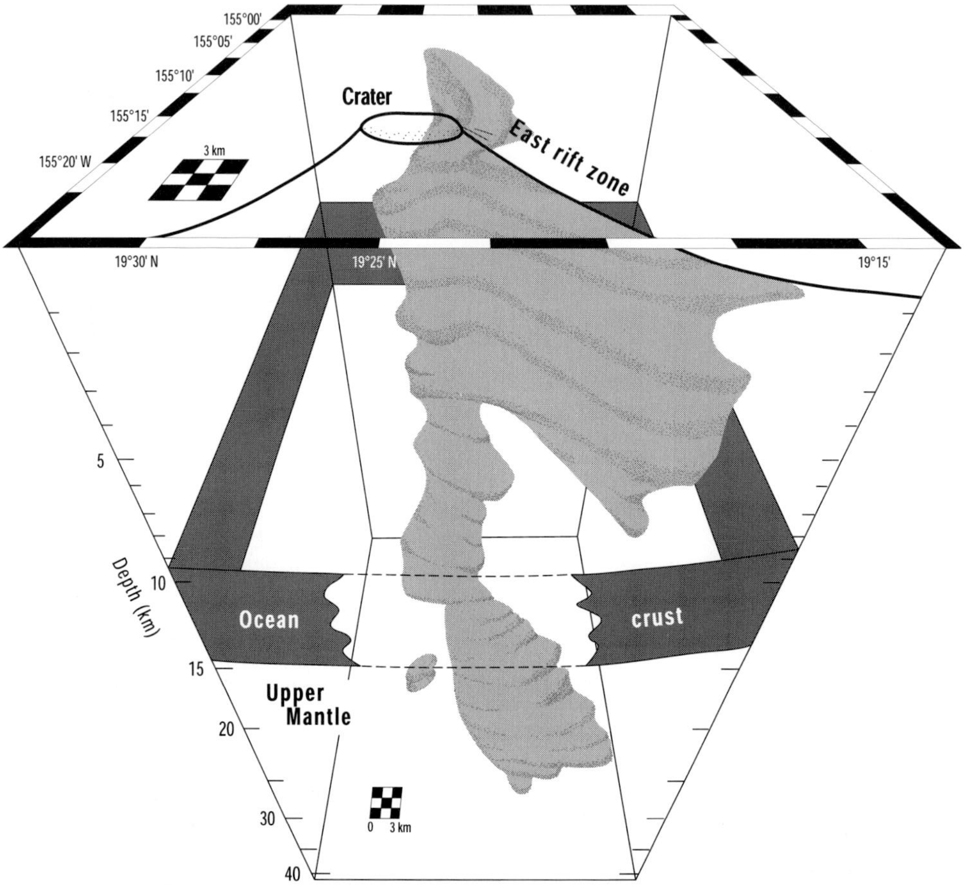

Fig. 10.6. The 'plumbing system' and magma reservoirs below Kilauea, Hawaii (from Ryan,1988). See text for explanation.

lithosphere moving away from the ridge and the hot spot (§ 1.3.1). The subsidence effect resulting from the overload exerted by the volcano itself is probably minor, as it occurs during its edification (Figure 10.5) and is thus likely compensated by further growth.

Tubuai island (Figures 10.7 and 10.8) is representative of numerous subsiding Polynesian volcanoes. Its shape is ovoid (9×6 km), and it is encircled by a barrier reef which isolates a 20-km-large lagoon of similar shape from the ocean (Figure 10.8). Behind the algal crest, which faces the open ocean, bioclastic sands composed of coral fragments are transported by wind-driven currents towards the western part of the lagoon, and they progressively bury the volcanic structures (Figure 10.8). The coast is delineated by a 5-m-high coral-sand beach ridge, which isolates swampy areas from the lagoon. The residual volcanic reliefs, which represent about one-half of the island surface, correspond to two edifices. In the eastern part, the main

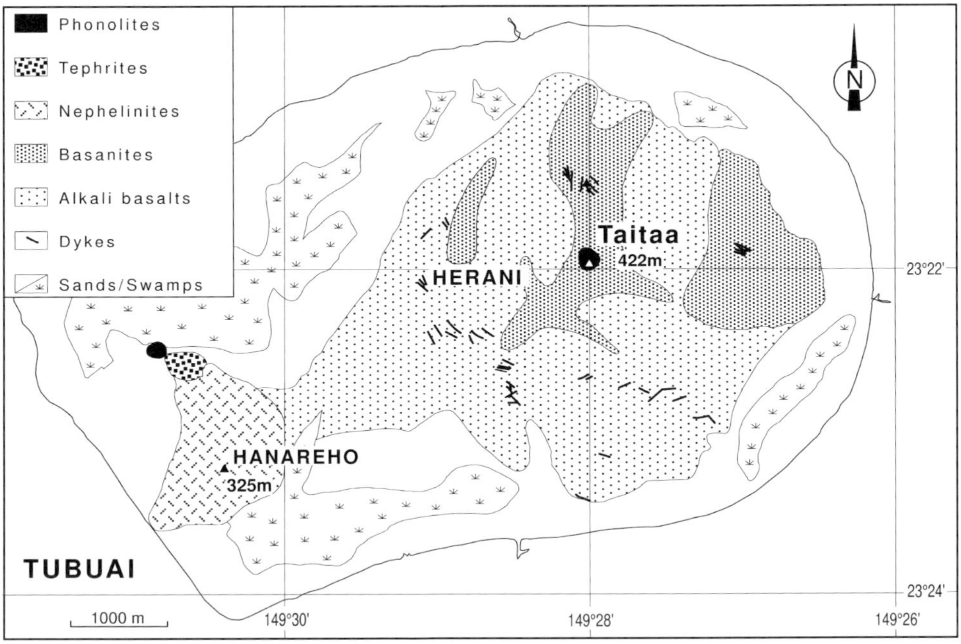

Fig. 10.7. Geology of Tubuai Island, Australs. Most of the two volcanic strutures (Herani and Hanareho) is buried below recent coral sands.

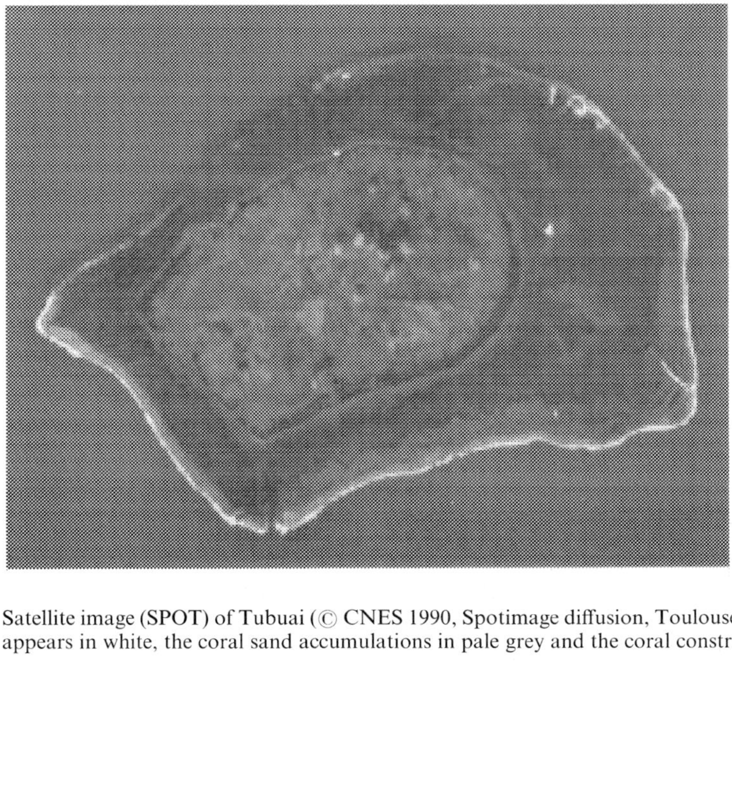

Fig. 10.8. Satellite image (SPOT) of Tubuai (© CNES 1990, Spotimage diffusion, Toulouse). The external algal crest appears in white, the coral sand accumulations in pale grey and the coral constructions in dark grey.

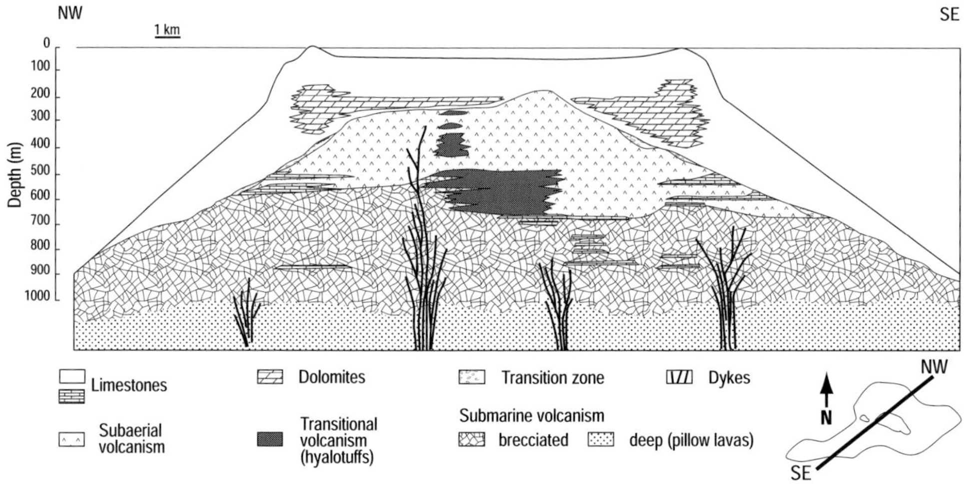

Fig. 10.9. Schematic NW–SE cross-section through the Mururoa atoll (from Buigues *et al.*, 1992). See text for explanation.

volcano of Herani was built between 12 and 10 Myr (Late Miocene). It is made up of a pile of alkali basalt flows overlain by basanitic flows (Table 2.4 and Figure 2.5). However, the summit of this volcano is a phonolitic plug, Mt Taitaa, whose chemical composition and age (9.5 Ma) are consistent with those of lava from the western volcanic centre, the Hanareho. Although younger (9.5 to 9 Myr) than the main volcano, the Hanareho remnants are limited to a crescent-shaped chain of hills, probably an ancient caldera wall. They are made of nephelinitic (Table 2.4) and tephritic lava flows intruded by phonolitic plugs.

10.1.6 The atoll stage: Mururoa (Pitcairn-Gambier)

The well-known Mururoa atoll (Figure 10.9; see also Guille *et al.*, 1995) has an elongated shape (28×10 km) with a general N80° trend parallel to the Australs Fracture Zone (Figure 10.1). Numerous drillings conducted through its upper 1000 m have greatly increased knowledge of its lithology and structure (Figure 10.9). The lower part of the drilled pile is composed of alkali basalt pillow flows. They are overlain, up to depths of 700–600 m, by submarine hyaloclastic breccias made up of glassy pillow fragments, which are in turn overlain by hydromagmatic Surtseyan-type breccias (hyalotuffs) derived from fragmentation of basaltic magmas within shallow sea water. This sequence is cut by numerous dykes grouped into four swarms which probably mark the position of previous rift zones. Subaerial alkali basalt flows locally interbedded with strombolian breccias coat the volcanic edifice, whose rather flat top is located at −300 m, and culminates at −200 m below the lagoon, where it is directly overlain by reef limestones. Below the barrier reef, the volcanic edifice is separated from the overlying carbonate rocks by a 'transition zone' composed of volcanic pebbles and debris in a carbonate matrix, which correspond to the ancient coastal line of the volcanic island. In addition, below the barrier, reef limestones have been converted to dolomites through interaction with briny waters.

Mururoa, although already an atoll, is only 12–10 Myr (Figure 10.2). Consequently, its upper carbonate pile is relatively thin: 300 m below the lagoon, 400–500 m below the barrier reef (Figure 10.9). Except in cases of abrupt variation in sea level (like those which led to its emersion at the peak of Quaternary glaciations) the coral reef barrier will continue to grow, compensating for progressive subsidence of the volcanic island. Geophysical data have shown that reef limestone piles of some Tuamotu atolls (Hao) are up to 2000 m thick.

10.2 ORIGIN AND EVOLUTION OF POLYNESIAN MAGMAS

The successive stages of building of shield volcanoes are marked by changes in magma composition which, although they remain within the range of OIB (Ocean Island Basalts) (Chapter 2), vary from tholeiitic to alkaline, with significant evolution of their isotopic signatures. Do these variations translate the structure and compositional variations of underlying plumes and hot spots, or alternatively do they mostly reflect the occurrence of small-size heterogeneities within the suboceanic mantle?

10.2.1 The example of Nuku Hiva

The composition of Nuku Hiva lavas, when plotted in an alkali-silica diagram (Figure 10.10a), shows two patterns of variation. The first corresponds to the alkali basalt–hawaiite–mugearite–benmoreite–trachyte sequence of the Taiohae volcano, a rather typical OIB series with a noticeable Daly gap at the level of mugearites. The other trend, almost perpendicular to the former in this diagram, traduces the transition from tholeiitic basalts dominant in the Tekao shield to transitional basalts, and then to alkali basalts and uncommon basanites in the Taiohae edifice and the products of the late eruptions. Figure 10.10b shows that transition from tholeiites to basanites corresponds to increases in incompatible element concentrations proportional to their degree of incompatibility, i.e. more accentuated on the left side of the multielement plot. Such variations, rather inconsistent with a fractional crystallization process which would lead to nearly parallel patterns, are usually taken as indicative of decreasing degrees in partial melting of an homogeneous source. Indeed, during batch partial melting, the most incompatible elements are transferred to the early-formed liquids, and later diluted when partial melting increases.

Other geochemical effects of combined partial melting and fractional crystallization processes in Nuku Hiva are shown in Figure 10.11. La and Nb, which have rather similar incompatibilies in basaltic systems, increase concomitantly from tholeiites to alkali basalts (decrease of partial melting degrees), then from alkali basalts to benmoreites by fractional crystallization (see also Table 2.7 and Figure 2.10). The points corresponding to trachytes are rather scattered due to the

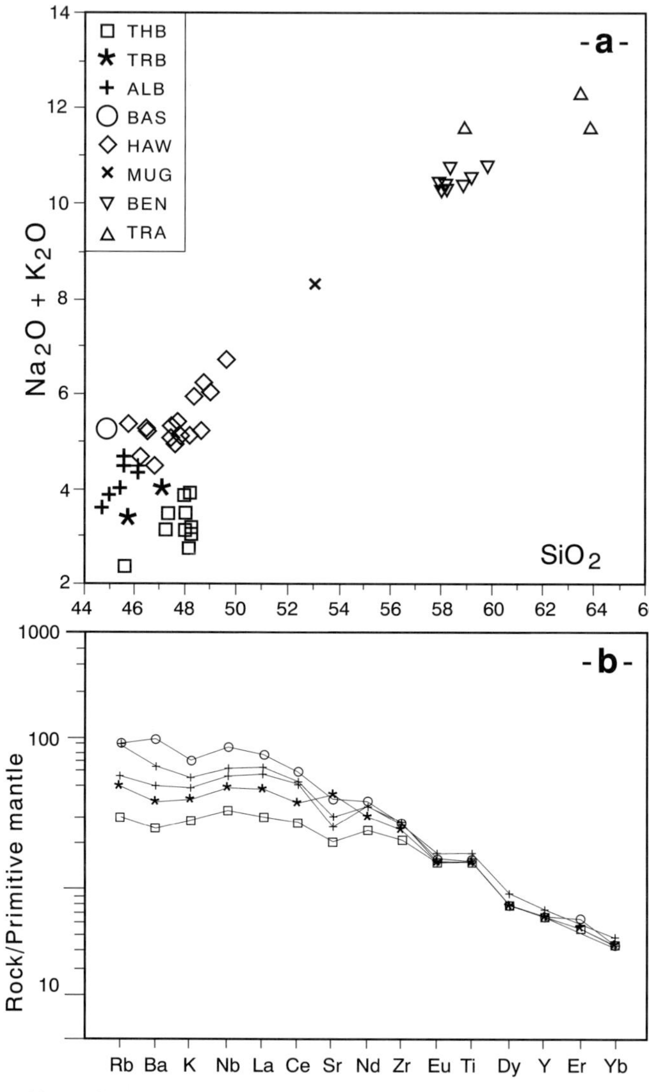

Fig. 10.10. Composition of Nuku Hiva lavas (from Le Dez *et al.*, 1996). (a) Alkali–silica plot; (b) primitive mantle-normalized multielement plots; THB: tholeiitic basalts; TRB: transitional basalts; ALB: alkali basalts; BAS: basanites; HAW: hawaiites; MUG: mugearites; BEN: benmoreites; TRA: trachytes.

incorporation of significant amounts of La into apatite and amphibole phenocrysts which fractionate in the benmoreitic magmas.

Fractional crystallization leads to a strong decrease in compatible elements and increase in incompatible elements. On the other hand, a decrease in partial melting degrees of an homogeneous lherzolitic source results in an increase in incompatible elements without noticeable variation of compatible elements, which are abundant in

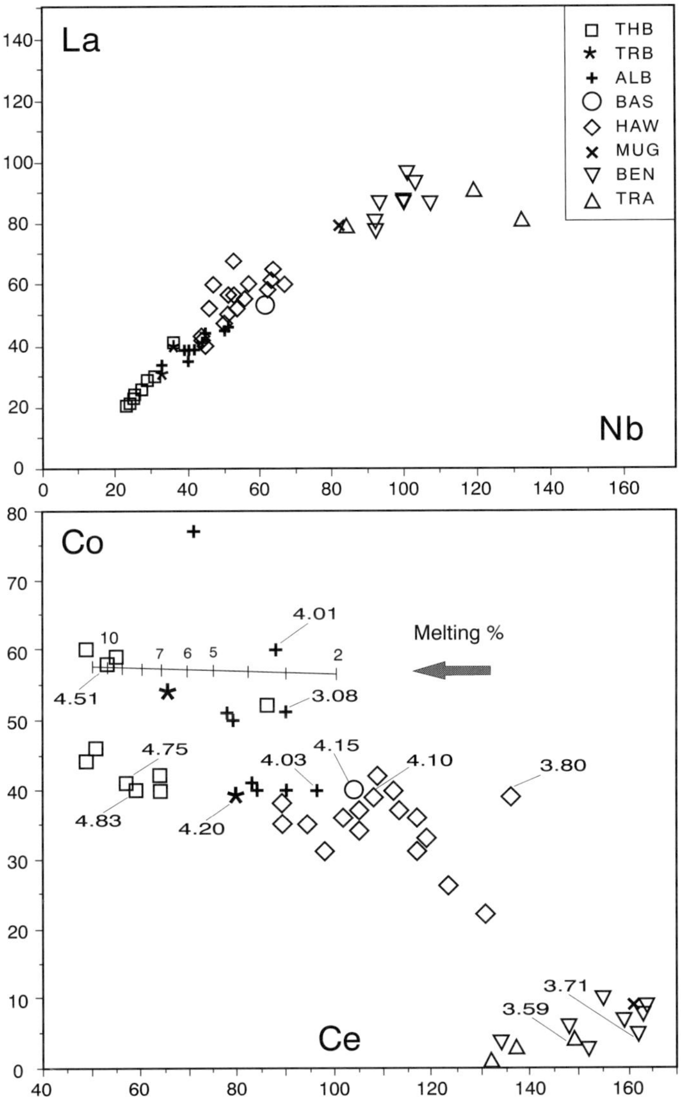

Fig. 10.11. Partial melting and fractional crystallization trends of Nuku Hiva magma (from Le Dez *et al.* 1996). Top: La/Nb plot showing the increase in these incompatible elements from basalts to trachytes. Bottom: Co/Ce plot showing the effects of partial melting (the numbers denote the ages of the samples, in Myr). Symbols as in Figure 10.10.

mantle minerals (Co and Ni in olivine, Cr in chromite and clinopyroxene). These two processes can be distinguished in a Co–Ce diagram (Figure 10.11, bottom): Co decreases and Ce increases through fractional crystallization from alkali basalts to trachytes, whereas Ce increases (at nearly constant Co contents) from tholeiites to alkali basalts and basanites. Calculated batch partial melting degrees of a lherzolitic

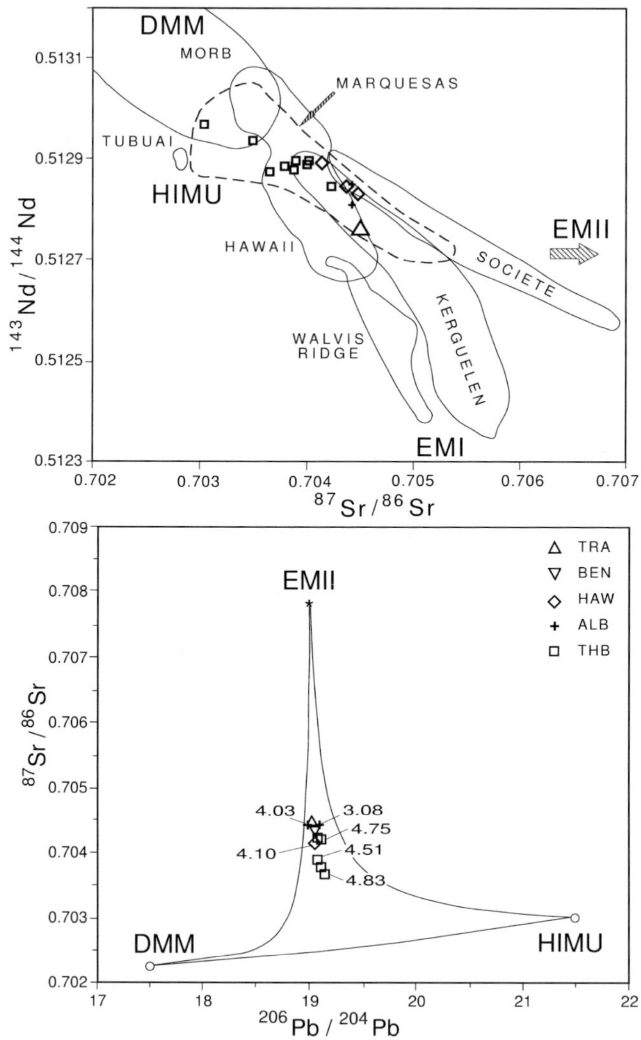

Fig. 10.12. Isotopic heterogeneity of Nuku Hiva lavas (from Le Dez *et al.*, 1996). Top: neodymium–strontium plot in comparison with other archipelagoes. Bottom: Evolution of isotopic compositions (numbers denote ages in Myr). Symbols as in Figure 10.10. See text for comments on DMM, EMI, EMII, HIMU.

source are also shown on this diagram: Nuku Hiva tholeiites correspond to degrees of 11% to 7%, and alkali basalts of 5%–2%. The decrease occurred between the main stage of building of the Tekao shield (4.8–4.5 Myr) and the end of its edification (4.2–4.1 Myr).

Nuku Hiva lavas are isotopically heterogeneous (Figure 10.12). In the Nd–Sr isotopic plot, they spread over the field of variation observed for lavas from the whole Marquesas archipelago. The heterogeneity observed within a single island is thus nearly as important as inter-island variations. The Nuku Hiva data are intermediate between three end-members (Zindler and Hart, 1986) corresponding respectively to: (i) the depleted mantle which represents the source of the MORB (DMM: Depleted MORB Mantle), (ii) a Tubuai-type enriched mantle pole with characteristically high U/Pb ratio (usually labelled μ, hence HIMU: HIgh MU), and finally (iii) another enriched mantle pole labelled EMII (Enriched Mantle II). The alkali basalts derive from a source enriched in EMII end-members with respect to the tholeiites which plot closer to DMM and HIMU end-members. However, the Sr–Pb plot in Figure 10.12 shows that isotopic variations occur rather independently of time and petrologic type. The early tholeiites emplaced at 4.83, 4.75 and 4.51 Myr are isotopically very heterogeneous, and the 4.75 Myr sample derived from a source rather similar to that of most alkali basalts.

10.2.2 Fractional crystallization effects

Intermediate and evolved lavas resulting from fractional crystallization processes occurring during tholeiitic stages of oceanic island activity are very uncommon. The very rapid growth of tholeiitic shields is probably responsible for this peculiarity, because of the very short residence time of basaltic magmas within the underlying reservoirs. On the other hand, intermediate and evolved lavas are commonly emplaced together with alkali basalts during the early and final stages of activity (pre-shield and post-shield). For instance, numerous mugearite dykes occur on Mehetia islet, and benmoreites and trachytes have been dredged from the Society hot spot. Trachytes and phonolites are common in Polynesian islands. The former are usually associated with alkali basalts (Moorea, Raiatea, Nuku Hiva, Eiao, Mururoa), and the phonolites to basanites or even nephelinites (Tubuai, Ua Pou in the Marquesas).

Fractional crystallization-related processes are quite diversified in Polynesian islands. In the apparently simplest cases, e.g. Mururoa (Maury *et al.*, 1992a), the lavas display constant initial isotopic Sr, Nd ratios which suggest limited interactions between evolving magmas and reservoir wallrocks. The case of the alkali basalt–trachyte series recovered from deep drillings in Eiao islet (Marquesas) is very different (Caroff *et al.*, 1995). Indeed, this series evolved by means of open-system fractional crystallization coupled with assimilation of materials from the ocean crust hosting the reservoirs. Some of its intermediate lavas (hawaiites and mugearites) have a lead isotopic signature closer to DMM than the associated basalts, which might indicate Assimilation coupled with Fractional Crystallization (AFC) processes involving assimilation of MORB-type basalts previously enriched in lead by hydrothermal sulphides.

Another case of rather complex evolution is shown in Figure 10.13. The lavas from Moorea, an island very close to Tahiti, range from alkali basalts to trachyphonolites and trachytes. There is a lot of petrographic and geochemical evidence for a 'regular' fractional crystallization process as shown by the abrupt decrease of MgO versus SiO_2 and Ni versus Th, driven by olivine fractionation (dotted curves). However, many intermediate lavas (hawaiites, mugearites) display

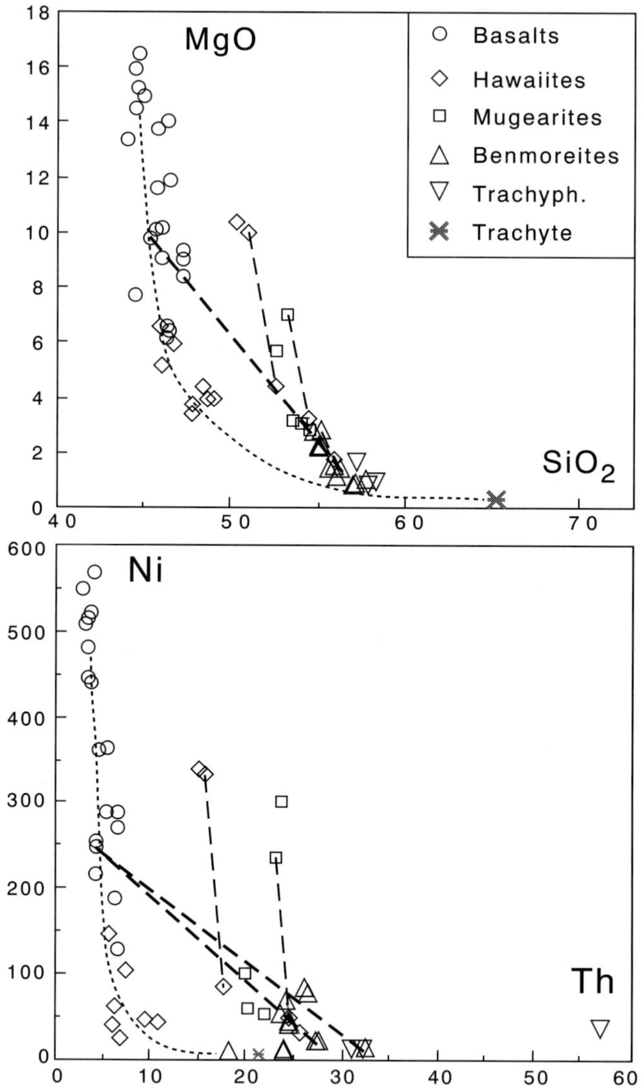

Fig. 10.13. Chemical variations in Moorea (Society) lavas illustrating the roles of fractional crystallization and magma mixing (from Le Dez, 1996).

abnormally high MgO and Ni content. These rocks contain numerous xenocrysts of Mg-rich olivine and clinopyroxene, often destabilized, which indicates that they derive from the mixing of evolved trachyphonolitic liquids with primitive basaltic magmas reinjected into the reservoir (see Figure 2.7). The bold dashed lines in Figure 10.13 indicate this process which, combined with mechanical accumulation of phenocrysts (normal dashed lines), accounts for the abnormal enrichment in compatible elements.

10.2.3 Partial melting effects

The example of Nuku Hiva, which is representative of most Polynesian islands containing both tholeiitic and alkali basalts (Eiao, Ua Pou, Marquesas; Fangataufa, Pitcairn-Gambier), shows that a significant decrease in degree of batch partial melting of a lherzolitic source accounts for the transition from tholeiites (11–7%) to alkali basalts (5–2%; Figure 10.11). Tubuai Island (Australs) illustrates variations linked to partial melting leading to alkali-rich magmas, from alkali basalts towards nephelinites. The Tubuai mafic lavas display a temporal increase of alkalinity (and silica-undersaturation (Figure 10.7). The main volcano (Herani) is composed of alkali basalt flows overlain by basanitic flows. This pile is crosscut by a number of later dykes, including highly silica-undersaturated trachybasanites (15 to 30% of normative feldspathoids) which are called analcitites because they contain phenocrysts of olivine and calcic clinopyroxene in a groundmass almost exclusively made up of amiboidal analcite patches. Finally, the mafic flows from the younger Hanareho volcano are nephelinites with more than 30% of normative nepheline plus leucite.

These four groups are easily distinguished in the alkali-silica diagram of Figure 10.14. The corresponding multielement patterns (Figure 2.5) show strong increases in concentrations of the most incompatible elements from alkali basalts to nephelinites, which suggests that the four groups might derive from variable degrees of partial melting. All the Tubuai lavas are isotopically homogeneous, and thus likely derive from the melting of a single source or type of source characterized by a strong HIMU signature. Figure 10.15 shows very good positive correlations between the highly incompatible elements Th, La and Nb. However, these correlations are

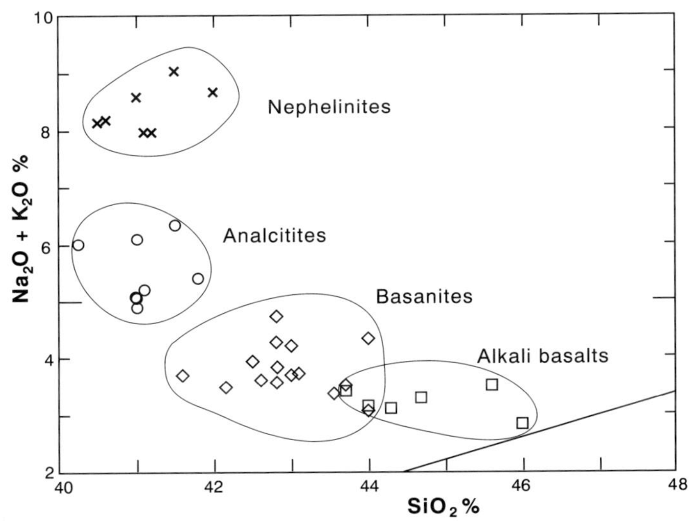

Fig. 10.14. Chemical variations in mafic magma from Tubuai (Australs) (alkali-silica plot from Caroff *et al.*, 1997). See text for explanations.

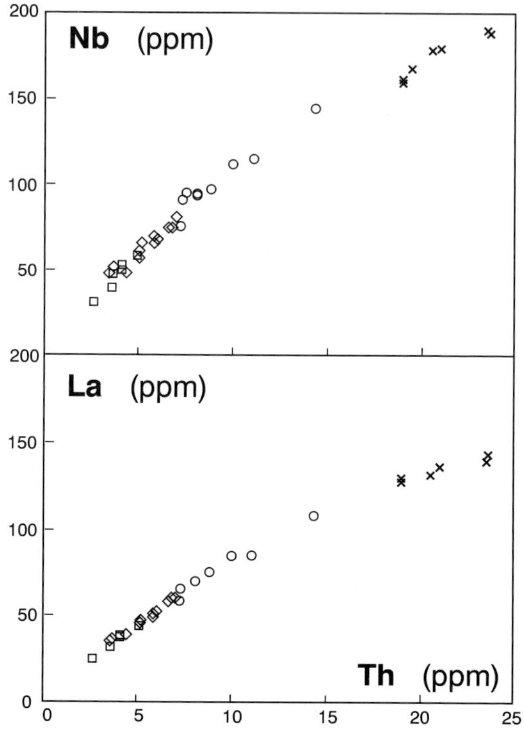

Fig. 10.15. Nb/Th and La/Th plots illustrating partial melting effects in Tubuai (from Caroff *et al.*, 1997). Symbols as in Figure 10.14.

not straight lines consistent with fractional crystallization but curves typical of variable partial melting degrees: Th, slightly more incompatible than Nb and La (Table 2.2), is selectively concentrated with respect to these elements in the liquids corresponding to the lowest degrees of partial melting, i.e. the nephelinites. A more sophisticated model than that shown in Figure 10.11 should propose a decrease of batch partial melting degrees from 8–5% for alkali basalts to 3–1,5% for the nephelinites (Caroff *et al.*, 1997). These latter magmas thus derive from very low partial melting degrees, close to the minimum values consistent with their extraction from the mantle source zone (§ 2.2c).

10.2.4 Mantle source heterogeneities

The example of Nuku Hiva demonstrates the occurrence of different isotopic signatures within mafic magmas from a single island, and even within those belonging to a single geological and petrologic unit (tholeiites from the Tekao shield). The isotopic variations observed for the young lavas from Polynesian archipelagoes have a range greater than those observed in the Hawaiian chain (Figure 10.12), allowing detailed study of corresponding signatures (Vidal *et al.*, 1987).

The ranges in variation of $^{87}Sr/^{86}Sr$ and $^{206}Pb/^{204}Pb$ in the main Polynesian linear chains are shown in Figure 10.16. As in the ($^{143}Nd/^{144}Nd$)/($^{87}Sr/^{86}Sr$) diagram of Figure 10.12, the fields clearly depart from that of the MORB-source mantle (labelled DMM end-member for Depleted MORB Mantle) and even from the field of enriched MORB like those from certain parts of the east-Pacific rise (EPR field). To account for the variability of these isotopic compositions (Figures 10.12 and 10.16), it is possible to define a minimum of three theoretical mantle end-members (Zindler and Hart, 1986): EMI (Enriched Mantle type I), EMII (Enriched Mantle type II) and HIMU (HIgh MU, μ symbolizing the U/Pb ratio).

The petrogenetic significance and even the actual occurrence as mantle rocks of these end-members are largely debated. Indeed, with the exception of some Austral-Cook islands characterized by a nearly pristine HIMU signature (Mangaia, the extreme case, Rimatara, Tubuai, Raivavae, and to a lesser extent Rurutu; Woodhead *et al.*, 1996), the Polynesian OIB derive from sources representing rather complex blends of DMM, EMI or EMII and HIMU end-members.

Mixes between DMM and materials rich in incompatible elements (e.g. the continental crust or detrital sediments) will get, with respect to DMM, higher Rb/Sr ratios (Rb being more incompatible than Sr) and U/Pb ratios, uranium being more incompatible than lead (Table 2.2). On the other hand, Nd being more incompatible than Sm, such mixes will have Sm/Nd ratios lower than that of DMM. With time, such materials will display stronger increases in their $^{87}Sr/^{86}Sr$ and $^{206}Pb/^{204}Pb$ ratios with respect to $^{143}Nd/^{144}Nd$. This applies rather well to the transition from DMM to EMII. The EMII end-member, characterized by very high (for mantle rocks) $^{87}Sr/^{86}Sr$ ratios together with rather low $^{143}Nd/^{144}Nd$, is usually interpreted as reflecting the ancient (>1.5 Gyr) incorporation of continental crust and/or derived sediments within depleted mantle. Precambrian subductions could be responsible for this incorporation, possibly through tectonic erosion of an active continental margin or subduction of detrital sediments. An alternative possibility includes the delamination of the continental lithosphere followed by its entrainment at depth by cold sinking plumes (McKenzie and O'Nions, 1995). This end-member contributes to the isotopic signatures of lavas from the Society and Marquesas archipelagoes (Figure 10.16). These lavas have relatively low Nb/U ratios, also consistent with a continental crust contribution to their source.

The EMI end-member differs from EMII by its lower $^{87}Sr/^{86}Sr$ ratios, extremely low $^{143}Nd/^{144}Nd$ and its lack of enrichment in $^{206}Pb/^{204}Pb$ with respect to DMM. This enriched pole, which plays a considerable role in the signature of the sources of Pitcairn and Rarotonga lavas (Figure 10.16), derives from the ancient incorporation into the mantle of materials with Rb/Sr, U/Pb and Sm/Nd lower than those of detrital sediments. Subducted Precambrian pelagic sediments could be at the root of the EMI signature, although the contribution of sinking subcontinental lithosphere has also been proposed, and the origin of this end-member is less constrained than those of the others (Hofmann, 1997).

The last enriched end-member, HIMU, presents the apparently contradictory features of low $^{87}Sr/^{86}Sr$ ratios, only slightly higher than those of DMM, together with extremely high $^{206}Pb/^{204}Pb$ ratios. It is hardly possible to decouple in such a

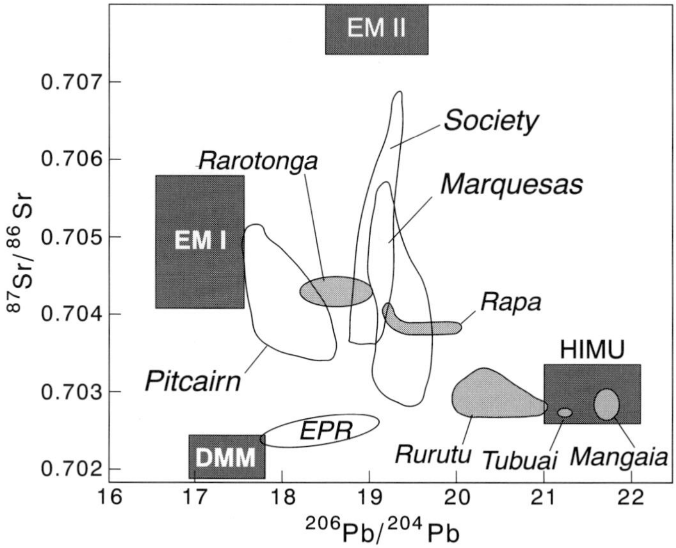

Fig. 10.16. Sr–Pb isotopic plot showing the heterogeneity of Polynesian OIB (from Le Dez, 1996). For comparison, EPR field of MORB from the east-Pacific rise.

way Rb/Sr from U/Pb during truly magmatic processes, because Rb and U on the one hand, Sr and Pb on the other, have amost similar levels of incompatibility (Table 2.2). Several hypotheses concerning the origin(s) of this decoupling have been proposed. The most widely admitted (Chauvel *et al.*, 1992; Woodhead, 1996; Kogiso *et al.*, 1997) involve a very old subduction (2 Gyr for Tubuai and Mangaia) of oceanic crust, which would have suffered either: (i) hydrothermal alteration leading to selective leaching of Rb and input of uranium, or (ii) dehydration or, more likely, partial melting during subduction (§ 11.3), resulting in Rb and Pb enrichment with respect to Sr and U, respectively. The Austral-Cook chain is famous for the HIMU signatures of its lavas, which are moreover variable from one island to another and even within a single island (from the old to the young volcanic episodes in Rurutu). The HIMU lavas are clearly enriched in Nb with respect to elements of similar incompatibility (Figure 2.5). This feature is consistent with the hypothesis of a contribution of subducted ocean crust, as Nb is not transferred in the arc mantle by hydrous fluids emanating from the subducted slab (§ 11.3.5).

Numerous authors have proposed geochemical models for hot spots, usually based on simple assumptions regarding the location of end-members within the mantle. For instance, postulating that DMM is located within the suboceanic lithospheric mantle and that EMII and HIMU are of deep-seated origin and hosted by the ascending plume, the changing isotopic signature of lavas from an island or even a chain might reflect either variable melting degrees of the mantle rocks hosting the end-member signature, or alternatively the stages of migration of the island or volcanic chain above an heterogeneous plume. The major problem with such models is that they are poorly constrained regarding the vertical and horizontal locations of

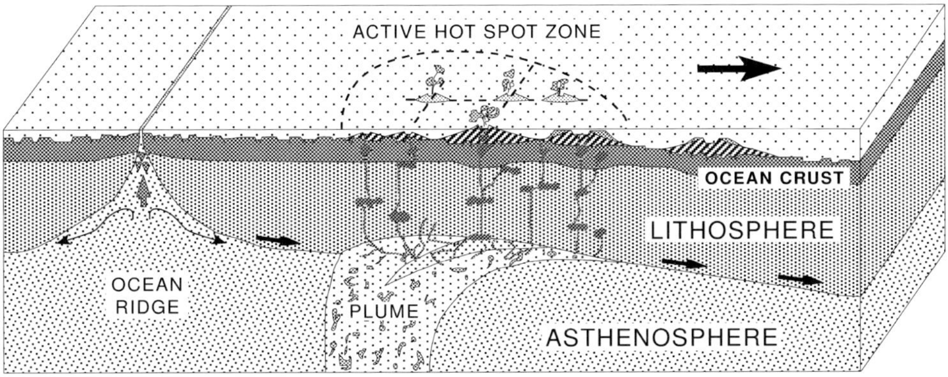

Fig. 10.17. Drawing of a plume–hot-spot system showing source heterogeneity and possible interactions with the lithosphere (from Le Dez, 1996).

the various end-members, and thus allow a prolific number of hypotheses. For instance, the origin of the DMM signature of OIB lavas may be related to: (i) the melting of the base of a DMM oceanic lithosphere above the plume, (ii) the melting of depleted DMM parts of an heterogeneous plume, which could be either of deep-seated origin or alternatively incorporated in the plume during its ascent, or finally (iii) the contamination of plume-derived magmas ascending through a DMM oceanic lithosphere. Similarly, the enriched end-members EMI/ EMII and HIMU are generally thought to be of deep-seated and plume-related origin, but their signature might also have been transferred to the oceanic lithosphere when it passed over another hot spot/plume. For instance, parts of the Pitcairn-Gambier lithospheric mantle might have been enriched when they passed over the Easter Island hot spot (Dupuy *et al.*, 1993). Various kinds of zoned plumes have also been proposed, with HIMU cores and EM rims, or EM cores and DMM rims, etc.

Besides their rather speculative character, geochemical hot-spot models usually rely on the hypothesis that heterogeneities in the mantle are of relatively large size (>10 km), especially within the plume: so that contemporaneous islands (Marquesas) or seamounts (Society) sample different sources, or that a given volcanic edifice migrating over the hot spot will collect successively contrasted signatures. On the other hand, recent data tend to show that lava isotopic signatures vary abruptly with time, sometimes independently of partial melting degrees (Nuku Hiva) which are basically responsible for petrologic changes (e.g. tholeiites/alkali basalts/nephelinites). Polynesian island signatures seem rather consistent with the contribution of plumes with small-scale heterogeneities as shown in Figure 10.17. An enriched plume including small-scale domains carrying EM and HIMU signatures rises through a depleted asthenosphere (DMM), incorporating parts of it. Its top is deflected by lithospheric motions, and suffers partial melting generating basaltic magmas that uprise though the DMM lithosphere, which contributes to accentuate their DMM signature. If partial melting degrees are higher in the central part of the plume (see Figure 10.22), the successive stages of edification of a theoretical island passing over

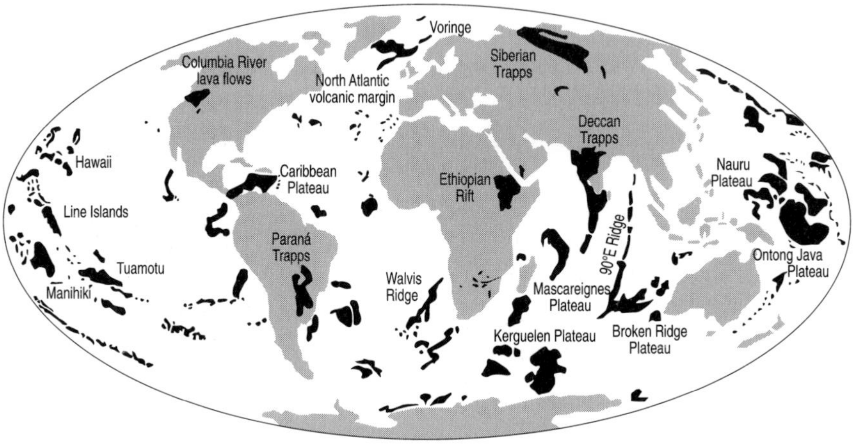

Fig. 10.18. Distribution of Large Igneous Provinces including oceanic plateaus (from Coffin and Heldholm, 1993a).

the hot spot should be initially alkalic (seamount) then tholeiitic (shield-building stage), and finally alkalic again (post-shield stage).

10.3 THE OCEANIC PLATEAUS: GEOLOGY AND PETROLOGY

Intraplate oceanic magmatism is not exclusively represented by Hawaiian or Polynesian-type linear chains of islands and seamounts. In some linear chains, volcanic edifices are numerous and close together (Tuamotu). Aseismic volcanic ridges like the Walvis Ridge—made up of alkali basalts—in the South Atlantic or the Ninetyeast Ridge in the Indian Ocean (Figure 10.18) can be considered as linear chains of almost coalescent seamounts formed by plume/hot-spot activity. Young basaltic fluid flows have been emplaced over the oceanic crust near Hawaii and in French Polynesia away from conventional hot spots. During the early Cretaceous, paroxysmal intraplate magmatic activity emplaced, mostly as fluid submarine lava flows, huge quantities of basalts as oceanic plateaus.

10.3.1 Location and ages of oceanic plateaus

The occurrence of Large Igneous Provinces (LIP) in continental intraplate domains is well known: they correspond to the basaltic 'traps' of Siberia, the Deccan in India, the Paraná in Brazil, the Columbia River in the western USA (Figure 10.18). Although of different ages, these large intracontinental basaltic provinces share a number of common features.

—Large volume: 1.3×10^6 km^3 for the Columbia River basalts, 6×10^6 km^3 for the dispersed fragments of the Paleocene North Atlantic Province ('Thulean

Province' from the west of the British Isles, Greenland and the Faroe Islands), 10^7 km^3 for the Deccan traps.

—High rates of magma production: the volumes of the individual fluid lava flows of the Columbia River reach up to several hundreds of km^3 (whereas the great 1783 fissural eruption of Laki in Iceland emplaced only 15 km3 of basalt).

—Very short periods of activity with respect to their sizes: 1.5 Myr for the Miocene Columbia River plateau, 1 Myr for the Deccan traps at the K-T boundary, less than 3 Myr for the Paleocene Thulean Province.

Over the last 20 years, progress in exploration of the oceans has shown the importance of a second type of LIP, which represent the submerged equivalent of the traps: the oceanic plateaus, the locations of which are shown in Figure 10.18.

The oceanic plateaus have sizes and volumes even larger than those of the traps (Coffin and Eldholm, 1993b): 2×10^6 km^2 and more than 36×10^6 km^3 for the Ontong Java plateau, 2×10^6 km^2 and 20×10^6 km^3 for the Kerguelen plateau, 800 000 km^2 for the Caribbean plateau before its fragmentation, 650 000 km^2 for the Manihiki plateau, 360 000 km^2 for the Nauru plateau.

Like the traps, oceanic plateaus are characterized by very high emplacement rates: 4.5 Myr for the Kerguelen plateau, and less than 3 Myr for the Ontong Java plateau, which in the latter case implies the emission of 12 to 15 km^3 of magma per year, i.e. almost as much as the world ridge system (16 to 26 km^3/yr since the Late Jurassic).

In contrast with the continental traps, most of the major oceanic plateaus were emplaced during a quite short period in the Early Cretaceous: 127–131 Myr for Nauru, 121–124 Myr for Ontong Java, 110–114 Myr for Kerguelen, 107–111 Myr for Manihiki and the start of the Caribbean Plateau activity. Their emplacement corresponds to a clear Early Cretaceous paroxysmal intraplate activity in the Pacific and Indian Oceans (Floyd, 1989; Coffin and Eldholm, 1993b).

The oceanic plateaus form reliefs 2500 to 3500 m higher than the surrounding abyssal plains (hence their name). This gives only a very minimal estimation of their thickness, because of local subsidence of the underlying oceanic crust in response to the overload. Seismic data show total crustal thickness of ca. 20 km on average, i.e. more than twice the normal thickness of oceanic crust. Their ages are often only slightly younger than those of their substratum: by 20 Myr for Nauru, 5 to 15 Myr for Ontong Java and Manihiki. Thus, they likely formed over young oceanic crust, in a near-ridge environment. After their formation, hot-spot-type activity persisted in a number of them, until 90 Myr in Ontong Java, 75 Myr in Nauru, and until the Quaternary in the Kerguelen archipelago.

10.3.2 A large oceanic plateau: Ontong Java

The Ontong Java plateau is located in the West Pacific (Figure 10.18). It covers an area of more than 2 million km^2, and its crustal thickness reaches up to 39 km. Although this last value exceeds the thickness of normal continental crust, seismic and gravimetric studies demonstrate its mafic nature: it corresponds to oceanic crust overthickened by addition of basaltic materials. The magmatic units of this huge structure, almost totally below sea level, are poorly known from three oceanic drill-holes (the deepest of which, ODP site 807, penetrates them for 149 m) and field studies of two islands from the Solomon archipelago, Santa Isabel and Malaita. The summital units of the plateau, 4 km thick, outcrop in Malaita as a pile of massive and pillow basaltic flows. Individual flow thickness are 20 m on average, and they reach up to 70 m. They are very fluid submarine lava flows, the glassy upper and lower parts are quenched by seawater. There are only a few sedimentary interbedded units, as thin clay-rich or pelagic limestone levels deposited near the carbonate compensation depth during the Valanginian, i.e. 2–3 km. All these features are typical of a very quick rate of emplacement. The submarine lava sequence is nearly fresh, having likely been protected from interaction with seawater by the high rate of growth of the pile. The basalts are subaphyric (<5 modal% phenocrysts), vesicle free, and their petrographic features and mineralogical composition are much like those of MORB.

The available ages of emplacement of the plateau lavas are surprisingly constant: 121 to 124 Myr (but of course only a minute part of the structure has been sampled!). It is obviously difficult to understand why the units from the top of the plateau were emplaced in a rather deep submarine environment given the total crustal thickness. Seismic data suggest that the lower parts of Ontong Java plateau are made of dense rocks, e.g. picrites, gabbros or garnet-bearing mafic granulites. It is possible that most of the crustal thickening occurred via accumulation of mafic magmas below the pre-existing oceanic crust (Saunders *et al.*, 1996).

10.3.3 The fragmented Caribbean Plateau

The distribution of mafic Cretaceous volcanic massifs in the Caribbean is shown in Figure 10.19. The Caribbean basin's oceanic crust is of probable Jurassic age. It formed in an East Pacific position (Farallon plate), as did the oceanic plateau overlying it, which likely derived from a precursor of the Galapagos hot spot. Then the corresponding plateau lithosphere collided with the Eastern Pacific subduction zone, and the 'Caribbean indenter' was inserted between the Americas, the Atlantic lithosphere subducting then below its eastern edge (the Antilles arc). Most of the Caribbean plateau is still in an ocean bottom position, especially in the Colombian basin and the Beata ridge where it has been drilled (DSDP sites 146 to 153, Figure 10.19), and possibly in the substratum of the Lesser Antilles island arc. Other parts of this plateau, the initial area of which is estimated at 800 000 km^2, were tectonically accreted to the peri-Caribbean chains. They occur (Figure 10.19) in the Colombian cordilleras and in Gorgona island, famous for its komatiites; in the Nicoya and Santa Elena peninsulas (Costa Rica); in Aruba and Curaçao Islands, north of the Venezuela coast; in Jamaica; and finally in Hispaniola, in the southern peninsula of Haiti (Dumisseau Formation) and the Dominican Republic (Duarte Complex). The ages of plateau formation are still debated because most of the rocks suffered hydrothermal or metamorphic alteration. However, the paroxysmal plateau-building stage occurred during the Turonian (90–88 Myr).

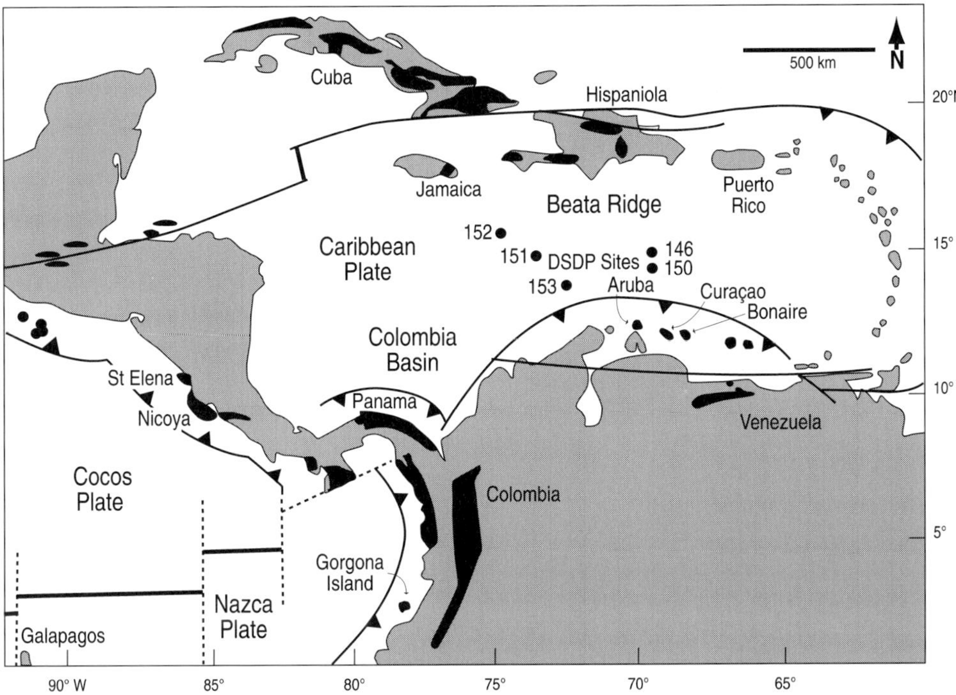

Fig. 10.19. The Cretaceous magmatic province of Central America, including the remnants of the Caribbean Plateau (from Kerr *et al.*, 1997). Magmatic units are shown in black. Numbers 146, 150, 151, 152, 153 refer to the DSDP drilling sites in the Caribbean Sea.

The Caribbean Plateau magmatic rocks are known from the study of DSDP samples and of the emerged fragments of the plateau, and indirectly from the interpretation of seismic data in the Caribbean Sea. Basalts from DSDP sites and from southern Haiti occur either as fluid or pillow-lava flows. Their petrologic and geochemical features are generally MORB-like. The lower units are incompatible element-depleted, and the upper ones enriched. Picritic basalts with flat REE patterns and rather homogeneous isotopic signatures intermediate between MORB and OIB are very common. Depleted and enriched lavas occur together in Gorgona island, where highly depleted picritic lavas (komatiites) occur (§ 2.4.4). The remnants of the Caribbean plateau found in the Colombian cordilleras and in Hispaniola (Duarte) enable reconstruction of its crustal structure (Fig 10.20). About 20 km thick, it contains thick gabbroic layers, perhaps inherited from the initial oceanic crust. They are overlain by a 3-km-thick komatiitic and picritic layer highly unusual for a normal oceanic crust, and finally by ca. 8 km of pillow submarine flows, cross-cut by numerous gabbroic intrusions (Kerr *et al.*, 1997).

Seismic refraction and reflexion data provide a complementary (and rather different) view of the Caribbean crust, the thickness of which seems extremely variable: 14 to 16 km in the Colombian basin and near Haiti (Fig 10.19), but less

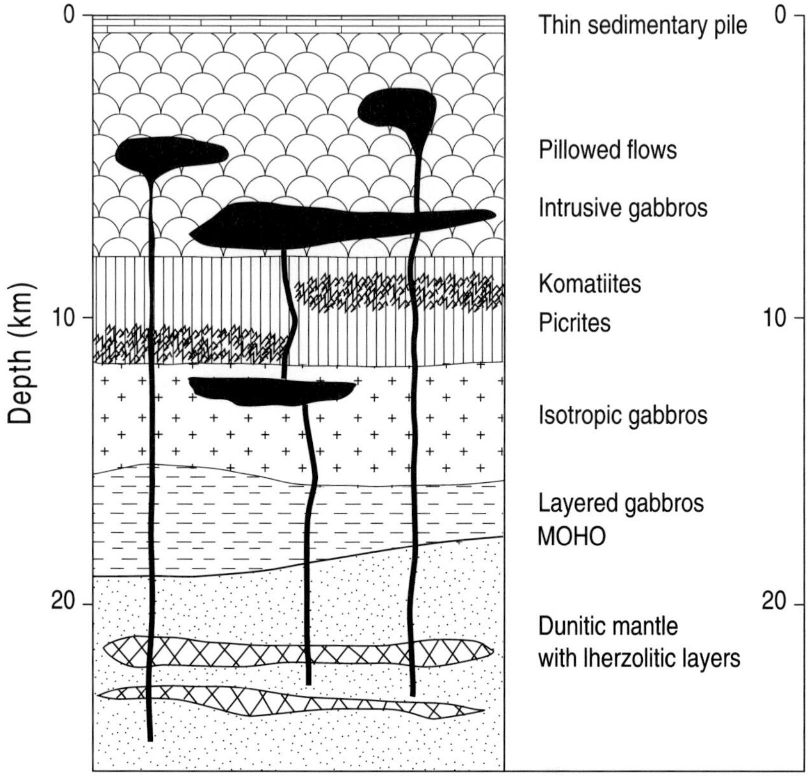

Fig. 10.20. Hypothetic structure of the Caribbean plateau crust (from Kerr *et al.*, 1997).

than 5 km north of the Venezuelan coast (Mauffret and Leroy, 1997). The basaltic layer 2 is rather thin (2–4 km), even at the level of the Beata Ridge high. Layer 3 (presumably gabbros and cumulates) shows major variation in thickness, which reaches up to 13-14 km near Nicaragua and below the Beata ridge. It contains a reflector suggesting the occurrence of dense rocks (picrites or komatiites?) in its lower part. It seems that the crustal thickening observed in oceanic plateaus may have several origins: piling up of fluid lava flows on the ocean bottom, gabbroic or picritic intrusions in the deep crust, or accumulation of these materials below the initial oceanic crust (magmatic underplating).

10.3.4 Origin of oceanic plateau magmas

Oceanic Plateau Basalts (OPB) share many N-MORB features (§ 2.4.3). They differ, however, from N-MORB only by : (i) the frequency of Mg-rich compositions (picrites, Gorgona komatiites), (ii) their relative enrichment in Rb, Ba, K, Th, LREE, Nb, Ta and Ti (Table 2.4, Figure 2.5), and (iii) their higher $^{87}Sr/^{86}Sr$ and lower $^{143}Nd/^{144}Nd$ ratios, which resemble those of OIB (Figure 10.21). The OPB are

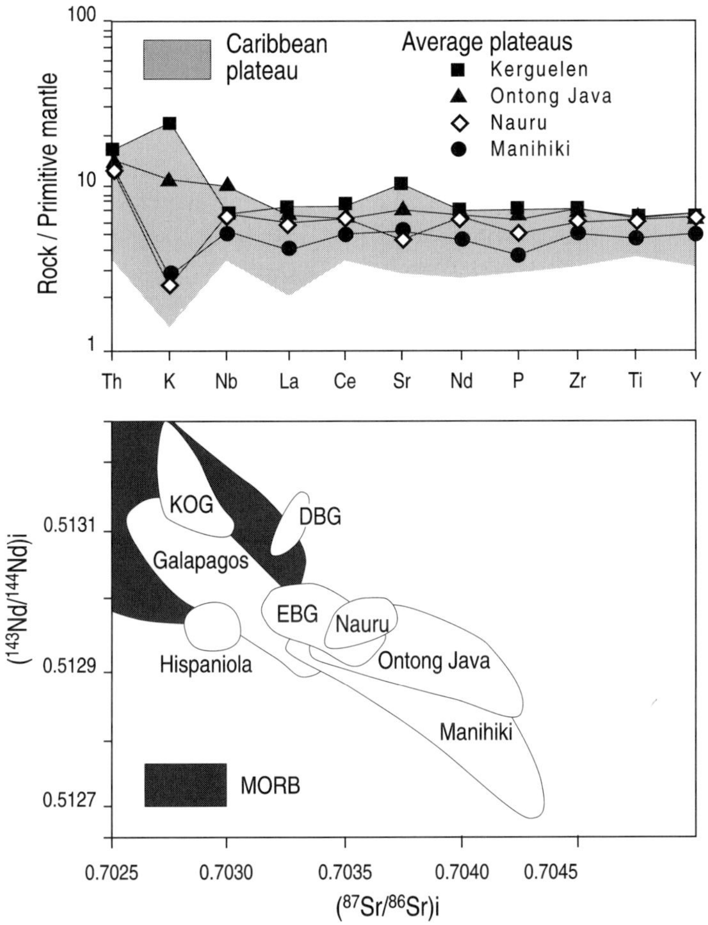

Fig. 10.21. Composition of magmatic rocks from oceanic plateaus (from Kerr *et al.*, 1997). KOG: Komatiites from Gorgona; DBG and EBG, depleted and enriched basalts from Gorgona.

much less homogeneous, however, than the MORB. Although their average composition usually displays flat multielement patterns (Figure 10.21, top), two groups of basalts can be distinguished in most cases: depleted basalts, picritic or not, usually at the base of the sequence, and, overlying them, enriched basalts and picrites with an OIB-type isotopic signature. Depleted basalts and komatiites from Gorgona derive from sources isotopically very close to those of N-MORB (Figure 10.21, bottom). MgO content of primitive picritic and komatiitic liquids is 16–18%, corresponding to partial melting degrees of mantle rocks ranging from 25 to 30% (Kerr *et al.*, 1996). Such melting degrees require, under pressures of 4 to 5 GPa, temperatures of ca. 1600°C, i.e. 300°C more than considered usual in asthenospheric mantle melting models. Taking into account the above two constraints (abnormally high temperatures and unusually heterogeneous geochemical signatures) leads us to consider that OPB derive from a deep and very hot mantle, presumably rising towards the surface as plumes. The tectonic settings of ocean plateaus and their intracontinental equivalents, the traps, are consistent with this hypothesis.

10.4 PLUMES, HOT SPOTS AND HOT LINES

The pattern of evolution of Pacific shield volcanoes toward atolls was described in 1842 by Charles Darwin. Dana (1849) interpreted the morphologies of Hawaiian islands as indicating a migration of volcanic activity toward the southeast. The basic ingredients of the presently widely admitted plume/hot-spot model have been discussed by Wilson (1963) and Morgan (1971, 1972).

10.4.1 A 'simple' hot-spot model

Hot spots represent the loci, at the top of the asthenosphere, of thermal and/or chemical transfers linked to the ascent of deep mantle plumes which result from thermal convection. Morgan (1972) proposed for mantle plumes a cylindric shape and a relatively narrow diameter of 100–200 km. These features are in very good agreement with the shape and size of the areas of active volcanism, which are often, and rather confusingly, also called hot spots. For instance, the so-called 'Society hot spot' (Figure 10.3), is a circular area 150 km in diameter, where present-day eruptions occur. These are the superficial effects of the activity of the genuine hot spot, which corresponds to the underlying partial melting zone located at the top of the asthenosphere. The existence of a topographic high (the bulge) over the hot spot, as an effect of heating of the oceanic lithosphere, is confirmed by bathymetric data (Hawaii, Society). Finally, seismic tomography has recently imaged a mantle plume below Iceland as a vertical cylinder 150 km in radius, extending down to at least 400 km (Wolfe *et al.*, 1997) and presumably originating from the lower mantle (Shen *et al.*, 1998).

Two kinds of plumes can be envisioned: exclusively thermal plumes (transferring heat but not materials from the deep mantle towards the surface), and chemical plumes (with coupled transfers of heat and matter). The chemical-plume model (Morgan, 1971, 1972) is now accepted because it accounts for the peculiarities of the OIB sources (EMI, EMII, HIMU end-members) with respect to composition of the suboceanic mantle. Morgan also proposed that plumes originate from the D″ layer, at a depth of 2900 km. This hypothesis, which implies whole-scale convection in the sublithospheric mantle, is also adopted in most recent Earth models (Chapter 12).

The conditions of magma genesis near the top of an ascending chemical plume are illustrated in Figure 10.22 (Wyllie, 1988). Melting results from the adiabatic uprise of plume materials: during decompression, their P-T path crosses the peridotite solidus (Figure 2.1). If the mantle contains traces of water ($<0.1\%$) and CO_2, they segregate as vapour at depths of ca. 300 km. Partial melting starts at the same level, in the hottest central part of the plume, and its temperature increases during ascent (dynamic melting). Magmas are only extracted from source peridotites at the end

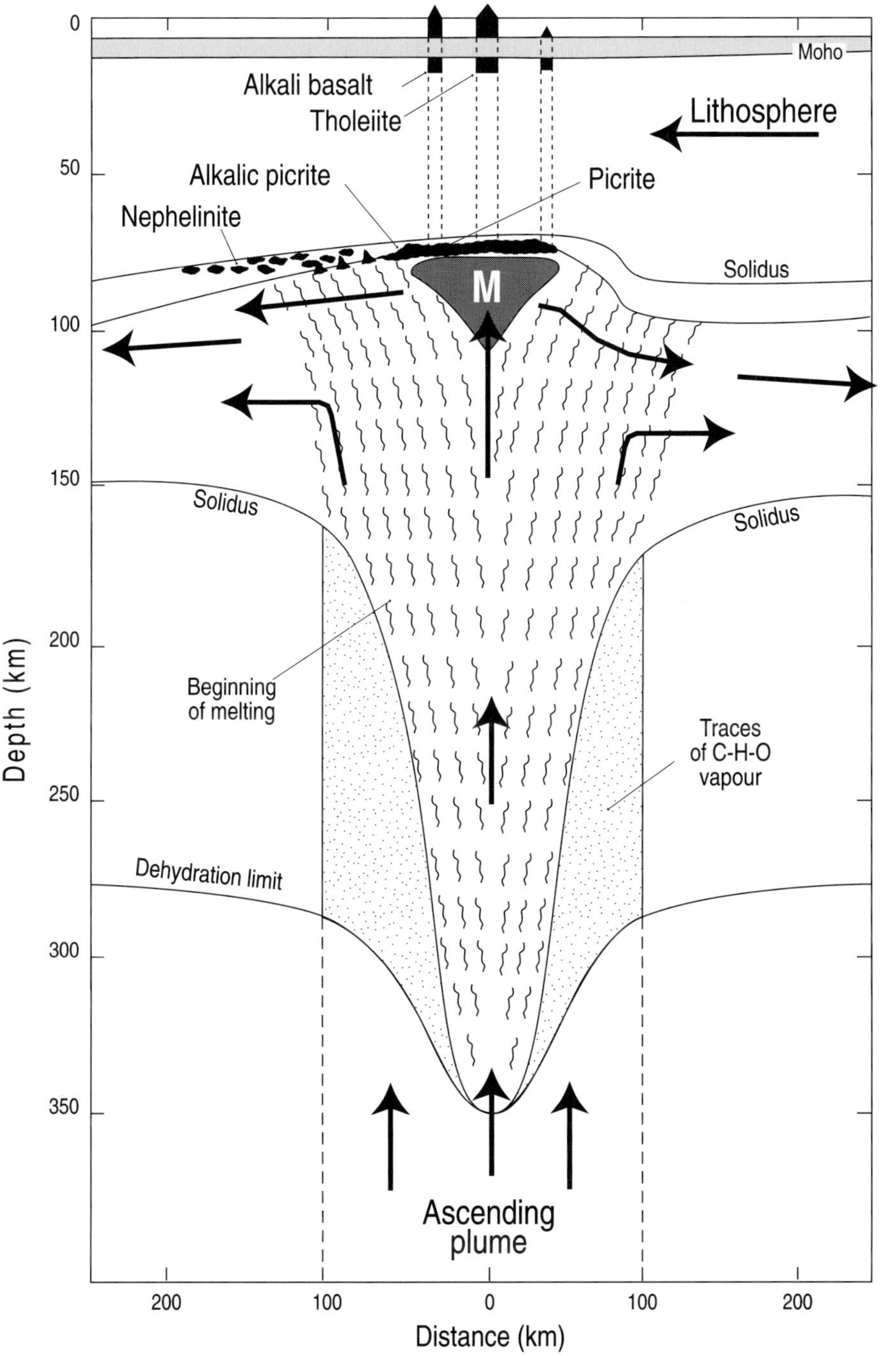

Fig. 10.22. Magma production in an ascending mantle plume (from Wyllie, 1988). M: zone of magma storage near the top of the plume.

of melting, which is thus adequately described by batch melting laws (§ 2.2.2). In the central part of the plume, they result from relatively high melting degrees (tholeiites), which decrease towards the periphery (generating alkali basalts, then nephelinites). These primary magmas are Mg-rich (picritic). They pond near the lithosphere–asthenosphere boundary, where early mineral fractionation occurs, before being emplaced at the surface (tholeiites above the central part of the hot spot, alkali basalts above its rims).

This model presents several original features. Most of the magma is stored in the central part of the hot spot (zone M): it will be used for the building of the shield. In addition, the peridotite solidus curve penetrates the basal part of the lithospheric mantle, which can thus contribute to magma genesis. Finally, the top of the plume is deflected by the overlying lithospheric motion: thus the early-emplaced magmas come from the central part of the plume, but the late alkalic volcanism can persist even after the island has migrated away from the hot spot-area (post-erosional nephelinites).

This general model, mostly based on the Hawaii example, cannot of course account for the petrologic and geochemical diversity of hot-spot magmatism. For instance, the Society linear chain lacks tholeiitic composition, even in its largest shield volcanoes (Tahiti Nui, Raiatea) and above its hot spot. Tholeiitic basalts are, on the other hand, rather common in the Marquesas. In the Pitcairn–Gambier linear chain, the active volcanoes overlying the hot spot and the neighbouring Pitcairn island are composed of alkali basalts, Gambier islands are exclusively tholeiitic, Mururoa is alkalic and Fangataufa shows a transition from tholeiitic to alkalic. The specific features of a given hot spot can thus significantly alter the general pattern depicted in Figure 10.22.

10.4.2 The complexity of hot spots

Numerous variants of the hot-spot model have been explored. Griffiths and Campbell (1990) performed analogue experiments on chemical plumes (Figure 10.23). The initial step (a) is the formation of a hot sphere (the plume 'head") which uprises through cold material but remains connected to its source by a narrow cylinder, its 'stem' or 'tail' which extends during head ascent. Thermal diffusion allows an important fraction of colder host materials to be incorporated within the plume head (b and c).

However, mixing between the hot and cold material is of limited extent, and internal convection within the plume's head leads to its chemical zonation. When the plume reaches the bottom of the overlying lithosphere, its top spreads out beneath it (d).

The popularity of this model derives partly from the elegant explanation that it provides for OIB source heterogeneity: enriched components would have a deep-seated origin and depleted components (DMM) would be incorporated within the ascending plume head. In addition, great variability of source blends can be expected, leading to the complex combination of EM, HIMU, DMM end-members evidenced from OIB isotopic studies. Another important point is the support

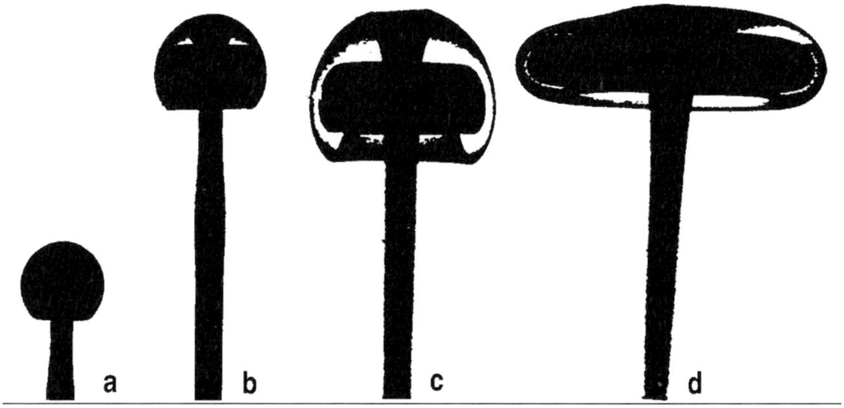

Fig. 10.23. Analogue model of the evolution of chemical plumes (from Griffiths and Campbell, 1990). Photographs of the ascent of coloured glucose syrup heated at 80°C within colourless glucose syrup at 20°C.

provided by this hypothesis to a unique mode of emplacement for oceanic plateaus and linear chains. The large size and high rate of emplacement of the former are consistent, for Griffiths and Campbell (1990) and their numerous followers, with the large outspread head of a plume reaching the base of the oceanic lithosphere. After the main stage of plateau formation, the plume tail would remain active, leading to a 'conventional' hot-spot volcanism creating a linear chain. Such a model is consistent with the geology and chronology of oceanic plateaus (§ 10.3.1).

Another interesting variant of the hot-spot model is the behaviour of a plume reaching the base of the oceanic lithosphere (Richards and Griffiths, 1989). It will experience transverse shear stress due to overlying lithospheric motion and will be deflected in the direction of this motion. Corresponding analogue models suggest that an important fraction of cold host material (sometimes more than 50%) can be incorporated within the plume head during this process. It will concentrate in the central part of the plume (Figure 10.17). This model adequately describes the final stages of the evolution of an ascending plume. It is also consistent with the isotopic heterogeneity of OIB and provides an explanation for the decreasing contribution of DMM, presumably incorporated in the plume during its ascent, from the shield-building step to the later stages (§ 10.2.4).

Another refinement of the Wilson–Morgan plume model concerns the interaction between the plume and the overlying oceanic lithosphere. The weakness or fracture zones of the latter may represent preferred ways for magma ascent. The rift-zone pattern of submarine volcanoes corresponds to the fracture trends of the underlying crust (§ 10.1.2). The Mururoa atoll (§ 10.1.2), located close to the Austral fracture zone (Figure 10.1), is elongated along its trend and prolongated along the same direction by a submarine volcanic plateau 130 km long and 30 km wide. McNutt *et al.*, (1989) have proposed an original explanation for the geological peculiarities of the Marquesas archipelago (lack of present-day volcanic activity, but concentration of young lavas near the southeast edge of the chain, close to the Marquesas fracture zone; random age patterns in the central part of the archipelago; unusual N140° trend of the latter, corresponding to the Paleocene Pacific motion at the time of formation of the underlying oceanic crust). For these authors, the Marquesas hot spot would be 'weak', its magmas being able only to reach the surface episodically, using the fracture network of the oceanic lithosphere. In such a case the surface pattern of intraplate magmatism would be mostly controlled by tectonic features of the underlying oceanic lithosphere.

10.4.3 Ridge–hot spot interactions

The activity of an ocean ridge can be significantly modified by a neighbouring or underlying plume/hot spot system, as demonstrated by the present-day example of Iceland. Such a situation was rather common during the Early Cretaceous, at the time of formation of the major oceanic plateaus. The 1800-km-long Foundation linear chain, southeast of Easter Island and west of the Pacific–Antarctic Ridge (PAR; Figure 10.24), was recently discovered using altimetric data. This linear chain extends to the ridge that it intercepts at 37°30′S. It includes 47 seamounts, the main ones shown in Figure 10.24 reaching depths of 300–600 m only. The Foundation volcanoes have been emplaced over a very young oceanic lithosphere (less than 10 Myr). Recent bathymetric data show significant variation of seamount shapes with their distance from the ridge. Far away from the PAR, they have typical star-shaped rift-zone patterns similar to those of the Society hot-spot area. Closer to the ridge, their shape changes towards that of elongated ridges oblique with respect to the direction of spreading. Finally, in the immediate vicinity of the PAR, these ridges

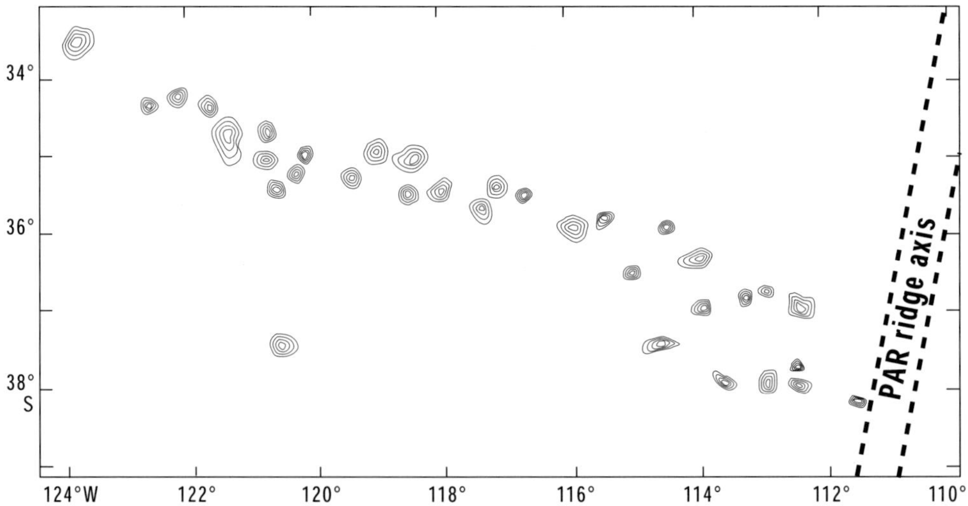

Fig. 10.24. The Foundation intraplate linear chain. Simplified bathymetric chart showing a 1800-km-long chain of seamounts close to the Pacific–Antarctic Ridge (PAR).

trend perpendicular to it and acquire a very smooth shape. These variations are connected with petrologic and geochemical changes. Far away from the PAR, the seamounts are made up of typical OIB, the sources of which contain HIMU and EMI end-members. At less than 200 km from the PAR axis, compositions intermediate between OIB and MORB start to occur. Finally, the MORB geochemical imprint becomes prominent closer to the axis, indicating the occurrence of asthenospheric fluxes connecting the hot spot with the neighbouring ridge.

10.4.4 Hot lines

The term 'hot line' was coined by Bonatti and Harrison (1976) to describe the seamount and island linear chain extending over 3500 km on the Nazca plate, from the east Pacific rise at the level of Easter Island towards San Felix and San Ambrosio, two islands located near the Chilean coast (Figure 10.25). This linear chain differs from the usual type by the occurrence of lavas younger than 3 Myr both at its western tip (Easter and Sala y Gomez islands), its eastern tip (San Felix and San Ambrosio), and its central part, over the fossil Nazca ridge (vertical arrows in Figure 10.25). Bonatti and Harrison proposed an explanation based on the occurrence in the asthenospheric upper mantle of convective cells perpendicular to

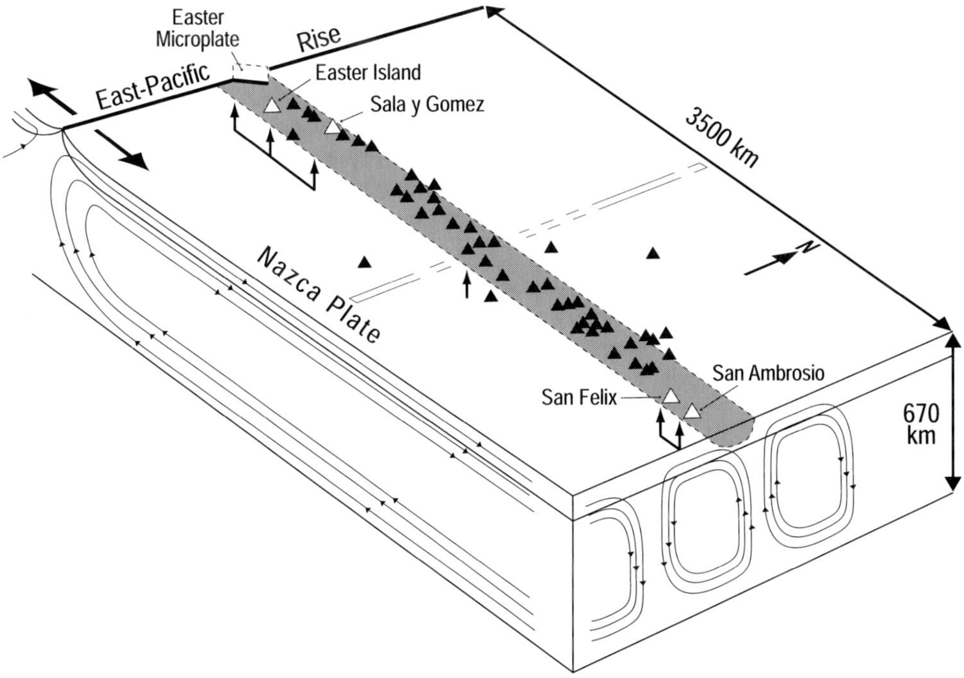

Fig. 10.25. The Easter Island hot line (from Bonatti *et al.*, 1977). Vertical arrows indicate young volcanic areas (< 3 Myr).

the ridge axis (Richter, 1973). Strike–slip constraints due to lithosphere motions would result into helicoidal convective motions alternatively clockwise and anti-clockwise. The ascending path located below Easter Island would allow episodic uprise of deep hot mantle from the base of the upper mantle (Figure 10.25), and its partial melting during adiabatic decompression.

The Austral–Cook linear chain has often been considered a hot line because of the occurrence of Quaternary volcanics at both its ends and in its central part (Rurutu; Figure 10.2). It differs from the Easter case, however, by its location far away from the East-Pacific Rise and the presence at its southeast end of a 'conventional' hot spot below the MacDonald seamount. The occurrence of two other hot spots located at 1000-km intervals along the linear chain derived from the MacDonald, approximately below Rurutu and Rarotonga, seems rather unlikely (McNutt *et al.*, 1997). It has, however, received some geochemical support: the young Rurutu lavas have a geochemical signature different from the old ones (likely formed at the level of the MacDonald hot spot), and owe their HIMU flavour to their contamination by the oceanic lithosphere previously enriched during the Miocene magmatic event (Chauvel *et al.*, 1997). The passage of a lithospheric plate over a hot spot creates a weakness zone within it, thus creating a preferred route for ascending magmas. If the South Pacific superswell overlies a large dying plume head (§ 10.4.5), melting zones too weak to have permanent magmatic surface effects could exist in some parts of the asthenosphere. The ascent of their magma towards the ocean bottom could be made easier, either by Richter's convective roll system, or by a lithospheric weakness zone (track of MacDonald hot spot). A somewhat different interpretation of the Austral–Cook linear chain is illustrated in Figure 10.26. According to this model, the young volcanic activities far away from the MacDonald would be due to the extraction of hot mantle materials from the top of the MacDonald plume, and especially from the part deflected by Pacific plate motion. These materials would

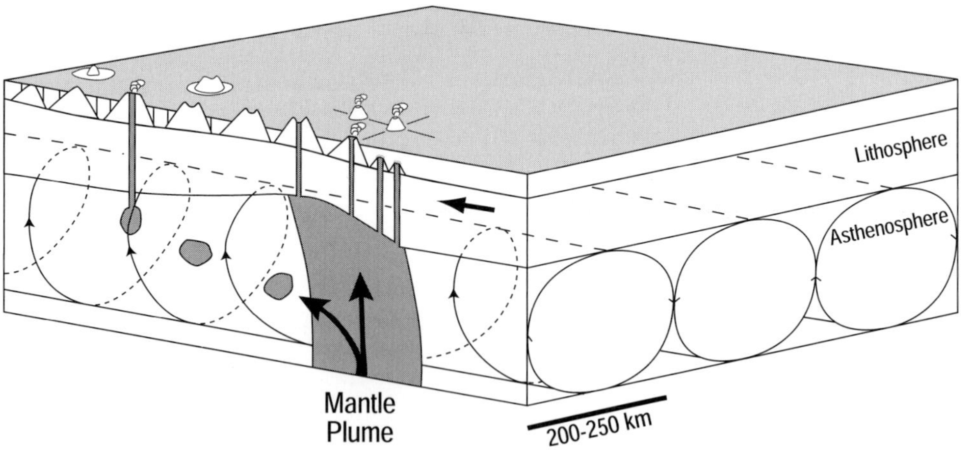

Fig. 10.26. A tentative model of a complex hot spot (from Diraison, 1991).

migrate below the chain, transported by Richter's convective rolls. Their partial melting would occur episodically, as in the case of the Easter hot line, when they are dragged upwards by ascending fluxes between two adjacent rolls.

10.4.5 Origin and evolution of hot spots

Present-day hot spots are not randomly distributed below the Earth's lithosphere. A plate kinematic reconstruction of Pangea during the Early Jurassic (Figure 10.27) shows that it fits quite well with the position of hot spots at that time and also with the Atlantic–Africa zone of positive residual geoid anomalies. Another large zone of positive geoid anomalies fits the location of hot spots in the Equatorial Pacific. For Anderson (1982) and Le Pichon and Huchon (1984), these two zones would result from heat accumulation within the mantle underlying the Pangea and Pacifica supercontinents which remained stable for ca. 100 Myr. This excess heat would have conditioned mantle convection, facilitating plume uprise and leading to the fragmentation of the supercontinents during the Middle and Late Jurassic. The Early Cretaceous magmatic crisis was concomitant with dispersion of the fragments of these supercontinents.

The existence of correlations between the positions of the active hot spots and those of LIP is largely accepted (Richards and Griffiths, 1989). For instance, the Réunion hot spot is connected to the Deccan trapps by a very long submarine

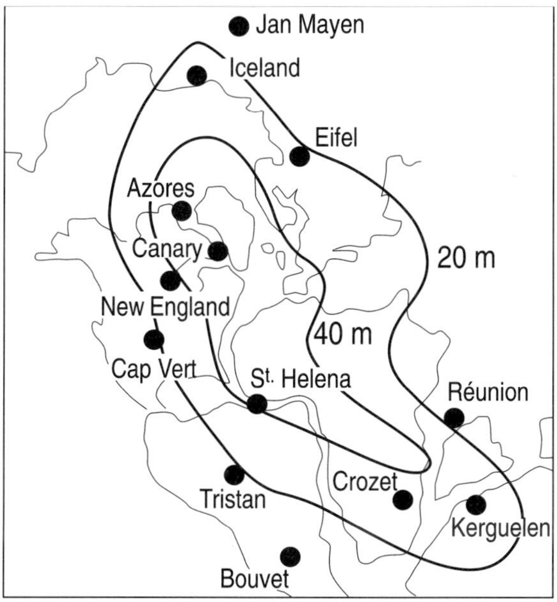

Fig. 10.27. Reconstruction of the Early Jurassic (200 Myr) position of Pangea, together with hot-spot locations (from Anderson, 1982). The curves denote positive residual geoid isoanomalies (+ 20 m and + 40 m) in the Atlantic–Africa sector.

volcanic linear chain including the Mascareigne plateau and the Chagos-Laccadive aseismic ridge (Richards *et al.*, 1989). The Caribbean plateau is, similarly, connected to the Galapagos hot spot, and the Ontong Java plateau to the Louisville hot spot. In the Central and Western Pacific, an intraplate volcanic trend consistent with Pacific plate motions since the Early Cretaceous connects the Polynesian archipelagoes with the Manihiki plateau. It is thus possible to consider the Polynesian hot spots and the associated 'superswell' as the remnants of a 'superplume' which contributed to the emplacement of the huge Pacific oceanic plateaus during the Early Cretaceous.

The problem of the depth of plume-source zones is discussed in Chapter 12 in connection with mantle convection models. Numerous geochemical arguments support a very deep-seated origin in the lower mantle or even the D″ layer. Indeed, the previously discussed mantle end-members (DMM, EMI, EMII, HIMU) evolve with time according to their respective Rb/Sr, Sm/Nd, U/Pb and Th/U ratios. These evolutions, recalculated back by 1 to 3 Gyr, converge towards a common composition called FOZO (FOcus ZOne; Hart *et al.*, 1992). The FOZO component, the signature of which is never prominent in OIB (like, for instance, the HIMU imprint in Tubuai and Mangaia; Figure 10.16), is characterized by low $^{87}Sr/^{86}Sr$ together with high $^{144}Nd/^{143}Nd$, $^{206}Pb/^{204}Pb$ and $^{3}He/^{4}He$ ratios. This end-member, which does not fit any known component of the crust and upper mantle, could be representative of the lower mantle (as well as another end-member known as 'C' for Common Component). It could be periodically transferred into the upper mantle by ascending plumes, especially during tectono-magmatic crises (Chapter 12).

11

The fate of the oceanic crust: from subduction to recycling

The active volcanic arcs which underline the converging plate boundaries have a total length of ca. 30 000 km, i.e. about half the total length of the active ocean ridges. The geochemical characteristics of their products, and especially their strontium, neodymium, lead and beryllium isotopic ratios, indicate that they carry back to the Earth's surface chemical components from the ocean basaltic crust and its associated sediments which have been buried down into the mantle by subduction processes. Although the occurrence of recycling processes responsible for these transfers is widely accepted, their modalities are still a matter of debate. Especially, the behaviour of major and trace elements during recycling and the corresponding chemical budgets remain largely unknown. The recycling processes are likely variable from one volcanic arc to another depending on subduction parameters, e.g. the age of the subducting slab. The ultimate fate of slab residues buried within the lower mantle and the effective contribution of arc magmatism to continental growth are frequently debated and remain largely speculative.

11.1 A TYPICAL VOLCANIC ARC: THE SUNDA ARC IN JAVA (WESTERN INDONESIA)

11.1.1 Tectonic setting

The Sunda Arc represents one of the most striking geological features of the Earth. It is some 5200 km long including its eastern prolongation (sometimes referred to as the Banda Arc) and it includes more than 500 active volcanoes, many of them reaching altitudes of ca. 3000 m. Its most famous historical pyroclastic eruptions are those of Krakatoa (Sunda Strait, 20 km^3 of magma emitted in 1883) and Tambora on Sumbawa Island (150 km^3 emitted in 1815). However, the most voluminous Quaternary eruption was that of the Toba Caldera ignimbrite which emplaced 2800 km^3 of rhyolitic pumices 74 000 years ago, whereas the famous 1902 pyroclastic

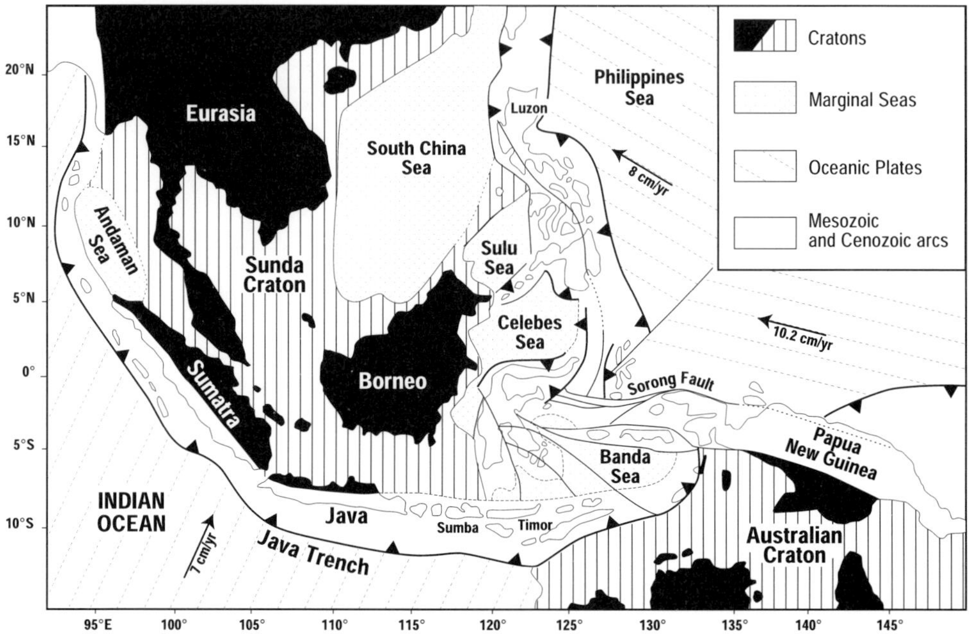

Fig. 11.1. Structural context of the Sunda Arc (from Simandjuntak and Barber, 1996).

eruptions of Mt Pelée in Martinique (Lesser Antilles) emitted nearly 0.2 km^3! The Sunda Arc results from the relatively fast (7 cm/year) subduction of the Indian Ocean lithosphere beneath the Eurasian margin (Figure 11.1). The latter corresponds, at the level of Sumatra and Northern Java, to the Sunda craton, which includes Permo-Triassic terranes and abundant Cretaceous granites. The age of the Indian oceanic crust entering the subduction zone increases progressively from Northern Sumatra (50 Myr) to the islands east of Java (160 Myr). East of Timor Island the Sunda Arc collides with the Australian Craton without interrupting its magmatic activity.

The Sunda Arc is seismically very active (Figure 11.2). Its Benioff–Wadati plane shows an average dip of 60° from the surface to a depth of 300 km beneath Java. A seismic gap is then observed between 300 and 600 km, and very deep earthquakes occur below the Java Sea north of Java. Despite this gap, tomographic studies have led to the conclusion that there is no hiatus in the descending slab below Java, whereas it is torn beneath Sumatra. Tomography also demonstrates the exceptional size of the Indian Ocean subducted slab. Indeed, high seismic wave speeds interpreted as reflecting 'cold' subducted material are observed down to depths of 1500 km in the lower mantle (Widiyantoro and van der Hilst, 1996). The subducted slab extends 2500 km north of the Sunda trench, and its total length is therefore about 3500 km. Postulating an average subduction rate of 7 cm/yr (70 km/Myr) similar to today's, such a length is consistent with 50 Myr of continuous subduction. Kinematic reconstruction of Southeast Asia suggests that the Sunda subduction was

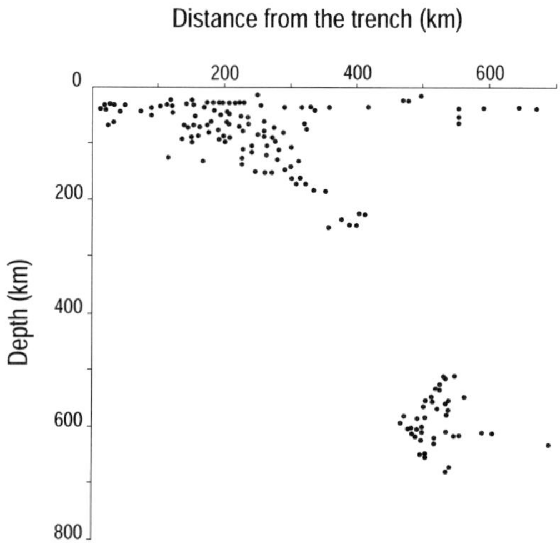

Fig. 11.2. Distribution of seismic foci below Central Java (from Eva *et al.*, 1988).

already active during the Eocene (Rangin *et al.*, 1990). In addition, the age of magmatic rocks of island-arc type range from 42 Myr (Java) and 55 Myr (Sumatra) to present without noticeable temporal gap.

11.1.2 Geology of Java

The 1000 km-long island of Java is located 300 km north of the Java Trench (Figure 11.3), from which it is separated by a relatively thin accretionary prism and by shallow fore-arc basins. No pre-Tertiary rocks crop out on the island, except as blocks within tectonic mélange breccias. The Java substratum is mostly composed of Paleogene sedimentary rocks overlain by Miocene reef limestones and thick Neogene volcanic sediments (mainly laharic breccia deposits). However, Cretaceous granitic bodies recognized in drill holes located in the northern part of the island as well as in the Java Sea mark the southern limit of the Sundaland Craton, which represents a dependance from the Eurasian Plate (Figure 11.1). Java overlies a Benioff plane the depth of which ranges from 100 km in the South to 400 km in the North. Island-arc magmatic formations outcrop as four stripes parallel to the Java Trench (Figure 11.3).

—In the south, volcanics locally referred to as the 'Old Andesites', which range in age from the Eocene (42 Myr) to the Late Miocene (12 Myr), are located about 50 km south of the younger units. They correspond to basaltic and andesitic pyroclastic breccias and to numerous intrusive bodies some of them also crosscut Paleogene sedimentary formations.

—Upper Miocene–Pliocene (11 to 2 Myr old) lava flows and intrusions outcrop north of the previous units, and they are often found in the substratum of large Quaternary volcanoes.

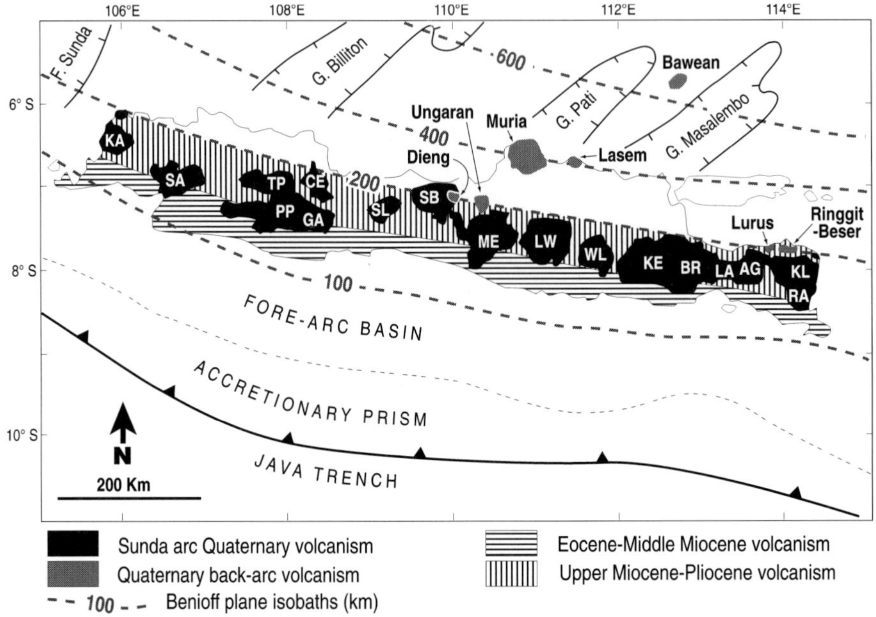

Fig. 11.3. The Sunda Arc at the level of Java (from Soeria-Atmadja *et al.*, 1994). Main active or recent volcanoes from the axis of the arc: KA, Karang; SA, Salak–Gede; TP, Tangkubanprahu; PP, Papandayan; GA, Galunggung; CE, Cereme; SL, Slamet; SD, Sundoro–Sumbing; ME, Merapi–Merbabu; LW, Lawu; WL, Wilis; KE, Kelud; BR, Bromo–Tengger–Semeru; LA, Lamongan; AG, Argopuro; RA, Raung; IJ, Kawah Ijen.

—These are rather regularly located along the central mountain range of Java, with an average interval of 80 km between two successive major eruptive centres. They overlie a Benioff plane 150–180 km deep, a rather high value compared with the average of about 100 km found for modern island arcs (Gill, 1981). Most active volcanoes in Java are andesitic cones with summit domes and predominant pyroclastic flow deposits.

—Finally, some Quaternary volcanoes occur in the back-arc domain, at variable distances from the axis of the Sunda Arc. They include, above the 200 km isobath, the Dieng and Ungaran Recent volcanic centres and the 1.2–0.5 Myr old Lurus, Ringgit and Beser volcanoes in Eastern Java. The Muria composite stratovolcano (1.1–0.4 Myr) and the Lasem andesitic cone overlie a 400 km deep Benioff plane, and the Bawean basanitic/phonolitic island (0.8–0.3 Myr) the deepest seismic foci (650 km).

Figure 11.3 shows that the position of the volcanic axis of the Sunda Arc in Java remained rather constant over its 42 Myr long history, which suggests little changes of the subduction parameters over time. The migration of this axis 50 km northward might be due to tectonic erosion of the Sunda plate margin by the downgoing slab. However, this migration occurred rather abruptly at the end of the Middle Miocene,

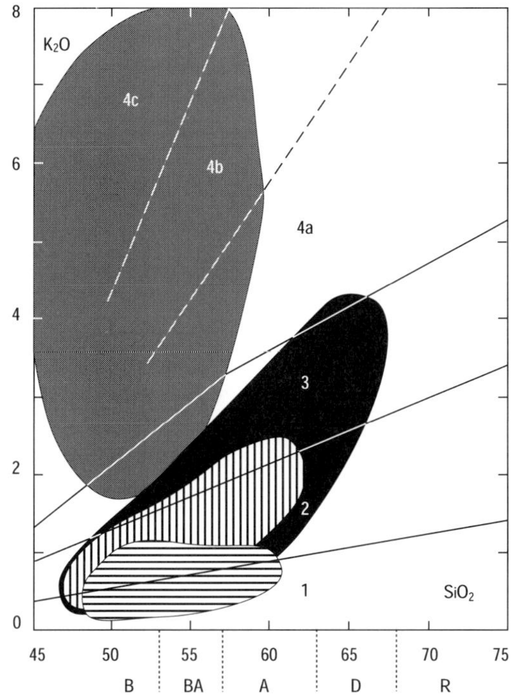

Fig. 11.4. Chemical composition of Java lavas. In the K_2O–SiO_2 plot, the fields correspond respectively to low-K calc-alkaline series (1), medium-K calc-alkaline series (2), high-K calc-alkaline series (3), shoshonitic series (4a), alkali potassic series (4b) and alkali ultrapotassic series (4c). Horizontal hatches: Eocene–Early Miocene lavas (42–15 Myr); vertical hatches: Late Miocene–Pliocene lavas (15–2 Myr); black: Quaternary lavas from the axis of the arc; shaded: Quaternary lavas from the back-arc domain. B: basalts; BA: basaltic andesites; A: andesites; D: dacites; R: rhyolites.

when major collisions involving continental blocks of Australian origin started in Eastern Indonesia, and it could thus simply result from a regional kinematic change.

11.1.3 Petrology and geochemistry of Java arc lavas

The petrographic compositions of lavas from the volcanic axis of Java do not differ significantly from other active arcs, e.g. Northern Japan, Lesser Antilles or Central Andes. They are highly porphyritic and rich in plagioclase phenocrysts. Andesites are more common than basalts, and the average composition of Java lavas is of a medium-K calc-alkaline andesite (see Chapter 2). The volcanoes in a back-arc position display rather different composition: their prominent rocks are K-rich or even ultrapotassic basalts containing abundant phenocrysts of magnesian olivine, calcic clinopyroxene (diopside) and sometimes leucite.

When plotted in a conventional K_2O–SiO_2 diagram (Figure 11.4), Java lavas display considerable variation. The most obvious pattern is a temporal increase of potassium content: Eocene–Middle Miocene lavas (42–15 Myr) are low to

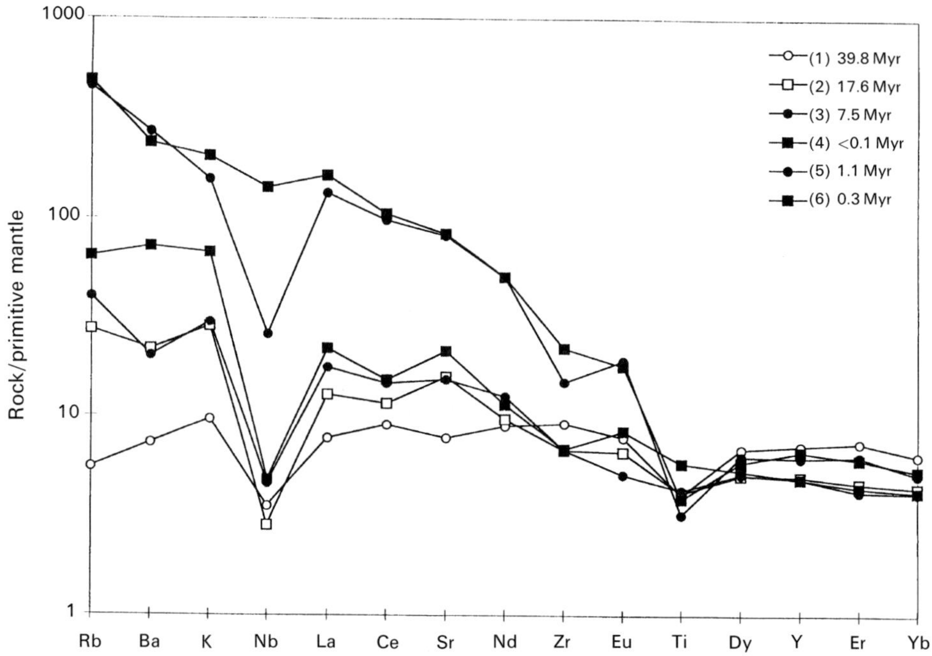

Fig. 11.5. Multielement plots of Java lavas. Normalization to the primitive mantle after Sun and McDonough (1989). (1) Eocene tholeiitic basaltic andesite from Karangsambung; (2) Middle Miocene calc-alkaline basaltic andesite from Pangandaran; (3) Late Miocene calc-alkaline basalt from Cianjur; (4) Recent potassic calc-alkaline basalt from the Merapi; (5) Quaternary potassic basanite from Muria; (6) Quaternary ultrapotassic (leucite-bearing) basanite from Bawean.

medium-K calc-alkaline and Upper Miocene-Pliocene lavas (15–2 Myr) are mostly medium-K calc-alkaline. The Quaternary lavas from the axis of the Sunda Arc range from medium-K to high-K calc-alkaline, whereas the lavas from the Quaternary back-arc volcanoes plot within the shoshonitic, alkali potassic and alkaline ultrapotassic (leucite-bearing lavas) fields of the K_2O–SiO_2 diagram.

The primitive mantle-normalized diagrams of Figure 11.5 show that these temporal variations do not only affect K but also most incompatible elements. Indeed, an Eocene representative lava (1) displays a rather flat pattern typical of low-K calc-alkaline (island-arc tholeiitic) series. The patterns of younger samples from the axis of Java (medium-K calc-alkaline samples 2 and 3, 17.6 and 7.5 Myr old, respectively; K-rich calc-alkaline basaltic andesite 4 from the Merapi active volcano) appear more and more enriched in the most incompatible elements (left part of the diagram) over time. The two samples from the back-arc alkalic volcanoes are much more enriched in highly incompatible elements than the former ones. Although the Muria potassic alkali basalt sample (5) pattern continues to show the negative niobium anomaly of typical arc-related magmas, that of the leucite-bearing ultrapotassic basanite from Bawean Island (6) is hardly distinguishable from those

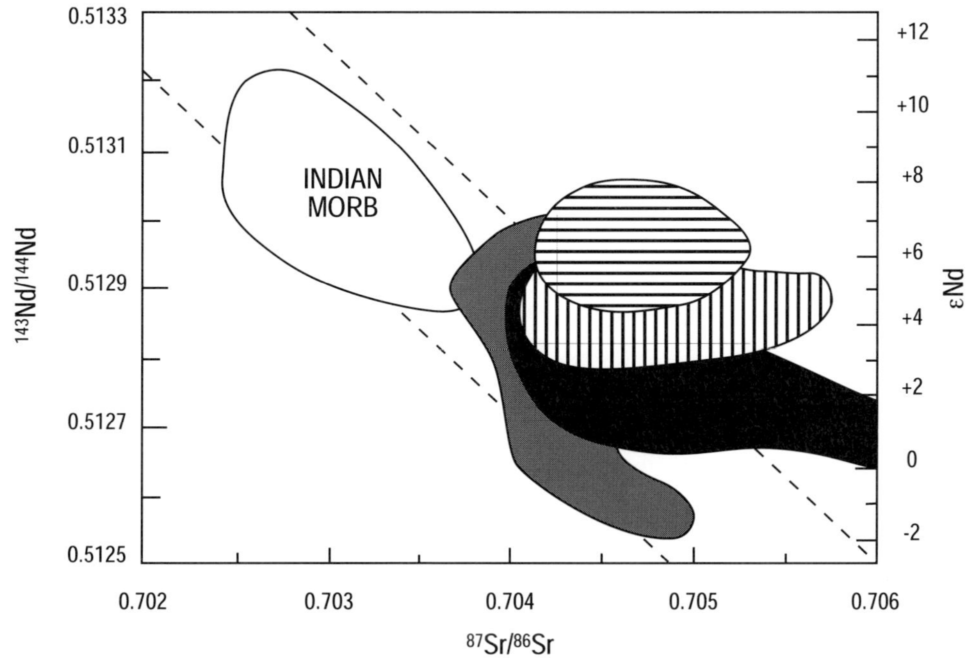

Fig. 11.6. Sr–Nd isotopic compositions of Java lavas (modified from Romeur, 1991). Symbols as in Figure 11.4.

of strongly silica-undersaturated magmas from oceanic and continental intraplate settings (nephelinites, leucitites).

The radiogenic Sr, Nd isotopic ratios of Java lavas are illustrated in Figure 11.6. Like those of the vast majority of island arc magmas, their positions in the (^{143}Nd/^{144}Nd)/(^{87}Sr/^{86}Sr) plot are intermediate between the field of MORB (mid-ocean ridge basalts) and those of oceanic sediments and the continental crust (^{143}Nd/^{144}Nd < 0.5126; ^{87}Sr/^{86}Sr > 0.7090), a feature which suggests that they contain some component of crustal origin. Such a crustal contribution is also consistent with the fact that in Figure 11.6 the fields of Java lavas deviate from the dashed stripe corresponding to the 'mantle array' (fields where plot magmas are considered as derived exclusively from melting of the mantle). Temporal variations can also be noticed in this plot. The oldest samples are characterized by relatively high ^{143}Nd/^{144}Nd ratios (0.5130) within the range of MORB, but their ^{87}Sr/^{86}Sr ratios are higher than 0.7040, as a result of either interaction with seawater (^{87}Sr/^{86}Sr $= 0.7091$) or contribution of pelagic sediments. The isotopic features of younger lava from the axis of the Sunda Arc (Upper Miocene–Pliocene and Quaternary) are characterized by evolution towards higher ^{87}Sr/^{86}Sr and lower ^{143}Nd/^{144}Nd ratios, consistent with increase in their sedimentary/crustal imprint (Figure 11.6). Such temporal variation towards higher contents in K and other incompatible elements together with an increasingly 'crustal' isotopic signature are often observed in arcs borne of long periods of subduction (e.g. the Lesser Antilles). They are frequently collectively referred to as the 'maturation' of the arc (Gill, 1981), a rather ambiguous term for the results of a highly complex combination of time-evolving petrogenetic processes.

The back-arc Quaternary magmas present significant isotopic differences compared with the others (Figure 11.6): most of them plot within the 'mantle array', and the most radiogenic among them display isotopic and trace-element features rather similar to those of OIB (Ocean Island Basalts).

11.1.4 Petrogenetic model of the Java Arc

A schematic cross-section of the Sunda Arc at the level of Central Java is presented in Figure 11.7 (without vertical exaggeration). The arc crust, 30–35 km thick, includes from south to north (see also Figures 11.1 and 11.3): (i) the modern sediment accretionary prism, (ii) the Mesozoic accretionary complex overlain by Tertiary sediments below Southern and Central Java, and (iii) the southern border of

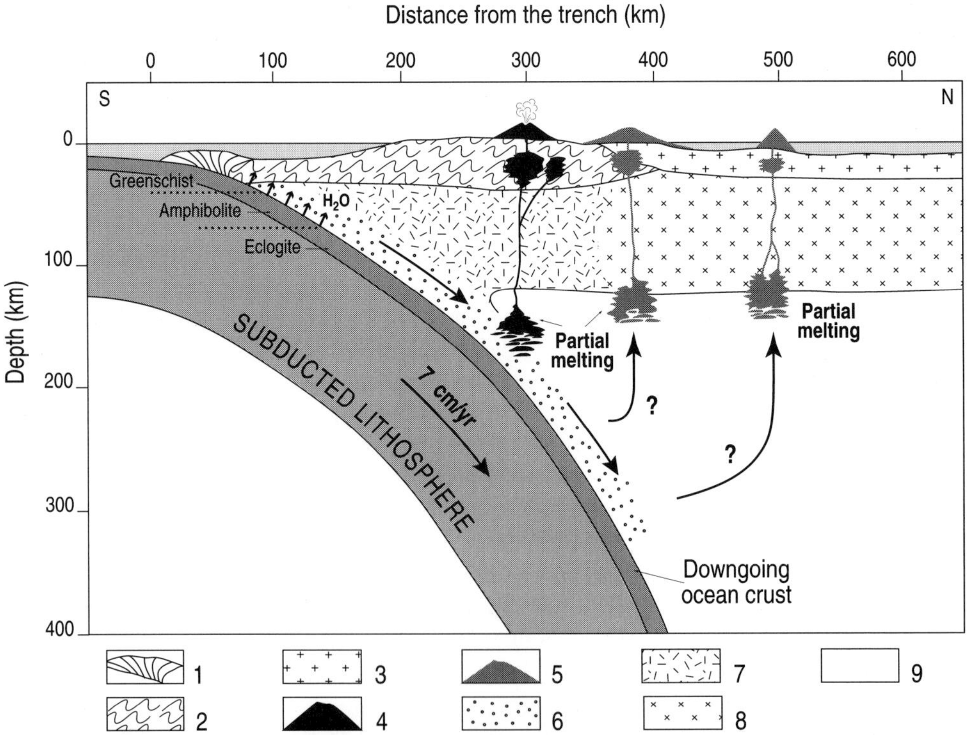

Fig. 11.7. Interpretative cross-section of the Sunda Arc at the level of Central Java (no vertical exaggeration except for volcanoes). (1) Accretionary prism; (2) Java crust (Mesozoic/Cenozoic accretionary complex); (3) crust of the back-arc domain (Sunda craton); (4) volcano from the axis of the arc; (5) back-arc volcano; (6) metasomatized mantle; (7) depleted lithospheric mantle; (8) enriched lithospheric mantle from the back-arc domain; (9) asthenospheric mantle.

the Sunda craton, which represents the crust of the northern part of Java and of the back-arc domain (Java Sea). The underlying upper mantle is also likely different: it could correspond to depleted suboceanic mantle below the accretionary complexes, and to enriched subcontinental mantle (possibly metasomatized during Mesozoic subduction events) below the Sunda craton. The downgoing oceanic crust is of Late Cretaceous age (110–120 Myr) south of Java. It was drilled during DSDP (deep-sea drilling project) Leg 22, which allowed recovery of Indian-type MORB overlain by a ca. 400-m-thick sedimentary pile, mostly pelagic in character. This ocean crust likely begins to lose its water as soon as it is underthrust below the accretionary prism, first through expelling interstitial water from the sediments, and then through dehydration effects resulting from metamorphic reactions. At increasing depths (and thus P-T conditions), these reactions lead from the diagenetic stage to zeolite facies rocks (not shown), and then successively to greenschist or blueschist facies, then amphibolite facies and finally to the anhydrous eclogite facies rocks. The water released into the mantle wedge is primarily seawater which contains dissolved large-ion lithophile elements (Rb, Ba, K, Sr) together with light rare-earth elements, thorium and uranium. These elements and the water carrying them will be incorporated within hydroxyl-bearing minerals (mostly pargasitic amphibole and phlogopite), resulting in the formation of a metasomatized mantle zone overlying the Benioff plane. The metasomatized peridotites carry an isotopic signature influenced by seawater and subducted sediments, as well as selective enrichment in the above-mentioned incompatible elements.

These metasomatic processes occur at depths lower than (or close to) 60–80 km (Figure 11.7), i.e. at relatively shallow levels compared with the considerable depth of the Benioff plane below the volcanic axis of the Sunda Arc (>150 km). It is thus unlikely that expulsion of water from the subducted slab into the mantle wedge could directly initiate melting of the latter, as was commonly supposed before the 1980s. Consequently, it seems necessary to postulate that the metasomatized mantle carrying the geochemical signatures of seawater and/or sediments is transferred to greater depths, most likely by mechanically coupled convection in the part of the mantle wedge immediately overlying the downgoing slab. This metasomatized mantle begins to melt at about 1000°C (Figure 2.1), usually through dehydration melting of pargasite-bearing peridotites following amphibole breakdown. The corresponding magmas rise within the lithospheric mantle and then the overlying crust. They are usually stored temporarily (or sometimes definitively) within intracrustal magma reservoirs in which they evolve by fractional crystallization sometimes coupled with chemical interactions with their host rocks (assimilation) before being emplaced at the surface (volcanic eruptions) or at depth (batholiths, plugs and other types of intrusions).

Such a petrogenetic model must also take into account the spatial and temporal patterns observed in magmatic arcs, e.g. the increase of incompatible element content and 'crustal' radiogenic imprint with time, or the occurrence of alkalic volcanism in back-arc position. A model involving a very limited recycling at the surface (arc volcanism) or at crustal depths (plutonic bodies) of arc magmas with respect to the flux of materials transferred from the subducting slab to the mantle wedge can account for such variations. For instance, a rate of subduction of 7 cm/yr of a crust overlain by a 400-m-thick sedimentary pile results in input at a depth of $(7 \times 10^{-5} \times (4 \times 10^{-1}) = 2.8 \times 10^{-5}\ \text{km}^3/\text{yr}$ of sediments per kilometer of arc (ignoring accretion of sediments into the fore-arc prism, which is justified in the case of the Sunda Arc but would be wrong for others, e.g. the Lesser Antilles). The amount of sediment contribution to arc magmas is usually estimated at less than 5 weight% (§ 11.3.1) as arc lavas derive mostly from the melting of mantle peridotites. The maximal sedimentary input to the Sunda arc mantle would thus theoretically allow annual production of $5.6 \times 10^{-4}\ \text{km}^3$ of magma per km of arc providing that all the sediments were recycled. There are no precise data regarding the rate of magma production along the Sunda Arc, but their range for other very active arcs (Aleutians, Mexico–Nicaragua, Peru–Chile) have been estimated as 1.3 to $3.7 \times 10^{-5}\ \text{km}^3$ per year and km of arc (Francis and Hawkesworth, 1994). It is thus clear that the proportion of subducted sediments effectively recycled by arc magmatism is very small (<10%). Most of the subducted sedimentary components remain at depth within the arc mantle, either as material buried within the deep mantle together with the rest of the subducted slab, or as a metasomatic fraction of the mantle wedge mingled with depleted peridotites by its convective motion. So temporal increase in crustal/sedimentary signature of such a mantle wedge over long subduction periods can be expected, and consequently temporal increase of this signature into its melting products, i.e. the arc lavas.

The petrogenetic relationship between magmas emplaced along the axis of the Sunda Arc and the K-rich volcanoes of the back-arc domain remain under debate. The location of the back-arc volcanoes seems relatively independent of the depth of the Benioff plane: it could trace a deep fracture system oblique to the trench, approximately parallel to those delineating the back-arc grabens (Figure 11.3). Moreover, the compositions of the back-arc magmas do not vary according to Benioff plane depths: the ultrapotassic volcanoes from Eastern Java (Lurus–Ringgit–Beser), very close to the axis of the Sunda Arc, are almost similar in composition to those of Bawean Island, located above a 600-km-deep slab. The case of Bawean is rather interesting, because the subducted Indian Ocean crust must be highly refractory at such depths (see Figure 11.26), and thus likely unable to contribute to magma genesis.

The numerous petrologic and geochemical similarities between the back-arc alkalic lavas from Java and those from intraplate settings suggest that the former derive mostly from the melting of enriched mantle, likely of subcontinental origin (see Figure 11.7). However, although the Java back-arc lavas plot within the 'mantle array' in Figure 11.6, most display negative Nb anomalies (with respect to neighbouring incompatible elements) which are characteristic of subduction-related magmas (Figure 11.5; see also Figure 2.6). The interpretation of these anomalies is still conjectural. They could reflect the contribution of subducted sediments through dehydration and transfer into the mantle wedge of hydrous fluids (hypothetic arrows in Figure 11.7), just as for the lavas from the axis of the arc. The occurrence of similar negative Nb and Ta anomalies in back-arc basin basalts (see § 11.4.3) indicates that extensive mingling of metasomatized and depleted mantles may arise from convection within the mantle wedge. However, back-arc lavas from Java do not contain appreciable amounts of ^{10}Be isotope, marker of an input of young sediments into their source, and so differ in this respect from the lavas of the axis of the Sunda Arc (Edwards *et al.*, 1993). This fact leads to favouring the hypothesis of an 'old' sediment contribution to the mantle source of back-arc lavas, possibly during subduction events that led to the emplacement of numerous Cretaceous granites within the Sundaland craton. Such calc-alkaline granites have been recovered from several drill holes in Northern Java.

The Javan example has been used to establish a once very popular correlation between K_2O content of arc lavas and depths of the Benioff plane (Dickinson and Atherton, 1967). This correlation was used in the 1970s to propose fossil subduction dips or even paleo-depths of vanished subduction planes. The lack of obvious petrogenetic relationships between the arc axis and back-arc lavas from Java suggests that across-arc K_2O variation is more indicative of the occurrence of distinct mantle domains (resulting from the long geological history of the Sundaland margin) rather than being linked to the polarity of subduction. In other arcs, similar 'positive correlations' have been shown to be based on rocks of very different ages, or even on lavas emplaced within different tectonic plates (New Guinea). Genuine examples of across-arc increase in K and other LILE (Large Ion Lithophile Elements) have been shown to occur in arcs (e.g. in Martinique, Lesser Antilles: Westercamp *et al.*, 1989; Coulon *et al.*, 1991), but this variation is mostly observed at a relatively small scale (10 to 30 km) corresponding to the usual width of volcanic-arc chains.

11.2 THE DIVERSITY OF MAGMATIC ARCS

There are 34 arcs including historically active volcanoes (Table 2.1 and Figure 11.8). Eight of them are installed on active continental margins, especially on the western America margin(s) which could be considered as the longest subduction zone on Earth, although it is interrupted by several hiatuses that divide it into six major magmatic arcs, some of them again being discontinuous at a smaller scale. The 26 other arcs lie on oceanic or at least partly submerged basements: they are island arcs. The average length of volcanic arcs is close to 100 km (Table 11.1). It varies from 100–500 km for 'small' (and usually ephemeral) arcs which generally occur within complicated tectonic settings involving mosaics of microplates to 1500–2500 km for huge (and long-lived) arcs installed along major tectonic plate boundaries (e.g. the Sunda Arc, or the Alaska–Aleutians Arc). The subduction rates vary from 1 to 12 cm/yr and their range is similar to that of accretion rates. Some arcs are apparently devoid of appreciable fore-arc sedimentary prisms (Marianas, Tofua, Peru–Chile). In the others, the size of these accretionary prisms is very variable, but they rarely reach the stage of emersion as in Barbados, which lies in fore-arc position with respect to the volcanic Lesser Antilles islands. Back-arc basins characterized by active extension and basaltic volcanism are rather uncommon (Table 11.1).

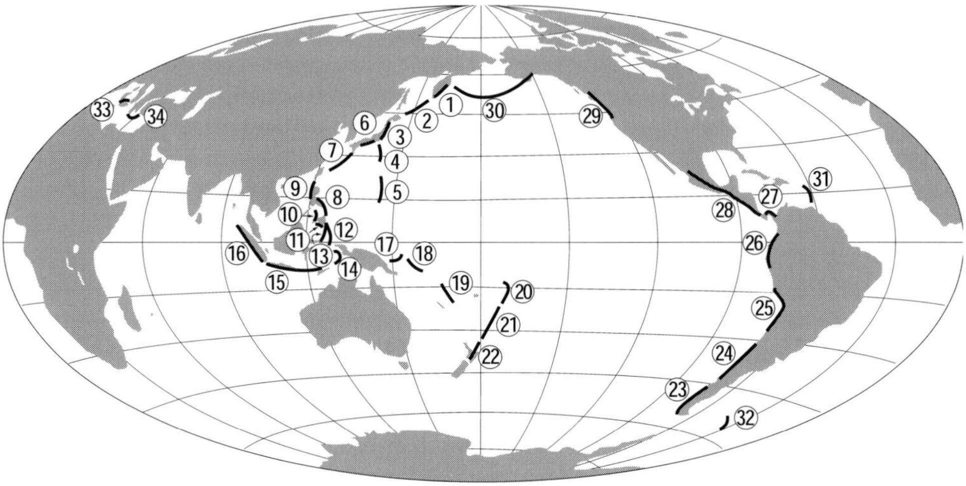

Fig. 11.8. The active volcanic arcs (numbers as in Table 11.1).

11.2.1 Arcs installed on thin crusts and on thick crusts

One of the most widely used classifications of volcanic arcs is based on the thickness of the arc crust, which ranges from typically oceanic values (10–12 km) up to 70 km below the Andean margin altiplano (set of values in Gill, 1981). Three major groups can be distinguished : (i) the island arcs installed on genuine oceanic crust, which are relatively uncommon (Tofua, South Sandwich, Izu-Bonin), (ii) the island arcs installed on 'intermediate' crust from viewpoint its thickness (15–35 km), which are the most common, and finally (iii) the volcanic arcs installed on continental crust (>30 km). The last category includes arcs installed on continental-crust fragments (New Zealand), continental peninsulas (Kamchatka) or active continental margins, the crust of which is either of 'normal' thickness (Cascades) or abnormally thick (Southern Peru and Northern Chile).

Many authors have documented correlations between crust thickness and petrologic or geochemical features of arc magma. For instance, the average silica content of magmatic rocks increases with thickness of arc crust (Coulon and Thorpe, 1981): basalts are the dominant rock type in arcs built on oceanic crust, whereas andesites predominate in arcs installed on continental crust. Large ignimbritic rhyolitic deposits typically occur in arcs built over continental crust, and very often derive from anatectic melting of this crust (altiplano ignimbritic flows). Finally, low-K (and more generally incompatible element-poor) magmatic series are mostly found in intra-oceanic arcs (Tofua, Kermadec, Marianas, South Sandwich: Figure 11.9), whereas K-rich series with strong 'crustal' isotopic signatures are rather common in continental arcs.

All these features are consistent with the occurrence of chemical interactions between basaltic to andesitic-arc magmas and the continental crust, leading to the enrichment of the former in silica, potassium and other LILE, and radiogenic

Table 11.1. The active volcanic arcs

Active volcanic arc			Associated subduction		Subducted ocean crust		Peculiarities
N°	Name	Length (Km)	Average dip (°)	Average speed (cm/yr)	Age (Myr)	Location [Ocean, Sea (S.), Basin (B.)]	AS: aseismic subd.; RS: ridge subd.; AD: adakite; BA: back-arc basin
1	S. Kamchatka	700	50	9	50–80	NW Pac.	
2	Kuriles	1100	45	9	80–100	NW Pac.	
3	NE Japan	1000	30	10	130–140	NW Pac.	
4	Izu-Bonin	1200	65	9	>140	NW Pac.	BA (Sumisu)
5	Marianas	500	80	9	>140	NW Pac.	BA (Marianas)
6	SW Japan	300	10	2	15–25	Shikoku B.	AS; AD
7	Ryu Kyu	900	40	2	50–65	Philippines S.	BA (Okinawa)
8	W Philippines	800	45	6	40–50	Philippines S.	AD
9	Luzon	1200	30	2	15–30	China S.	AD
10	Negros	400	15	2	10–20	Sulu B.	AS; AD
11	SE Mindanao	200	45	2	30–40	Celebes B.	AD
12	Sangihe–Sulawesi	500	60	8	25–35	Molucca S.	
13	Halmahera	200	45	9	25–35	Molucca S.	
14	Banda	1200	40	7	>120	Timor S.	BA (Banda S.)
15	Sunda (Java–Flores)	2400	60	7	100–160	NE Indian	
16	Sunda (Sumatra)	1600	35	8	50–90	NE Indian	
17	New Britain	500	30	9	25–35	Salomon S.	BA (Manus); AD
18	Salomon	500	30	10	<10	Woodlark B.	RS
19	Vanuatu	1700	70	12	55–65	Coral S.	BA (Coriolis)
20	Tofua (Tonga)	800	45	9	100–135	SW Pac.	BA (Lau)
21	Kermadec	900	60	9	100–135	SW Pac.	BA (Havre)
22	New Zealand	300	50	6	100–135	SW Pac.	
23	Austral Andes	900	10	2	<20	SE Pac.	AS; RS; AD
24	Central Chile	2000	15	9	15–40	E Pac. (Nazca)	AD
25	Peru–Chile (N)	1600	15	10	40–5-	E Pac. (Nazca)	AD
26	Colombia–Ecuador	900	35	8	10–20	E Pac. (Nazca)	AD
27	Panama–Costa Rica	400	20	8	10–20	E Pac. (Cocos)	AS; AD
28	Mexico–Nicaragua	1700	55	7	40–50	E Pac. (Cocos)	AD
29	Cascades	1100	15	2	<15	E Pac. (J. Fuca)	AS; RS; AD
30	Alaska–Aleutians	2500	55	8	55–65	N Pac.	AD
31	Lesser Antilles	800	55	2	70–100	W Atl.	
32	S. Sandwich	300	50	2	55–65	A Atl.	
33	Aeolian	100	55	1	80–120	E Medit. S.	
34	Aegean	400	35	3	80–120	E Medit. S.	

strontium. These processes, collectively referred to as crustal contamination, mostly occur during the ascent and storage of mantle-derived magmas within the silica-rich continental crust (see Table 2.1 for average analysis). The main mechanism leading to this contamination is Assimilation (of country rock) coupled with Fractional Crystallization (AFC: De Paolo, 1981). The occurrence of AFC is a direct consequence of Bowen's Reaction Principle : the energy provided by the exothermic

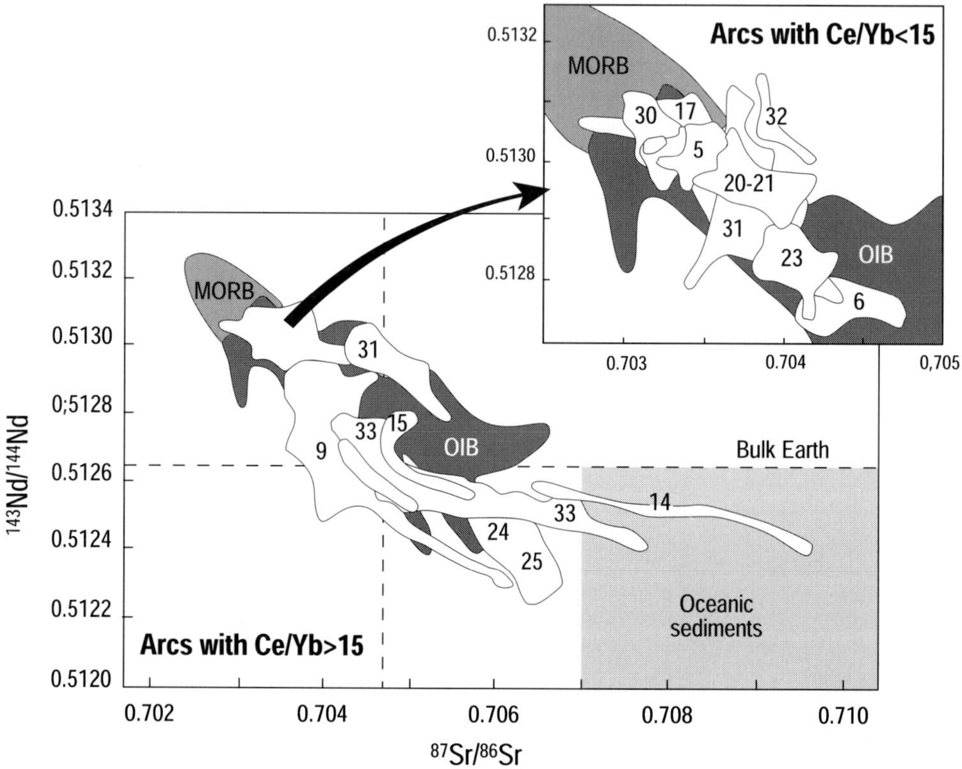

Fig. 11.9. Sr–Nd isotopic compositions of arc lavas (numbers as in Table 11.1 and Figure 11.8), adapted from Hawkesworth *et al.* (1994).

crystallization of the phenocrysts during the cooling of an igneous body is partly (at least) consumed by the endothermic reactions linked to the melting of its wallrocks and xenoliths. The anatectic liquids derived from continental crust can then mix with mantle-derived liquids (hybridation). Contamination may occur though a combination of complex processes (MASH: Melting, Assimilation, Storage, Homogenization) when mafic mantle-derived arc magmas are stored within thick continental crust, e.g. below the Andes (Hildreth and Moorbath, 1988). Crustal assimilation by mafic magmas may be documented using a combination of Sr (or Nd) and O isotopes, as it generates mixing curves different from those traducing incorporation of crustal material into the mantle (Figure 2.3).

Despite its obvious importance, the thickness and composition of the arc crust is not the only, or even the major, parameter controlling the composition of arc magmas. For instance, the crust of the Alaska–Aleutians Arc is continental in the eastern part of the arc and oceanic in its western part, but this major change is apparently not correlated with variation in the isotopic signature of basaltic magmas. However, as discussed above, andesites are the prominent rock type in

Alaska, where anatectic rhyolites are emplaced as ignimbritic flows. Some magmatic arcs installed over continental crust (Austral/Andes, Figure 11.9) are characterized by low strontium isotopic ratios, very different from those of neighbouring sections of the Andes active continental margin (Central Chile, Peru–North Chile). K-rich lavas occasionally occur in arcs considered as built on oceanic crust (Vanuatu). Highly radiogenic lavas (Figure 11.9) occur in the Banda Arc, which is installed on a relatively thin crust, but below which materials of continental (Australian) origin are subducting. This example suggests that characteristics of the subducted materials can also control the compositions of arc magmas.

The geological meaning of arc crustal thicknesses is not yet fully understood. It seems likely that the crusts of 'intermediate' thickness characteristic of many arcs were initially oceanic in character, either made up of typical ocean-crust rock sequences or ocean-crust already thickened by the adjunction of plateau materials (see § 10.3). Then, during long subduction periods, intrusion of arc magmas would result in further thickening of these crusts, mainly due to batholiths and plugs and possibly also to storage and cooling of arc magmas below the ocean crust (magmatic underplating). The Lesser Antilles arc crust is an example of such a complex evolution: according to recent data it was originally a Late Jurassic ocean crust thickened up to 20 km by the emplacement of the Caribbean Plateau lavas and intrusives during the Turonian (90–88 Myr) (see § 10.3.3). Then it became the locus of arc magmatism during Late Cretaceous, Oligocene–Early Miocene and Late Miocene to Present subduction events. This crust reaches thicknesses up to 35 km despite lacking rocks older than the Late Jurassic. Thus, although it is widely admitted that differences in thickness linked to plateau magmatism can control the location and geometry of incipient intra-oceanic subductions, adjunction of arc magmas can play a major role in determining the final thickness (and chemical composition) of arc crusts.

The origin of extreme thickening (up to 70 km) documented for the Andean crust is highly debated. Current explanations include or even combine: (i) emplacement of large intracrustal calc-alkaline batholiths since the Mesozoic, (ii) storage of huge quantities of arc magmas in the Moho region (magmatic underplating), (iii) accumulation below the initial Moho of slices of subducted ocean crust (tectonic underplating) and finally (iv) thickening due to major intracontinental thrusts, e.g. the underthrusting of the western edge of the Brazilian craton below the Andes.

11.2.2 Enriched and depleted arcs

Some active arcs, e.g. the Sunda Arc, are mostly made up of calc-alkaline and/or potassic calc-alkaline magmatic rocks stongly enriched in highly incompatible elements (including LREE (Light Rare Earth Elements)) with respect to moderately incompatible elements (e.g. HREE (Heavy Rare Earth Elements)). Contrarily, other arcs bear a majority of low-K magmatic rocks with nearly flat or even depleted (i.e. characterized by positive slopes) rare-earth patterns and spidergrams. These two categories can be differentiated using LREE/HREE ratios like La/Yb or Ce/Yb.

Hawkesworth *et al.* (1994) used Ce/Yb ratios to distinguish unenriched arcs (Ce/Yb < 15) from enriched arcs with Ce/Yb > 15. The boundary between the two groups corresponds to a primitive mantle-normalized Ce/Yb ratio of 4 (Table 2.2) and thus to nearly flat REE spectra: for instance, in Figure 11.5, the only Java sample with Ce/Yb < 15 is the Eocene low-K andesite (39.8 Myr).

The main arcs with low and high Ce/Yb ratios, respectively, are shown in Figure 11.9 (where the numbers are the same as in Table 11.1). There is an obvious general correlation between LREE enrichment and Sr and Nd isotopic ratios. The magmas from arcs with Ce/Yb < 15 display isotopic compositions rather similar to those of MORB and do not deviate significantly from the MORB–OIB mantle array. Contrarily, the magmas from LREE-enriched arcs are almost as enriched in radiogenic Sr as the OIB, or even more, and some of them are characterized by a very strong sedimentary/crustal imprint.

The major depleted or mildly enriched arcs (Ce/Yb < 15) are the Aleutians, Izu-Bonin, Marianas, New Britain, Vanuatu, Tofua, Kermadec, South Sandwich, but also the Northern Lesser Antilles and the Austral/Andes. These arcs are characterized by highly variable crustal thickness, as well as by very different ages for subducting crusts (Table 11.1). The most obvious feature they share in common is that the sedimentary pile on the top of the subducting crust is thin (a few hundred meters) and mostly pelagic in character. These characteristics may reflect either subduction of young ocean crust (Austral/Andes, New Britain) or subduction of older crust devoid of detritric sediments of continental origin (Marianas, Vanuatu, Tonga, Kermadec, South Sandwich). Contrarily, enriched arcs usually correspond either to examples of subduction of ocean crust topped by thick piles of detrital sediments, or alternatively to cases of long subduction history (Java).

11.2.3 Subduction of old crust versus subduction of young crust

Table 11.1 data show that the ages of plates subducting below modern magmatic arcs are extremely variable: they range from Jurassic for the oldest ocean crusts still in the position on the seafloor to Present for the three active subducting ocean ridges. Indeed, the Chile Ridge is presently subducting beneath the Austral/Andes at the level of Taitao Peninsula, the Juan de Fuca Ridge below the Cascades chain and the Woodlark Basin Ridge below the Solomon arc. About 15 arcs overlie Mesozoic downgoing slabs, and 12 of them are associated to subduction of Neogene ocean crusts.

Figure 11.10 data show that relatively 'young' downgoing crusts of post-Eocene age (<35 Myr) usually have low dipping angles (<30°), whereas those of Mesozoic slabs are higher than, or close to, this value. Indeed, old (and thus dense) lithosphere tends to sink within the mantle, whereas young, and therefore still hot, downgoing lithosphere is buoyant and tends to remain at shallow depths immediately below the arc lithosphere. Such a situation is favourable to tectonic underplating of slices of ocean crust, which may contribute to the crustal thickening process. Subduction of very young, and thus still plastic, lithosphere (<20 Myr), is nearly aseismic: tomographic data rather than locating of deep seimic foci, have led to the

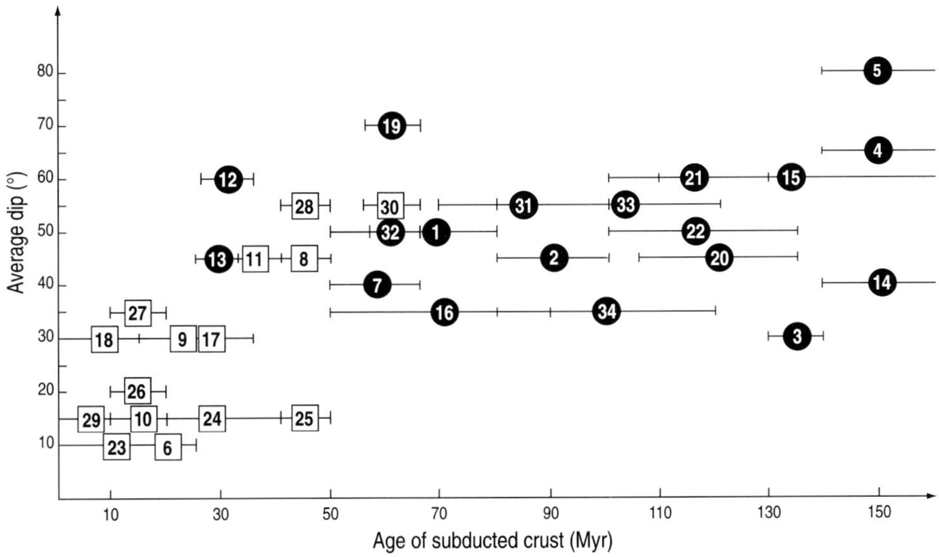

Fig. 11.10. Correlation between dips and ages of the subducted crust (numbers as in Table 11.1 and Figure 11.8). Squares: arcs containing adakites; circles: other arcs.

recognition of Benioff planes below the Cascades and the Panama Arc. Seisms of high magnitude (7 and more) are usually associated with subduction of slabs of Paleogene age (Peru and North Chile). Finally, high-dipping subduction of Mesozoic crust (Mariana's type of Uyeda and Kanamori, 1979) usually occurs without major seisms because of the mechanical decoupling of the two plates during gravity-driven sinking of the old subducting slab. In such cases, extension in the upper plate often results in the formation of a back-arc basin and in normal faulting in the fore-arc domain (Marianas). Five out of the eight back-arc basins mentioned in Table 11.1 are linked to the subduction of an oceanic plate older than 100 Myr.

There are also interesting correlations between the age of the subducting slab and the petrology and geochemistry of arc magmatic rocks. Indeed, the chemical interactions between subducted lithosphere and arc lithosphere vary according to the temperature of downgoing ocean crust. Young and therefore still hot crusts are able to suffer partial melting, generating silica-rich magmas referred to as adakitic, which may either be emplaced at the surface or within the crust (volcanic and plutonic adakites) or alternatively metasomatize the arc mantle. Indeed, all but one of the adakite occurrences indicated in Table 11.1 correspond to active subduction zones involving descending ocean crusts younger than 50 Myr. Subduction of active oceanic ridges leads to emplacement of unusual rocks including adakites and associated rocks (magnesian andesites), boninites and MORB contaminated by arc crust. Contrarily, an old (and cold) subducting crust will mainly, as in the Java example, liberate its water and the elements dissolved in it, which will be transferred into the arc mantle.

11.3 CONTRIBUTION OF THE SUBDUCTING OCEANIC CRUST TO ARC MAGMAS

11.3.1 Evidence for crustal contribution

It is presently widely accepted that the vast majority of arc magmas originate from melting of the depleted mantle of the upper plate on which arcs are built (Arculus, 1994). Indeed, almost all active arcs bear basaltic rocks containing up to 8 to 12 weight% of MgO (magnesian basalts; Table 2.5). Given the range of melting temperatures consistent with present-day oceanic and continental geotherms (Figure 2.1), such magmas are likely to derive from low (<20%) melting degrees of sources which are thus necessarily much richer in MgO than these basaltic magmas: the upper mantle peridotites fit this requirement (Table 2.1). In addition, the low-K basalts common in intra-oceanic island arcs, and often referred to as island-arc tholeiites, in common with the MORB share so many petrologic and chemical features that both types should derive from rather similar sources.

On the other hand, there is impressive amount of evidence for contribution of the subducting oceanic crust (and especially of its sediments) to chemical composition of arc magmas; such as the occurrence in many arc lavas of the cosmogenic isotope ^{10}Be, which forms in the high atmosphere by spallation of nitrogen and oxygen induced by cosmic rays. It is quickly trapped within clay-rich oceanic sediments. The radioactive decay of ^{10}Be leads to boron (^{10}B), after a fairly short period (1.5 Myr). Consequently, ^{10}Be cannot be detected in rocks older than ca. 7.5 to 9 Myr. This isotope has not been shown to occur in recent MORB or OIB, but is frequently detected in arc lavas, including the case where both magmatic rocks from the subducted crust and sediments from the arc crust are older than 10 Myr. The source of these lavas must therefore contain a young sedimentary component. The behaviour of ^{10}Be also demonstrates that the recycling process can be very short: less than 9 Myr from the formation of the oceanic sediment to the eruption of arc lava including the intermediate stages of sediment subduction, mantle metasomatism and partial melting.

Many other stable or radiogenic isotopes (Sr, Nd, Pb, Th, O) can be used to demonstrate a sedimentary contribution to arc-magma geochemistry. In most plots using two different isotopic ratios, the field of arc magmatic rocks lies between those of N-MORB, considered as genuine melting products of depleted suboceanic mantle, and those of oceanic sediments, respectively: typical examples are shown in Figure 11.9 (Nd–Sr plot) and Figure 11.11 (Nd–Pb plot).

In these two diagrams, magmatic rocks from enriched arcs with Ce/Yb<15 plot closer to the sediment field than those from depleted to mildly enriched arcs, which of course suggests a more important sedimentary contribution to the former group. This contribution can be modelled using mixing curves that allow calculation of sediment mass proportions (Figure 2.3). The data shown in Figures 11.9 and 11.11 could lead to the conclusion of very high sedimentary contributions in some cases where the isotopic signature of arc magmas is nearly identical to that of sediments, with the consequence that the bulk chemical composition of such rocks should be

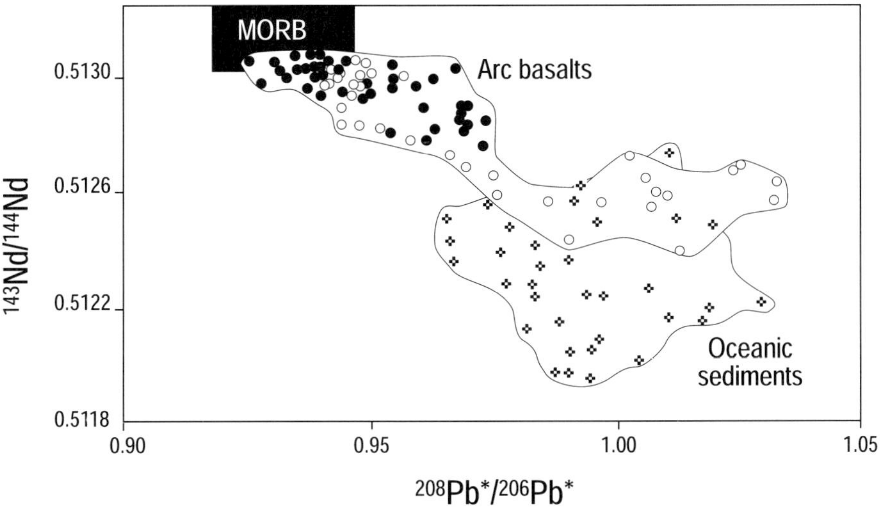

Fig. 11.11. Sr–Pb isotopic compositions of arc lavas, compared with that of MORB (hatched area) and oceanic sediments (crosses), from Hawkesworth *et al.* (1994). $^{208}Pb^*/^{206}Pb^*$: ratio between the amounts of radiogenic isotopes ^{208}Pb and ^{206}Pb produced since the Earth's accretion through the decay of ^{232}Th and ^{238}U. Filled circles: lavas from arcs with Ce/Yb < 15; empty circles: lavas from arcs with Ce/Yb > 15.

sedimentary rather than igneous. However, such simple mass balance calculations apply only to bulk mixing processes. Bulk sediment addition to the mantle does not adequately describe metasomatic processes which likely involve the addition to the mantle of a metasomatic agent (hydrous fluid or melt) the composition of which is very difficult to constrain. Postulating that Sr, Nd and Pb content is high in the metasomatic agents with respect to their concentration in depleted mantle peridotites, the data of Figures 11.9 and 11.11 can be modelled with only low percentages of sediments (usually less than 5 weight%).

11.3.2 Evolution of genetic models

The origin of arc magmas has been the subject of much speculation, and a number of models have been proposed with the aim of explaining their geological relationships and geochemical characteristics. Three of the most famous of these models are shown in Figure 11.12. In Ringwood's (1974) model (Figure 11.12a), metamorphic reactions lead to a progressive dehydration of the downgoing slab. The hydrous fluids thus released are incorporated into the mantle at depths ranging from 50 to 100 km. The melting of this mantle produces island arc tholeiitic magmas. At greater depths (100–150 km), the partial melting of subducted ocean crust at temperatures higher than 750°C produces silica-rich magmas which metasomatize the overlying mantle, which becomes the source of calc-alkaline magmas. Wyllie and Sekine (1982) proposed several refinements of Ringwood's model. The release of hydrous fluids from the downgoing crust results in the formation of metasomatic amphibole in the

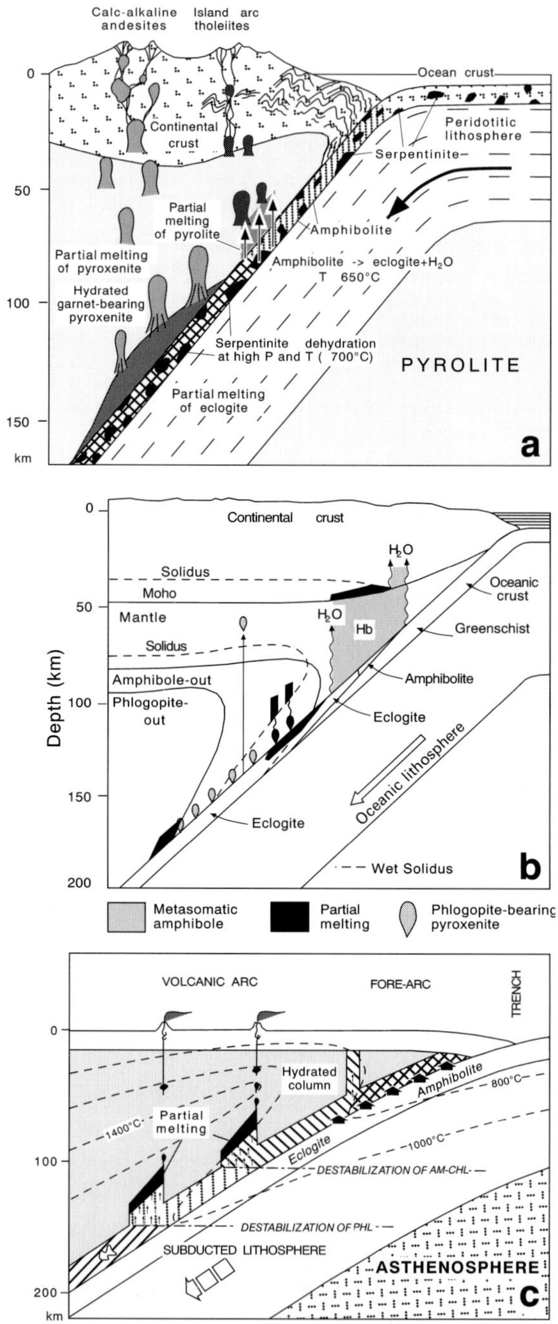

Fig. 11.12. Evolution of genetic models for arc magmas (from Maury *et al.*, 1996). (a) From Ringwood (1974); (b) from Wyllie and Sekine (1982); (c) from Tatsumi (1989). Hb: hornblende; AM: amphibole; CHL: chlorite; PHL: phlogopite.

arc mantle (Figure 11.12b), which reaches its solidus at depths of ca. 90–100 km. The subducted ocean crust melts partially at depths of 100–120 km, producing silica-rich magmas which—as in the former model—metasomatize the mantle but do not reach the surface. Tatsumi's (1989) model, which does not consider any melting of the subducted crust (Figure 11.12c), involves three successive steps. The first one is progressive dehydration of the downgoing crust, which is achieved at depths of 60–80 km with the transition from amphibolites to eclogites, and the migration of the released hydrous fluids into the mantle wedge where they form metasomatic minerals (amphibole and phlogopite). The following step is the convection-induced descent of this metasomatized mantle down to 110–120 km (below the volcanic arc) where amphibole breaks down and then to ca. 170 km (below the back-arc domain) where phlogopite becomes unstable. The breakdown of these hydroxyl-bearing minerals triggers the partial melting (dehydration melting) of the mantle wedge, producing volcanic arc and back-arc magmas, respectively.

11.3.3 The variable modalities of crustal/sedimentary contribution: the example of the Luzon Arc (Philippines)

According to recent studies, the crustal/sedimentary imprint can be acquired in many ways, which may change not only from one arc to another but also in space and time within a single arc: from one island to another or even from one volcano to another on a single small island.

The Luzon Arc (Figures 11.1 and 11.13) resulted from the subduction of the South China Sea oceanic plate below the Philippine plate. This oceanic basin opened from 30 to 15 Myr within the continental Eurasiatic margin (Figure 11.1). Its fossil ridge is presently subducting below the northern part of Luzon Island (Figure 11.13). The main peculiarity of the Luzon Arc is that both its northern and southern edges correspond to active collision zones between the arc domain and the China continental margin in the north (Taiwan), and the Mindoro continental block in the south. Volcanic activity is extinct in these two collision zones but the central part of the Luzon Arc possesses a number of very active volcanoes (Pinatubo, Taal). In its northern half (Figure 11.13), there are only four active/dormant volcanic centres north of Luzon (Batan, Babuyan de Claro, Camiguin, Mt Cagua) and magmatic activity started at ca. 9 Myr in Batan.

The Neogene lavas from the northern part of the Luzon Arc display clear along-arc geochemical variations (Figure 11.13): from south to north they become more radiogenic in Sr and Pb and less radiogenic in Nd. These variations parallel those of Recent sediments cored off the Manila Trench, which contain an increasing amount of detrital particles close to the China continental margin.

Geochemical studies of lavas from these areas suggest that the origin of the above-mentioned isotopic variations is linked to an increase in the sediment contribution towards the north of the arc. Coupled O–Sr and O–Nd isotopic data indicate that the assimilation of materials from the arc crust is mainly effective in the northern part of Luzon (Baguio), but rather negligible in the volcanic-island chain from Camiguin to Batan (Fourcade *et al.*, 1994). The lavas from this last group of islands also show

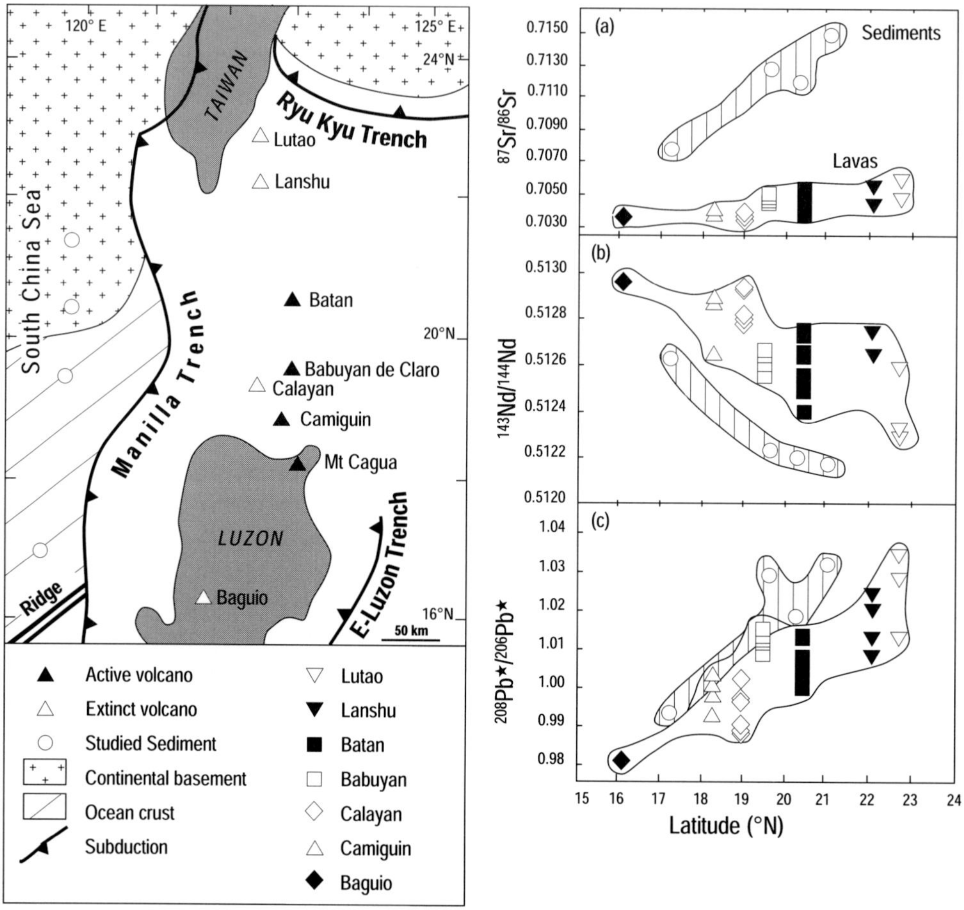

Fig. 11.13. The northern part of the Luzon Arc: location and isotopic variations in lavas and sediments (from McDermott *et al.*, 1993). $^{208}Pb^*/^{206}Pb^*$: see Figure 11.11.

temporal compositional variations (Figure 11.14). K_2O content increases over time from low-K series (Calayan) to medium-K calc-alkaline and finally to high-K calc-alkaline (young Batan lavas). These changes are observed either from one island to another (inter-island variation: Babuyan and Camiguin lavas are richer in K, with equivalent silica content, than those of Calayan) or within the same island (Batan). Most incompatible elements follow similar patterns of temporal enrichment. However, with reference to a moderately incompatible element (Yb), Ba, Th and Ta increases over time are more spectacular for Batan than for the other islands. Th/Yb, Ta/Yb and La/Yb ratios have been plotted against Sr and Nd isotopic ratios in Figure 11.15. The increase in sedimentary imprint (higher $^{86}Sr/^{87}Sr$, lower $^{143}Nd/^{144}Nd$) corresponds to an increase in these trace element ratios which again is more spectacular in the case of Batan than for the other islands.

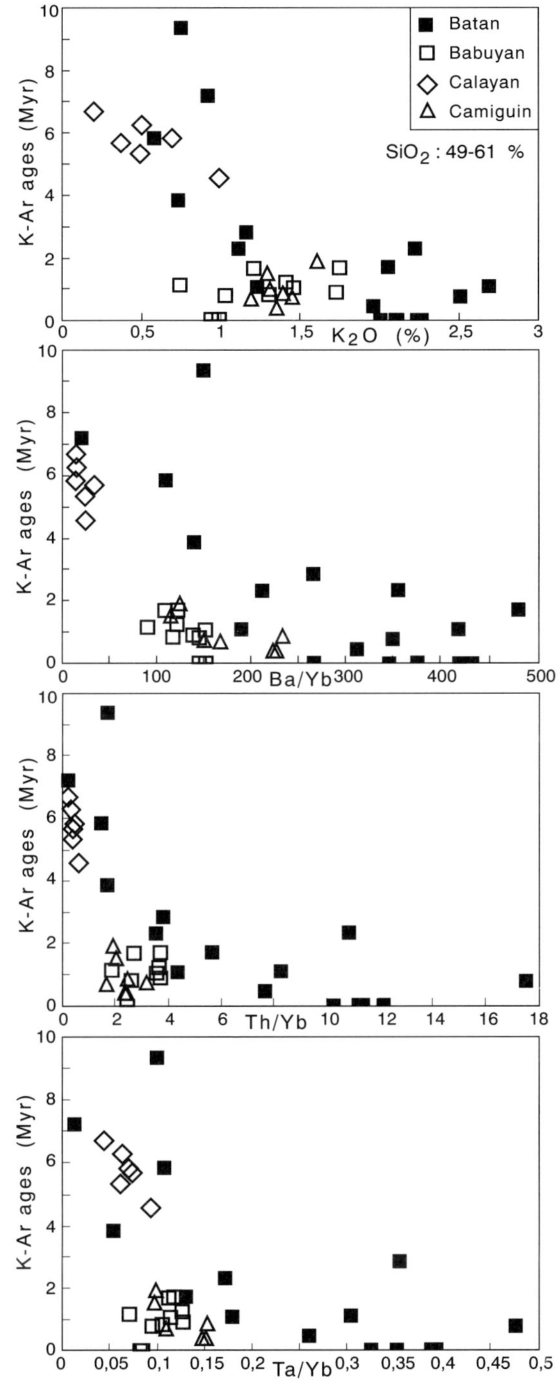

Fig. 11.14. Temporal evolution of Northern Luzon lavas (from Maury *et al.*, 1998).

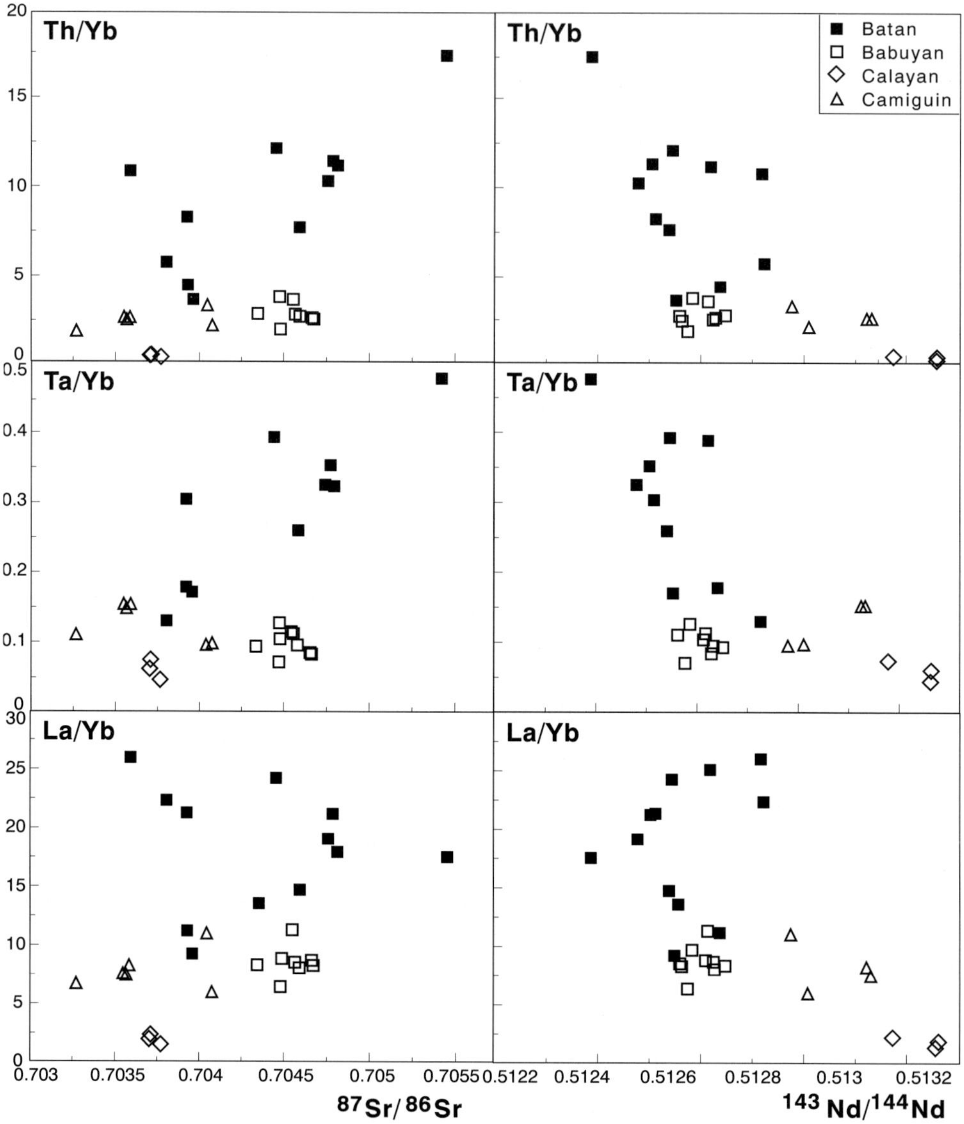

Fig. 11.15. Coupled trace element and isotopic enrichments in Northern Luzon lavas (from Maury *et al.*, 1998).

These variations suggest that the mantle source of arc magmas was metasomatized much more efficiently beneath Batan (especially considering the youngest lavas from this island) than below Babuyan, Camiguin and Calayan. The corresponding metasomatic agent was likely able to carry not only LILE but also light REE, Th and even HFSE like Ta, which are not soluble in hydrous fluids. Young Batan lavas are noteworthy for containing metasomatized peridotitic mantle nodules. Their primary (deformed) minerals are typical of highly residual harzburgites, depleted in all incompatible elements by successive partial melting events. These peridotites contain veinlets of metasomatic (undeformed) phlogopite and pargasite. Their bulk geochemical signature (incompatible element patterns and Sr, Nd, O isotopes) is similar to that of potassic calc-alkaline magmas and they can represent an analogue of the source of their host lavas (Vidal *et al.*, 1989; Maury *et al.*, 1992). Batan peridotites provide evidence for two kinds of metasomatic agents (Schiano *et al.*, 1995b): hydrous fluids still found as inclusions, which probably transported most of the LILE (Rb, K, Ba, Sr), and acidic magmas (adakites), also found as glassy inclusions, which carried most of REE, Th, U, Pb and HFSE (including Ta). However, adakites have not yet been found in the island, despite the young age of the subducting crust. It is thus possible that melting products from the subducted crust contribute towards mantle-wedge metasomatism without reaching the surface, as envisioned by Ringwood (1974) and Wyllie and Sekine (1982).

11.3.4 Hydrous fluid metasomatism and partial melting of the mantle wedge

The chemical composition of an arc basalt of low-K calc-alkaline or 'island arc tholeiitic' type (Table 2.4 and Figure 2.6) differs mainly from that of N-MORB (Table 2.3 and Figure 2.4) by the higher content of the former in LILE (Rb, Ba, K, Sr) and LREE (La, Ce, Nd). The content of other incompatible elements (Nb and Ta, Zr, Ti, Y and the HREE) is either similar in both types or lower in the arc basalt than in the N-MORB. These features support a commonly accepted hypothesis that arc basalts derive from the partial melting of harzburgitic upper mantle at least as residual as the source of common MORB (regarding Y, Zr, Nb and HREE). Prior to melting, such a depleted mantle would have been selectively enriched in LILE and LREE originating from the subducted slab, and carrying a 'sedimentary' Sr–Nd isotopic imprint. The most likely metasomatic agent responsible for transferring of these elements from the slab to the overlying mantle wedge is the water expelled from the downgoing slab. Indeed:

(i) the sediments and even the altered basalts from the top of the ocean crust are relatively water-rich (1.7 weight% for the average of basalts from Hole 504B);
(ii) large ion alkali and alkali-earth elements (Rb, Ba, K, Sr) are easily soluble in water whereas Nb, Ta, Zr, Ti, Y which are insoluble cannot thus be transferred into the arc mantle by hydrous fluids;
(iii) the very uncommon xenoliths of mantle peridotite found in arc lavas (Batan) contain hydroxyl-bearing metasomatic minerals (pargasite-type amphibole, phlogopite);
(iv) at temperatures close to 1000°C which are usually reached at depths of ca. 100 km in the arc-mantle wedge, melting may occur only in the presence of water (Figure 2.1);
(v) whatever the dip of the Benioff plane, the volcanic axes of most arcs overlie it for depths ranging from 90 to 120 km;

(vi) finally, arc magmas are more water-rich (1 to 5 weight% at the pre-eruptive stage) than MORB and OIB (<0.5%), as evidenced by their frequently explosive eruptions.

Considering hydrous fluids as the principal metasomatic agents in the arc mantle (Tatsumi and Kogiso, 1997) provides a simple explanation for the most general and obvious geochemical feature of arc magmas, i.e. their depletion in insoluble incompatible elements (HFSE) (High Field Strength Elements) with respect to water-soluble ones (LILE) of similar incompatibility. This hypothesis has also received experimental support. Tatsumi *et al.*, (1986) have shown that a trace-element-spiked serpentinite dehydrated under a pressure of 1.2 GPa selectively loses its LILE (Cs, Rb, K, Ba) as well as La, whereas Nb and HREE do not migrate in hydrous fluids. Other experiments in similar conditions (Brenan *et al.*, 1994) as well as amphibolite dehydration studies (Kogiso *et al.*, 1997) have demonstrated the mobility of U, Pb, Ce, and the immobility of Zr and Ti. Such apparently simple behaviour should make it easy to recognize hydrous fluid-driven metasomatism from the lack of transfer of HFSE and HREE to the source of arc magmas. Unfortunately, later-occurring complex petrogenetic processes, e.g. crustal contamination and AFC, result in geochemical scatters which often prevent a precise characterization of the initial sources.

The water from the subducted crust is transferred progressively into the arc crust, then into the underlying mantle, during the initial stages of descent. The final stage of dehydration occurs at depths of ca. 55–70 km (1.5–2 GPa) with the transition from amphibolite facies to eclogite facies. The expelled water and its dissolved incompatible elements are incorporated within metasomatic Mg-amphibole (pargasite) and Mg-mica (phlogopite). The 1000°C isotherm corresponding to the dehydration melting of mantle triggered by pargasite breakdown is encountered at depths of 90–120 km, i.e. below the volcanic-arc axis. This implies that the metasomatized part of the mantle wedge is dragged 20–30 km down or even more by convection mechanically coupled with the motion of the descending slab. Examples of incipient subduction or initiation of subduction suggest that arc magmatic activity begins as soon as the subducted slab has reached a depth of 100 to 120 km, suggesting that the rate of descent of metasomatized mantle is more or less similar to the rate of slab subduction.

11.3.5 Melting of the subducted ocean crust

Defant and Drummond (1990) coined the word 'adakite' (from Adak Island, Aleutians) for intermediate/acidic arc-related magmatic rocks ($SiO_2 > 56\%$), of various petrographic types: andesite/diorite, dacite/granodiorite or rhyolite/granite. Adakites are usually sodium-rich (Table 2.5), and mainly differ from their 'normal' calc-alkaline equivalents by their very low Y and HREE contents (Figure 11.16), usually less than twice the primitive mantle value ($Y < 10$ ppm; $Yb < 1$ ppm). They represent the modern equivalents of the Archean TTG (Tonalites–Trondhjemites–Granodiorites) as described by Martin (1986, 1993). They are usually richer in Sr

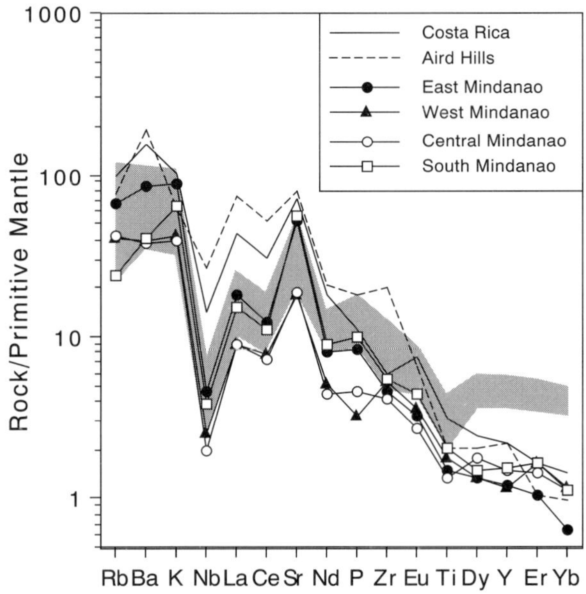

Fig. 11.16. Composition of adakites compared with 'normal' calc-alkaline lavas from Leyte, Philippines (shaded field), from Maury *et al.*, (1996). Normalization to the primitive mantle after Sun and McDonough (1989).

than their 'normal' calc-alkaline counterparts (Figure 11.16), from which they can be readily distinguished using a (Sr/Y)/Y plot (Figure 11.17). They have a typically MORB-type isotopic signature, although some of them show a minor sedimentary or continental crust imprint. More than 20 Plio-Quaternary adakitic provinces are presently known all around the Pacific (Figure 11.18). They belong to 15 of the 34 active arcs mentioned in Table 11.1. An equivalent number of older (Miocene, Paleogene or Mesozoic) adakite occurrences have been found, and new ones are described every year.

The very low Y and HREE content of adakites strongly suggest that they derive from the partial melting of a source containing residual garnet which concentrates these elements. Their isotopic compositions indicate this source is of MORB-type. Given the fact that they usually occur in active subduction zones where they overlie 50–80 km-deep Benioff planes, their most likely source is the downgoing basaltic ocean crust metamorphosed into garnet amphibolite or eclogite facies (Figures 11.7 and 11.12).

Experimental data on melting of basaltic materials under pressures consistent with depths lower than 150 km are shown in Figure 11.19. In the absence of water (eclogite facies) melting begins only at ca. 1100°C (dry solidus of Green, 1982), a temperature very difficult to reach at moderate depths in modern subduction zones. In the presence of excess water, the solidus temperatures are lowered down to 650–700°C, but this may seem unrealistic because of the progressive expulsion of water

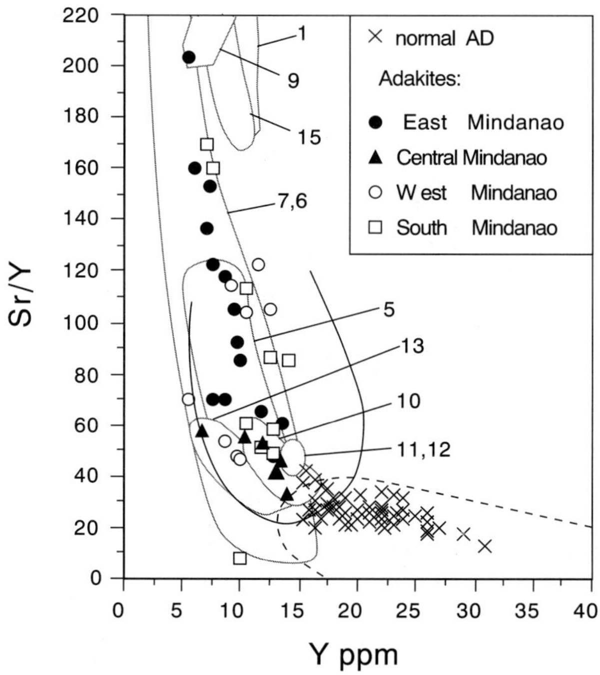

Fig. 11.17. Plot of Sr/Y against Y (from Maury *et al.*, 1996) showing the fields of adakites (full line) and mantle-derived magmas (dashed line). Circles and squares: adakites from Mindanao; ×: mantle-derived andesites and dacites from Leyte; dotted line: adakites from other areas (numbers as in Figure 11.18).

from the descending crust. Dehydration melting due to amphibole breakdown can begin between 800 and 1000°C according to data of Sen and Dunn (1995) or Rapp *et al.* (1991). The liquids corresponding to partial melting degrees of 10–30% are acidic and their compositions are consistent with those of natural adakites. The residues of this partial melting are plagioclase-free but amphibole- and garnet-rich, which is consistent both with the high content in Sr (concentrated in plagioclase) and low Y and HREE (concentrated in garnet) contents typical of adakitic magmas.

Adakitic magmas are thus likely to form through partial melting of MORB-type crust within a set of P-T conditions defined by the lack of plagioclase and the presence of residual garnet and amphibole (garnet amphibolite residue). The corresponding field (shaded in Figure 11.19 according to the results of Green, and Sen and Dunn) corresponds to the 'adakitic window' which has to be crossed by the P-T paths of subducting crust to generate adakites.

These P-T paths vary considerably according to the age of the downgoing crust, the subduction rate and the average dip of the Benioff plane (Table 11.1). An additional factor which is still poorly known resides in the thermal effects of shear stresses, which are negligible according to some authors and for others may reach up to 300°C in cases of rapid and oblique subduction (shear heating: Peacock *et al.*,

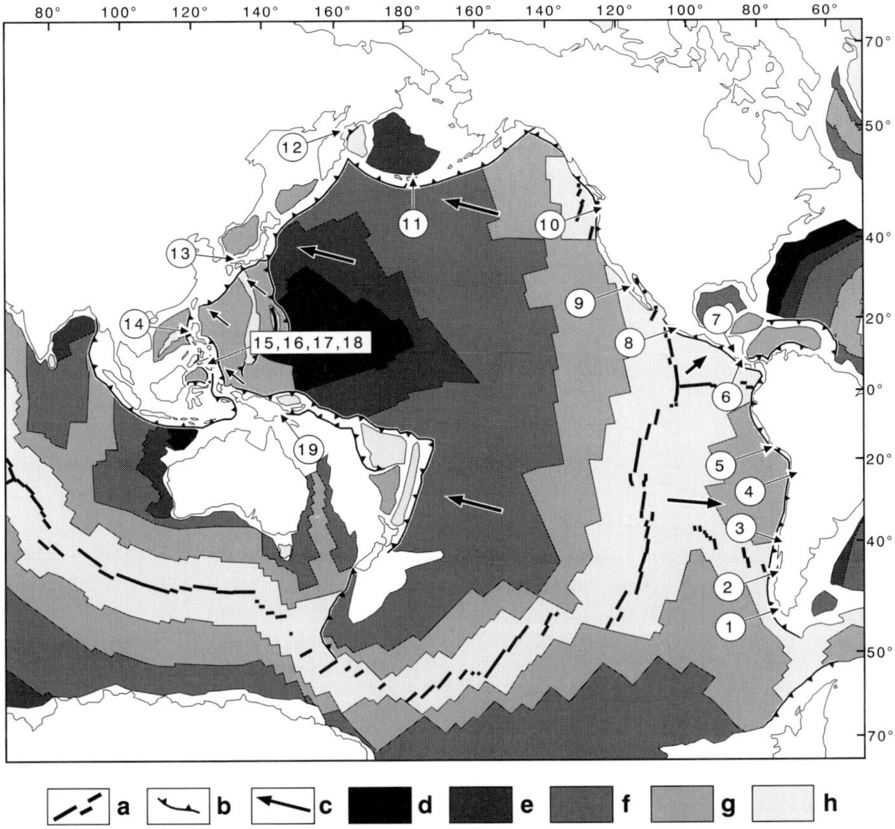

Fig. 11.18. Occurrences of recent adakites (from Maury *et al.*, 1996). (1) Austral/Andes; (2) Taitao and Cerro Pampa, Chile; (3) Central Chilean Andes; (4) El Negrillar, Central Andes; (5) Cordillera Blanca, Peru; (6) Panama; (7) Costa Rica; (8) Guatemala; (9) Baja California; (10) Cascades (Mount St Helens); (11) Adak Island and Aleutians; (12) Kamchatka; (13) Shikoku, Japan; (14) Luzon, Philippines; (15) East Mindanao, Philippines; (16) West Mindanao and Negros; (17) Southeast Mindanao; (18) Central Mindanao; (19) Aird Hills, New Guinea. a: Spreading ridges; b: subduction zones; c: plate motions. Ages of subducting crust: d: Jurassic; e: Early Cretaceous; F: Late Cretaceous; g: Paleogene; h: Neocene.

1994). The P-T paths (geotherms) of subducting crust reported in Figure 11.19 according to Peacock et al. do not usually pass through the adakitic window with the exceptions of curve 1 (subduction of an active ridge), curve 4 (high shear heating) and of course curve 10 (Archean geotherm, effective for the genesis of the TTG; see § 11.5.3). However, adakites do occur in most arcs where the subducting crust is younger than 50 Myr (Figure 11.10) although the corresponding theoretical geotherms (curves 3, 6, 7, 8) seem to preclude their melting.

Several explanations have been proposed for this apparent discrepancy (see Maury *et al.*, 1996, for a review). The thermal regime of the upper part of the subducting crust may be significantly enhanced by a number of factors including high shear heating (rapid and/or oblique subduction, initiation of subduction or end

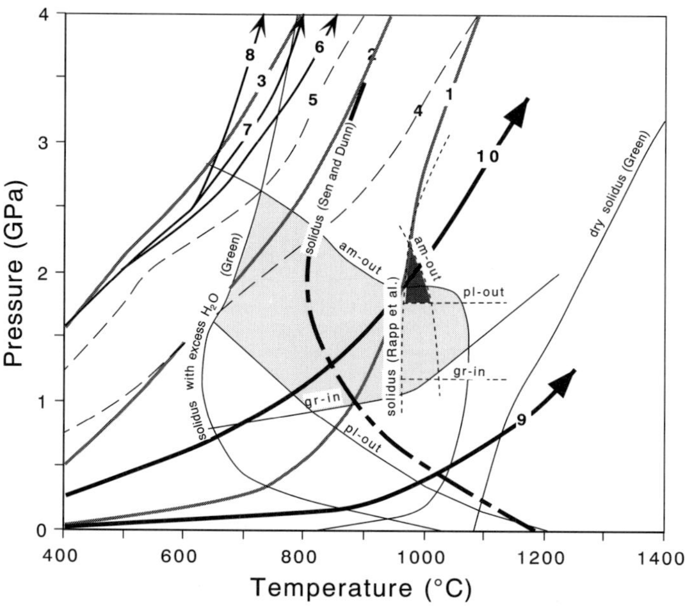

Fig. 11.19. Experimental constraints relevant to adakite genesis (from Maury *et al.*, 1996). This pressure–temperature diagram shows: the solidi of basalt in dry conditions and in the presence of excess water (Green, 1982); the dehydration melting of amphibole/eclogite in the absence of vapour (Rapp *et al.*, 1991) and the dehydration melting of amphibole (Sen and Dunn, 1994). Abbreviations: am, amphibole; pl, plagioclase; gr, garnet. Numerical models of the various P-T paths of subducted ocean crust (Peacock *et al.*, 1994) are also shown. Curves 1, 2 and 3 correspond to P-T paths of crusts aged up to 0 Myr (active ridge subduction), 5 Myr and 10 Myr, respectively, for a subduction rate of 3 cm/yr. Curves 4 and 5 correspond to cases of high-shear heating (100 and 67 MPa) during fast subduction (10 cm/yr). Curves 6, 7 and 8 correspond to incipient subduction of 50-Myr-old crust with subduction rates of 10, 3 and 1 cm/yr, respectively. Finally, curves 9 and 10 represent the P-T paths of Early Archean and Middle Archean mantle, respectively (Martin, 1987; Rapp *et al.*, 1991).

of subduction. Additional water may be provided by the dehydration of mantle serpentinites underlying the ocean crust or occurring within fracture zones, and the release of this water can substantially lower the solidus of MORB crust. Anyway, adakites are clearly much less rare than initially supposed, and remain rather good geodynamic markers of subduction of young ocean crust (two-thirds of the known occurrences correspond to slab ages younger than 20 Myr). There is increasing evidence that they could also trace melting, in post-collisional settings, of slivers of ocean crust incorporated within the thickened lithosphere of collision zones (Papua–New Guinea, Mindanao).

The experimental data discussed above apply only to the melting of basaltic/gabbroic rocks. What is the fate of the associated downgoing sediments? The isotopic signature of adakites is usually MORB-like, sometimes with additional contamination by continental crust (adakites from the Andes), but this may traduce the melting of young crust with a thin (or even lacking) sedimentary pile. High-pressure melting experiments of pelagic clay-rich sediments under 4 GPa (corresponding to depths of ca. 120 km) are consistent with a relatively low solidus at 650–700°C in presence of excessive water (Nichols *et al.*, 1994). These temperatures are nearly identical to those of Green's basalt solidus under equivalent conditions (Figure 11.19). So it seems likely that, if the young downgoing ocean crust contains a significant sediment component, it will melt in the same conditions as the associated basalts, and the two melts will probably mix. A 'sedimentary imprint' can thus be carried into the mantle wedge by sediment melts (Shimoda *et al.*, 1998) or by adakitic magmas having incorporated them; if these melts metasomatize the mantle, this process can quite adequately be described as bulk sediment addition from viewpoint of Sr, Nd and Pb isotopes.

In addition to being emplaced at the surface or within the arc crust, do adakitic magmas contribute to mantle wedge metasomatism, as envisioned by Ringwood and suggested by the data on Batan ultramafic nodules? Many potassic calc-alkaline magmas are, like the young Batan lavas, enriched in Nb and Ta with respect to island-arc tholeiites (although their multi-element patterns still present negative anomalies in these elements). This feature could indicate that their source has been metasomatized by adakitic magmas which contain significant amounts of Nb and Ta (Stolz *et al.*, 1996). Moreover, highly unusual arc basalts referred to as high-niobium basalts are often found in spatial and temporal association with adakites. They display a MORB-like isotopic signature together with Nb and Ta contents much higher than typical subduction-related magma values (Table 2.5 and Figure 2.6). They might derive from the melting of pargasite-rich mantle formed through extensive reaction of harzburgite with adakitic melts, because pargasitic amphibole is able to admit in its structure fairly high amounts of Nb and Ta.

11.3.6 Simplified petrogenetic models

The genesis of arc magmas is illustrated very schematically in Figure 11.20. When the subducted slab is old (on right) the major or even only metasomatic process is the migration of hydrous fluids from the subducted crust to the mantle wedge (small arrows). The corresponding metasomatic pargasite-phlogopite-bearing mantle (pattern 7) is dragged at depth by convection (large arrows). When it reaches ca. 100 km (corresponding to the 1000°C isotherm) the breakdown of pargasite triggers melting. On their way to the surface, the magmas interact (I) with the upper mantle and then the crust, modifying their wallrocks and being in turn contaminated by them. Within the crust, they are usually stored within magma reservoirs where they evolve by fractional crystallization (C) often coupled with assimilation (AFC) before being erupted on the surface or crystallizing at depth.

In arcs resulting from the subduction of young crust (left side of Figure 11.20) the same processes happen, resulting in the emplacement of 'normal' calc-alkaline magmas. In addition, adakitic magmas derived from the melting of garnet amphibolites from the subducting crust at depths of 60–75 km can migrate towards the surface, interacting during their ascent with the lithospheric mantle and the arc crust (I). They can also be consumed through these reactions, leading to 'clusters' (pattern 8) of adakite-metasomatized mantle. These areas, geochemically different

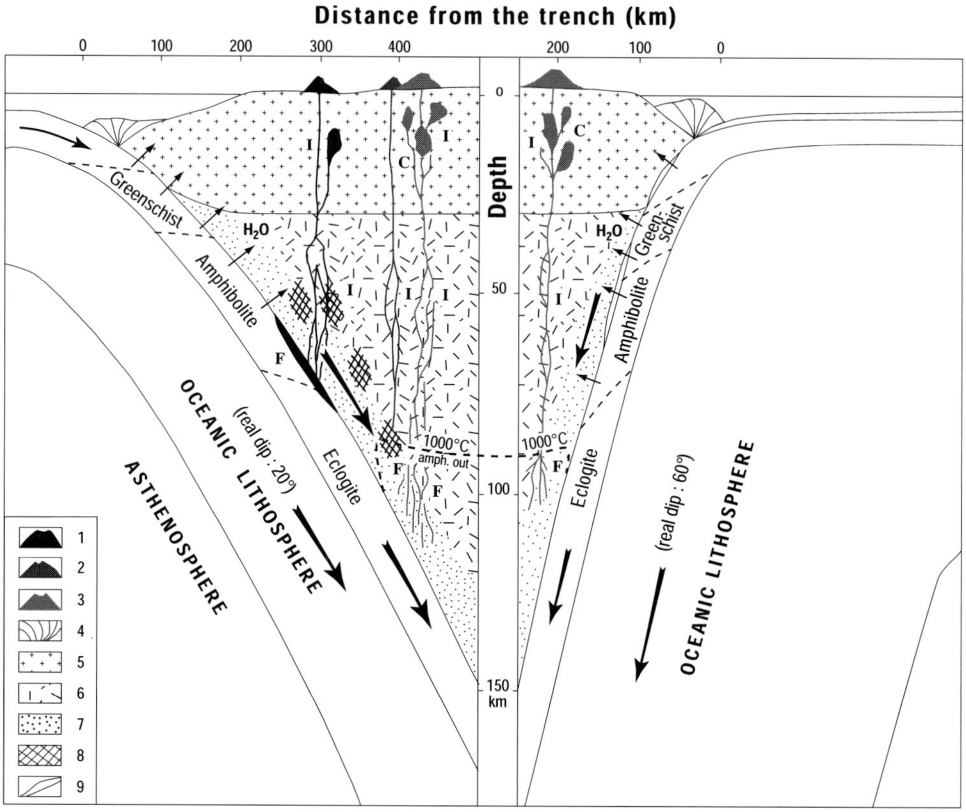

Fig. 11.20. Petrogenetic processes below volcanic arcs (vertical exaggeration ×5). Right; subduction of a 120-Myr-old oceanic lithosphere (based on the Sunda and Lesser Antilles examples); left: subduction of a 10-Myr-old lithosphere (based on the Panama–Costa Rica and Mindanao cases). (1) Adakitic volcano; (2) Nb–Ta-rich basaltic volcano; (3) 'normal' calc-alkaline volcano; (4) accretionary prism; (5) arc crust; (6) depleted lithospheric mantle; (7) hydrous fluid-metasomatized mantle; (8) adakite-metasomatized mantle; (9) sediments. F: Partial melting zones; I: sites of interactions between magmas and mantle or crust (contamination, assimilation, metasomatism); C: sites of fractional crystallization.

from hydrous-fluid-metasomatized mantle, can also be dragged down by convection and melt when they reach the 1000°C isotherm to form basaltic magmas rich in Nb and Ta or at least less depleted in these elements than common calc-alkaline basalts.

Recent studies suggest that these two petrogenetic models implying contrasted metasomatic agents (hydrous fluids and adakitic silica-rich magmas, respectively) are complementary rather than mutually exclusive. There is increasing experimental evidence that adakitic magmas are water-rich (>5 weight% at depth). At 1000°C and 1 GPa, a continuity is observed from hydrous fluid to water-rich silica melt (Shen and Keppler, 1995): the hydrous fluid and adakitic inclusions occurring together in Batan peridotites may derive from low-pressure demixing of such a 'hybrid' metasomatic agent.

11.4 FORE-ARC AND BACK-ARC MAGMAS

The example of Java shows that magma emplacement may occur off the volcanic axis of arcs, especially in the back-arc domain. When this area is affected by an extensional tectonic regime (Mariana-type arcs), back-arc magmatism often results in the formation of a new oceanic domain, the back-arc basin, characterized by an active spreading axis similar to the main ocean ridges. Emplacement of magmatic rocks can also occur in the fore-arc domain, where it is usually linked to subduction of an active ridge or of a very young crust.

11.4.1 Magmatic effects of active ridge subduction

There are three present-day examples of subducting active ridges (Table 11.1). The kinematics and the magmatic effects of the Late Miocene subduction of the east Pacific rise below California are also well known.

The subduction of an active ocean ridge has three kinds of consequences. First, the production of MORB at the ridge axis is not immediately interrupted, and these basaltic magmas uprise through the crust of the arc where they can be stored before being erupted at the surface. For instance, the subduction of the Chile Ridge beneath the Taitao Peninsula (South Chile) resulted in the emplacement of Pliocene volcanics in a shallow marine environment, only 10 km away from the trench (Figure 11.21). Their compositions range from those of MORB, similar to those emplaced along the ridge axis off the trench, to those of calc-alkaline andesites and dacites. The La/Nb ratios of these rocks increase with silica content, and are also correlated with an

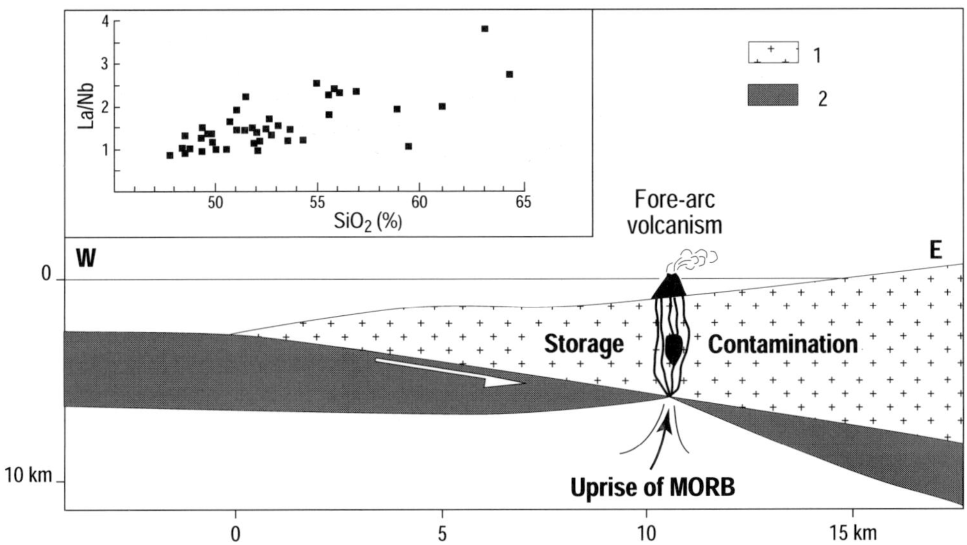

Fig. 11.21. Drawing depicting the subduction of the active Chile Ridge below the South American margin (Taitao Peninsula) during the Pliocene (from Lagabrielle *et al.*, 1994).

increase in their $^{87}Sr/^{86}Sr$ ratios. These features are consistent with an AFC process involving the progressive assimilation of the Chile margin continental crust by MORB magmas.

A second likely consequence of subduction of an active ridge is a shallow partial melting of the downgoing ocean crust below the fore-arc domain, generating adakitic magmas. Pliocene adakites occur in the Taitao Peninsula, and are considered as formed by melting, at depths of ca. 30 km, of the very young (and thus very hot) Chile ridge ocean crust. Neogene magnesian andesites having an adakitic signature ('bajaites') occur in Baja California, where they formed after the subduction of very young ocean crust, close to a ridge axis.

The heating of the crust and lithospheric mantle of the arc by the subducting ridge may generate their partial melting, generating either crustal melts (anatectic rhyolites and granites) or boninitic melts (see below) if the mantle wedge is harzburgitic. These effects are enhanced when the downgoing ridge is cut by transform faults: in the case of oblique convergence, parts of the arc lithosphere may be successively heated by two (or even three) active ridge segments. When the production of MORB at the ridge axis ceases, an asthenospheric window may open because of the divergence of the two sides of the former ridge. Such a window connects the hot asthenospheric mantle of the ridge with the overlying lithospheric arc mantle. If the latter is enriched (subcontinental-type), its melting may lead to the emplacement of highly alkalic magmas (absarokites, minettes, lamprophyres) which have been recognized for instance in California and in the Andes.

11.4.2 Boninites and their origins

Boninites are rather unusual fore-arc lavas (Table 2.5). More silica-rich than basalts ($SiO_2 = 53$–60%), they are characterized by high MgO content (8–20%). They are typically rich in hydrated glass, plagioclase-free and contain phenocrysts of Cr-spinel (magnesiochromite), olivine, Mg-rich orthopyroxene and calcic clinopyroxene, and sometimes clinoenstatite, a rare calcium-free magnesian clinopyroxene. Boninites have a very low content in incompatible elements, usually less than ten times the primitive mantle (Figure 2.6). Their REE patterns are either flat (Figure 11.22) or LREE-depleted, and REE concentrations much lower than those of MORB and IAB. Like the latter, their spidergrams display negative anomalies in Nb and Ta and selective enrichment in Rb, K, Ba and Sr with respect to other highly incompatible elements (Figure 11.22).

Such features suggest a rather unusual set of conditions for their genesis. Their very low incompatible element concentrations indicate high degrees of melting of depleted suboceanic mantle. In addition, the high water content and selective enrichment in LILE of boninitic magmas are consistent with a metasomatism of their source by hydrous fluids. Their lack of depletion in HREE relative to LREE (contrarily to adakites) suggest a garnet-free melting residue, which implies relatively shallow melting at depths lower than 50–80 km (Figure 11.23).

Experimental data on depleted peridotites indicate that their dry solidus temperatures increase from 1200 to 1400°C with depth (Figure 11.23; see also Figure

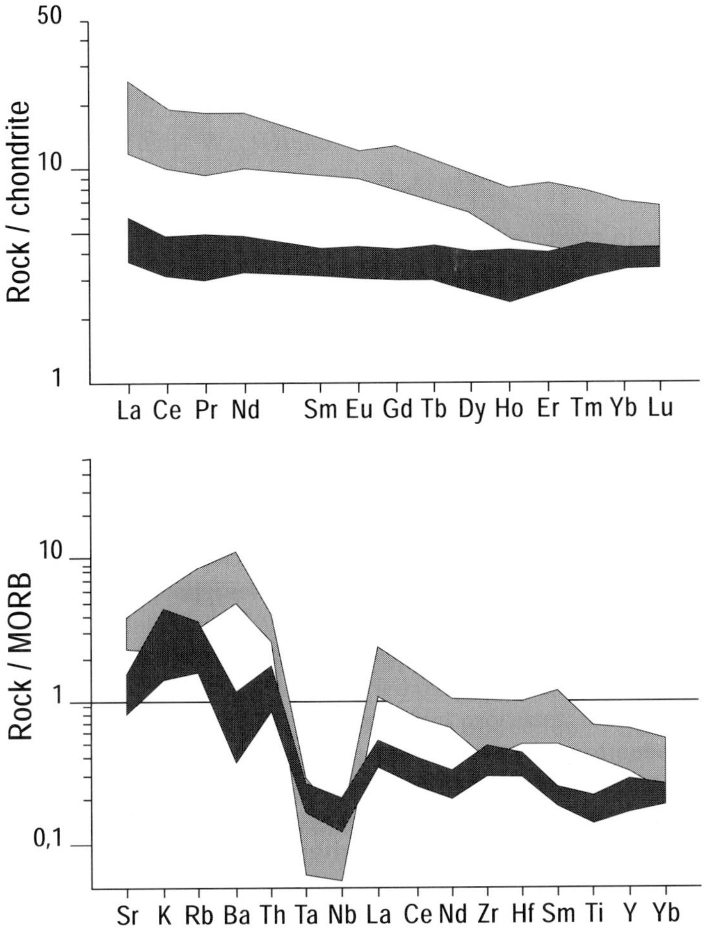

Fig. 11.22. Composition of boninites compared with that of associated calc-alkaline basalts. Example of the Marianas–Bonin from Pearce *et al.* (1992).

2.1): such thermal conditions are unlikely to occur in modern subduction zones. In the presence of 2% water, melting begins at 1200°C whatever the pressure. As a temperature of 1000°C is only usually reached at depths of 100 km in the arc mantle (Figure 11.20), the genesis of boninites requires a P–T path 200°C (or even 300°C to get high melting degree) hotter than usual. For instance, the adiabatic uprise (R in Figure 11.23) of very hot depleted asthenospheric mantle (1280°C) will allow it to remain significantly above solidus until it reaches the surface if the ascent is instantaneous, or until it reaches depths of 15 km and 40 km for uprise times of 2 Myr and 4 Myr, respectively.

Boninites should thus occur within tectonic settings involving subduction of active ridges or very young crust. However, the interpretation of their occurrences is

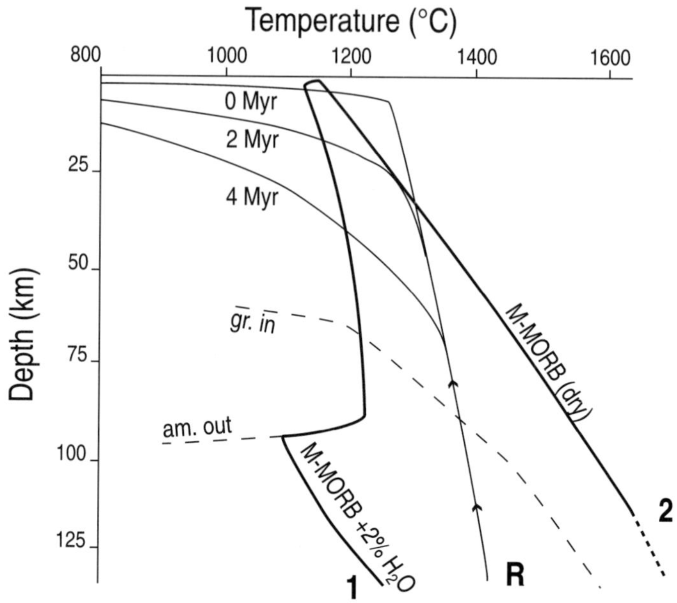

Fig. 11.23. Conditions of boninite genesis (after Pearce *et al.*, 1992). M-MORB: depleted mantle source of MORB; R: adiabatic rise of a diapir of M-MORB; 1: melting of M-MORB in presence of 2% water; 2: dry melting of M-MORB; am: amphibole: gr: garnet.

sometimes difficult, because most of them correspond to relatively old rock sequences. The type occurrences are Eocene (47–42 Myr) lava flow and pyroclastic sequences, located in three small islands from the Bonin fore-arc domain. They have also been found in the ODP holes drilled in this fore-arc area. At the time of their emplacement the subducting crust of the West Philippines basin (47–48 Myr) was very young, as well as that of the Shikoku basin (25–17 Myr) when its subduction led to the emplacement, at 13 Myr, of the Southern Japan boninites (Tatsumi and Maruyama, 1989). The occurrence of magnesian andesites intermediate between boninites and adakites suggests that these rock types can be emplaced in similar tectonic settings. In addition, boninites occur in back-arc basins which opened close to the volcanic-arc axis, e.g. in the Northern Marianas basin and the Lau basin. They have often been described in ophiolitic series considered as formed in back-arc basins (SSZ: Supra-Subduction Zone ophiolites; see § 6.2.4).

Although most authors agree that boninites are derived from the shallow high-degree melting of garnet-free hydrated residual mantle, the corresponding possible tectonic settings remain debated. The two most common models (Figure 11.24) respectively imply: (i) subduction of a young lithosphere below a ridge (the water from the subducted crust migrates into the depleted asthenospheric mantle below the ridge, increases its LILE content and triggers its melting), and (ii) subduction of an active ridge below a young lithosphere. In the latter case, the subducting slab provides water (presumably from serpentinites from transform faults) and LILE,

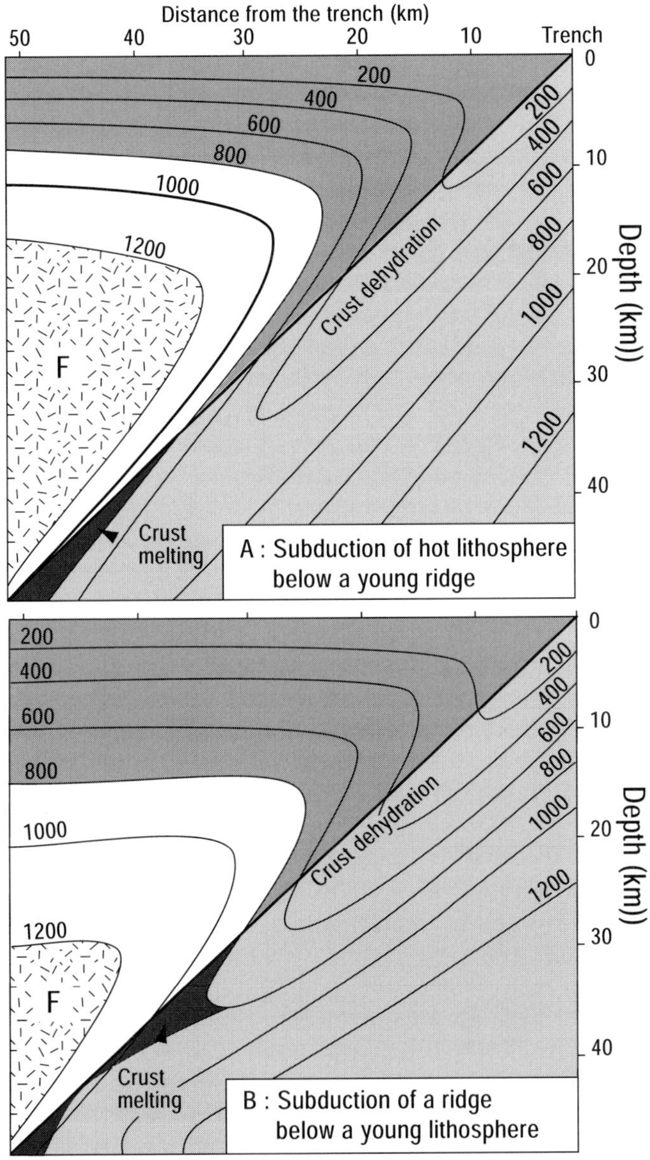

Fig. 11.24. Theoretical thermal models relevant to boninite genesis (after Pearce *et al.*, 1992). F: sites of partial melting of depleted mantle. Isotherm temperatures in °C.

and the 'blowtorch effect' of the downgoing ridge triggers melting of the arc mantle. It is noteworthy that although melting of subducted crust (shaded areas) should occur together with melting of depleted mantle in these two models, field associations of adakites with boninites have not yet been described.

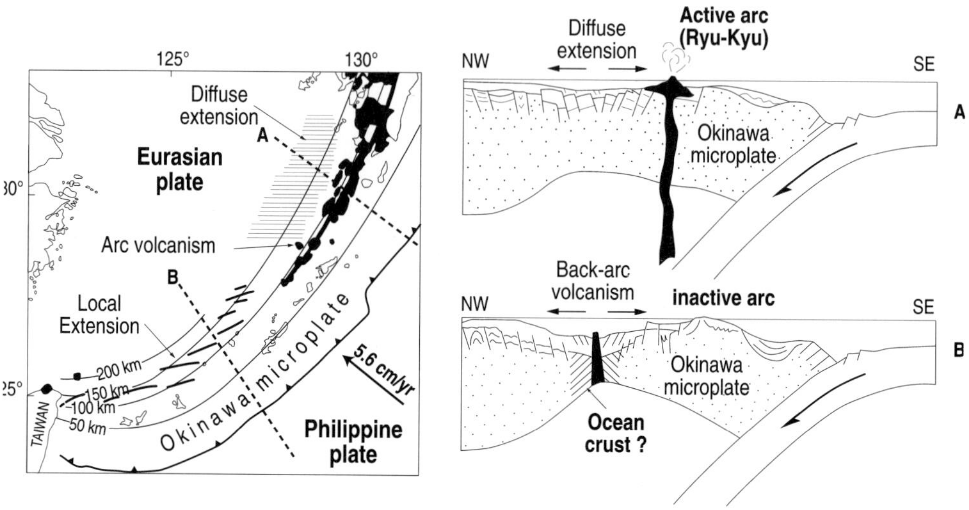

Fig. 11.25. Early opening of a back-arc basin: the Okinawa Trough (from Sibuet *et al.*, 1987).

11.4.3 Back-arc basin magmas

The example of the Okinawa basin illustrates the initial stages of opening of back-arc basins (Figure 11.25). The Ryu Kyu arc results from the subduction of the Paleocene–Eocene Philippines plate below the Okinawa microplate. At the rear of the active arc (in the north) back-arc extension occurs in a rather diffuse way, and is traduced by a series of grabens devoid of young magmatic rocks. In the southern Ryu Kyu, on the contrary, the volcanic arc is extinct. In the corresponding back-arc domain, overlying a 100 to 160-km-deep Benioff plane, occur a set of grabens *en échelon* characterized by very narrow (<1 km wide) central volcanic ridges. They are made up of vacuolar (presumably water-rich) basalts and subaphyric dacites. The Okinawa back-arc basalts are roughly similar to E-MORB from which they differ, however, by their more radiogenic Sr signature and by their negative Nb anomalies similar to those of arc magmas (Figure 11.26). These features indicate that their petrogenesis was strongly influenced by the subducting plate. The Okinawa basin began to open only 2 Myr ago (Sibuet *et al.*, 1987), and the volcanic axes of the grabens are not yet connected to form a single back-arc ridge.

Tertiary and Quaternary back-arc basins are common in the Southwest Pacific where some of them reach large sizes (South China Sea, Celebes Sea, Sulu Sea). The successive opening, at the rear of the Bonin and Mariana trenches, of the West Philippines basin, then of the Parece-Vela and Shikoku basins, followed by the Marianas and Sumisu basins, have led to the progressive motion of these active trenches away from the Eastern Asia margins.

All the transitions exist between: (i) 'genuine' back-arc basins, the opening of which reflects an extensional tectonic regime due to the decoupling between the upper plate and the subducting plate, and (ii) basins which may be located in back-

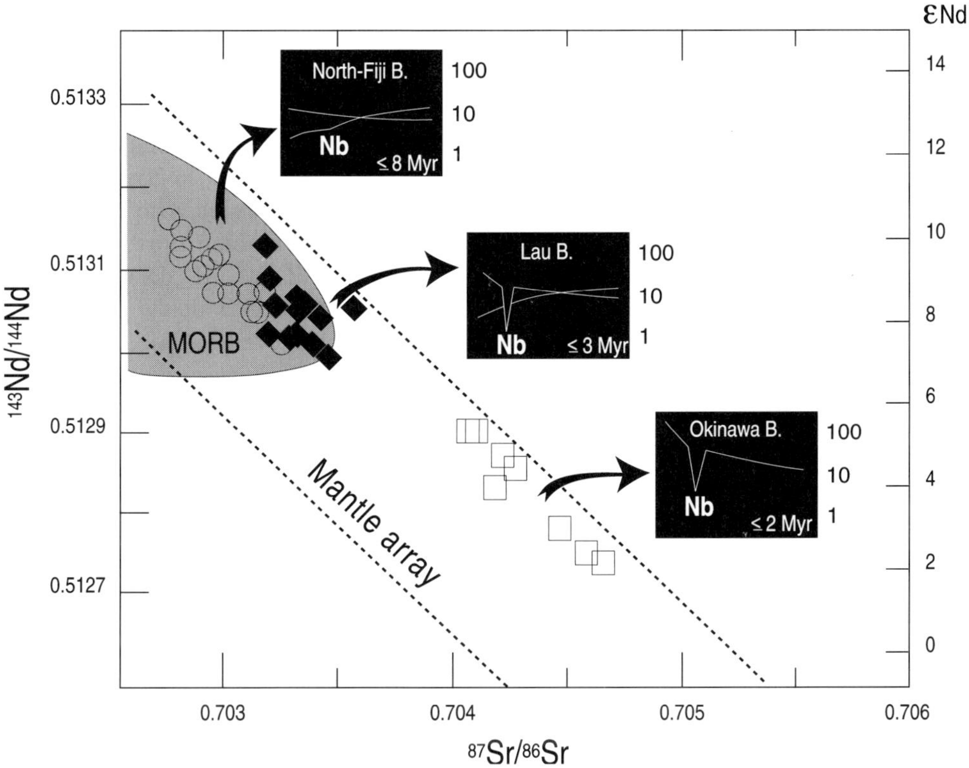

Fig. 11.26. Geochemical evolution of back-arc basin basalts (from Auzende *et al.*, 1990).

arc position, but the opening of which was controlled by other tectonic processes (these latter have not been included in Table 11.1). For instance, the shape of the Japan Sea, often regarded as a typical back-arc basin, is mostly controlled by major dextral strike-slip faults which delineate its western and eastern margins. The Andaman basin, north of Sumatra, is a typical pull-apart basin, and the Woodlark basin currently opening within the New Guinea margin can be considered as a submerged continental rift (Benes *et al.*, 1994). The opening of the Tyrrhenian Sea is only partly due to the Eolian arc subduction. The North Fiji Basin is the largest (350 000 km^2) and the most active of the genuine Quaternary back-arc basins with five active ridges the total length of which is ca. 3000 km (Lagabrielle *et al.*, 1996). It is located in a particularly complex tectonic setting, being framed by three subduction zones corresponding to the Vanuatu (in the west), Vitiaz in the north (inactive) and Tonga (in the east) trenches. Although the initial stages of its opening, during the Late Miocene, seem mostly related to the Vanuatu subduction, it is presently located above the deep part of the Pacific plate subducting beneath Tofua/ Tonga.

The composition of back-arc basin basalts vary with the distance that separates the back-arc active basaltic ridge from the trench and thus reflects in some respects the history of the opening of these basins (Figure 11.26). During the initial stages of opening (Okinawa), the geochemical imprint of subduction is strong, as indicated by the radiogenic Sr isotopic ratios, important negative Nb and Ta anomalies and the water-rich character of basalts. With the progress of oceanization, this imprint becomes weaker (e.g. in the lavas from the Lau basin, active for 3 Myr), and N-MORB occur together with Nb-depleted basalts (Figure 11.26). Finally, in large back-arc basins where the axial ridge is located far away from any trench (e.g. the North Fiji basin, active for 8 Myr) the composition of basalts becomes indistinguishable from that of N-MORB from large ocean ridges (Figure 2.4).

11.5 THE FATE OF SUBDUCTED OCEANIC CRUST

The oceanic crust created at ocean ridges and involved in plate motions will not remain in its original position on the ocean floor for more than 150 to 200 Myr. Its possible fates include: (i) subduction involving dehydration and/or melting followed by partial recycling into the arc crust, (ii) incorporation as ophiolitic slivers into the continental crust through obduction processes generally linked to the early stages of collision, (iii) tectonic underplating at the bottom of the continental crust of collision zones or the thickened crust of Andean-type active continental margins, and finally (iv) recycling at depth into the asthenospheric upper mantle or even the lower mantle.

11.5.1 From subduction to collision

The end of the activity of a magmatic arc results either from a general or regional change in plate kinematics like that which occurred during the Eocene (43 Myr) in the Pacific, or, more frequently, because of a collision. For instance, in the north of the Luzon Arc (Figure 11.13), the interruption of volcanic activity seems linked to the collision between the arc and the China continental margin. The subduction of an aseismic ridge (D'Entrecasteaux Ridge in the Vanuatu trench) or of a fossil spreading ridge (like that of the South China Sea in front of northern Luzon, Figure 11.13) usually results in a temporary interruption or paucity of volcanic activity. Within the mosaic of microplates of southeast Asia, the arrival at the trench of a continental microblock or even of an oceanic crust area thickened by plateau magmatism often seems enough to induce the blocking of subduction followed by a local kinematic reorganization. The Banda Arc in Eastern Indonesia (Figure 11.1) is apparently the only present-day example of continuation of volcanic activity during the subduction of a continental passive margin (the Australian shelf).

The magmatic series from collision zones display petrologic and geochemical characteristics remarkably similar to those from Andean-type active continental margins (Harris *et al.*, 1986): prominence of andesites, occurrence of anatectic rhyolitic magmas usually emplaced as ignimbritic flows, strong crustal isotopic imprint generally traducing the occurrence of AFC, MASH and related processes. After the end of subduction-related calc-alkaline activity, the syn-collisional stage is usually marked by the emplacement of anatectic magmas as granites and ignimbrites. The post-collisional stage corresponds to the emplacement of potassic calc-alkaline series showing strong evidence of crustal contamination. These magmas can post-date the end of collision by as much as 25 Myr (Tibet: Coulon *et al.*, 1986). They are followed by distension-related intraplate alkali basalts which still retain for several Myr negative Nb and Ta anomalies that may traduce either their contamination by continental crust or the persistence at depth of the mantle metasomatized during previous subduction events.

The sequence of magmatic activities linked to the Alpine collision, however, follow a rather distinct pattern. The emplacement of anatectic granites, of strongly contaminated potassic calc-alkaline granitoids and of leucite-bearing ultrapotassic dykes likely derived from subcontinental mantle occurred almost concomitantly from 42 to 25 Myr. Such a short magmatic history could reflect a detachment of the deep part of the subducting lithosphere, resulting in the opening of an asthenospheric window and the subsequent heating and melting of the subcontinental mantle and the overlying crust (von Blanckenburg and Davies, 1995).

The sources of post-collisional magmatism are not well known because the original geochemical signatures have usually been obliterated by the strong continental imprint resulting from AFC and MASH processes. The behaviour and fate of the downgoing oceanic crust after the end of subduction remains problematic. Geophysical data (regional seismicity patterns, seismic tomography, deep crustal seismic profiles) often lead to suggest the occurrence, below young collision chains, of ocean crust slivers. These fragments could either accumulate at the bottom of the crust and contribute to its thickening (tectonic underplating) or pond into the underlying lithospheric mantle. As discussed above, the occurrence of adakites in collision zones suggests that these slivers could melt either at the end of subduction or during a post-collisional stage of adiabatic uplift.

11.5.2 Deep recycling of oceanic crust

Numerous arguments for the reality of deep recycling of subducted ocean crust have been developed. Arc magmatism accounts only for crustal recycling of a small fraction of the downgoing materials (§ 11.1.4). The most largely accepted interpretation of the enriched mantle end-members EMI (Enriched Mantle I), EMII (Enriched Mantle II) and HIMU (HIgh MU) are that they derive from 1–2-Ga-old subducted basaltic crusts and their associated sediments (§ 10.3; see also Hofmann, 1997). Isotopic studies of rare gas from mantle rocks and plume-related basalts also suggest the occurrence of sediment recycling into the deep mantle (Allègre *et al.*, 1987).

Deep mantle rocks are mainly known from the study of: (i) peridotitic xenoliths brought to the surface by alkali basalt flows and kimberlite pipes, the latter containing diamonds which are only stable at depths in excess of 150 km, and (ii) the high-temperature peridotitic intrusions from Alpine chains, which have been exhumated from depths corresponding to pressures up to 5 GPa. In all these occurrences layers or xenoliths of eclogitic composition (garnet pyroxenites) are often found. They are thought to represent fragments of ocean crust buried at great depths, sometimes within the diamond stability field. The Sr, Pb and O isotopic compositions of some of these pyroxenites are different from those of depleted mantle and N-MORB. For instance, the diamond-bearing garnet pyroxenites from Beni Bousera (Morocco) are more radiogenic and display rather high $\delta^{18}O$ values consistent with a contribution of seawater and/or pelagic sediments (Pearson *et al.*, 1993).

Many diamonds are rich in mineral inclusions which provide useful indications of their origin. The most common are microcrysts of garnet, pyroxenes and olivine, the compositions of which are consistent with those of upper mantle minerals at depths ranging from 150 to 400 km (Figure 11.27). However, garnet inclusions in diamonds are by far the most common, and their compositions are similar to those of garnets from eclogites rather than from peridotites. These features suggest that host diamonds crystallized within eclogitic rocks from subducted basaltic crust.

Extremely rare but fascinating minerals have been found as inclusions within diamonds (Sautter and Gillet, 1994). Majorite is a garnet containing excess silicium in octahedral sites, which forms from pyroxenes plus garnet at depths of 400 to 660 km corresponding to the transition zone between the upper and lower mantle (Figure

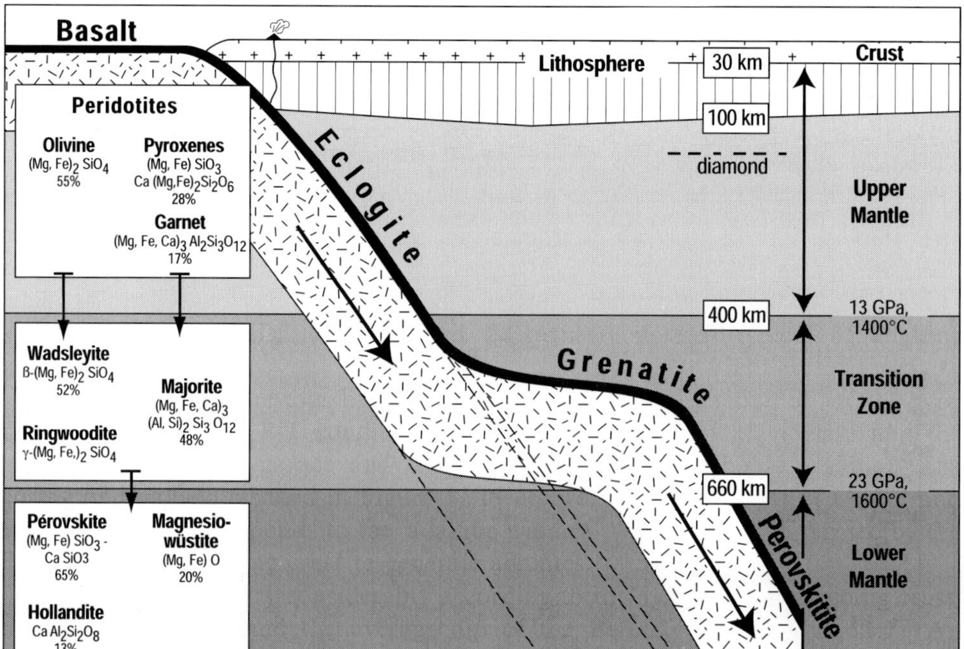

Fig. 11.27. Deep mantle evolution of subducted oceanic lithosphere (from Sautter and Gillet, 1994). See text for explanation.

11.27). Some majorite inclusions have an eclogitic-type composition, which leads us to consider their formation from subducted ocean crust converted to a nearly massive majorite garnet aggregate in the transition zone. Finally, perovskite and magnesiowüstite (Figure 11.27) have recently been found as inclusions within diamonds, suggesting an origin from the lower mantle of their host minerals.

Can the subducted ocean crust reach the mantle-core boundary? Some authors (Haggerty, 1994) consider that diamonds form in the D″ layer at depths of ca. 2900 km. Tomographic data suggest that the subducted lithosphere can reach depths of 1500 km or more below the major subduction zones (Wysession, 1996). Although the Indian slab subducting beneath Java does not seem to be interrupted in the transition zone, other mechanical behaviours are suspected to occur, e.g. tearing of the slab beneath Sumatra and sinking of its deepest part into the lower mantle (Widiyantoro and van der Hilst, 1996), or thickening and accumulation (ponding?) of the slab materials in the transition zone below the Tonga–Kermadec arc (van der Hilst, 1995). Such cases may be more favourable than that of Java to the mingling and dispersion of ocean crust remains within the mantle.

11.5.3 Subduction and continental growth

The average composition of continental crust (Table 2.1) is very close to that of a calc-alkaline andesite or even (with the exception of the HREE) of an adakite. A major geochemical role of arc magmatism is thus to extract continental crust-type materials from the oceanic crust: either directly through partial melting followed by emplacement of adakitic magmas, or indirectly through mantle metasomatism followed by melting, ascent of basaltic magmas and their differentiation towards acidic liquids within the arc crust (Figure 11.20). Some (or even sometimes most) of the oceanic sediments involved in this process are derived from the alteration of continental materials. In such a case arc magmatism only recycles (and rejuvenates) continental materials within the crust. At the same time the subduction process reinjects ocean crust residues into the mantle, together probably with part of the associated detrital sediments. It is thus unclear whether the bulk budget of subduction leads only to recycling of already existing continental crust or alternatively contributes to the progressive growth of continents at the expense of the oceanic crust and of the mantle from which it derives.

Most of the oldest rocks known on Earth have been found in the North Atlantic craton, which has been much more studied than many others. The oldest preserved remnants of continental crust are 3.9-Gyr-old gneisses from Canada. Komatiites (3.8 Gyr-old) display LREE depletion which may be inherited from their mantle source, previously depleted through extraction of oceanic or continental crust. The first TTG (Tonalites–Trondhjemites–Granodiorites, equivalent to modern adakites) appear at 3.6 Gyr from the melting of basaltic crust converted to garnet amphibolite (Martin, 1986). Typical MORB (3.3–3.1 Gyr old) occur in the Pilbara craton in Eastern Australia, but differ from modern ones by their less depleted character (Table 2.3). Bimodal komatiite–TTG associations are prominent in all cratons up to the end of the Archean at 2.5 Gyr. Magmatic associations more or less similar to modern ones appear during the Proterozoic, around 2.0–1.8 Gyr, in the Canadian shield. They include 'true' ophiolites similar to the Alpine ones, calc-alkaline arcs and ocean island-type basalts and associated rocks. However, melting of subducted crust remained a very common process until the end of the Proterozoic: for instance, in the Arabic shield of Upper Proterozoic age (860–560 Myr), 25% of the granitoids still present chemical characteristics consistent with those of TTG and adakites (Harris *et al.*, 1993).

Sr, Nd, Pb isotopic data can be used to constrain the models of continental growth, although there are a lot of discrepancies between the published models (for a review see Taylor and McLennan, 1985). Most continental crust (75–80% of the surface of continents) seems to have formed before the end of the Archean. Continental growth would then have slowed until the end of the Paleozoic (245 Myr), when a nearly 'steady state' of subduction-driven recycling was reached. On a global scale, the growth of continental crust seems to reflect the temporal decrease of TTG–adakite production through melting of subducted oceanic crust in the progressively decreasing thermal regime of the Earth's mantle.

12

Conclusion: the oceanic crust cycle

Two main problems arise from the study of the successive stages of evolution of the oceanic crust from its formation at the ocean ridge axes to its recycling within the mantle that underlies subduction zones.

—Which part of the Earth is involved in this cycle? Does it mainly affect the crust and the upper mantle, or does it concern the bulk mantle down to the D″ layer?

—Does this cycle occur in roughly similar conditions within the supercontinent cycle (Wilson's cycle) or is it interrupted by major tectono-magmatic crises, like the Early Cretaceous one, separated by nearly stable periods of magmatic production at the Earth's scale?

12.1 PETROLOGIC AND GEOCHEMICAL CONSTRAINTS

The conclusions arising from the former chapters lead to the following key points.

(a) Oceanic crust magmatic rocks concentrate incompatible elements, stored within that part of the depleted suboceanic mantle submitted to partial melting, below the axes of oceanic ridges.

(b) The bulk quantities of these elements stored within the oceanic crust increase during its travel from the ridge to the trench by means of: (i) weathering of basalts by interaction with seawater, (ii) deposition of incompatible element-rich sediment of variable but sometimes important thickness over the top of the basaltic sequence, and finally (iii) thickening of the oceanic crust by intraplate magmatic activity which adds to it OIB (Ocean Island Basalts) and OPB (Ocean Plateau Basalts) that are richer in incompatible elements than the MORB (Mid-Ocean Ridge Basalts), and which recycle components from the deep mantle.

(c) A fraction of these incompatible elements is extracted from the downgoing ocean crust in subduction zones and then recycled on the surface or within the arc crust by subduction-related magmatism. This recycling process involves either direct emplacement of melting products from the ocean crust (adakites) or mantle-wedge metasomatism followed by partial melting which produces calc-alkaline magmas.

(d) An important fraction of the incompatible elements is not extracted from the descending ocean crust but stored in the deep mantle for 1–2 Gyr, and then transferred again to the surface by plume/hot-spot magmatic activity. They contribute to the enriched mantle end-members EMI (Enriched Mantle I), EMII (Enriched Mantle II) and HIMU (HIgh MU).

The occurrence in some oceanic basalts, especially from the Indian Ocean and surrounding areas, of isotopic signatures involving an enriched EMI/EMII mantle component in their source (DUPAL anomaly) indicates that subducted sediment components can also be recycled within the ocean crust. This means that the Indian Ocean crust, the top of which is sedimentary, also contains ancient sediment-derived components both in its MORB section (basalts and related gabbros) and in the plume-derived oceanic plateaus and islands which overlie it.

Are the enriched mantle components EMI, EMII and HIMU exclusively derived from subducted ocean crust? The continental crust is also likely recycled into the arc mantle by two processes at least: the subduction of a continental margin (a rather uncommon case, but which is presently occurring beneath the Banda Arc) and the tectonic erosion of a continental margin by the downgoing crust, leading to dissemination of crustal materials within the upper mantle. The occurrence of this last process is probably very common, but its geochemical consequences have not yet been fully evaluated. The possible contribution to the EM (especially the EMI) signatures of the enriched mantle underlying the continents (subcontinental mantle) also remains largely a matter of speculation (Hofmann, 1997).

Are the plumes carrying the EM and HIMU signatures ascending only from the base of the upper mantle or from the D″ layer? Geochemical studies have not presently given a clear answer to this question, although there are reasons for presuming a very deep origin. Indeed, the composition of the lower mantle is still poorly known (FOZO (FOcus ZOne) and C hypotheses, § 10.4.5). The high $^{3}He/^{4}He$ and low $^{21}Ne/^{22}Ne$ ratios from active hot-spot basalts suggest a strong contribution from primordial helium and neon originating from a deep source, most probably the lower mantle. However, their amounts seem too low to account for the noble gas content of the core and lower mantle, except if the degassing of the deep Earth occurs in an irregular way, with peaks associated to tectono-magmatic crises.

12.2 DEEP STRUCTURE OF THE EARTH AND CONVECTION

The deep recycling of crustal materials requires convective mingling of materials originating from the lithosphere, the asthenospheric upper mantle and the lower mantle, respectively. Convective motions are strongly dependent on the distribution of densities and viscosities in the bulk mantle. Two kinds of models are currently proposed. In the first one, the upper mantle–lower mantle boundary, at depths of

Plate I. Map of the world ocean floors (copyright Hachette-Guides Bleues, Paris, all rights reserved).

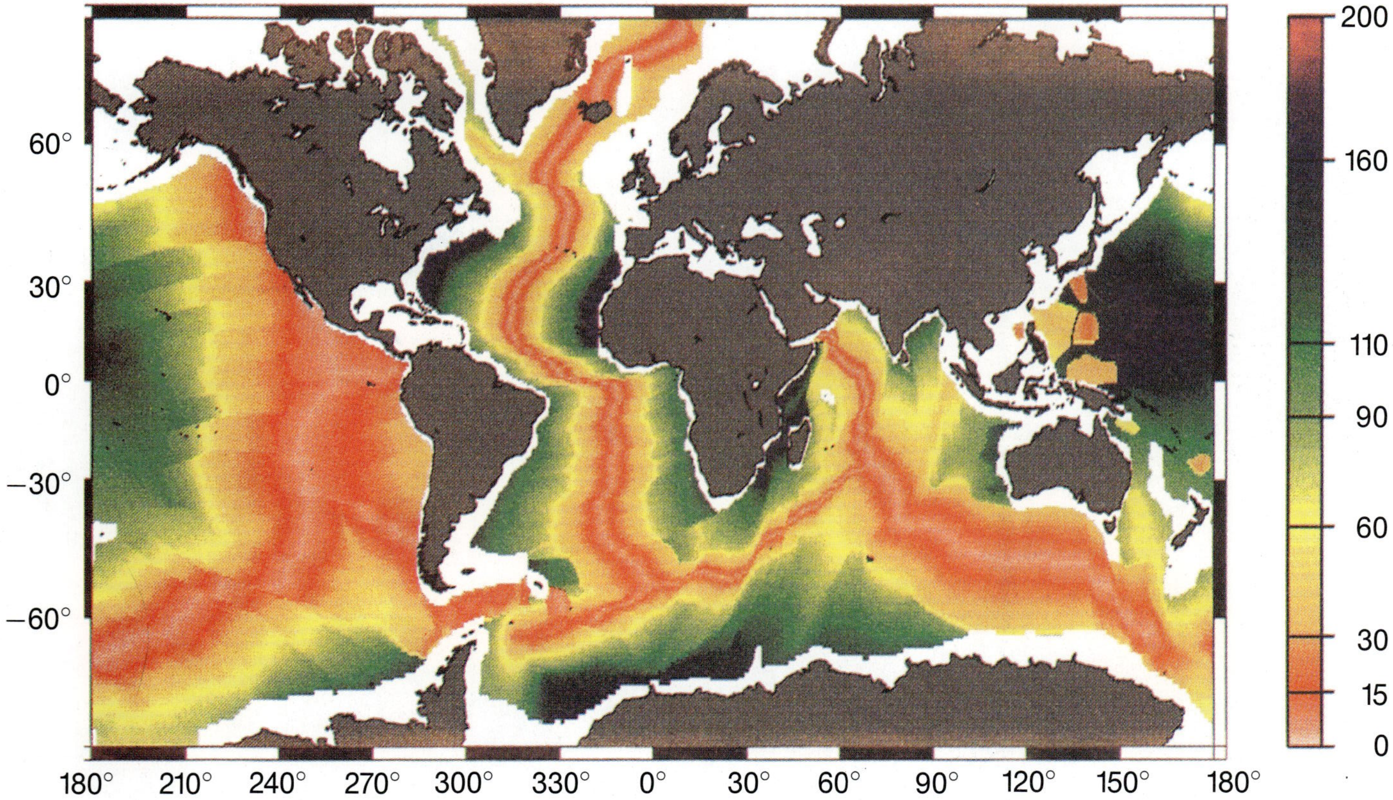

Plate II. Map of the ages of the world ocean floors, based on the distribution of the magnetic anomalies in the oceanic basins, and on the data from the deep

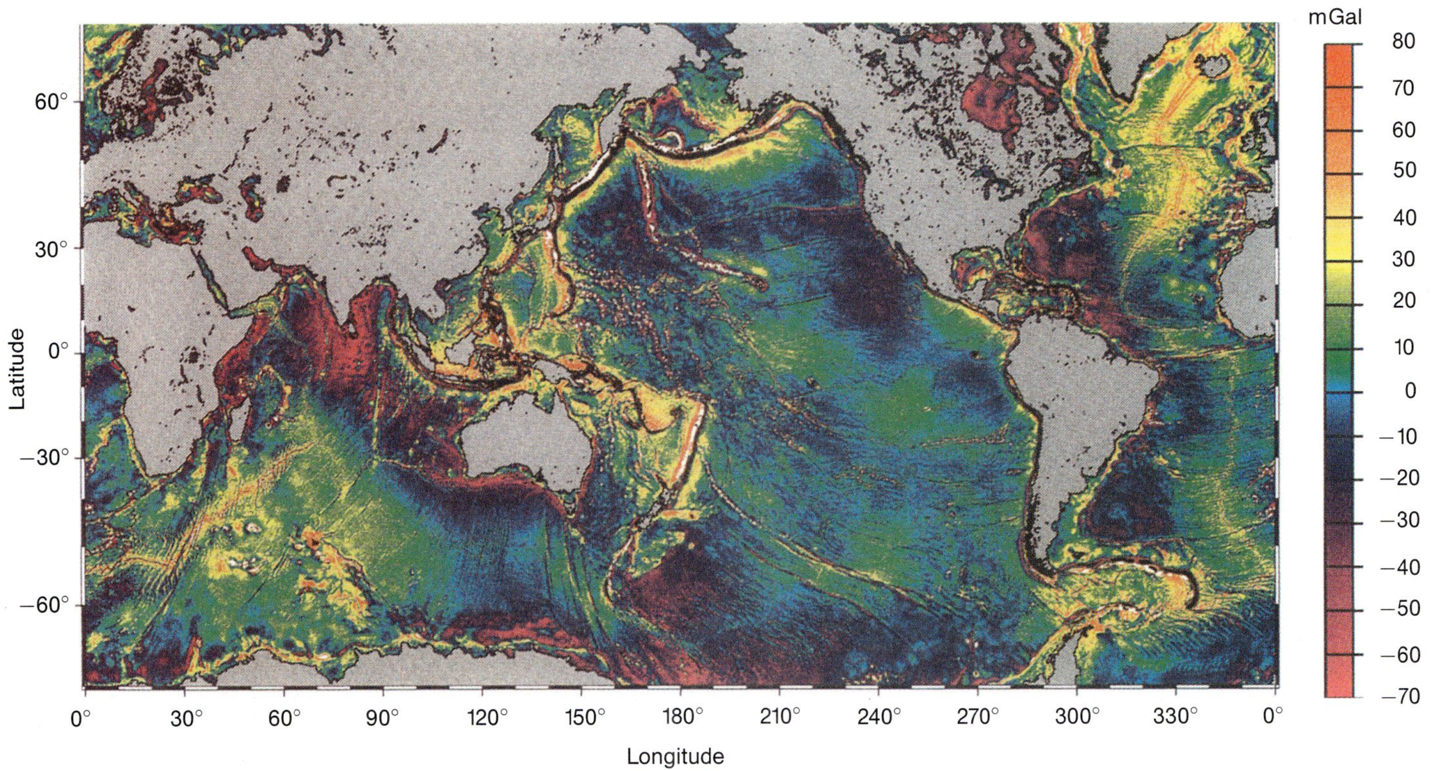

Plate III. Map of the free-air gravity anomalies from satellite altimetry (from Sandwell and Smith, 1995. Copyright 1995 David T. Sandwell). The altimetric radar aboard the satellite determines the distance separating the satellite from the ocean surface (using the double travel-time of the emitted radar waves). Simultaneously, the satellite orbit is calculated with respect to a reference ellipsoid, by laser telemetric or Doppler effect tracking stations on land. The height of the Geoid with respect to the ellipsoid is obtained by difference between these two measurements. After correction of the tide, currents, wind and other non-permanent perturbation effects, the topography of the geoid reflects directly the gravitational influence of sea-bottom reliefs, within a wide range of spatial scales.

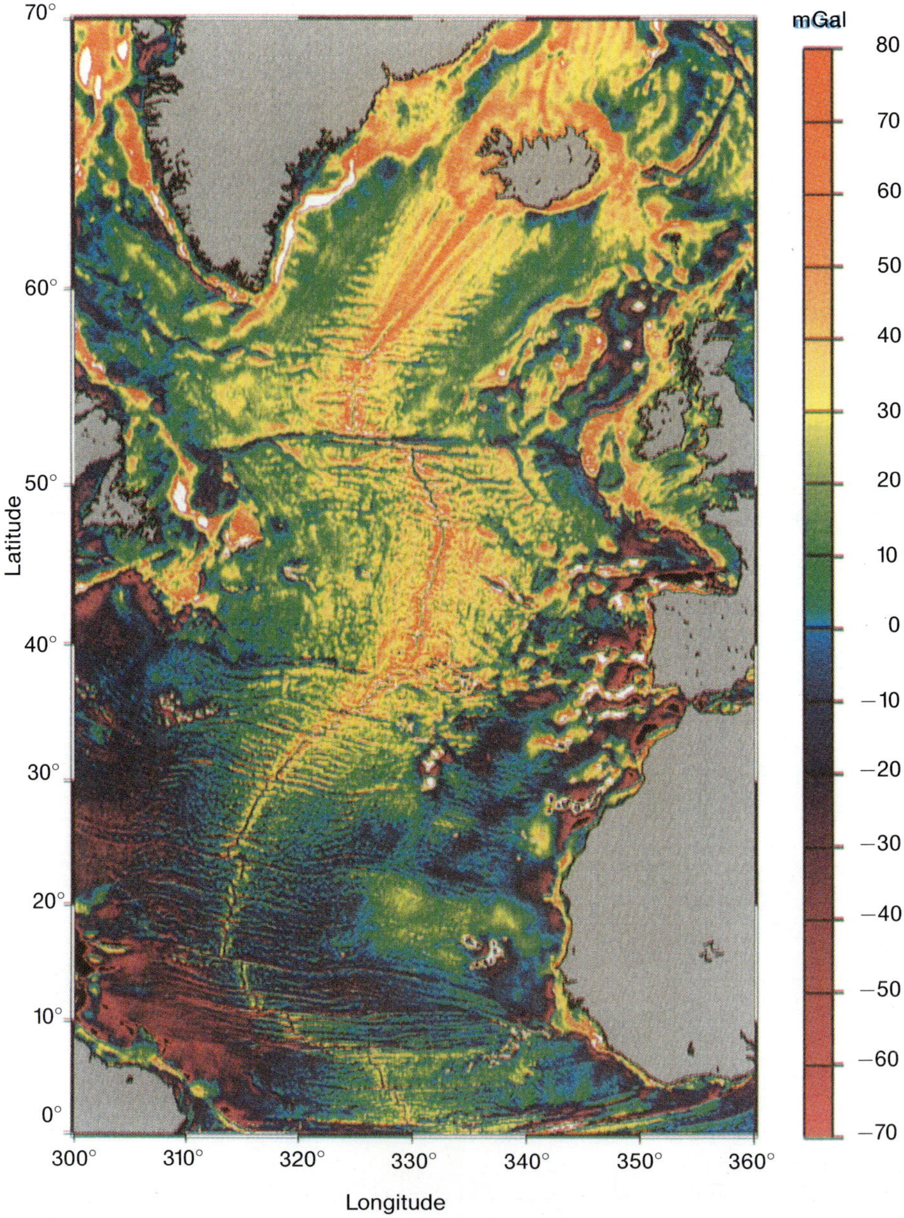

Plate IV. Free-air gravity map of the Northern and Central Atlantic Ocean between 15°N and 50°N, from satellite altimetric data (after Sandwell *et al.*, 1994).

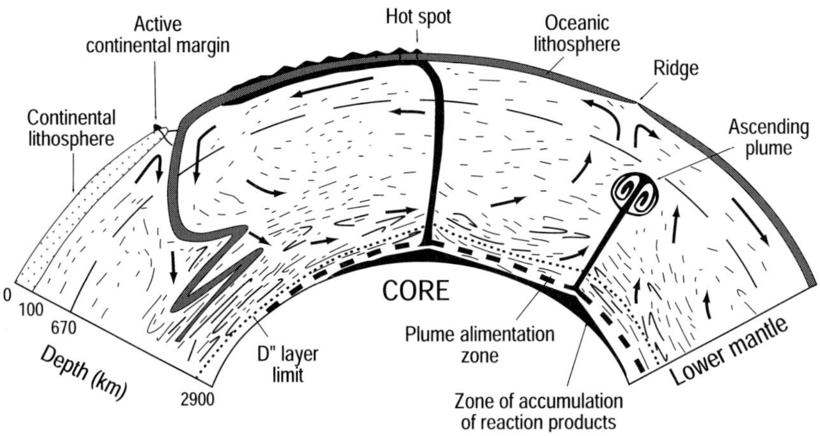

Fig. 12.1. Model of whole-scale mantle convection (from Davies and Richards, 1992). Explanations in the text.

660–670 km, corresponds to abrupt density and viscosity changes which make this the limit for convective currents. As a consequence, lithospheric materials would mingle only with the asthenospheric upper mantle without reaching the lower mantle. This model is not consistent with tomographic data suggesting that subducted slabs might reach depths of 1500 km. Inclusions within diamonds and noble gas isotopes also support the occurrence of (at least episodic) transfers of materials between the upper and the lower mantle.

The second model postulates whole-scale convection in a viscosity-stratified mantle (Figure 12.1). The viscosity increases by a factor of 1000 from the top of the mantle to the D″ layer (Davies, 1990; Davies and Richards, 1992). The upper mantle–lower mantle boundary, corresponding to mineralogical phase transitions (Figure 11.27), does not act as a barrier preventing convective motions, although it is marked by a viscosity increase. Consequently, the descending lithospheric plates will go across this boundary with or without deflection. Within the lower mantle, they will suffer folding and delamination because of the very high viscosities. Their fragments will either pond at the base of the mantle (D″ layer) or be progressively disseminated within its lower part (black strokes in Figure 12.1), creating mantle heterogeneities.

Convective motions independent of the sinking of dense and cold oceanic lithosphere originate in the two (relatively) low-viscosity layers of the mantle. One of them is the upper part of the asthenosphere, from which the mantle diapirs start ascending below the ridges (convective cells parallel to the ridge axis) and the hot lines (Richter's convective cells perpendicular to the ridge: Figure 10.25). The second low-viscosity layer corresponds to the core–mantle boundary. The D″ layer, heated by the outer core, is 150 to 500 km thick (Davies, 1990), and the plumes could start from its hot basal part, 30 to 100 km thick. In addition to the accumulation of subducted lithospheric fragments, this layer could concentrate reaction products

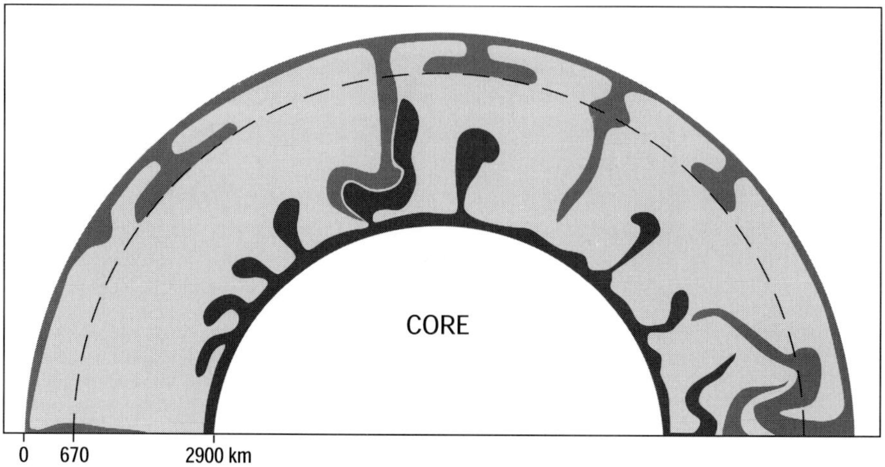

Fig. 12.2. Numerical simulation of mantle convection (from Powell, 1991). Downgoing cold materials are shown in light grey, uprising hot materials in dark grey. Other explanations in the text.

(wüstite FeO) from interactions between mantle rocks and liquid iron from the outer core. The dispersion of the thermal energy of the core would initiate instabilities within the D'' layer, triggering the uprise of its materials as mantle plumes. The ascent of these plumes through 2800 km of the mantle would allow their summital sphere (the 'head') to develop and reach diameters of ca. 2000 km consistent with the sizes of oceanic plateaus.

Numerical simulations of mantle convection have been conducted (Figure 12.2). Their results provide some support both to the one-layer and two-layer convective models. The cold materials, analogues of subducted oceanic lithosphere and possibly also of delaminated subcontinental mantle, usually pond over the 670 km boundary. Episodically, they cross this boundary, and then sink down to the core–mantle boundary. After a long period of heating (more than 1 Gyr) they uprise as deep mantle plumes.

12.3 FROM THE EXTRACTION OF OCEANIC CRUST TO ITS RECYCLING

The main steps of the ocean crust cycle are shown schematically in Figure 12.3 at the scale of a 900-km-wide ocean domain (top) and at that of the Earth (bottom).

In the upper sketch, the ocean crust is formed at the axis of a slow-spreading ridge (1 cm/yr) by cooling of MORB magmas derived from the partial melting of an adiabatically ascending diapir of depleted asthenospheric upper mantle. As it moves away from the ridge, sedimentary piles of increasing thickness cover its top, and the bulk lithospheric thickness also increases because of its progressive cooling. On the left of the sketch, this still young crust (20 Myr) subducts below another (and older)

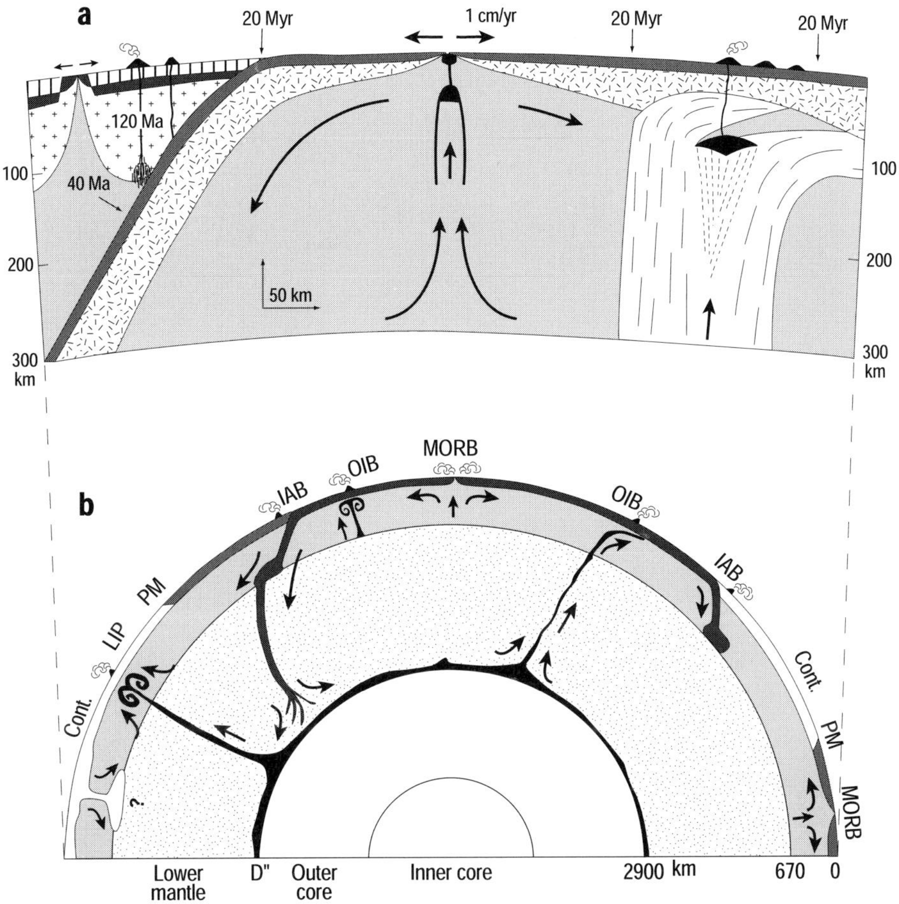

Fig. 12.3. The ocean crust cycle: a summary.

oceanic lithosphere, the thickness of which doubled during the Early Cretaceous (120 Myr) by the formation of an oceanic plateau. The partial melting of subducted oceanic crust at a depth of 70 km leads to the emplacement of adakites in the fore-arc domain. The metasomatized arc mantle melts at 100–110 km, producing calc-alkaline magmas, and a new oceanic basin opens in the back-arc domain. On the right, the oceanic lithosphere is heated again and thinned when it passes over a chemical plume, the top of which is deviated by the shear stresses induced by lithospheric motion. The magmas derived from the adiabatic decompression-induced melting of the top of this plume are erupted at the surface, where they form a hot-spot island chain.

In the lower sketch, derived from Figures 12.1 and 12.2, MORB derive from melting linked to ascending convective motions of the upper mantle, and IAB (Island Arc Basalts) from the subduction of oceanic lithosphere beneath another oceanic plate or a continental margin. The deep subducted lithosphere either ponds at the base of the upper mantle or penetrates the lower mantle into which it delaminates and contributes to the growth of D″ layer and to the creation of mantle heterogeneities. Cold, enriched and delaminated continental roots might also sink just like the oceanic lithosphere into the deep mantle, although this point is still very conjectural. Deep mantle plumes rise up from the D″ layer and supply magmas which build OIB hot-spot chains or Large Igneous Provinces (here a continental flood basalt province, or trap). OIB might also derive from shallower plumes formed at the base of the upper mantle.

12.4 THE GREAT TECTONO-MAGMATIC CRISES IN THE OCEANS

The temporal distribution of the main Cretaceous and Cenozoic intraplate magmatic provinces is shown in Figure 12.4. Most of the large oceanic plateaus formed between 135 Myr (onset of activity in Nauru) and 90–88 Myr (end of the Caribbean

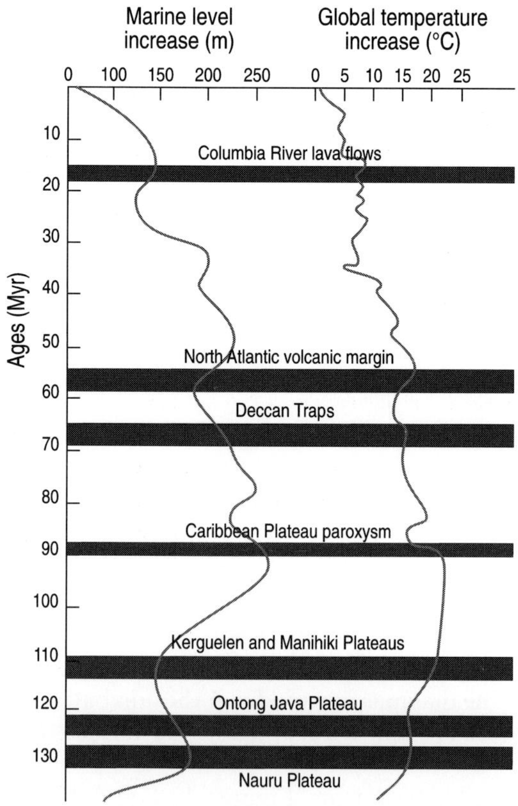

Fig. 12.4. Correlations between marine level changes, climate warming and emplacement of oceanic plateaus (from Coffin and Heldholm, 1993a).

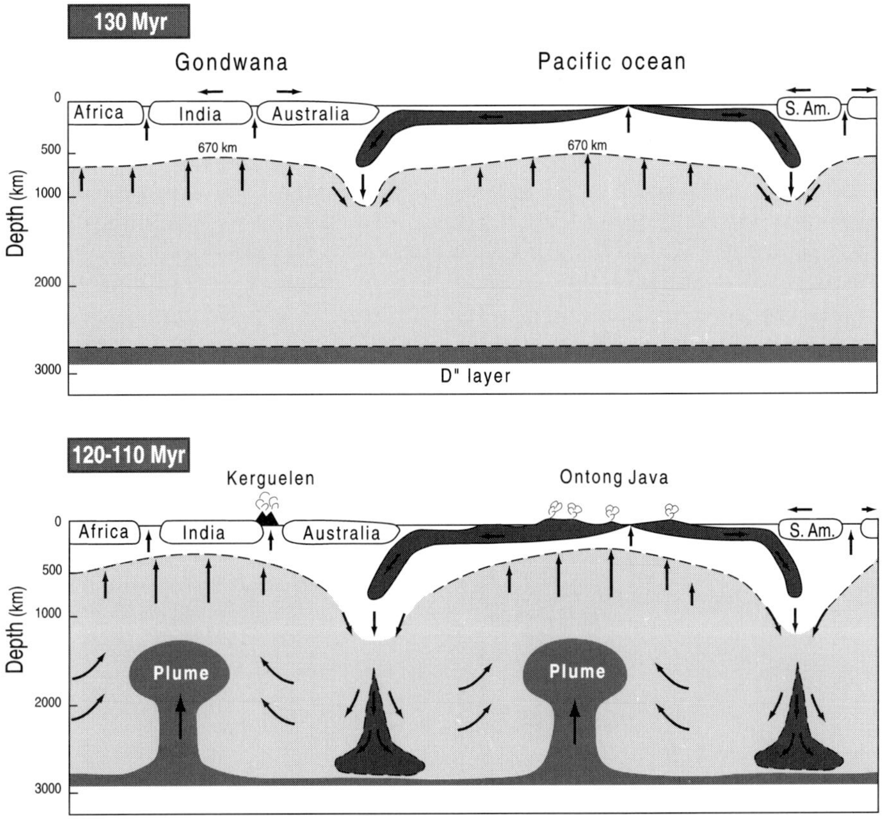

Fig. 12.5. An interpretation of the Early Cretaceous magmatic crisis in the Pacific (from Larson and Kincaid, 1996).

paroxysmic event). This time span corresponds to the Early Cretaceous period of high marine level and warming of climate. It can be estimated that the mainly submarine magmatic crisis of the Early Cretaceous emplaced 10^8 km^3 of basaltic materials, corresponding to 5 or 6 Myr of 'normal' ridge activity. Such an event was obviously able to generate significant changes at the planetary scale, and moreover had an equivalent effect on the continents where voluminous traps were emplaced, e.g. the Paraná basalts in Brazil (paroxysmal activity from 138 to 126 Myr; Garland *et al.*, 1996) and the Rajmahal traps in north-eastern India (115 Myr). Such a magmatic crisis did not happen again until the Late Cretaceous. It is thus possible to envision planetary geodynamics as an alternance of 'normal' or 'quiet' periods, sometimes referred to as Wilsonian periods (from J.T. Wilson), and of magmatic or tectono-magmatic crises.

What could the origin(s) of such crises be? Numerical simulations of mantle behaviour suggest that convective motions are not steady-state. 'Quiet' periods of two-level convection could alternate with crises during which whole-scale convection operates into the mantle (Powell, 1991). Such tectono-magmatic crises could efficiently enhance the degassing processes of the lower mantle. They have been referred to as MOMO periods (for Mantle Overturn Major Orogeny). However, the features of convection can, during a given period, differ from one area to another. For instance, during the Cretaceous, the Atlantic and Alpine areas remained relatively 'cold' and 'quiet' with respect to a hot and very active Pacific sector.

A tentative interpretation of the Early Cretaceous tectono-magmatic crisis in the Pacific is shown in Figure 12.5. An acceleration, the origin of which remains entirely hypothetical, of subduction rates would lead to a generalized sinking of oceanic plates into the lower mantle down to the D″ layer. In return, numerous ascending plumes would generate the Large Igneous Provinces shown in Figure 12.4. Besides the problem of the triggering factor(s), the initiation of such a crisis which leads to an abrupt change of convective regime and motions in about 10 Myr obviously requires a major decrease in viscosities in the deep mantle with respect to usual estimations.

References

Alabaster T., Pearce J.A., Malpas J. (1982). The volcanic stratigraphy and petrogenesis of the Oman Ophiolite Complex. *Contrib. Mineral. Petrol.*, **81**, 168–183.

Allègre C. J., Hamelin B., Dupré B. (1984). Statistical analysis of isotopic ratios in MORB: the mantle blob cluster model and the convective regime of the mantle. *Earth Planet. Sci. Lett.*, **71**, 71–84.

Allègre C.J., Staudacher T., Sarda P. (1987). Rare gas systematics: formation of the atmosphere, evolution and structure of the Earth's mantle. *Earth Planet. Sci. Lett.*, **81**, 127–150.

Alt J.C. (1995). Subseafloor processes in mid-ocean ridge hydrothermal systems. In: S. Humphris *et al.* (eds), *Seafloor Hydrothermal Systems: Physical, Chemical, Biological and Geological Interactions.* Geophysical Monograph 91, American Geophysical Union, Washington, DC, pp. 85–114.

Anderson D.L. (1982). Hotspots, polar wander, Mesozoic convection and the geoid. *Nature*, **297**, 391–393.

Anonymous (1972). Penrose Field Conference on Ophiolites. *Geotimes*, **17**, 24–25.

Arculus R.J. (1994). Aspects of magma genesis in arcs. *Lithos*, **33**, 189–208.

ARCYANA (1977). Rocks collected by bathyscaph and diving saucer in the FAMOUS area of the Mid-Atlantic Rift valley: petrological diversity and structural setting. *Deep-Sea Research*, **24**, 565–589.

ARCYANA (1978). *Atlas photographique de la Dorsale Médio-Atlantique: rift et faille transformante par 3000 mètres de fond.* Bordas éd., Paris, 128 pp.

Aubouin J. (1965). *Geosynclines.* Elsevier, Amsterdam, 335 pp.

Auzende J.M., Bideau D., Bonatti E., Cannat M., Honnorez J., Lagabrielle Y., Malavieille J., Mamaloukas-Frangoulis V., Mevel C. (1988a). Une coupe complète de la croûte océanique sur le mur Sud de la zone de fracture Vema (Atlantique central): résultats préliminaires de la campagne Vemanaute. *C. R. Acad. Sci. Paris*, **307**, Sér. II, 2061–2067.

Auzende J.M., Boespflug X., Bougault H., Dosso L., Foucher J.P., Joron J.L., Ruellan E., Sibuet J.C. (1990). From intracratonic extension to mature spreading in back arc basins: examples from the Okinawa, Lau and North Fiji basins. *Oceanologica Acta*, **10**, 153–163.

Auzende J.M., Cannat M., Gente P., Henriet J.P., Juteau T., Karson J., Lagabrielle Y., Mevel C., Tivey M. (1994.). Observation of sections of oceanic crust and mantle cropping out on the southern wall of Kane FZ. *Terra Nova*, **6**, 143–148.

Auzende J.M., Eissen J.P., Lafoy Y., Gente P., Charlou J.L. (1988b). Sea-floor spreading in the North Fiji Basin (SW Pacific). *Tectonophysics*, Spec. Issue, **146**, Wezel F.C. édit., 317–351, .

Auzende J.M., Tivey M., Cannat M., Gente P., Henriet J.P., Juteau T., Karson J., Lagabrielle Y. (1993). Première exploration in situ de la dorsale médio-Atlantique au sud de la zone de fracture Atlantis (29°06′N–43°11′W). *C. R. Acad. Sci. Paris*, **316**, 1415–1422.

Averous P. (1981). *Chercheurs sur l'océan*. Hachette, Paris.

Ayadi M., Pezard P., De Larouzière F.D. (1996). Fracture distribution from downhole electrical images at the base of the sheeted dike complex in Hole 504B. In: Alt, J.C., Kinoshita, H., Stokking; L.B. and Michael, P.J. (eds), *Proc. ODP Sci. Results*, **148**: College Station, Texas (Ocean Drilling Program), 307–315.

Ballard R.D., Francheteau J., Juteau T., Rangin C., Normark W. (1981). East Pacific Rise at 21°N: the volcanic, tectonic and hydrothermal processes of the central axis. *Earth Planet. Sci. Lett.*, **55**, 1–10.

Ballard R.D., Moore J.G. (1977). *Photographic Atlas of the Mid-Atlantic Ridge Rift Valley*, Springer-Verlag, New York, 114 pp.

Bard J.P. (1980). *Microtextures des roches magmatiques et métamorphiques*. Masson, Paris, 192 pp.

Benes V., Scott S.D., Binns R.A. (1994). Tectonics of rift propagation into a continental margin: Western Woodlark Basin, Papua New Guinea. *J. Geophys. Res.*, **99**, 4439–4455.

Benn K., Nicolas A., Reuber I. (1988). Mantle-crust transition zone and origin of wehrlitic magmas: evidence from the Oman ophiolite. *Tectonophysics*, **151**, 75–85.

Bideau D., Hekinian R., Francheteau J. (1977). Orientation of ocean floor basaltic rocks at time of cooling: a general method. *Contrib. Mineral. Petrol.*, **65**, 19–28.

Biju-Duval B. (1994). *Océanologie*. Dunod édit., Paris, 247 pp.

Binard N., Maury R.C., Guille G., Talandier J., Gillot P.Y., Cotten J. (1993). Mehetia Island, South Pacific: geology and petrology of the emerged part of the Society hot spot. *J. Volc. Geoth. Res.*, **55**, 239–260.

Bird J.M. (1987). Wilson cycle. In: Seyfert C.K. (ed.), *The Encyclopedia of Structural geology and Plate tectonics*, Van Nostrand Reinhold, New York, pp. 836–838.

Bischoff J.L., Dickson F.W. (1975). Seawater basalt interaction at 200°C and 55 bars: implications for origin of sea-floor heavy metal deposits and regulation of seawater chemistry. *Earth Planet. Sci. Lett.*, **25**, 285–297.

Bischoff J.L., Rosenbauer R.J. (1989). Salinity variations in submarine hydrothermal systems by layered double-diffusive convection. *J. Geol.*, **97**, 613–623.

Blanckenburg H. von, Davies J.H. (1995). Slab breakoff: a model for syncollisional magmatism and tectonics in the Alps. *Tectonics*, **14**, 120–131.

Bonatti E., Harrison C.G.A. (1976). Hot lines in the Earth's mantle. *Nature*, **263**, 402–404.

Bonatti E., Harrison C.G.A. (1988). Eruption styles of basalt in oceanic spreading ridges and seamounts: effects of magma temperature and viscosity. *J. Geophys. Res.*, **93** (B4), 2967–2980.

Bonatti E., Harrison C.G.A., Fisher D.E., Honnorez J., Schilling J.G., Stipp J.J., Zentilli M. (1977). Easter volcanic chain (southeast Pacific): a mantle hot line. *J. Geophys. Res.*, **82**, 2457–2478.

Bonatti E., Honnorez J. (1976). Sections of the earth's crust in the equatorial Atlantic. *J. Geophys. Res.*, **81**, 4104–4116.

Boudier F., Nicolas A. (1985). Harzburgite and lherzolite subtypes in ophiolitic and oceanic environments. *Earth and Planet. Sci. Lett.*, **76**, 84–92.

Boudier F., Le Sueur E., Nicolas A. (1989). Structures in the Trinity Complex (Eastern Klamath Mountains). An atypical ophiolite. *Geol. Soc. Amer. Bull.*, **101**, 820–833.

Boudier F., Nicolas A., Ildefonse B. (1996). Magma chambers in the Oman ophiolite: fed from the top and the bottom. *Earth Planet Sci. Lett.*, **144**, 239–250.

Bougault H., Joron J.L., Treuil M., Maury R. (1985). Local versus regional mantle heterogeneities: evidence from hygromagmaphile elements. In: Bougault H., Cande S.C. *et al.*, *Initial Reports of the Deep Sea Drilling Project*, **82**, U.S. Government Printing Office, Washington DC, pp. 459–482.

Bowen N.L. (1927.). *The evolution of the igneous rocks*. Dover Publ., New York, 332 pp.

Bowen N.L. (1949). The origin of ultrabasic and related rocks. *Amer. J. Sci.*, **14**, 89–108, .

Brenan J.M., Shaw H.R., Phinney D.L., Ryerson F.J. (1994). Rutile-aqueous fluid partitioning of Nb, Ta, Hf, Zr, U and Th: implications for high field strength element depletions in island arc basalts. *Earth Planet. Sci. Lett.*, **128**, 327–339.

Brongniart A. (1813). Essai d'une classification minéralogique des roches mélangées. *J. des Mines, Paris*, **199**, 5–48

Brousse R., Barsczus H.G., Bellon H., Cantagrel J.-M., Diraison C., Guillou H. & Léotot C. (1990). Les Marquises (Polynésie française): volcanologie, géochronologie, discussion d'un modèle de point chaud. *Bull. Soc. géol. France.*, **8**, VI, No. 6, 933–949.

Brunn J.H. (1960). Mise en place et différenciation de l'association pluto-volcanique du cortège ophiolitique. *Rev. Géogr. phys. Géol. dyn.*, **3** (3), 115–132.

Bryan W.B. (1972). Morphology of quench crystals in submarine basalts. *J. Geophys. Res.*, **77**, 5812–5819.

Bryan W.B. (1983). Systematics of modal phenocrysts assemblages in submarine basalts: petrologic implications. *Contrib. Mineral. Petrol.*, **83**, 62–74.

Bryan W.B. (1992). From pillow-lava to sheet flow. Evolution of deep sea volcanology. *Oceanus, Woods Hole Ocean. Inst.*, **34** (4), 42–50.

Bryan W.B., Moore J.G. (1977). Compositional variations of young basalts in the Mid-Atlantic Ridge rift valley near lat. 36°49′N. *Bull. Geol. Soc. Am.*, **88**, 556–570.

Buigues D., Gachon A., Guille G. (1993). L'atoll de Mururoa (Polynésie française). I. Structure et évolution géologique. *Bull. Soc. géol. France*, **163**, 645–657.

Cann J.R.(1970). New model for the structure of the ocean crust. *Nature*, **226**, 928–930.

Cann J.R. (1974). A model for oceanic crustal structure developed. *Geophys. J. R. Astron. Soc.*, **39**, 169–187.

Cann J.R. (1979). Metamorphism of the oceanic crust. In: Talwani M., Harrison C.G. and Hayes D.E. (eds), *Deep Drilling Results in the Atlantic Ocean: Ocean Crust*, Maurice Ewing Symposium Series 2, American Geophysical Union, Washington, DC, **48**, 230–238.

Cann J.R. (1981). Basalts from the ocean floor. In: *The Sea*, Vol. 7 *The Oceanic Lithosphere*, John Wiley & Sons, New York, pp. 363–390.

Cannat M. (1990). Mécanisme de l'accrétion océanique et mise l'affleurement de gabbros et de péridotites à l'axe des dorsales océaniques. 'Habilitation' thesis, Université de Bretagne Occidentale, Brest.

Cannat M. (1993). Emplacement of mantle rocks in the seafloor at mid-ocean ridges. *J. Geophys. Res.*, **98**, 4163–4172.

Cannat M. (1996). How thick is the magmatic crust at slow spreading oceanic ridges? *J. Geophys. Res.*, **101**, 2847–2857.

Cannat M., Mevel C., Stakes D.S. (1991). Stretching of the deep crust at the slow spreading Southwest Indian Ridge. *Tectonophysics*, **190**, 73–94.

Cannat M., Mevel C., Maia M., Deplus C., Durand C., Gente P., Agrinier P., Belarouchi A., Dubuisson G., Humler G., Reynolds J. (1995). Thin crust, ultramafic exposures, and rugged faulting patterns at the Mid-Atlantic Ridge (22°–24°N). *Geology*, **23**, 49–52.

Caroff M., Maury R.C., Guille G., Cotten J. (1997). Partial melting below Tubuai (Austral islands, French Polynesia). *Contrib. Mineral. Petrol.*, **127**, 369–382.

Caroff M., Maury R.C., Vidal Ph., Guille G., Dupuy C., Cotten J., Guillou H., Gillot P.-Y. (1995). Rapid temporal changes in Ocean Island Basalt composition: evidence from a 800 m-deep drill hole in Eiao shield (Marquesas). *J. Petrol.*, **36** (5), 1333–1365.

Caron J.M., Gauthier A., Schaaf A., Ulysse J., Wozniak J. (1989). *Comprendre et enseigner la Planète Terre*, Ophrys édit., Paris, 271 pp.

Carroll, M.J., Wyllie, P.J. (1989). Experimental phase relations in the system tonalite-peridotite-H_2O at 15 kbar, implications for assimilation and differentiation processes near the crust-mantle boundary. *J. Petrol.*, **30**, 1351–1382.

Ceuleneer G. (1986.). Structure des ophiolites d'Oman: flux mantellaire sous un centre d'expansion océanique et charriage à la dorsale. Ph.D. thesis, Université de Nantes, 217 pp.

Ceuleneer G., Nicolas A., Boudier F. (1988). Mantle flow patterns at an oceanic spreading center: the Oman peridotites record. *Tectonophysics*, **151**, 1–26.

Chadwick W.W., Embley R.M., Fox C.G. (1991.). Evidence for eruption on the southern Juan de Fuca ridge between 1981 and 1987. *Nature*, **350**, 416–418.

Chauvel C., Hofmann A.W., Vidal P. (1992). HIMU-EM: The French Polynesian connection. *Earth Planet. Sci. Lett.*, **110**, 99–119.

Chauvel C., McDonough W., Guille G., Maury R.C., Duncan R. (1997). Contrasting old and young volcanism in Rurutu Island, Austral chain. *Chemical Geol.*, **139**, 125–143.

Choukroune P., Francheteau J., Hekinian R. (1984). Tectonics of the East Pacific Rise near 132°50′N: a submersible study. *Earth Planet. Sci. Lett.*, **68**, 115–127.

Christensen N.I. (1966). Elasticity of ultrabasic rocks. *J. Geophys. Res.*, **71**, 5921–5931.

Christensen N.I., Salisbury M.H. (1972). Seafloor spreading, progressive alteration of layer 2 basalts and associated changes in seismic velocities. *Earth Planet. Sci. Lett.*, **56**, 367–375.

Christensen N.I., Salisbury M.H. (1975). Structure and constitution of the lower oceanic crust. *Rev. Geophys. Space Phys.*, **13**, 57–86.

Coffin M., Eldholm O. (1993a). Les grandes provinces volcaniques. *Pour la Science*, **194**, 76–83.

Coffin M., Eldholm O., (1993b). Scratching the surface: estimating dimensions of large igneous provinces. *Geology*, **21**, 515–518.

Coleman R.G. (1971). Plate tectonic emplacement of upper mantle peridotites along continental edges. *J. Geophys. Res.*, **76**, 1212–1222. .

Coleman R.G. (1981). Tectonic setting for ophiolite obduction in Oman. *J. Geophys. Res.*, **86**, 2497–2508.

Constantin M. (1995). Pétrologie des gabbros et péridotites de la dorsale Est-Pacifique: la transition croûte-manteau aux dorsales rapides. Ph.D. thesis, Université de Bretagne Occidentale, Brest, 286 pp.

Coulon C., Dupuy C., Dostal J., Escalant M. (1991). Spatial and temporal evolution of the volcanism of Martinique (Lesser Antilles). Petrogenetic implications. *Bull. Soc. géol. France*, **162**, 1037–1047.

Coulon C., Maluski H., Bollinger C., Wang S. (1986). Mesozoic and Cenozoic volcanic rocks from central and southern Tibet: ^{39}Ar–^{40}Ar dating, petrological characteristics and geodynamical significance. *Earth Planet. Sci. Lett.*, **79**, 281–302.

Coulon C., Thorpe R.S. (1981). Role of continental crust in petrogenesis of orogenic volcanic associations. *Tectonophysics*, **77**, 79–93.

Cousens B.L., Allan J.F., Leybourne M.I., Chase R.L., Van Wagoner N. (1995). Mixing of magmas from enriched and depleted mantle sources in the northeast Pacific: West Valley segment, Juan de Fuca ridge. *Contrib. Mineral. Petrol.*, **120**, 337–357.

Cox K.G., Bell J.D., Pankhurst R.J. (1979). *The Interpretation of Igneous Rocks*, Allen and Unwin, London, 450 pp.

Crane K., Ballard R.D. (1981.). Volcanics and structure of the FAMOUS-Narrowgate Rift: evidence for cyclic evolution. *J. Geophys. Res.*, **86**, 5112–5124.

Cross W., Iddings J.P., Pirsson L.V., Washington H.S. (1903). *Quantitative Classification of Igneous Rocks*, University of Chicago Press, Illinois.

Dana J. D. (1849). Geology. *The United States Exploring Expedition During the Years 1838–1842*, Sherman and Co, Philadelphia, **10**, 756 pp.

Darwin C. (1842). *The Structure and Distribution of Coral Reefs*. London, 2nd edn 1874, 3rd edn 1889.

Davies G.F. (1990). Mantle plumes, mantle stirring, and hot-spot chemistry. *Earth Planet. Sci. Lett.*, **99**, 94–109.

Davies G.F., Richards M.A. (1992). Mantle convection. *J. Geol.*, **100**, 151–206.

Davis B.T.C., Schairer J.F. (1965). Melting relations in the join diopside-forsterite-pyrope at 40 kilobars and at one atmosphere. *Carnegie Inst. Washington Yearbook*, **64**, 123–126.

Decandia A., Elter P. (1972). La zona ofiolitifera del Bracco nel settore fra Levanto e la Val Graveglia (Apennino ligure). *Mem. Soc. Geol. It.*, **11**, 503–530.

Decker R.W., Hadikusumo D. (1961). Results of the 1960 expedition to Krakatoa. *J. Geophys. Res.*, **66**, 3497–3511, .

Defant, M.J., Drummond, M.S. (1990). Derivation of some modern arc magmas by melting of young subducted lithosphere. *Nature*, **347**, 662–665.

Defant, M.J., Jackson, T.E., Drummond, M.S., De Boer, J.Z., Bellon, H., Feigenson, M.D., Maury, R.C., Stewart, R.H. (1992). The geochemistry of young volcanism throughout Western Panama and Southeastern Costa Rica: an overview. *J. Geol. Soc. London*, **149**, 569–579.

Delesse A. (1846). Sur les roches globuleuses. *Mém. Société géologique de France,* 2ème série, Vol. 4.

De Paolo D.J. (1981). Trace element and isotopic effects of combined wallrock assimilation and fractional crystallization. *Earth Planet. Sci. Lett.*, **53**, 189–202.

De Roever W.P. (1957). Sind die alpinotypen Peridotitmassen vielleicht tektonisch verfrachtete Bruchstücke der Peridotitschale? *Geol. Rundschau*, **46**, 137–146.

Detrick R., Collins J., Stephen R., Swift S. (1994). In situ evidence for the nature of seismic layer 2/3 boundary in oceanic crust. *Nature*, **370**, 288–290.

Dewey J.F., Bird J.M. (1970). Mountain belts and the new global tectonics. *J. Geophys. Res.*, **75**, 2625–2647.

Dewey J.F., Kidd W.S.F. (1977). Geometry of plate accretion. *Geol. Soc. Amer. Bull.*, **88**, 960–968.

Dick H.J.B., Meyer P.S., Bloomer S., Kirby S., Stakes D., Mawer C. (1991). Lithostratigraphic evolution of an in-situ section of oceanic layer 3. In: Von Herzen R.P., Robinson P.T. *et al.* (eds), *Proc. ODP Sci. Results*, 118, College Station, 439–537.

Dickinson W.R., Atherton T. (1967). Andesitic volcanism and seismicity around the Pacific. *Science*, **157**, 801–803.

Diraison C. (1991). Le volcanisme aérien des archipels polynésiens de la Société, des Marquises et des Australes-Cook. Téphrostratigraphie, datation isotopique et géochimie comparées. Contribution à l'étude des origines du volcanisme intraplaque du Pacifique central. Ph.D. thesis, Université de Bretagne Occidentale, Brest, 413 pp.

Donaldson C.H. (1976). An experimental investigation of olivine morphology. *Contrib. Mineral. Petrol.*, **57**, 187–213.

Donnelly T.W., Pritchard R.A., Emmermann R. (1979). The aging of oceanic crust: synthesis of mineralogical and chemical results of DSDP Legs 51–53. *Init. Reports DSDP*, **51–53**, Part 2, 1563–1577

Dosso L., Bougault H., Joron J.L. (1993). Geochemical morphology of the North Mid-Atlantic Ridge, 10°–24°N: Trace element-isotope complementarity. *Earth Planet. Sci. Lett.*, **120**, 443–462.

Dubertret L. (1953). Géologie des roches vertes du Nord Ouest de la Syrie et du Hatay (Turquie). *Mus. Nat. Hist. nat. Paris, Notes Mém. Moyen-Orient*, **6**, 179.

Dungan M.A., Long P.E., Rhodes J.M. (1978a). The petrography, mineral chemistry and one-atmosphere phase relations of basalts from site 395, Leg 45 DSDP. In: Melson W.G., Rabinowitz P.D. *et al.* (eds), *Initial Reports of the Deep Sea Drilling Project*, Vol. 45, U.S. Government Printing Office, Washington, D.C., pp. 461–478.

Dungan M.A., Rhodes J.M., Long P.E., Blanchard D.P., Brannon J.C., Rodgers K.V. (1978b). Petrology and geochemistry of basalts from site 396, Legs 45 and 46 of the Deep Sea Drilling Project. In: Dmitriev L.V., Heirtzler J.R. *et al.* (eds), *Initial Reports of the Deep Sea Drilling Project*, Vol. 46, U.S. Government Printing Office, Washington, D.C., pp. 89–114.

Dupré B., Allègre C.J. (1983). Pb-Sr isotopic variations in Indian Ocean basalts and mixing phenomena. *Nature*, **303**, 142–146.

Dupuy C., Vidal Ph., Maury R.C., Guille G. (1993). Basalts from Mururoa, Fangataufa and Gambier islands (French Polynesia): geochemical dependence on the age of the lithosphere. *Earth Planet. Sci. Lett.*, **117**, 89–100.

Durand C., Gente P., Dauteuil O. (1995). Caractéristiques morphologiques des segments axiaux le long de la dorsale médio-Atlantique. *C. R. Acad. Sci., Paris*, **320**, 411–418.

Edgar N.T. (1974). Acoustic stratigraphy in the deep oceans. In: C.A. Bur, C.L. Drake (eds), *The Geology of Continental Margins,* Springer-Verlag, Berlin, pp. 243–246

Edwards C.M.H., Morris J.D., Thirlwall M.F. (1993). Separating mantle from slab signatures in arc lavas using B/Be and radiogenic isotope systematics. *Nature*, **362**, 530–533.

Eissen J.P., Bideau D., Juteau T. (1981). Présence de basaltes porphyriques dans les zones de fracture de la dorsale est-Pacifique. *C. R. Acad. Sci. Paris*, **293**, 61–64.

Elter G. (1971). Schistes lustrés et ophiolites de la zone piémontaise entre Orco et Doire Baltée. Hypothèse sur l'origine des ophiolites. *Géol. Alpine, Grenoble*, **47**, 147–169

Ernewein M. (1987). Histoire magmatique d'un segment de croûte océanique téthysienne: pétrologie de la séquence plutonique du massif ophiolitique de Salahi (nappe de Semail, Oman). Ph.D. thesis, Université Louis Pasteur, Strasbourg, 205 pp.

Ernewein M., Pflumio C., Whitechurch H. (1988). The death of accretion zone as evidenced by the magmatic history of the Sumail ophiolite (Oman). *Tectonophysics*, **151**, 247–274.

Escalant M. (1988). Evolution spatiale et chronologique du volcanisme de l'île de la Martinique (Petites Antilles): implications pétrogénétiques. Ph. D. thesis, Université d'Aix-Marseille III, 188 pp.

Eva C., Cattaneo M., Merlanti F. (1988). Seismotectonics of the central segment of the Indonesian arc. *Tectonophysics*, **146**, 241–259.

Filmer P.E., McNutt M.K., Webb H.F., Dixon D.J. (1994). Volcanism and archipelagic aprons in the Marquesas and Hawaiian Islands. *Marine Geophys. Res.*, **16**, 385–406.

Floyd P.A. (1989). Geochemical features of intraplate oceanic plateau basalts. *Magmatism in the Ocean Basin*, Geol. Soc. Sp. Publ. 42, pp. 215–230.

Fourcade S., Maury R.C., Defant M.J., McDermott F. (1994). Mantle metasomatic enrichment versus arc crust contamination in the Philippines: oxygen isotope study of Batan ultramafic nodules and northern Luzon arc lavas. *Chem. Geol.*, **114**, 199–215.

Francheteau J., Armijo R., Cheminée J.L., Hékinian R., Lonsdale P., Blum N. (1990). 1 Myr East Pacific Rise oceanic crust and uppermost mantle exposed by rifting in Hess Deep (equatorial Pacific Ocean). *Earth Planet. Sci. Lett.*, **101**, 281–295.

Francheteau J., Armijo R., Cheminée J.L., Hékinian R., Lonsdale P., Blum N. (1992). Dyke complex of the East Pacific Rise exposed in the walls of Hess Deep and the structure of the upper oceanic crust. *Earth Planet. Sci. Lett.*, **111**, 109–121.

Francheteau J., Ballard R.D. (1983). The East Pacific Rise near 21°N, 13°N and 20°S: inferences for along-strike variability of axial processes of the Mid-Ocean Ridge. *Earth Planet. Sci. Lett.*, **64**, 93–116.

Francheteau J., Juteau T., Needham D., Rangin C. (1980). *CYAMEX—Naissance d'un océan*. CNEXO éd., Paris, 86 pp.

Francheteau J., Juteau T., Rangin C. (1979). Basaltic pillars in collapsed lava pools on deep ocean floor. *Nature*, **281**, 209–211.

Francis P.W., Hawkesworth C.J. (1994). Late Cenozoic rates of magmatic activity in the Central Andes and their relationship to continental crust formation and thickening. *J. Geol. Soc., London*, **151**, 845–854.

Gansser A. (1966). The Indian ocean and the Himalayas, a geological interpretation. *Eclogae geol. Helv.*, **59**, 832–848.

Garland F., Turner S., Hawkesworth C. (1996). Shifts in the source of the Paraná basalts through time. *Lithos*, **37**, 223–243.

Geist D., Howard K.A., Larson P. (1995). The generation of oceanic rhyolites by crystal fractionation: the basalt-rhyolite association at Volcán Alcedo, Galapagos archipelago. *J. Petrol.*, **34**, 965–982.

Gente P. (1987). Etude morphostructurale comparative de dorsales océaniques à taux d'expansion variés. Ph.D. thesis, Université de Bretagne Occidentale, Brest, 371 pp.

Gente P. (1995). Segmentation et formation des reliefs des dorsales. Implications sur les processus de l'accrétion. 'Habilitation' thesis, Université de Bretagne Occidentale, Brest, 136 pp. plus appendices.

Gente P., Auzende J.M., Renard V., Fouquet Y., Bideau D. (1986). Detailed geological mapping by submersible of the East Pacific Rise axial graben near 13°N. *Earth Planet. Sci. Lett.*, **78**, 224–236.

Gente P., Pockalny R.A., Durand C., Deplus C., Maia C., Ceuleneer G., Mevel C., Cannat M., Laverne C. (1995). Characteristics and evolution of the segmentation of the Mid-Atlantic ridge between 20°N and 24°N during the last 10 million years. *Earth Planet. Sci. Lett.*, **129**, 55–71.

Gill J. (1981). *Orogenic Andesites and Plate Tectonics*. Springer-Verlag, Berlin, 390 pp.

Gillot P.Y., Lefèvre J.C., Nativel P.E. (1994). Model for the structural evolution of the volcanoes of Réunion Island. *Earth Planet. Sci. Lett.*, **122**, 291–302.

Girardeau J., Mercier J.C., Wang Xibin (1985). Petrology of the mafic rocks of the Xigaze ophiolite, Tibet. Implications for the genesis of the oceanic lithosphere. *Contr. Mineral. Petrol.*, **90**, 309–321.

Goslin J., Beuzart P., Francheteau J., Le Pichon X. (1972.). Thickening of the oceanic layer in the Pacific Ocean. *Marine Geophys. Res.*, **1**, 418–427.

Green D.H., Ringwood A.E. (1967). The genesis of basaltic magmas. *Contrib. Mineral. Petrol.*, **15**, 103–190.

Green T.H. (1982). Anatexis of mafic crust and high pressure crystallization of andesite. In: Thorpe R.S. (ed.), *Andesites*. Wiley, New York, pp. 465–486.

Griffiths R.W., Campbell I.H. (1990). Stirring and structure in mantle starting plumes. *Earth Planet. Sci. Lett.*, **99**, 66–78.

Guille G., Goutière G., Sornein J.F., Buigues D., Gachon A., Guy C. (1995). *The atolls of Mururoa and Fangataufa (French Polynesia). I. Geology-Petrology-Hydrogeology. From volcano to atoll*, Louis-Jean, Gap, 175 pp.

Haggerty S.E. (1994). Superkimberlites: a geodynamic diamond window to the Earth's core. *Earth Planet. Sci. Lett.*, **122**, 57–69.

Hall A. (1987). *Igneous Petrology*. Longman, Harlow, 573 pp.

Harris N.B.W., Hawkesworth C.J., Tindle A.G. (1993). The growth of continental crust during the Late Proterozoic: geochemical evidence from the Arabian shield. In: Prichard H.M. *et al.*, (ed.). *Magmatic Processes and Plate Tectonics*, Geol. Soc. Sp. Publ. 76, 363–371.

Harris N.B.W., Pearce J.A., Tindle A.G. (1986). Geochemical characteristics of collision-zone magmatism. In: Coward M.P., Ries A.C. (ed.). *Collision Tectonics*, Geol. Soc. Sp. Publ. 19, 67–81.

Harris P.G., Reay E.A., White I.J. (1967). Chemical composition of the upper mantle. *J. Geophys. Res.*, **72**, 6359–6369.

Hart R.A (1973). A model for chemical exchange in the basalt-seawater system of oceanic layer 2. *Can. J. Earth Sci.*, **10**, 799–816.

Hart S.R., Hauri E.H., Oschmann L.A., Whitehead J.A. (1992). Mantle plumes and entrainment: isotopic evidence. *Science*, **256**, 517–520.

Hawkesworth C.J., Gallagher K., Hergt J.M., McDermott F. (1994). Destructive plate margin magmatism: geochemistry and melt generation. *Lithos*, **33**, 169–188.

Haymon R. and Adventure Team (1991). Active eruption seen on East Pacific Rise. *EOS Trans. Amer. Geophys. Union*, **72** (No. 46, 12 Nov.), 505–507.

Hékinian R. (1982). *Petrology of the Ocean Floor*, Elsevier Scientific, Amsterdam, 393 pp.

Hékinian R., Hoffert M. (1975). Rate of palagonitization and manganese coating on basaltic rocks from the rift valley in the Atlantic ocean near 36°50'N. *Mar. Geol.*, **19**, 91–109.

Hess H. H. (1962). History of ocean basins. In: E.J. Engel, H.L. James and B.F. Leonard (eds), *Petrological Studies: A Volume in Honor of A. F. Buddington,* Geological Society America, pp. 559–620.

Hildreth W., Moorbath S. (1988). Crustal contribution to arc magmatism in the Andes of Central Chile. *Contrib. Mineral. Petrol.*, **98**, 455–489.

Hilst R.D. van der (1995). Complex morphology of the subducted lithosphere in the mantle beneath the Tonga trench. *Nature*, **374**, 154–157.

Hofmann A.W. (1988). Chemical differentiation of the Earth: the relationship between mantle, continental crust, and oceanic crust. *Earth Planet. Sci. Lett.*, **90**, 297–314.

Hofmann A.W. (1997). Mantle geochemistry: the message from oceanic volcanism. *Nature*, **385**, 219–229.

Honnorez J. (1981). The aging of the oceanic crust at low temperature. In: C. Emiliani (ed.), *Oceanic Lithosphere*, J. Wiley, London, pp. 525–587.

Honnorez J., Kirst P. (1975). Submarine basaltic volcanism: morphometric parameters for discriminating hyaloclastites from hyalotuffs. *Bull. Volcanol.*, **39–3**, 1–25.

Hören H., Zamora M., Dubuisson G. (1996). Seismic waves velocities and anisotropy in serpentinized peridotites from Xigaze ophiolite: abundance of serpentine in slow spreading ridge. *Geophys. Res. Lett.*, **23**, 9–12.

Houtz R., Ewing J. (1976). Upper crustal structure as a function of plate age. *J. Geophys. Res.*, **81**, 2490–2498.

Irvine T.N. (1980). Magmatic density currents and cumulus processes. *Amer. J. Sci.*, **280**, 1–58.

Irvine T.N. (1982). Terminology for layered intrusions. *J. Petrol.*, **23**, 127–162.

James D.E. (1981). The combined use of oxygen and radiogenic isotopes as indicators of crustal contamination. *Ann. Rev. Earth Planet. Sci.*, **9**, 311–344.

Jaques A.L., Green D.H. (1980). Anhydrous melting of peridotite at 0–15 kb pressure and the genesis of tholeiitic basalts. *Contrib. Mineral. Petrol.*, **73**, 287–310.

Juteau T. (1975). Les ophiolites des nappes d'Antalya (Taurides occidentales, Turquie). Pétrologie d'un fragment de l'ancienne croûte océanique téthysienne. Thèse Doct. d'Etat, Université de Nancy I, Sciences de la Terre, Mém. No. 32, Nancy, 692 pp.

Juteau T. (1993a). Le volcanisme des dorsales océaniques. *Mém. Soc. géol. France*, No. 163, 81–98.

Juteau T. (1993b). *La naissance des océans*. Payot and Rivages édit., 382 pp.

Juteau T., Whitechurch H. (1989). La séquence plutonique de la nappe ophiolitique de Semail en Oman: dualité du magmatisme et segmentation axiale. *Bull. Soc. géol. France*, 8th sér., V, 189–204.

Juteau T. & CYAMEX Scientific Team. (1980). Homogeneous basalts from the East Pacific Rise at 21°N: steady state magma reservoirs at moderately fast spreading centers. *Oceanol. Acta*, **3**, 487–503.

Juteau T., Eissen J.P., Monin A.S., Zonenshain L.P., Sorokhtin O.G., Matveenkov V.V., Almukhamedov A.I. (1983). Structure et pétrologie du rift axial de la Mer Rouge vers 18°N. *Bull. Cent. Rech. Explor. Prod. Elf Aquitaine*, **7**, 217–231.

Juteau T., Ernewein M., Reuber I., Whitechurch H., Dahl R. (1988a). Duality of magmatism in the plutonic sequence of the Sumail nappe, Oman. *Tectonophysics*, **151**, 107–135.

Juteau T., Beurrier M., Dahl R., Nehlig P. (1988b). Segmentation at a fossil spreading axis. The plutonic sequence of the Wadi Haymiliyah area (Haylayn Block, Sumail Nappe, Oman). *Tectonophysics*, **151**, 167–197.

Juteau T., Cannat M., Lagabrielle Y. (1990). Serpentinized peridotites in the upper oceanic crust away from transform zones: a comparison of the results of previous DSDP and ODP Legs. In: Detrick R., Honnorez J., Bryan W.B., Juteau T. *et al.* (eds), *Proc. ODP, Sci. Results*, 106/109, College Station, Texas (Ocean Drilling Program), pp. 415–431.

Juteau T., Bideau D., Dauteuil O., Manac'h G., Naidoo D., Nehlig P., Ondreas H., Tivey MA.A., Whipple K.X., Delaney J.R. (1995). De la propagation à l'accrétion océanique: étude par submersible du mur nord de la fosse Ouest-Blanco (zone de fracture Blanco, Pacifique NE). *Bull. Soc. géol. France*, **166**, 105–121.

Karson J.A. (1992). Tectonic of slow spreading ridges. *Oceanus, Woods Hole Ocean. Inst.*, **34** (4), 51–59.

Karson J.A., Hurst S.D., Lonsdale P. (1992). Tectonic rotations of dikes in fast-spread oceanic crust exposed near Hess Deep. *Geology*, **20**, 685–688.

Kay, R.W. (1978). Aleutian magnesian andesites: melts from subducted Pacific Ocean crust. *J. Volcanol. Geotherm. Res.*, **4**, 117–132.

Keary P., Vine F.J. (1990). *Global Tectonics*. Blackwell Scientific Publ., Oxford, 302 pp.

Kelemen P.B., Koga K., Shimizu N. (1997). Geochemistry of gabbro sills in the crust-mantle transition zone of the Oman ophiolite: implications for the origin of the oceanic lower crust. *Earth Planet. Sci. Lett.*, **146**, 475–488.

Kelley D.S., Robinson P.T., Malpas J.G. (1992). Processes of brine generation and circulation in the oceanic crust: fluid inclusion evidence from the Troodos ophiolite, Cyprus. *J. Geophys. Res.*, **97**, 9307–9322.

Kerr A.C., Marriner G.F., Arndt N.T., Tarney J., Nivia A., Saunders A.D., Duncan R.A. (1996). The petrogenesis of Gorgona komatiites, picrites and basalts: new field, petrographic and geochemical constraints. *Geology*, **23**, 843–846.

Kerr A.C., Saunders A.D., Tarney J., Beryn H., Hards V.L. (1995). Depleted mantle-plume geochemical signatures: no paradox for plume theories. *Lithos*, **37**, 199–221.

Kerr A.C., Tarney J., Marriner G.F., Nivia A., Saunders A.D. (1997). The Caribbean-Colombian cretaceous igneous province: the internal anatomy of an oceanic plateau. In: Coffin M., Mahoney J. J. (eds), *Large Igneous Provinces: Continental, Oceanic and Planetary Flood Volcanism*, Geophys. Monogr. 100, American Geophysical Union, pp. 123–144.

Kirkpatrick R.J. (1978). Processes of crystallization in pillow basalts. Hole 396B, DSDP Leg 46. In: Dmitriev L., Heirtzler J., *et al.* (eds), *Initial Reports of the Deep Sea Drilling Project*, **46**, 271–282.

Klitgord K.D., Schouten H. (1986). Plate kinematics of the central North Atlantic. In: Vogt, P.R., Tucholke B.E. (eds), *The Geology of North America. The western north Atlantic region*, Geological Society of America, Boulder, Colorado, Vol. M, pp. 351–404.

Kogiso T., Tatsumi Y., Nakano S. (1997). Trace element transport during dehydration processes in the subducted crust: 1. Experiments and implications for the origin of ocean island basalts. *Earth Planet. Sci. Lett.*, **148**, 193–205.

Kohlstedt D. (1992). Structure, rheology and permeability of partially molten rocks at low melt fractions. In: *Mantle Flow and Melt Generation at Mid-ocean Ridges*, Geophysical Monograph 71, American Geophysical Union, Washington, D.C., pp. 103–121.

Kong L.S.L., Detrick R.S., Fox P.J., Mayer L.A., Ryan W.B.F. (1988). The morphology and tectonics of the Mark Area from Sea Beam and Sea MARC I observations (Mid-Atlantic Ridge 23°N). *Marine Geophys. Res.*, **10**, 59–90.

Lacroix A. (1927). La constitution lithologique des îles volcaniques de la Polynésie australe. *Mém. Acad. Sci., Paris*, **59**, 1–82.

Lagabrielle Y. (1987.). Les ophiolites: marqueurs de l'histoire tectonique des domaines océaniques. Thèse Doctorat d'Etat, Université de Bretagne Occidentale, Brest, 350 pp.

Lagabrielle Y., Cannat M. (1990). Alpine jurassic ophiolites resemble the modern central Atlantic basement. *Geology*, **18**, 319–322.

Lagabrielle Y., Fudral S., Kienast J.R. (1989). La couverture océanique des ultrabasites de Lanzo (Alpes occidentales): arguments lithostratigraphiques et pétrologiques. *Geodinamica acta*, **3** (2), 43–55.

Lagabrielle Y., Le Moigne J., Maury R.C., Cotten J., Bourgois J. (1994). Volcanic record of the subduction of an active spreading ridge, Taitao Peninsula (Southern Chile). *Geology*, **22**, 515–518.

Lagabrielle Y., Polino R., Auzende J.M., Blanchet R., Caby R., Fudral S., Lemoine M., Mevel C., Ohnenstetter M., Robert D., Tricart P. (1984). Les témoins d'une tectonique intraocéanique dans le domaine téthysien: analyse des rapports entre les ophiolites et leurs couvertures métasédimentaires dans la zone piedmontaise des Alpes franco-italiennes. *Ofioliti*, **9** (1), 67–88.

Lagabrielle Y., Ruellan E. and 15 co-authors (1996). Active oceanic spreading in the northern North Fiji basin: results of the NOFI cruise of R/V l'Atalante (Newstarmer Project). *Marine Geophys. Res.*, **18**, 225–247.

Langmuir C.H., Klein E., Plank T. (1992). Petrological systematics of mid-ocean ridge basalts: constraints on melt generation beneath ocean ridges. In: Morgan J.P., Blackman D.K., Sinton J. (eds.), *Mantle Flow and Melt Generation at Mid-Ocean Ridges*, Geophysical Monograph 71, American Geophysical Union, Washington D.C., pp. 183–280.

Larson R.L., Kincaid C. (1996). Onset of mid-Cretaceous volcanism by elevation or the 670 km thermal boundary layer. *Geology*, **24**, 551–554.

Laubscher H., Bernouilli D. (1980). Cross-section from the Rhine graben to the Po Plain. Excursion No. 031A. *26th Internat. Geol. Congress*, Paris, Livret-guide des excursions, Switzerland, Schweizerische geologische Kommission, Wepf and Co., Basle, pp. 183–209.

Laverne C. (1987). Les interactions basalte-fluides en domaine océanique. Minéralogie, pétrologie et géochimie d'un système hydrothermal: le Puits 504B, Pacifique oriental. Ph.D. thesis, Univ. Aix-Marseille (St Jérome), 315 pp.

Laverne C., Vanko D.A., Tartarotti P., Alt J.C. (1995). Chemistry and geochemistry of secondary minerals from the deep sheeted dike complex, DSDP/ODP Hole 504B. In: Erzinger J., Becker K., Dick H.J.B., and Stocking I.B. (eds), *Proc. ODP Sci. Results*, 137–140, College Station, Texas, pp. 167–189.

Le Dez A. (1996). Variations pétrologiques et géochimiques associées à l'édification des volcans-boucliers de Polynésie française: exemples de Nuku Hiva et Hiva Oa (Marquises) et de Moorea (Société). Ph.D. Thesis, Université de Bretagne Occidentale, Brest, 407 pp.

Le Dez A., Maury R.C., Vidal P., Bellon H., Cotten J., Brousse R. (1996). Geology and geochemistry of Nuku Hiva, Marquesas: temporal trends in a large Polynesian shield volcano. *Bull. Soc. géol. France*, **167**, 197–209.

Le Douaran S., Francheteau J. (1981). Axial depth anomalies from 10 to 50° north along the mid-Atlantic ridge: correlation with other mantle properties. *Earth Planet. Sci. Lett.*, **54**, 29–47.

Lemoine M. (1980). Serpentinites, gabbros and ophicalcites in the piemont-ligurian domain of the western Alps: possible indicators of oceanic fracture zones and associated serpentinite protrusions in the Jurassic-Cretaceous Tethys. *Archives des Sciences, Genève*, **33**, 103–116.

Le Pichon X., Francheteau J., Bonnin J. (1973). *Plate Tectonics*. Developments in Geotectonics 6, Elsevier, Amsterdam.

Le Pichon X., Huchon P. (1984). Geoid, Pangea and convection. *Earth Planet. Sci. Lett.*, **67**, 123–135.

Leroy S. (1995). Structure et origine de la plaque caraïbe. Implications géodynamiques. Ph.D. thesis, Université Pierre et Marie Curie, Paris, 249 pp.

Le Sueur F., Boudier F. (1986). Structures du complexe basique et ultrabasique de Trinity, Californie: genèse d'une ophiolite atypique. *Bull. Soc. géol. France*, **6**, 1007–1014.

Lewis B.T.R., Snydsman W.E. (1979). Fine structure of the lower oceanic crust on the Cocos Plate. *Tectonophysics*, **55**, 63–86.

Lewis J.V. (1914). Origin of pillow lavas. *Bull. Geol. Soc. Amer.*, **25**, 591–654.

Lippard S.J., Shelton A.W., Gass I.G. (1986). The ophiolite of Northern Oman. *Geol. Soc. London Mem.*, **11**, 178 pp.

Lister C.R.B. (1982). 'Active' and 'passive' hydrothermal systems in the ocean crust. Predicted physical conditions. In: K.A. Fanning, F.T. Manheim (eds), *The Dynamic Environment of the Ocean Floor*, D. C. Heath, Lexington, Massachusetts, pp. 441–470, .

Lofgren G. (1971). Spherulitic textures in glassy and crystalline rocks. *J. Geophys. Res.*, **76**, 5635–5648.

Lofgren G. (1974). An experimental study of plagioclase crystal morphology: isothermal crystallization. *Amer. J. Sci.*, **274**, 243–273, .

Lofgren G. (1980). Experimental studies on the dynamic crystallization of silicate melts. In: Hargraves R. (ed.), *Physics of Magmatic Processes*, Princeton University Press, New Jersey, pp. 487–551.

Lonsdale P.F. (1977). Deep-tow observations at the mounds abyssal hydrothermal field, Galapagos rift. *Earth Planet. Sci.. Lett.*, **36**, 92–110.

Maaloe S. (1985). *Principles of Igneous Petrology*. Springer-Verlag, Berlin, 374 pp.

Maaloe S., Aoki K. (1977). The major element composition of the upper mantle estimated from the composition of lherzolites. *Contrib. Mineral. Petrol.*, **63**, 161–173.

Macdonald G.A. (1972). *Volcanoes*. Prentice-Hall, Englewood Cliffs, New Jersey, USA, 510 pp.

Macdonald K.C. (1982). Mid-ocean ridges: fine scale tectonic, volcanic and hydrothermal processes within the plate boundary zone. *Annual Review of Earth Planet. Sciences*, **10**, 155–190.

Macdonald K.C. (1989). Anatomy of the magma reservoir. *Nature*, **339**, 178–179.

Macdonald K.C., Fox P.J., Perram L.J., Eisen M.F., Haymon R.M., Miller S.P., Carbotte S.M., Cormier M.H., Shor A.N. (1988). A new view of the Mid-Ocean Ridge from the behaviour of ridge-axis discontinuities. *Nature*, **335**, 217–225.

Macdonald K.C., Sempere J.C., Fox P.J. (1986). Reply: the debate concerning overlapping spreading centers and mid-ocean ridge processes. *J. Geophys. Res.*, **91**, 10 501–10 511.

Marion E. (1992). Interactions croûte océanique profonde-eau de mer dans les gabbros de la zone MARK (dorsale médio-Atlantique/zone de fracture Kane). Ph.D thesis, Univ. P. et M. Curie, Paris, 159 pp.

Martin H. (1986). Effect of steeper Archaean geothermal gradient on geochemistry of subduction zone magmas. *Geology*, **14**, 753–756.

Martin H. (1993). The mechanisms of petrogenesis of the Archaean continental crust. Comparison with modern processes. *Lithos*, **30**, 373–388.

Mauffret A., Leroy S. (1997). Seismic stratigraphy and structure of the Caribbean igneous province. *Tectonophysics*, **283**, 61–104.

Maury R.C. (1984). Les conséquences volcaniques de la subduction. *Bull. Soc. géol. France*, **26** (7), 489–500.

Maury R.C. (1993). Les séries volcaniques. *Mém. Soc. géol. France, 'Pleins feux sur les volcans'*, **163**, 39–55.

Maury R.C., Caroff M., Achard S., Guille G., Joron J.L., Gachon A., Rocaboy A., Leterrier J. (1992a). L'atoll de Mururoa (Polynésie française). II: La série magmatique. *Bull. Soc. géol. France*, **163**, 659–679.

Maury R.C., Defant M.J., Bellon H., Jacques D., Joron J.L., McDermott F., Vidal P. (1998). Temporal geochemical trends in northern Luzon arc lavas (Philippines): implications on metasomatic processes in the island arc mantle. *Bull. Soc. géol. France*, **169**, 69–80.

Maury R.C., Defant M.J., Joron J.L. (1992b). Metasomatism of the sub-arc mantle inferred from trace elements in Philippine xenoliths. *Nature*, **360**, 661–663.

Maury R.C., El Azzouzi M., Bellon H., Liotard J.M., Guille G., Barsczus H.G., Chauvel C., Diraison C., Dupuy C., Vidal Ph., Brousse R. (1994). Géologie et pétrologie de l'île de Tubuai (Australes, Polynésie française). *C. R. Acad. Sci. Paris*, **318**, II, 1341–1347.

Maury R.C., Sajona F.G., Pubellier M., Bellon H., Defant M.J. (1996). Fusion de la croûte océanique dans les zones de subduction/collision récentes: l'exemple de Mindanao (Philippines). *Bull. Soc. géol. France*, **167**, 579–595.

Maynard G.L. (1970). Crustal layer of seismic velocity of 6.9–7.6 kilometers per second under deep oceans. *Science*, **168**, 120–121.

McBirney A.R. (1963). Factors governing the nature of submarine volcanism. *Bull. Volcanol.*, **26**, 455–469.

McCulloch, M.T., Gamble, J.A. (1991). Geochemical and geodynamical constraints on subduction zone magmatism. *Earth Planet. Sci. Lett.*, **102**, 358–374.

McDermott F., Defant M.J., Hawkesworth C.J., Maury R.C., Joron J.L. (1993). Isotope and trace element evidence for three component mixing in the genesis of the N. Luzon arc (Philippines). *Contrib. Mineral. Petrol.*, **113**, 9–23.

McKenzie D.P. , Bickle M. (1988). The volume and composition of melt generated by extension of the lithosphere. *J. Petrol.*, **29**, 625–679.

McKenzie D., O'Nions R. (1995). The source regions of Ocean Island Basalts. *J. Petrol.*, **36**, 133–159.

McNutt M. K., Fischer K. (1987). The south Pacific superswell. In: Keating B. H. *et al.* (ed.), *Seamounts, Islands and Atolls*, American Geophysical Union Monograph 43, pp. 25–34.

McNutt M.K., Caress D.W., Reynolds S., Sordahl K.A., Duncan R.A. (1997). Failure of plume theory to explain midplate volcanism in the southern Austral islands. *Nature*, **389**, 479–482.

McNutt M.K., Fischer K., Kruse S., Natland J. (1989). The origin of the Marquesas fracture zone ridge and its implications for the nature of hot spots. *Earth Planet. Sci. Lett.*, **91**, 381–393.

Mercier J.C., Nicolas A. (1975). Textures and fabrics of upper mantle peridotites as illustrated by basalt xenoliths. *J. Petrol.*, **16**, 454–487.

Mével C., Caby R., Kienast J.R. (1978). Lower amphibolite facies conditions in the oceanic crust: example of amphibolitized flaser-gabbros and amphibolites from an alpine ophiolite massif (Chenaillet massif, Hautes Alpes, France). *Earth Planet. Sci. Lett.*, **39**, 98–108.

Mével C., Cannat M. (1991). Lithospheric stretching and hydrothermal processes in oceanic gabbros from slow-spreading ridges. In: Ophiolite genesis and evolution of the oceanic lithosphere, Peters Tj. *et al.* (ed.), Ministry of Petroleum and Minerals, Sultanate of Oman, pp. 293–312.

Miyashiro A. (1973). The Troodos Complex was probably formed in an island arc. *Earth Planet. Sci. Lett.*, **25**, 217–222.

Miyashiro A. (1978). Nature of alkalic volcanic rock series. *Contrib. Mineral. Petrol.*, **66**, 91–104.

Monzier M., Danyushevsky L.V., Crawford A.J., Bellon H., Cotten J. (1993). High-Mg andesites from the southern termination of the New-Hebrides island arc (SW Pacific). *J. Volc. Geoth. Res.*, **57**, 193–217.

Moore J.G. (1975). Mechanism of formation of pillow lava. *Am. Scientist*, **63** (3), 269–277.

Morgan W.J. (1971). Convection plumes in the lower mantle. *Nature*, **230**, 203–213.

Morgan W.J. (1972). Deep mantle convection plumes and plate motions. *Am. Ass. Petrol. Geol. Bull.*, **56**, 42–43.

Morris J.D., Leeman W.P., Tera F. (1990). The subducted component in island arc lavas: constraints from Be isotopes and B–Be systematics. *Nature*, **344**, 31–36.

Natland J.H. (1978). Crystal morphologies in basalts from DSDP Site 395, 23°N, 46°W, Mid-Atlantic Ridge. In. Melson W.G., Rabinowitz P.D. *et al.*, *Initial Reports of the Deep Sea Drilling Project*, **45**, U.S. Government Printing Office, Washington DC, pp. 424–445.

Natland J.H. (1991). Mineralogy and crystallization of oceanic basalts. In: Floyd P.A. (ed.), *Oceanic Basalts*, Blackie and Son, London, pp. 63–93.

Nehlig P. (1989). Etude d'un système hydrothermal océanique fossile: l'ophiolite de Semail (Oman). Ph.D. thesis, Université de Bretagne Occidentale, Brest, 2 vol., 600 pp.

Nehlig P. (1993). Interactions between magma chambers and hydrothermal systems: oceanic and ophiolitic constraints. *J. Geophys. Res.*, **98**, 19 621–19 633.

Nehlig P., Juteau T. (1988). Flow porosities, permeabilities and preliminary data on fluid inclusions and fossil thermal gradients in the crustal sequence of the Samail Ophiolite (Oman). *Tectonophysics*, **151**, 199–221.

Nehlig P., Juteau T., Bendel V., Cotten J. (1994). The root zones of oceanic hydrothermal systems: constraints from the Semail ophiolite (Oman). *J. Geophys. Res.*, **99**, 4703–4713.

Nichols G.T., Wyllie P.J., Stern C.R. (1994). Subduction zone melting of pelagic sediments constrained by melting experiments. *Nature*, **371**, 785–788.

Nicolas A. (1989). *Structures of Ophiolites and Dynamics of Oceanic Lithosphere*, Kluwer Acad. Publ., Dordrecht, 367 pp.

Nicolas A. (1990). *Les montagnes sous la mer*. Editions du BRGM, Orléans, 188 pp.

Nicolas A. (1995). *The Mid-Oceanic Ridges: Mountains Below the Sea Level*, Springer-Verlag, Berlin.

Nicolas A. (1992). Kinematics in magmatic rocks with special reference to gabbros. *J. Petrol.*, **33**, 891–915.

Nicolas A., Girardeau J., Marcoux J., Dupré B., Wang Xibin, Cao Yougong, Zheng Haixiang, Xiao Xuchang (1981). The Xigaze ophiolite (Tibet): a peculiar oceanic lithosphere. *Nature*, **294**, 414–417.

Nicolas A., Jackson E.D. (1972). Répartition en deux provinces des péridotites des chaînes alpines longeant la Méditerranée: implications géotectoniques. *Bull. Suisse Minéral. Pétrol.*, **53**, 385–401.

Nicolas A., Reuber I., Benn K. (1988). A new magma chamber model based on structural studies in the Oman ophiolite. *Tectonophysics*, **151**, 27–56.

Nisbet E.G., Fowler C.M.R. (1978). The Mid-Atlantic Ridge at 37° and 45°N: some geophysical and petrological constraints. *Geophys. J. R. astr. Soc.*, **54**, 631–660.

Ohta H., Maruyama S., Takahashi E., Watanabe Y., Kato Y. (1996). Field occurrence, geochemistry and petrogenesis of the archean mid-oceanic ridge basalts (AMORBs) of the Cleaverville area, Pilbara Craton, Western Australia. *Lithos*, **37**, 199–221.

Pallister J.S., Hopson C.A. (1981). Samail ophiolite plutonic suite: field relations, phase variation, cryptic variation and layering, and a model of a spreading ridge magma chamber. *J. Geophys. Res.*, **86**, 2593–2644.

Parsons B., Sclater J.G. (1977). An analysis of the variation of ocean floor bathymetry and heat flow with age. *J. Geophys. Res.*, **82**, 803–827.

Peacock, S.M. (1990). Numerical simulation of metamorphic pressure-temperature-time paths and fluid production in subducting slabs. *Tectonics*, **248**, 329–337.

Peacock, S.M., Rushmer, T. and Thompson, A.B. (1994). Partial melting of subducting oceanic crust. *Earth Planet. Sci. Lett.*, **121**, 227–224.

Pearce J.A. (1980). Geochemical evidence for the genesis and eruptive setting of lavas from Tethyan ophiolites. In: Panayiotou A. (ed.), *Ophiolites: Proceedings of the International Ophiolite Symposium*, Nicosia, Cyprus, 1979, Geol. Survey Dept, Cyprus, pp. 261–272.

Pearce J.A., Cann J.R. (1973). Tectonic setting of basic volcanic rocks determined using trace element analysis. *Earth Planet. Sci. Lett.*, **19**, 290–300.

Pearce J.A., Lippard S.J., Roberts S. (1984). Characteristics and tectonic significance of supra-subduction zone ophiolites. In: Kokelaar B.P., Howell M.F. (eds), *Marginal Basin Geology*, Geol. Society of London Spec. Publ. 16, 77–94.

Pearce J.A., van der Laan S.H., Arculus R.J., Murton B.J., Ishii T., Peate D.W., Parkinson I.J. (1992). Boninite and harzburgite from Leg 125 (Bonin-Mariana forearc): a case study of magma genesis during the initial stages of subduction. In: Fryer P., Pearce J.A. *et al.* (eds), *Proc. ODP Sci. Res.*, **125**, 623–659.

Pearson D.G., Davies G.R., Nixon P.H. (1993). Geochemical constraints on the petrogenesis of diamond facies pyroxenites from the Beni Bousera peridotite massif, North Morocco. *J. Petrol.*, **34**, 125–172.

Pflumio C. (1988). Histoire volcanique et hydrothermale du massif de Salahi: implications sur l'origine et l'évolution de l'ophiolite de Semail (Oman). Ph.D. thesis, Ecole Nat. Sup. des Mines de Paris, Paris, France.

Pomerol C., Renard M. (1995). *Éléments de géologie*, Arman Colin Édit., Paris, 10th edn., 629 pp.

Powell C. (1991). Les structures profondes. *Pour la Science*, **166**, 90–97.

Purdy G.M., Detrick R.S. (1986). Crustal structure of the Mid-Atlantic Ridge at 23°N from seismic refraction studies. *J. Geophys. Res.*, **91**, 3739–3762.

Raitt R.W. (1963). The crustal rocks. In: Hill M.N. (ed.), *The Sea*, Vol. 3, Wiley Interscience, New York, pp. 85–102.

Rangin, C., Jolivet, L., Pubellier, M. and the Tethys Working Group (1990). A simple model for the tectonic evolution of Southeast Asia and Indonesia regions for the past 43 Myr. *Bull. Soc. Géol. France*, **6**, 887–905.

Rapp, R.P., Watson, E.B., Miller, C.F. (1991). Partial melting of amphibolite/eclogite and the origin of Archean trondhjemites and tonalites. *Precambrian Res.*, **94**, 4619–4633.

Renard V., Hekinian R., Francheteau J., Ballard R.D., Backer H. (1985). Submersible observations at the axis of the ultrafast spreading East Pacific Rise (17°30′ to 21°30′S). *Earth Planet. Sci. Lett.*, **75**, 339–353.

Reuber I. (1988). Complexity of the crustal sequence in northern Oman ophiolite (Fizh and southern Aswad block): the effect of early slicing? *Tectonophysics*, **151**, 137–165.

Richards M.A., Duncan R.A., Courtillot V.E. (1989). Thermal entrainment by deflected mantle plumes. *Science*, **246**, 103–107.

Richards M.A., Griffiths R.W. (1989). Flood basalts and hot-spot tracks: plume heads and tails. *Nature*, **342**, 900–902.

Richardson C.J., Cann J.R., Richards H.G., Cowan J.G. (1987). Metal-depleted root-zones of the Troodos ore-forming hydrothermal systems, Cyprus. *Earth Planet. Sci. Lett.*, **84**, 243–253.

Richter F.M. (1973). Convection and the large-scale circulation of the mantle. *J. Geophys. Res.*, **78**, 8735–8745.

Ringwood, A.E. (1974). The petrological evolution of island arc systems. *J. Geol. Soc. London*, **130**, 183–204.

Rittmann A. (1963). *Les volcans et leur activité*. Masson et Cie édit., Paris, 458 pp.

Romeur M. (1991). Séries magmatiques arc et arrière-arc de la Sonde: nature des sources impliquées (éléments en trace et isotopes Sr-Nd-Pb). Ph. D. Thesis, Université de Bretagne Occidentale, Brest, 447 pp.

Rona P.A., Scott S.D. (1993). Seafloor hydrothermal mineralization: new perspectives. *Econ. Geol.*, **88**, 1933–1975.

Routhier P. (1946). Vulcano-plutons sous-marins du cortège ophiolitique. *C. R. Acad. Sci. Paris*, **222**, 192–194.

Ryan M.P. (1988). The mechanics and three-dimensional internal structure of active magmatic systems: Kilauea volcano, Hawaii. *J. Geophys. Res.*, **93**, 4213–4248.

Ryerson, F.J., Watson, E.B. (1987). Rutile saturation in magmas: implications for Ti-Nb-Ta depletion in island-arc basalts. *Earth Planet. Sci. Lett*,, 86, 225–239.

Sajona, F.G. (1995). Fusion de la croûte océanique en contexte de subduction/collision: géochimie, géochronologie et pétrologie du magmatisme plioquaternaire de Mindanao (Philippines). Ph.D. Thesis, Université de Bretagne Occidentale, Brest, 223 pp.

Sajona F.G., Bellon H., Maury R.C., Pubellier M., Cotten J., Rangin C. (1994). Magmatic response to abrupt changes in geodynamic settings: Pliocene-Quaternary calc-alkaline and Nb-enriched lavas from Mindanao (Philippines). *Tectonophysics*, **237** (1–2), 47–72.

Sajona, F.G., Maury, R.C., Bellon, H., Cotten (1996). High field strength element enrichment of Pliocene-Pleistocene island arc basalts, Zamboanga Peninsula, Western Mindanao (Philippines). *J. Petrol.*, **37**, 693–726.

Sajona F., Maury R.C., Bellon H., Cotten J., Defant M.J., Pubellier M. (1993). Initiation of subduction and the generation of slab melts in western and eastern Mindanao, Philippines. *Geology*, *21*, 1007–1010.

Sandwell D.T., Smith W.H.F., Yale M.M. (1994). ERS1 geodetic mission reveals new tectonic fabric. *EOS Trans. American Geophysical Union*, **73**, 133.

Sandwell D.T., Smith W.H.F. (1995). *Marine Gravity Anomaly from Satellite Altimetry*. Geological Data Center, Scripps Institution of Oceanography, La Jolla, California.

Saunders A.D., Tarney J., Kerr A.C., Kent R.W. (1996). The formation and fate of large oceanic igneous provinces. *Lithos*, **37**, 81–95.

Sautter V., Gillet P. (1994). les diamants, messagers des profondeurs de la Terre. *La Recherche*, **25**, 1238–1245.

Schiano P., Clocchiatti R., Shimizu N., Maury R.C., Jochum K.P., Hofmann A.W. (1995). Hydrous, silica-rich melts in the sub-arc mantle and their relationship with erupted arc lavas. *Nature*, **377**, 595–600.

Schiffman P., Smith B.M., Varga R.J., Moores E.M. (1987). Geometry, conditions and timing of off-axis hydrothermal metamorphism and ore deposition in the Solea Graben. *Nature*, **325**, 423–425.

Schilling J.G., Zajac M., Evans R., Johnston T., White W., Devine J.D., Kingsley R. (1983). Petrologic and geochemical variations along the Mid-Atlantic Ridge from 29°N to 73°N. *Amer. J. Sci.*, **283**, 510–586.

Schmincke H.U., Bednarz U. (1990). Pillow, sheet flow and breccia flow volcanoes and volcano-tectonic hydrothermal cycles in the extrusive series of the northern Troodos ophiolite, Cyprus. In: *Ophiolites. Oceanic Crustal Analogues, Proceedings of the Symposium Troodos '87*, Cyprus Geological Survey Department, Nicosia, Cyprus, pp. 185–206.

Schmincke H.U., Rautenschlein M. (1987). Volcanology along the Akaki River Canyon. In: Xenophontos C., Malpas J.G. (eds), *Troodos '87, Ophiolites and Oceanic Lithosphere. Field Excursion Guidebook*, Cyprus Geological Survey Department, Nicosia, Cyprus, pp. 36–91.

Scott S.D. (1992). Polymetallic sulfide riches from the deep: act or fallacy? In: Hs K.J. and Thiede J. (eds), *Proceedings of Dahlem Conference on Use and Misuse of the Seafloor*, J. Wiley & Sons, New York, pp. 87–115.

Scott S.D. (1995). Polymetallic massive sulfide deposits of the modern ocean floor: a review. In: *Modern seafloor hydrothermalism and metallogeny*, Course notes of selected reading, Université de Bretagne Occidentale, Brest., 64 pp.

Sekine, T., Wyllie, P.J. (1982). Experimental simulation of mantle hybridization in subduction zones. *J. Geol.*, **91**, 511–528.

Sempéré J.C., Lin J., Brown H.S., Schouten H., Purdy G.M. (1993). Segmentation and morphotectonic variations along a slow spreading center: the Mid-Atlantic Ridge (24°00N–30°40′N). *Marine Geophys. Res.*, **15**, 153–200.

Sempéré J.C., Purdy G.M., Schouten H. (1990). Segmentation of the Mid-Atlantic Ridge between 24°N and 30°40′N. *Nature*, **344**, 427–431.

Sen, C., Dunn, T (1994). Dehydration melting of a basaltic composition amphibolite at 1.5 and 2.0 GPa: implications for the origin of adakites. *Contrib. Mineral. Petrol., 117*, 394–409.

Sen, C., Dunn, T (1995). Experimental modal metasomatism of a spinel lherzolite and the production of amphibole-bearing peridotite. *Contrib. Mineral. Petrol.*, **119**, 394–409.

Seyfried W.E., Bischoff J.L. (1979). Low temperature basalt interaction with seawater: an experimental study at 70°C and 150°C. *Geochim. Cosmochim. Acta*, **43**, 1937–1947.

Shen A., Keppler H. (1995). Infrared spectroscopy of hydrous silicate melts to 1000°C and 10 kbar: direct observation of H_2O speciation in a diamond-anvil cell. *Amer. Mineral.*, **80**, 1335–1338.

Shen Y., Solomon S.C., Bjarnason I.T., Wolfe C.J. (1998). Seismic evidence for a lower-mantle origin of the Iceland plume. *Nature*, **395**, 62–65.

Shimoda G., Tatsumi Y., Nohda S., Ishikaza K., Jahn B.M. (1998). Setouchi high-Mg andesites revisited: geochemical evidence for melting of subducted sediments. *Earth Planet. Sci. Letters*, **160**, 479–492.

Sibuet J-C., Letouzey J., Barbier F., Charvet J., Foucher J-P., Hilde T.W.C., Kimura M., Ling-yun C., Marsset B., Muller C., Stephanj F. (1987). Back-arc extension in the Okinawa trough. *J. Geophys. Res.*, **92**, 14 041–14 062.

Sinton J.M., Detrick R.S. (1992). Mid-ocean ridge magma chambers. *J. Geophys. Res.*, **97**, 197–216.

Simandjuntak T.O., Barber A.J. (1996). Contrasting tectonic styles in the Neogene orogenic belts of Indonesia. In: Hall R., Blundell D. (ed.), *Tectonic evolution of Southeast Asia*, Geol. Soc. Sp. Publ., **106**, 185–201.

Smewing J.D. (1981). Mixing characteristics and compositional differences in mantle-derived melts beneath spreading axes: evidence from cyclically layered rocks in the ophiolite of North Oman. *J. Geophys. Res.*, **86**, 2645–2550.

Smith D.K., Cann J.R. (1990). Hundreds of small volcanoes on the median valley floor of the Mid-Atlantic Ridge at 24–30°N. *Nature*, **348**, 152–155.

Smith D.K., Cann J.R. (1993). Building the crust at the Mid-Atlantic Ridge. *Nature*, **365**, 707–715 .

Soeria-Atmadja R., Maury R.C., Bellon H., Pringgoprawiro H., Polvé M., Priadi B. (1994). Tertiary magmatic belts in Java. *J. Southeast Asia Earth Sci.*, **9**, 13–27.

Spudich P., Orcutt J. (1980). A new look at the seismic velocity structure of the oceanic crust. *Rev. Geophys. Space Phys.*, **18**, 627–645.

Steinmann G. (1927). Die ophiolitischen Zonen in den mediterranen Keltengebirgen. *14th Internat. Geol. Congress*, Madrid, Vol. 2, pp. 638–667.

Stolz A.J., Jochum K.P., Spettel B., Hofmann A.W. (1996). Fluid- and melt-related enrichment in the subarc mantle: evidence from Nb/Ta variations in island-arc basalts. *Geology*, **24**, 587–590.

Storzer D., Selo M. (1976). Uranium and fission track ages of some basalts from the FAMOUS area. *Bull. Soc. Géol. Fr.*, **18** (4), 807–810.

Sun, S.S., McDonough W.F. (1989). Chemical and isotopic systematics of oceanic basalts: Implications for mantle composition and processes. *Magmatism in the Ocean Basin*, Geol. Soc. Sp. Publ. 42, pp. 313–345.

Sutton G.M., Maynard G.L., Hussong D.M. (1971). In: Heacock J.G. (ed.), *Widespread Occurrence of a High-velocity basal Layer in the Pacific Crust Found with Repetitive Sources and Sonobuoys.* Geophysical Monograph Series, American Geophysical Union, **14**, 193–209.

Tatsumi, Y. (1989). Migration of fluid phases and genesis of basalt magmas in subduction zones. *J. Geophys. Res.*, **94**, 4697–4707.

Tatsumi, Y., Hamilton, D.L., Nesbitt, R.W. (1986). Chemical characteristics of fluid phase released from a subducted lithosphere and origin of arc lavas: evidence from high-pressure experiments and natural rocks. *J. Volcanol. Geotherm. Res.*, **29**, 293–309.

Tatsumi Y., Kogiso T. (1997). Trace element transport during dehydration processes in the subducted oceanic crust: 2. Origin of chemical and physical characteristics in arc magmatism. *Earth Planet. Sci. Lett.*, **148**, 207–221.

Tatsumi Y., Maruyama S. (1989). Boninites and high-Mg andesites: tectonics and petrogenesis. In: Crawford A.J. (ed.), *Boninites and Related Rocks,* Unwin Hyman, London, pp. 50–71.

Taylor S.S., McLennan S.M. (1985). *The Continental Crust: Its Composition and Evolution.* Blackwell Scientific, Oxford, 312 pp.

Tazieff H. (1958). L'éruption 1957–1958 et la tectonique de Faial (Açores). *Soc. Belg. Geol. Bull.*, **67**, 13–49.

Teall J.J.H. (1894). On greenstones associated with radiolarian cherts. *Trans. Roy. Geol. Soc. Cornwall*, **11**, 560–565.

Tepley L., Moore J.G. (1974). *Fire Under the Sea: the Origin of Pillow Lava*, 16-mm movie film, Moonlight Productions, Mountain View, California.

The Open University (1989.). *The Ocean Basins: Their Structure and Evolution*, Pergamon Press, Oxford, 171 pp.

Thompson G. (1991). Metamorphic and hydrothermal processes: basalt-seawater interactions. In: P.A. Floyd (ed.), *Oceanic Basalts*, Blackie and Son, Glasgow, pp. 148–173.

Thompson G., Bryan W.B., Ballard R.O., Hzamuro K., Melson W.G. (1985). Axial processes along a segment of the East Pacific Rise 10°–12°N. *Nature*, **318**, 429–433.

Thorarinsson S. (1964). *Surtsey, the New Island in the North Atlantic*, Almenna Bokafelagid, Reykjavik, 64 pp.

Thornton C.P., Tuttle O.F. (1960). Chemistry of igneous rocks. I: Differentiation Index. *Amer. J. Sci.*, **258**, 665–684.

Tilling R.I., Dvorak J.J. (1993). Anatomy of a basaltic volcano. *Nature*, **363**, 125–133.

Toksöz, M.N., Hsui, A.T. (1982). Numerical studies of back-arc convection and the formation of marginal basins. *Tectonophysics*, **50**, 177–196.

Toomey D.R., Purdy G.M., Solomon S., Wilcox W. (1990.). The three dimensional seismic velocity structure of the East Pacific Rise near latitude 9°30'N. *Nature*, **347**, 639–644.

Uyeda S., Kanamori H. (1979). Back-arc opening and the mode of subduction. *J. Geophys. Res.*, **84**, 1049–1061.

Van Andel T.H. (1985). New views on an old planet. *Continental Drift and the History of the Earth*, Cambridge University Press, Cambridge, 324 pp.

Van Andel T.H., Ballard R.D. (1979). The Galapagos Rift at 86°W: 2. Volcanism, structure, and evolution of the rift valley. *J. Geophys. Res.*, **84**, 5390–5406.

Vera E.E., Buhl P., Mutter J.C., Harding A.J., Orcutt J.A., Detrick R.S. (1990). The structure of 0–0.2-My-old oceanic crust at 9°N in the East Pacific Rise from expanded spread profiles. *J. Geophys. Res.*, **95**, 15 529–15 556.

Vidal P. (1994). *Géochimie*, Dunod, Paris, 190 pp.

Vidal P., Dupuy C., Barsczus H.G., Chauvel. C. (1987). Hétérogénéités du manteau et origine des basaltes des Marquises (Polynésie). *Bull. Soc. géol. France*, 4, 633–642.

Vidal, P., Dupuy, C., Maury, R. C., Richard, M. (1989). Mantle metasomatism above subduction zones: trace element and radiogenic isotope in xenoliths from Batan island (Philippines). *Geology*, **17**, 1115–1118.

Vidal P., Le Guen de Kerneizon M., Maury R.C., Dupré B., White M. (1991). Large role of sediments in the genesis of some Lesser Antilles andesites and dacites (Soufrière, St. Lucia): isotopic constraints. *Bull. Soc. géol. France*, **162**, 993–1002.

Vine F.J., Matthews D.H. (1963). Magnetic anomalies over oceanic ridges. *Nature*, **199**, 947–949.

Wager L.R., Brown G.M. (1968). *Layered Igneous Rocks*. Oliver and Boyd, Edinburgh, 588 pp.

Wager L.R., Brown G.M., Wadsworth W.J. (1960). Types of igneous cumulates. *J. Petrol.*, **1**, 73–85.

Watson E.B., Brenan J.M. (1987). Fluids in the lithosphere, 1. Experimentally-determined wetting characteristics of CO_2–H_2O fluids and their implications for fluid transport, host-rock physical properties, and fluid inclusion formation. *Earth Planet. Sci. Lett.*, **85**, 497–515.

Westercamp D., Andreieff P., Bouysse P., Cottez S., Battistini R. (1989). *Carte géologique 1/50 000. Martinique*. Notice explicative. BRGM, Orléans, 246 pp. plus two maps.

White R.S. (1984). Atlantic oceanic crust: seismic structure of a slow spreading ridge. In: Gass I.G., Lippard S.J., Shelton A.W. (eds), *Ophiolites and Oceanic Lithosphere*, Geol. Soc. London Spec. Publ. 13, pp. 101–111.

White R.S. (1991). Structure of the oceanic crust from geophysical measurements. In: Floyd P.A. (ed.), *Ocean Basalts*, Blackie & Son Ltd, Glasgow, pp. 30–48.

White R.S., Detrick R.S., Sinha M.C., Cormier M.H. (1984). Anomalous seismic crustal structure of oceanic fracture zones. *Geophys. J. R. Astron. Soc.*, **79**, 779–798.

Whitechurch H. (1993). Les ophiolites téthysiennes de la chaîne du Taurus (Turquie): de l'accrétion océanique à l'obduction. Thèse Doct. d'Etat, Université Louis Pasteur, Strasbourg, 366 pp.

Widiyantoro S., van der Hilst R. (1996). Evidence from islands on the spreading of ocean floors. *Nature*, **197**, 536–538.

Wilson J.T. (1963). Evidence from islands on the spreading of ocean floors. *Science*, **271**, 1566–1570.

Wilson J.T. (1966). Did the Atlantic close and then re-open? *Nature*, **211**, 676–681.

Wilson M. (1989). *Igneous Petrogenesis. A Global Tectonic Approach.* Chapman & Hall, London, 466 pp.

Wolery T.J., Sleep N.H. (1976). Hydrothermal circulation and geochemical flux at mid-ocean ridges. *J. Geol.*, **84**, 249–276.

Wolfe C.J., Bjarnason I.T., Van Decar J.C., Solomon S.C. (1997). Seismic structure of the Iceland mantle plume. *Nature*, **385**, 245–247.

Woodhead J.D. (1996). Extreme HIMU in an oceanic setting: the geochemistry of Mangaia island (Polynesia), and temporal evolution of the Cook-Austral hotspot. *J. Volc. Geoth. Res.*, **72**, 1–19.

Wyllie P.J. (1988). Magma genesis, plate tectonics and chemical differentiation of the Earth. *Rev. Geophysics*, **26**, 370–404.

Wyllie, P.J., Sekine, T. (1982). The formation of mantle phlogopite in subduction zone hybridization. *Contrib. Mineral. Petrol.*, **79**, 375–380.

Wysession M.E. (1996). Imaging slabs at the base of the mantle: the sometimes fate of slabs? In: Bebout G.E., Scholl D.W., Kirby S.H., Platt J.P. (eds), *Subduction: Top to Bottom*, Geophys. Monogr. 96, American Geophysical Union, pp. 369–384.

Yamagishi H. (1985). Growth of pillow lobes—Evidence from pillow lavas of Hokkaido, Japan, and North Island, New Zealand. *Geology*, **13**, 499–502.

Yoder H.S Jr. (1976). *Generation of Basaltic Magma.* National Academy of Scienes, Washington D.C., 264 pp.

Yoder H.S. Jr., Tilley C.E. (1962). Origin of basalt magmas: an experimental study of natural and synthetic rock systems. *J. Petrol.*, **3**, 342–532.

Zindler A., Hart S.R. (1986). Chemical geodynamics. *Ann. Rev. Earth Planet. Sci.*, **14**, 493–571.

Index